Chemically Modified Carbon Nanotubes
for Commercial Applications

Chemically Modified Carbon Nanotubes for Commercial Applications

Edited by Jeenat Aslam, Chaudhery Mustansar Hussain, and Ruby Aslam

WILEY-VCH

Editors

Dr. Jeenat Aslam
Taibah University
College of Science
Department of Chemistry
Yanbu 30799
Al-Madina, Saudi Arabia

Prof. Chaudhery Mustansar Hussain
New Jersey Institute of Technology (NJIT)
Department of Chemistry and Environmental Sciences
Newark
New Jersey 07102
USA

Dr. Ruby Aslam
Aligarh Muslim University
Department of Applied Chemistry
202002 Aligarh
India

Cover Image: © Maciej Frolow/ Photographer's Choice RF/Getty Images

Library of Congress Card No.: applied for

British Library Cataloguing-in-Publication Data
A catalogue record for this book is available from the British Library.

Bibliographic information published by the Deutsche Nationalbibliothek
The Deutsche Nationalbibliothek lists this publication in the Deutsche Nationalbibliografie; detailed bibliographic data are available on the Internet at <http://dnb.d-nb.de>.

Print ISBN: 978-3-527-35072-8
ePDF ISBN: 978-3-527-83881-3
ePub ISBN: 978-3-527-83880-6
oBook ISBN: 978-3-527-83879-0

Typesetting Straive, Chennai, India
Printing and Binding CPI Group (UK) Ltd, Croydon, CR0 4YY

C121235_151222

Contents

Preface

Carbon nanotubes, 1D allotropes of carbon, have fascinated significant research interest since their innovation in 1991 due to their low mass density, large aspect ratio, and unique physical, chemical, and electronic properties that offer exciting possibilities for nanoscale applications. Chemically modified treatments can improve carbon nanotubes' poor dispersion in solvents and their interactions with other materials, thereby increasing their industrial applications. Chemically modified carbon nanotubes are relatively flexible and interact with cell membranes and penetrate different biological tissues. This book covers a broad range of topics related to carbon nanotubes, from synthesis to chemical modification to commercial applications. In general, this book summarizes the recent progress and developments in commercial applications of chemically modified carbon nanotubes at both experimental and theoretical model scales.

To capture the inclusive portrait of commercial applications of chemically modified carbon nanotubes and to offer readers a logical and eloquent design of the topic and concentrated up-to-date references, the book is divided into six parts, with each part comprising several chapters. Part 1 explores the overview, commercialization, and economic aspects of chemically modified carbon nanotubes. Topics covered in Chapters 1–4 include the overview and fundamentals of chemically modified carbon nanotubes, surface modification strategies for carbon nanotubes, and latest developments in commercial-scale fabrications for chemically modified carbon nanotubes and their economic aspects. Part 2 discusses the energy and environmental applications of chemically modified carbon nanotubes. Topics covered in Chapters 5–9 are chemically modified carbon nanotubes in energy production and storage, pollutant adsorption, removal of textile effluents, membrane separation, and water purification systems. Part 3 debates electronics and electrical applications of chemically modified carbon nanotubes. Topics covered in Chapters 10–12 include chemically modified carbon nanotubes for electronics and photonic applications, electrochemical sensors, and lab-on-a-chip devices. Part 4 talks about biomedical applications of chemically modified carbon nanotubes. Topics covered in Chapters 13–16 are chemically modified carbon nanotubes in cancer therapy, drug delivery, tissue engineering, and tissue engineering. Part 5 describes the construction applications of chemically modified carbon nanotubes. The topic covered in Chapter 17 is chemically modified carbon nanotubes in cement and concrete field. Lastly, Part 6 discovers the emerging applications of chemically modified carbon nanotubes. Topics covered in Chapters 18–21 are chemically modified carbon nanotubes in 3D and 4D printing, tribology applications, corrosion protection and chemically modified carbon nanotubes and their sustainability.

The aim of this book is to deliver the recent advancements in chemically modified carbon nanotubes for commercial applications arena. This book is intended for a very wide-ranging audience working in the fields of advanced materials science, chemistry, chemical engineering and technology, etc. This book will be an invaluable reference source for the libraries in universities and industrial institutions, government and independent institutes, individual research groups, and scientists. Overall, this book is planned to be a useful source for advanced undergraduates, graduate students, researchers, and scientists who are looking for chemically modified carbon nanotubes for commercial applications to meet modern research demands. The editors and contributors of all chapters are well-known researchers, scientists, and experts from academia and industry. On behalf of Wiley, we are very much pleased to thank the contributors of all chapters for their exceptional and wholehearted efforts in the making of this book. Invaluable thanks to Dr. Martin Preuss (Associate Publisher, Academic Publishing), Miss Daniela Bez (Managing Editor), and the Editorial Team at Wiley for their wholehearted support and help during this project. In the end, all appreciation to Wiley for publishing the book.

Dr. Jeenat Aslam
Department of Chemistry, College of Science
Taibah University, Yanbu-30799
Al-Madina, Saudi Arabia

Prof. Chaudhery Mustansar Hussain
Department of Chemistry and
Environmental Science
New Jersey Institute of Technology
Newark, N J 07102, USA

Dr. Ruby Aslam
Department of Applied Chemistry
Aligarh Muslim University
Aligarh, India

About the Editors

Jeenat Aslam, PhD, is currently working as an associate professor at the Department of Chemistry, College of Science, Taibah University, Yanbu, Al-Madina, Saudi Arabia. She obtained her PhD in surface science/chemistry at the Aligarh Muslim University, Aligarh, India. Her research is mainly focused on the materials and corrosion, nanotechnology, and surface chemistry. Dr. Jeenat has published several research and review articles in peer-reviewed international journals of ACS, Wiley, Elsevier, Springer, Taylor & Francis, and Bentham Science. She has authored and edited many books and has contributed to twenty-seven book chapters.

Chaudhery Mustansar Hussain, PhD, is an adjunct professor and director of laboratories in the Department of Chemistry and Environmental Science at New Jersey Institute of Technology (NJIT), Newark, New Jersey, United States. His research is focused on the applications of nanotechnology and advanced materials, environmental management, analytical chemistry, smart materials and technologies, and other various industries. Dr. Hussain is the author of numerous papers in peer-reviewed journals as well as a prolific author and editor of around 100 books, including scientific monographs and handbooks in his research areas. He has published with Elsevier, American Chemical Society, Royal Society of Chemistry, Springer, Wiley, and CRC Press.

Ruby Aslam, PhD, is currently a research associate fellow under CSIR-HRDG, New Delhi, in the Department of Applied Chemistry, Aligarh Muslim University, Aligarh, India. She received her MSc, MPhil, and PhD degrees from the same university. She has authored/coauthored several research papers in international peer-reviewed journals of wide readership, including critical reviews and book chapters.

Part I

Chemically Modified Carbon Nanotubes: Overview, Commercialization, and Economic Aspects

1

A Detailed Study on Carbon Nanotubes: Properties, Synthesis, and Characterization

Nishant Tripathi[1], Prachi Sharma[1,2], Vladimir Pavelyev[1], Anastasiia Rymzhina[1], and Prabhash Mishra[1,3,4]

[1] *Nanoengineering Department, Samara National Research University, 34, Moskovskoye Shosse, Samara, 443086, Russia*
[2] *Department of Electrical, Electronics and Communication Engineering, GITAM School of Technology, Bengaluru Campus, GITAM (Deemed to be University) NH 207, Nagadenehalli, Doddaballapura, Karnataka 561203, India*
[3] *Centre for Nanoscience and Nanotechnology, Jamia Millia Islamia, Jamia Nagar, New Delhi, 110025, India*
[4] *Center for Photonics & 2D Materials, Moscow Institute of Physics and Technology (MIPT), Dolgoprudny 141700, Russia*

1.1 Introduction

The present chapter deals with the evolution of carbon allotropes, especially with carbon nanotubes (CNTs). CNTs are one of the most important nanomaterials of the carbon family. The invention of CNTs brought a real revolution in the field of nanoscience and nanotechnology. The astonishing properties of CNTs make them a suitable material for the development of applications in electronics, optoelectronics, medical, and in many other fields. Worldwide researchers are doing aggressive work on CNTs-based gas sensors, optical detectors, heat detectors, humidity sensors, transistors, nanoelectronics devices, and display applications [1, 2]. It has been observed that different types of applications based on CNTs require different types of CNTs. For example, sensors, especially for gas and optical applications, required horizontal aligned/network type of CNTs while field emission devices required vertically aligned CNTs. So, it has become very important to grow CNTs with selective structure and orientation [3]. Generally, CNTs can be grown by chemical vapor deposition (CVD), arc discharge, and laser ablation technique but among all the techniques CVD is the most preferable due to its capability to grow CNTs with selective properties. In general, the growth process of CNTs by using CVD system, a catalyst film deposited substrate is loaded into the chamber of CVD system, and then the chamber is heated around 800 °C in presence of a carrier gas; at 800 °C along with carrier gas, a hydrocarbon gas is supplied in CVD chamber for CNTs growth. CVD has lots of parameters to tune to decide CNTs' structural quality as well as morphology. The major parameters are the type of catalyst, catalyst

Chemically Modified Carbon Nanotubes for Commercial Applications, First Edition.
Edited by Jeenat Aslam, Chaudhery Mustansar Hussain, and Ruby Aslam.

deposition technique, catalyst engineering, growth temperature, growth duration as well as the type of carrier and hydrocarbon gas [3–5].

After synthesis of CNTs, its analysis became an important task. To analyze CNTs, researchers utilized various nanomaterial characterization tools such as scanning electron microscope (SEM), transmission electron microscope (TEM), atomic force microscope (AFM), Raman spectroscopy, and X-ray diffraction (XRD). SEM and AFM are useful for analyzing morphology of CNTs while TEM, Raman spectroscopy, and XRD are useful for the characterization of structural quality of CNTs.

All of the discussions mentioned motivated us to write a present book chapter. The chapter deals with evolution of Carbon from graphite, diamond, graphene to CNTs. It covers in detail various structures of CNTs, defects in CNTs, and their applications. It also covers the various synthesis and characterization techniques of CNTs. Finally, we made a broad discussion on the synthesis of selective CNTs.

1.2 Evolution of Carbon: Graphite to CNTs

The fundamental constituent or the building block of CNTs is Carbon. This group IV element of the periodic table is well known for its incredible ability to form crystalline solids and a variety of other compounds. It is placed in sixth position in the periodic table. Two out of its six electrons lie in 1s orbital, remaining electrons form sp^3, sp^2, or sp hybrid orbital. These four valence electrons constitute allotropes of Carbon such as CNT, diamond, graphite, graphene, and fullerenes [6]. It is an important fact that existing electronic bonds are very poor in the outer two orbitals as compared to the first orbital. Due to the weaker attraction of outer shell electrons than inner shell electrons, outer orbital electrons participate in electron hybridization. Small energy gradient between outer two orbitals (i.e. 2s and 2p) aids the overlapping of orbital wave functions favoring electron hybridization. There are three accessible mixing of atomic orbitals in carbon atoms making hybrid orbitals typically referred to as sp, sp^2, and sp^3 hybridization. In sp hybridization, one 2s and one 2p electrons participate in mixing (forming σ-bonds) and leaving two 2p electrons free of mixing (π-bonds). sp^2 hybridization involves the mixing of one's orbital electron and two 2p electrons. In sp^3 hybridization, all outer shell electrons of carbon atoms take part in mixing. The orbitals are focused on corners of a tetrahedron restricted to carbon atoms. According to hybridization, a carbon atom makes bonds with minimum one to maximum of four partners to produce compounds. Structural quality of carbon-based compounds and allotropes also depend on type of hybridization. For sp^1 hybridization, $(l + 1)$ σ bonds take charge to form generally one-dimensional local structure.

The linear chain compound of carbon i.e. "Carbyne" is an example of sp hybridization. sp^2 hybridization leads to the formation of two-dimensional structures such as graphene. 3D-structures such as diamond are formed by sp^3 hybridization. It is also noticed that in sp and sp^2 mixing one or two p orbital does not involve hybridization, instead showing their presence in the form π-bonds. Depending on orbital mixing carbon has many allotropes and among these amorphous carbon, diamond, graphite, graphene, and CNTs have received a great amount of attention [7]. These allotropes of carbon exhibit a part or completely different properties in nature. For their commendable and extraordinary multidimensional properties be it CNTs, graphene or diamond, and others, allotropes of carbon have become a hot cake for research investigations.

1.2.1 Graphite

The word graphite is taken from a Greek word "graphein" i.e. "write." Pencil, which is an attractive and widely used tool for writing and drawing, is made from it. It was invented by Debye et al. in 1917 [8]. Graphite is a combined structure of layers. Each layer is called Graphene [9]. In graphite structure, carbon atoms are situated in hexagonal fashion on the *XY* plane [10]. The distance between carbon atoms in a single layer of graphite is 0.142 nm. The separation distance between layers of graphite is approximately 0.335 nm [11].

The graphene layers are held together by weak van der Waals forces to form solid structure graphite. Each carbon atom is situated on edge of hexagon having three σ-bond in sp^2 hybridization form in which three valence electrons participate in hybridization and one valence electron exists in p_z orbital creating two-dimensional electron gas in the form of π-bond or cloud. It is spread all over on individual *XY* planes of graphite. Due to the mobility of said electron gas, graphite shows electrical conductivity. It is more reactive than diamond. The separation distance between the nearest two carbon atoms on an individual layer of graphite is around 0.142 nm that is approximately equal to band order of 1.5 and two times larger than the aromatic carbon atom's covalent radius. The separation gap among the nearest two layers of graphite is approximately two times of van der Waals radius. The weak van der Waals interaction within layers of graphite causes the layers of graphite to easily move along the *XY* plane. Generally, two types of graphite are found in nature abundantly: a stable form that is hexagonal or α-graphite and rhombohedral or β-graphite. β-graphite is more abundant in nature than α-graphite. By heat treatment, β-graphite can be changed into α-graphite. Other than these two, a few more types of graphite are found in nature. Sometimes, due to some disorder in stacking, no more attraction exists between layers and hence individual layers of graphene or graphite randomly turn around the *z*-axis and move in the *XY* plane resulting in a turbostratic structure.

1.2.2 Diamond

Diamond is an allotrope of carbon with sp^3 hybridization. Every carbon atom in diamond is bonded to four nearest carbon atoms. Each diamond cell has a tetrahedron structure with a bond length of 154.45 pm. Mainly, two types of structures of diamond are found: cubic and hexagonal. Cubic structure is more available than hexagonal [12, 13]. Hexagonal-shaped diamonds are very precious and rare to find. It was first found in 1967 in a meteorite. Still, there is a possible way to synthesize artificially hexagonal type diamond from graphite by heat treatment of graphite under high pressure with ambient temperature along the vertical axis. Some other structures are also found in which nitrogen molecules are mixed with carbon atoms. Some noticed nitrogen-contained diamonds are Ia, Ib, IIa, and IIb. If nitrogen contamination is like platelets then it is Ia structure of C_3N composition. In Ib type nitrogen, molecules are homogeneously spread all over the structure. And, if the diamond structure has null nitrogen then it is called IIa. The last structure has a very small probability of being in nature. The last type, i.e. IIb, is a semiconducting diamond. In that case, nitrogen is completely absent. Instead of nitrogen, IIb type of the diamond structure has some contamination of aluminum. In a normal diamond unit cell, eight atoms exist having face-centered cubic packing. The lattice constant of diamond lattice is 356.68 pm [14–18]. Another noticeable point about diamond's structure is that it's packed in a sphalerite kind instead of dense packing. Its structure is a penetration of two faces. Its centered cubic lattice can be moved along unit cell space diagonal. Another thing is that if diamond is heated with specific conditions, i.e. 3750 °C temperature and 1840 psi pressure, it is converted to graphite. It is also possible to make a hexagonal structure like its cubic type structure from tetrahedrons of carbon with different configurations. In the specified configuration, the cell is made from four carbon atoms with lattice parameters a_0 = 252 pm and c = 412 pm [15].

1.2.3 Graphene

As described in Section 1.2.1, if a single layer of graphite is peeled-off from the bulk material, then it is called graphene. It is a single planar structure containing covalent bonded carbon atoms [19] thus also known as planar allotrope of carbon. Buckyballs, nanotubes, and graphite are allotropes of graphene. If graphene is folded in a shape of sphere then it becomes the buckyball or if rolled along any of its axes then it takes the form of a nanotube. Stacked layers of graphene through its *XY* plane are graphite. Therefore, graphene may be called a generator of three allotropes of carbon. Thus it is vital to study the properties of graphene to understand the properties of CNTs and the other two allotropes. The 2s, $2p_x$, and $2p_y$ electrons interact in hybridized fashion giving graphene its characteristics such as physical and electrical ones [19].

sp^2 hybridization results in the formation of very stable covalent bonds (σ-bonds) on graphene surface which is the reason for graphene and CNTs being the strongest material with superior mechanical properties among other materials known. Not hybridized atomic $2p_z$ orbital forms π-bonds normal to *XY* plane due to the electron gas delocalization. These π-bonds provide graphene with distinctive electronic properties [20]. Structurally, graphene has two types of lattice arrangements: direct and reciprocal lattice.

1.2.3.1 Direct Lattice

The honeycomb structure of direct lattice graphene can be explained by ball and stick model. In the ball and stick model, carbon atoms are denoted by ball and sticks represent σ-bonds. σ-bonds of graphene have a bond length around 1.42 Å. Such type of lattice may be considered as a Bravais lattice. The Bravais lattice basis is 2 given as two π electrons per unit cell and depicts enhanced electronic properties of graphene [21]. The primitive unit cell of mentioned honeycomb lattice may be taken as an equilateral parallelogram. The parallelogram side length is as following:

$$a = \sqrt{3}\,(\text{length of a } \sigma\text{-bond}) = 2.46\ \text{Å} \tag{1.1}$$

The vectors of primitive unit cell are as following:

$$a_1 = \left(\frac{\sqrt{3}a}{2}, \frac{a}{2}\right) \tag{1.2}$$

$$a_2 = \left(\frac{\sqrt{3}a}{2}, \frac{a}{-2}\right) \tag{1.3}$$

where the vector $|a_1| = |a_2| = a$ describes the separation distance between two nearest carbon atoms.

$$R_1 = \left(\frac{a}{\sqrt{3}}, 0\right) \tag{1.4}$$

$$R_2 = (-a_2 + R_1) \tag{1.5}$$

$$R_2 = \left(-\frac{a}{2\sqrt{3}}, -\frac{a}{2}\right) \tag{1.6}$$

$$R_3 = (-a_1 + R_1) \tag{1.7}$$

with $|R_1| = |R_2| = |R_3| =$ length of σ-bond.

1.2.3.2 The Reciprocal Lattice

The rotated lattice with an angle of 90° with reference to the direct lattice is called a reciprocal lattice of graphene [17]. Similar to the direct lattice, the reciprocal lattice also has carbon atoms arranged in a honeycomb structure. The vectors of said lattice are given below:

$$b_1 = \left(\frac{2\pi}{\sqrt{3}a}, \frac{2\pi}{a}\right) \tag{1.8}$$

$$b_2 = \left(\frac{2\pi}{\sqrt{3}a}, -\frac{2\pi}{a}\right) \tag{1.9}$$

where $|b_1| = |b_2| = \dfrac{4\pi}{\sqrt{3}a}$.

1.2.4 Carbon Nanotubes

The history of Carbon tubular structures start after the invention of electron microscopes in 1950 [22]. In 1952, a hollow tube of carbon was invented by a Russian researcher, Radushkevich and Lukyanovich, however, the article did not gather much attention being published in a local language [23]. Baker et al. again investigated related structure in 1972 [24]. In 1976, Oberlain et al. described single cylinder hubs in a graphite structure of carbon fibers [25]. Finally, in 1991 Sumio Iijima, an electron microscopist, discovered the cylindrical structure of carbon while analyzing fullerenes, another allotrope of carbon [26]. His discovery caught the attention of worldwide researchers in the field of nanotechnology. Almost two years later, the same group of researchers achieved another milestone in the field of nanotechnology with the invention of single wall carbon nanotubes (SWNTs). CNT is one of the essential allotropes of graphene and is formed when a graphene sheet is rolled into a seamless hollow tube. Practically, it is impossible to roll a graphene sheet into a cylindrical structure because of its ultra-small dimensions. It is only said to understand the structure of CNTs. Generally, diameter of CNTs lies between 0.5 and 5 nm thus termed as nanotubes. The length of CNTs, however, falls from a few mm to several cm [7]. CNTs are also called tubical fullerenes. Generally, CNTs are found in two structures: (i) SWNTs (see Figure 1.1a) and (ii) multi wall carbon nanotubes (MWCNTs) (see Figure 1.1b). If single graphene sheet is rolled to make CNTs then it is called SWNTs and if more than one graphene sheet is concentrically rolled in the "Russian dolls model" then the tubes are called MWNTs. SWNTs are discussed below followed by an introduction about the MWNTs.

1.2.4.1 SWNTs: Types and Structure

Based on structure, SWNTs are divided in three categories:

i. Zigzag
ii. Armchair
iii. Chiral

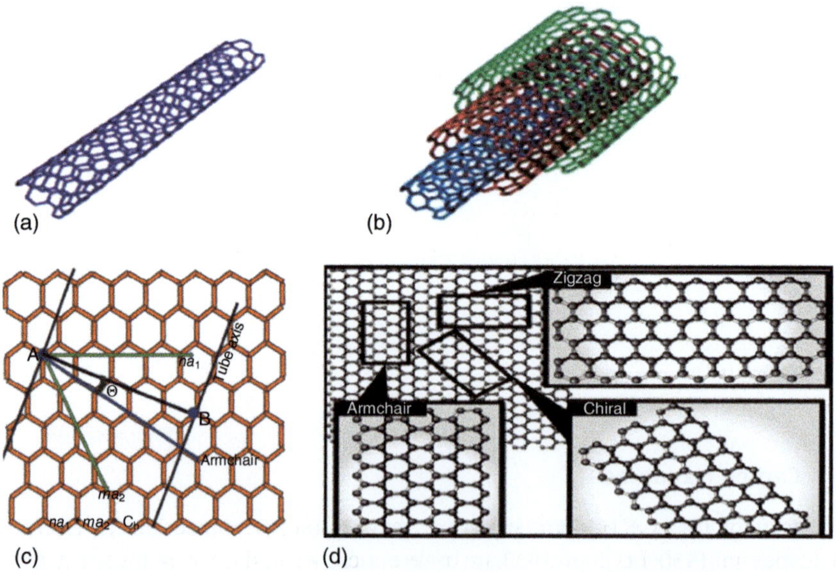

Figure 1.1 Figure (a) and (b) show the schematic diagram for SWNTs and MWNTs, respectively. Figure (c) shows the honeycomb structure of carbon atoms; (d) shows different types of CNTs structure. Source: Ahmad Aqel et al. [27]/with permission of Elsevier.

The types of SWNTs depend on the type of rolling of graphene sheets, which are the basis of the CNTs tubular structure. It is possible to calculate the chirality (helicity) as well as the diameter of SWNTs from a vector, called the chiral vector C_h. It is defined as

$$C_h = na_1 + ma_2$$

where $n, m = 1, 2, 3, \ldots$ with notation as (n, m).

The diameter of SWNTs according to the said model is calculated by the following equation:

$$d = \frac{\text{Chiral Vector}}{\pi} \tag{1.10}$$

$$d = \frac{\sqrt{3} * b\sqrt{n * n + nm + m * m}}{\pi} \tag{1.11}$$

where b has the value of 0.142 nm, which represents the carbon–carbon bond length. It is observed that the diameter of CNTs strongly depends on its chirality. A CNT having (5,5) chirality tends to have a smaller diameter than a CNT with (10,10) chirality. The steps for calculating chirality are as mentioned below [27, 28]:

I. Plot the tube axes. The tube axes are the nanotubes edges. If the tube axes join with each other in the form of a cylinder then it becomes nanotubes (see Figure 1.1c).
II. Mark a point A along the tube axis where the tube axis interacts with carbon atoms.
III. Draw the armchair line (thin green line) by finding any point along the tube axis that travels across each hexagon, separating them into two equal halves.
IV. Take point B on the second tube axis at the intersection of the tube axis and carbon atoms. It should be the nearest intersection point from the armchair line.
V. Draw a line AB representing the chiral vector C_h. The chiral vector is equivalent to the CNTs circumference.
VI. Angle θ between the armchair line and chiral vector decides the types of SWNTs.
VII. If the armchair line and chiral vector overlap each other, i.e. when $\theta = 0°$ $(n = m)$, then the resulting SWNTs are called the armchair SWNTs (see Figure 1.1d).
VIII. If the said wrapping angle is $\theta = 30°$ $(m = 0)$, then zigzag nanotubes are formed (see Figure 1.1d).
IX. If the wrapping angle θ is between 0° and 30°, then the fabricated SWNTs are called chiral.

Among all types of CNTs, chiral type CNTs have the superior place. Because chirality is an elemental theory to investigate different configured CNTs and their relative electronic band structures. Therefore, it is important to understand the concept and its application to identify CNTs structure. Table 1.1 shows parameters and their relative equations for CNTs.

Table 1.1 Various parameters for SWNTs.

S. no.	Symbol	Name	Chiral CNT	Armchair CNTs	Zigzag CNTs
1.	C_h	Chiral vector	$C_h = na_1 + ma_2 = (n,m)$	$C_h = (n,m)$	$C_h = (n,0)$
2.	C_h	Length of chiral vector	$C_h = \lvert C_h \rvert = a\sqrt{n^2 + nm + m^2}$	$C_h = a\sqrt{3n}$	$C_h = an$
3.	d_t	Diameter	$d_t = \frac{a}{\pi}\sqrt{n^2 + nm + m^2}$	$d_t = \frac{an}{\pi}\sqrt{3}$	$d_t = \frac{an}{\pi}$
4.	Θ	Chiral angle	$\cos\theta = \frac{2n + m}{2\sqrt{n^2 + nm + m^2}}$	$\theta = 30°$	$\theta = 0°$
5.	N	Number of hexagons/cells	$N = \frac{2C_h^2}{a^2 g_d}$	$N = 2n$	$N = 2n$

1.2.4.2 Chirality

The word chirality comes from the Greek language that means HAND. It is used to represent reflection symmetry between an object and its mirror image. In general, a chiral object is anti-symmetric to its mirror image. For example, if we take the mirror image of our left hand and try to superimpose it on our left hand, it does not accurately superimpose, indicating that hands are chiral objects. Similarly, CNTs that are superimposed to their mirror image are called achiral CNTs. Zigzag and armchair CNTs are examples of achiral CNTs.

1.2.4.3 Electronic Properties of CNTs

Electronic properties of CNTs provide great opportunities in nano-electronic research applications. Ultra-small dimensions and ultra-symmetric structures create remarkable quantum effects and electronic properties of nanotubes. Due to the circumferential confinement effect on tubes, SWNTs and MWNTs [28–33] show the quality of a quantum wire. Experimental investigations have proved that MWNTs and the rope of SWNTs behaved like a parallel assembly of single SWNTs. The electronic conductance for said assembly is given by

$$G = G_o M = \left(\frac{2e^2}{h}\right) M \tag{1.12}$$

where, the calculated value of the quantized conductance G_o is 12.9 kΩ^{-1}. Another parameter used in the equation, i.e. M is the measurement of exact conducting channels, and its exact value for an ideal defect-free SWNT is 2. The value of M is determined by the intrinsic properties of nanotubes, coupling between tubes impurities, defects in the structure of tubes, interaction with the substrate, and on contacts made for electronic connections. Therefore, the experimental value of conductance is less than the quantized value [34].

Many research groups reported about graphite's resistivity. It depends strongly on the quantity of graphite taken for analysis. The best quality graphite has the resistivity around 0.4 μΩm at room temperature [35]. In the case of CNTs, the MWNTs, as well as the rope of SWNTs, have much higher resistivity than the best quality graphite. It is also reported that the resistivity of the nanotubes mentioned above decreases with temperature. These results were found due to a random orientation of nanotubes on the substrate. The same measurements were made for purified CNTs aligned between four electrodes and were found to be much less than 0.4 μΩm [34, 36]. In defect-free nanotubes, π electrons are more distributed as compared to graphite. It happens due to σ–π re-hybridization. These conditions made graphene more conductive as compared to graphite [34]. Therefore, CNTs are called 1-D conductors.

1.2.4.4 Optical Properties of CNTs

Defect-free SWNTs have a direct band gap. They also have a well-defined structure of the sub band. Such type of structure is considered ideal for optical applications. Optical spectra of SWNTs may be obtained by resonant Raman spectroscopy [35] and by fluorescence spectroscopy [36, 37]. An optical spectrum obtained for grouped SWNTs, as well as for graphite under compression, is shown in Figure 1.2. Three important peaks that are found in SWNTs are invisible in graphite and attributed to symmetric transitions between the lowest sub-bands in semiconducting (A and B) and metallic (C) tubes. Generally, it is found that grown CNTs area a mixture of semiconducting and metallic tubes. In CNTs, the measured peak position and intensity of optical spectra provide information about the electronic structure or tube chirality or (D, Θ). Thus, for a detailed composition investigation of CNTs, optical spectra become indispensable. To understand the optical properties of CNTs, it is necessary to understand their band structure and density of states (DOS), which is already discussed in the electronic dispersion Section 1.2.4.3.

The optical transportation is only possible when electrons or holes are triggered from a lower energy level to a higher energy level. The energy required for optical

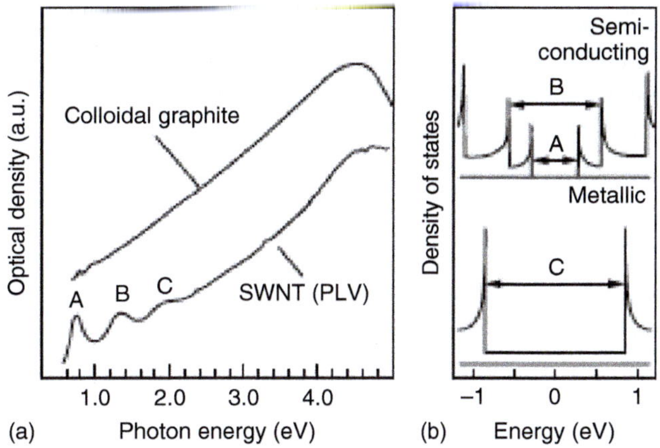

Figure 1.2 Figure shows the UV–vis–NIR analysis for various SWNTs samples. Figure (a) shows the absorbance spectra of SWNTs as well as of colloidal graphite. Figure (b) shows plots of the density of states for semiconducting and metallic CNTs. Source: Hagen and Hertel [37]/with permission of American Chemical Society.

transactions is donated by E_{pq}. There are two selection rules that are defined for that. First, if $p - q = 0$, it is defined for inter-band transitions. It is symmetric to the Fermi energy of the polarized light along the tube axis. Second, it requires light that is normal to the tube axis. However, it does not appear in optical spectra. It happens due to very weak transitions [37].

1.2.4.5 Chemical Properties of CNTs

Due to ultra-small dimensions and ultra-large specific area with σ–π re-hybridization, CNTs are highly sensitive to chemicals and environmental interactions. It is important to study the chemical properties of CNTs such as opening, wetting, filling, adsorption, charge transfer, doping, and intercalation for application purposes.

Opening It is reported by several research groups that the ends of CNTs are more reactive than their complete structure. CNTs' ends have metallic catalyst particles with a large curvature on their open end. Therefore, many approaches have been focused on opening CNTs ends such as treatment with acids, plasma treatment, etc. [38, 39].

Wetting and Filling CNTs are hydrophobic in nature. Mostly aqueous solvents are unable to wet CNTs. However, some organic elements, HNO_3, S, Cs, Rb, and Se as well as some oxides Bi_2O_2 are able to wet CNTs, if they are pressurized with capillary pressure. Using this technique, some other nonwetting elements may also be injected into CNTs. The capillary pressure for nanotubes is proportional to $\left(\frac{1}{D}\right)$ [40].

Adsorption and Charge Transfer It is expected that CNTs have enhanced molecular adsorption as well as charge transfer. Researchers have reported that CNTs show very good adsorption and charge transfer from oxygen to CNTs at room temperature. The localized sites on CNTs where pores exist, interstitials in tube bundles, the surface of CNTs as well as a groove between two attached CNTs are the perfect places for charge transfer and adsorption. In fact, the mentioned sites show the capability of CNTs for adsorption and charge transfer.

The CNTs property discussed above is successfully used in NO_2, $C_6H_5NO_2$, C_6H_6, NH_3 as well as in CH_4 sensing. When these molecules interact with CNTs, a change in CNTs resistance is observed. It is possible to design electronic device for sensing application on that principle [41, 42].

Chemical Doping, Intercalation, and Modification The discussed adsorption method may be used for noncovalent bonding doping of CNTs to make p-type as well as n-type CNTs to enhance their electronic conductivity. Intercalation of alkali metals is also used to enhance electronic conductivity. Researchers reported on behalf of the experimental evidence that alkali metals diffuse into inter tube space or defects that exist in CNTs, and hence enhance the CNTs charge transfer ability. To enhance the electrochemical capacity of CNTs, CNTs itself can be used as electrodes. In CNT-based electronic devices for energy production or harvesting, the flow of electrons is generated by reduction and oxidation reactions occurring at the electrodes [43–47].

1.2.4.6 Defects in CNTs

Defects, being a deterministic factor for the physical and chemical properties of a nanomaterial, are considered significant. Although nanomaterials like CNTs are perfect and have unique electrical and mechanical properties which change due to presence of certain defects such as vacancies, heptagon–pentagon pairs type transformations, doping, and interstitials, edges and adatoms (see Figure 1.3) [49], resulting CNTs can be used for further applications, such as sensing, hydrogen storage, and drug delivery systems.

Structural Defects Structural defects in CNTs arise due to their topology distortion caused by the introduction of pentagonal, heptagonal, and octagonal rings into the hexagonal carbon network, affecting nanotubes electrical properties resulting in a conical structure with sharp tips due to the presence of a pentagonal ring at the nanotubes peaks (as shown in Figure 1.3a,b). Also, different vertex angles are obtained by a pentagon insertion into a hexagonal carbon graphitic structure leading to 30° bend dictating that pentagon and hexagon are separated maximally that is they are placed opposite to each other (as shown in Figure 1.3c) and 0° bend is obtained when both structures are combined. Thereby, we can conclude that the vertex angles can indicate the count of hexagon–pentagon pairs present in the CNTs.

A spiral CNT (as shown in Figure 1.3d) is an out-of-plane structure obtained by introducing pentagonal and heptagonal rings in a perfect graphitic sheet and then rolling it with a rotation angle. Atoroidal CNT (as shown Figure 1.3e) is an in-plane structure obtained by connecting CNTs of various diameters by inserting pentagons and hexagons [50].

Bond Rotation Thrower stone wales (TSW) type defects produced by bond rotation in the nanotube are due to 90° rotation of C—C bond in a hexagonal network such as in fullerenes, resulting in a translation of two heptagons and pentagons (as shown in Figure 1.3f) [48–51]. Unlike the conical-end structure formed in structural defects, this TSW type defect does not result in large curvature deviations of the nanotube. TSW type defects cause a plasticity failure at elevated temperatures and can alter the chirality of CNTs. Further elongation of the tube at the defect location can lead to the nanotube collapse. Its significance lies in the innovation of nanoelectronic devices [52].

Doping-Induced Defects In order to increase the conductivity of a CNT, dopant atoms can be inserted (see Figure 1.3g) in the carbon lattice. Another method consists of the functionalization of nanotubes, which makes it suitable for biochemical and gas sensing purposes. The boron (p-type dopant) and nitrogen (n-type dopant) atoms in the internal CNT structure are used to sense carbon monoxide and water molecules. The boron (B) atom doping is carried out by arc discharge method using BN-rich as an anode and the nitrogen (N) atom doping is carried out by ferrocene-melamine mixture pyrolysis at high temperatures [53]. As a result of doping, both nanotubes become metallic in nature with no band-gap in comparison with the undoped lattice structure. Other atoms used for doping purposes are P, S, and Si in addition to

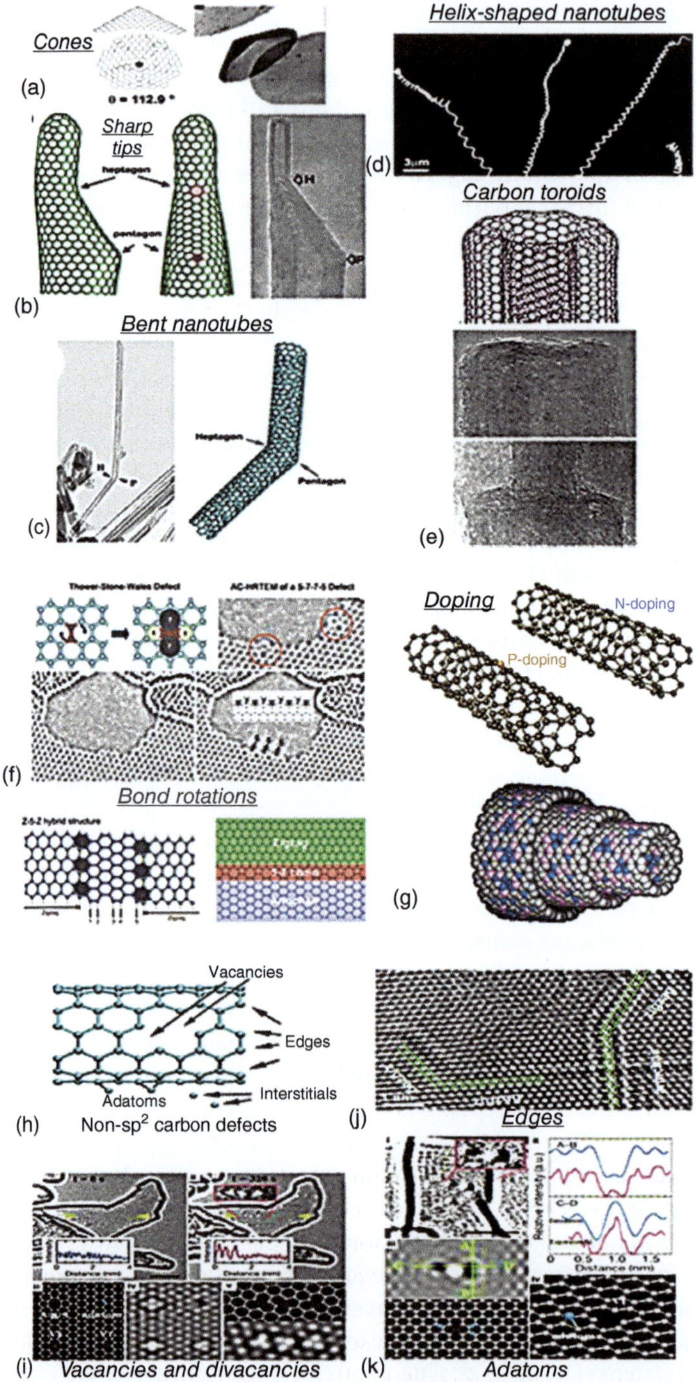
Cones
(a)
Sharp tips
(b)
Bent nanotubes
(c)
Helix-shaped nanotubes
(d)
Carbon toroids
(e)
(f)
Bond rotations
Doping
N-doping
P-doping
(g)
Vacancies
Edges
Adatoms
Interstitials
(h)
Non-sp2 carbon defects
(j)
Edges
(i)
Vacancies and divacancies
(k)
Adatoms

Figure 1.3 Figure shows the various defects involved in CNTs. Figure (a) shows the graphitic conical shape fabricated on the CNT tip. Figure (b) shows the presence of pentagon and heptagon impurities on the CNT tip. Figure (b) and (c) show bending and helices types of CNTs. Figure (e) represents the joining of two concentric CNTs by a hemitoroidal cap. Figure (f) shows the TEM analysis as well as molecular model of the TSW defect for the conversion of four hexagons into two pentagons and two heptagons. Figure (g) demonstrates nitrogen and phosphorous doped CNTs. Figure (h) shows simulated images of various non-sp^2 defects that occur in CNTs while figure (i–k) show experimentally produced vacancies, edges, and adatoms defects, respectively. Source: Lehman et al. [48]/with permission of Elsevier.

B and N atoms; all of them alter the reactivity of the nanotube, enhancing the binding energy of the sensing molecule with the doped species [54–57].

Non-sp^2 Carbon Defects These non-sp^2 hybridized defects in the CNT structure are formed due to dangling bonds, other than carbon atoms found in the lattice structure, or can be caused by vacancies, adatoms, and opened or closed nanotubes as shown in Figure 1.3h. Adatoms remain as dopants on the surface of nanotubes, in which the lattice structure stays undisturbed. Depending on the ratio between the lower and upper levels of occupied molecular orbitals of adatoms and the Fermi level of the tube, the electrons of dopant atoms are transmitted or absorbed.

1.2.4.7 CNTs Properties Modification by Chemical Functionalization Process

CNTs functionalization helps to improve their processibility and solubility as well as to mix the unique features of materials attached to CNTs with their own properties [58–63]. Chemical bounds can tailor the CNT interaction with other systems like biopolymer matrices, solvents, polymers, and other nanotubes. Approaches to CNTs functionalization are divided into noncovalent (physical) and covalent (chemical) as interactions between CNTs and attached elements as can be seen in Figure 1.4a–e [64, 66]. A distinction that is more detailed can be made if we divide the functionalization of MWNTs and SWNTs, as well as that of tube bundles and individual tubes [67].

Covalent functionalization constructs from the functional entities covalent linkage on the surface of CNT. It can be further divided into indirect covalent functionalization with carboxylic groups and direct covalent side wall functionalization. Direct covalent side wall functionalization appears due to hybridization change from sp^2 to sp^3 along with a conjugation loss. Indirect covalent functionalization (also known as defect functionalization) utilizes defect sites that already occur in the CNT. These sites can appear in a form of open ends as well as in the form of holes in the walls with attached, for example, carboxylic groups. Oxygenated sites also can be considered as defects [65]. Acid chloride, in major cases converted from carboxylic acid groups, is utilized to increase the CNTs reactivity with further treatment via amidation or esterification reaction [68].

Noncovalent functionalization is founded primarily on secondary interactions like van der Waals' forces or π-stacking. All the mentioned above functionalization methods are exohedral derivatizations. Endohedral CNTs functionalization (the

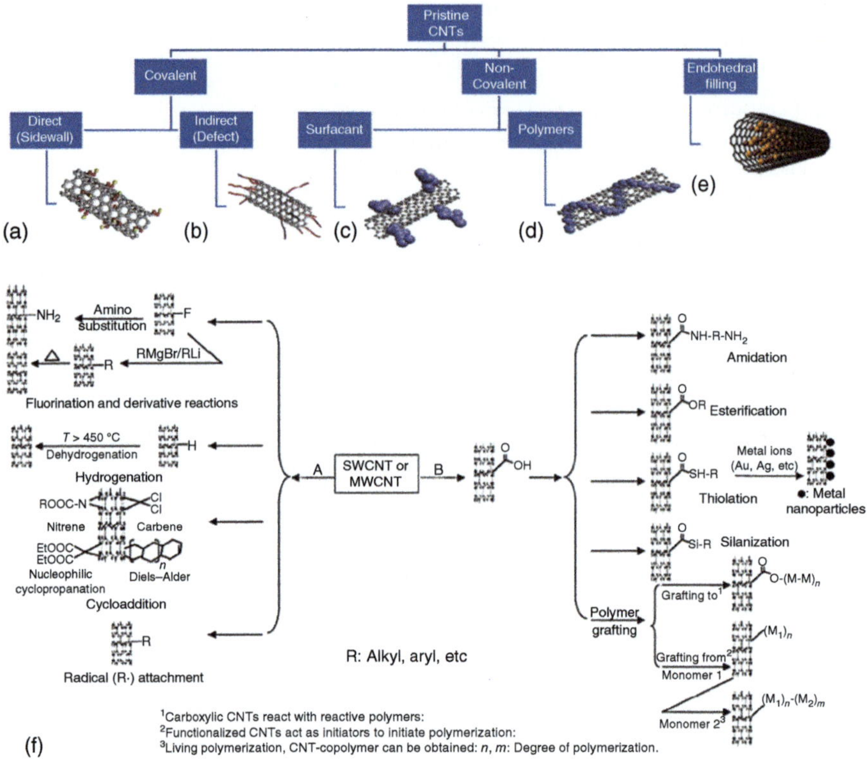

Figure 1.4 Functionalization types of CNTs: (a) chemical side wall functionalization, (b) defect-group functionalization, (c) physical exohedral functionalization with surfactants, (d) physical exohedral functionalization with polymers, (e) endohedral functionalization. For A–C approaches, the nanotubes are idealized, however, defects can be observed in reality. Source: Muhammad-Ahson Aslam et al. [64]/MDPI/Public Domain CC BY. Figure (f) CNTs covalent functionalization strategies: A-route represents side wall functionalization, B-route demonstrates defect functionalization. Source: Ma et al. [65]/ with permission of Elsevier.

tubes filling with atoms or small molecules) is a special case of functionalization [69, 70].

Covalent Functionalization Covalent functionalization can be implemented onto CNTs end caps or on their side walls that have numerous defects. Strategies for covalent functionalization are presented in Figure 1.4f [65]. Direct covalent side wall functionalization can be performed via reaction with particular molecules which possess high chemical reactivity. Initially, CNTs fluorination became popular due to the supposition that CNTs side walls are inert [71, 72]. CNTs that were functionalized via the fluorination method, possess weaker C—F bonds in comparison with alkyl fluorides [73]. Therefore, fluorinated CNTs provide replacement sites for additional functionalization [74]. Fluorine atoms substitutions by hydroxyl, alkyl, and amino groups were successfully achieved to date [75]. Plasma treatment can be utilized for MWNTs after functionalization. For example, maleic anhydride joint with MWNTs after plasma treatment of the latter improves the obtained material mechanical properties as well as electrical conductivity [76]. Moreover, the grafting approach can be used to synthesize SWNTs which will possess high polymer layers covalent bonding density [77]. In [77] styrene copolymers benzyl chloride groups reacted with the SWNTs alkyne groups which lead to a greater efficiency of polymer chains grafting.

The chains of polymer grafting on the surface of CNTs are associated with the establishment of strong chemical bonds appeared between the nanotubes and polymer. Plasma treatment provides an increase in the CNTs ability to disperse when mixed with the corresponding polymer. Moreover, it increases structural properties if implemented in bulk composites. The drawback of this method consists in the limitation of grafted polymer value due to macromolecules high steric hindrance and comparatively low reactivity. It can result in degradation of nanocomposites features [78].

Noncovalent Functionalization Methods of chemical functionalization have two main disadvantages which consist in a creation of a huge defect number onto CNTs side walls as well as in some cases due to ultrasonication treatment, CNTs breakage into small pieces. These drawbacks lead to disruption of π-electron system in CNTs and significant deterioration in its mechanical and electrical properties. Physical functionalization overcomes these issues. This method is based on supramolecular chemistry [79]. It comprises van der Waals forces, electrostatic effects, hydrogen bonding, π–π interactions, ionic bonding, and hydrophobic forces [80]. The modeling of functionalization materials can be performed with the usage of density functional theory (DFT). Thus, CNTs functionalized with free-based tetraphenyl porphyrin molecules possess biding energies from 1.1 up to 1.8 eV and demonstrate a strong π–π interaction between CNT and tetraphenyl porphyrin surface [81]. CNTs solution with polymers like polystyrene, polyphenylene, polyvinylene result in the polymer wrapping around CNTs forming supermolecular CNTs complexes [82]. Surfactants are also very popular materials for CNTs functionalization [83–89]. Many studies were reported on nonionic surfactants like polyoxyethylene octyl phenyl ether, nonylphenol ethoxylate as well as polyoxyethylene 8 lauryl [83–85]. The nonionic surfactants treatment is grounded on a strong hydrophobic attraction between the tail group and the surfactant solid surface. After surfactant adsorption on the surface of the material for functionalization, the molecules of the surfactant are self-assembled in micelles beyond the critical micelle concentration [85]. Furthermore, researchers also investigated numerous anionic surfactants such as poly(styrene sulfate), sodium dodecylbenzene sulfonate, and sodium dodecylsulfate [86, 87]. Moreover, cationic surfactants like cetyltrimethylammounium 4-vinylbenzoate and dodecyl tri-methyl ammonium bromide were studied as well [88]. The mechanism of aggregates formation preventing is based on the lowering CNTs surface tension caused by surfactant physical absorption on the CNTs surface. By steric or electrostatic repulsive forces, CNTs treated with surfactants overcome the van der Waals attraction. This method effectiveness is influenced by the surfactants properties, polymer matrix as well as by medium chemistry [89].

Moreover, some researchers report a SWNT functionalization technique, which includes both physical and chemical approaches. For instance, in [90] scientists present cyanate ester (CE) reinforcement. Authors of this work used a dispersant hydroxyl polyimide-graft bisphenol to enable usage of the both above-mentioned functionalization methods. Then, epoxidized SWNTs were used with diglyceryl acrylate (PIOH-BDA). Thus, SWNTs were functionalized physically with the help of PIOH-BDA dispersant. This approach led to a better SWNTs dispersion due to the interaction caused by 1,4-butanediamine (BDA) side chain steric hindrance and the main chain. In addition, epoxidized SWNTs are also functionalized chemically by PIOH-BDA due to PIOH-BDA reaction with SWNTs epoxide groups and CE.

Applications of Functionalized CNTs Applications of functionalized CNTs cover many spheres of human life. Thus, in [91] researchers developed an immunosensor of voltammetric factor-alpha tumor necrosis (TNF-α). This sensor comprises Au nanoparticles included in functionalized with thiol MWNTs (AuNPs/S-MWCNTs) as well as bimetallic Ni/Cu-MOFs. AuNPs/S-MWNTs in this construction work as a platform for sensor when Ni/Cu-MOFs works as an amplifier of the sensor. First of all, a mixture of S-MWNTs and AuNPs was used to create the platform of the sensor on glassy carbon electrode. The sensor platform plays two roles in the device: enhancement in conductivity on the electrode surface and binding sites formation for capture antibody 1-TNF-α. After that, via amino-Au affinity, capture TNF-α antibodies were paired with the platform of the sensor. Then, through immune reaction that occurs between Ni/Cu-MOFs which are conjugated with secondary TNF-α antibodies and AuNPs/S-MWNTs, the immunosensor was created. Due to bimetallic MOFs' big internal surface and its porosity secondary antibody 2-TNF-α conjugate without difficulty to Ni/Cu-MOFs through electrostatic and π–π interactions leading to electrochemical performance improvement [92]. Such immunosensor possesses a small border of detection which is 2.00 fg ml^{-1} as well as a linear range that lies from 0.01 up to 1.0 pg ml^{-1} [91].

A membrane for desalination based on CNTs functionalized with polytetrafluoroethylene/oleic acid was developed in [93]. First, CNTs were functionalized with oleic acid. After that, the composite membrane was developed by solution coating of polyvinylidene fluoride that contains functionalized CNTs onto support membrane made from polytetrafluoroethylene. Polytetrafluoroethylene support membrane has hydrophobic pores that help to pass water vapor. The active layer coating of CNTs functionalized with polytetrafluoroethylene/oleic acid onto the polytetrafluoroethylene support membrane leads to an increase in salt rejection by about 30%. CNTs functionalized with oleic acid may decrease the oxidized CNTs hydrophilicity. The polyvinylidene fluoride polymer existence on the functionalized CNTs surface and between the CNTs improves the membrane quality of operation in the vacuum membrane distillation process and increases the membrane hydrophobicity. In contrast, the polyvinylidene fluoride polymer presence leads to the CNTs interconnection in the coated active layer. The optimum fabrication conditions consist in concentration of 10 000 ppm, temperature of 70 °C as well as the flow rate of 150 LPH (liters per hour). The membrane based on polytetrafluoroethylene/oleic acid-functionalized CNTs composite allows to achieve high salt rejection of 99.45% as well as water flux of 89 LMH (liters per square meter per hour) in the vacuum membrane distillation process.

Moreover, functionalized SWNTs can be used for visible-light photocatalysis [94]. In [94] researchers performed endohedral functionalization of SWNTs with 10-phenylphenothiazine (PTH). The method used in this work is based on PTH solution refluxing in the cap-opened SWNTs presence. This technique made it possible to encapsulate inside the SWNTs as much as 8% of PTH. The obtained PTH/SWNTs structure demonstrates intense photoemission under visible light irradiation that makes it useful as a heterogeneous photocatalyst in reactions of single-electron transfer dehalogenation. Additionally, this material can be used for hydrodehalogenation reactions and more complex reactions like synthesis of tetrahydroisoquinolines as well as aldehydes α-alkylation when exposed to 420 nm light irradiation. This material possesses a high photocatalytic activity, showing turnover numbers of 3200. Furthermore, PTH/SWNTs structure demonstrates stability and recyclability. The physical process leading to these properties consists of the following. Under radiation exposure, an excited PTH electron is transported from the molecule π system to the delocalized SWNT π-cloud, therefore justifying the enhanced photocatalytic activity. Therefore, injection of electrons with subsequent electron–hole separation leads to the photocatalytic activity under visible light exposure, which initiates reactions of single-electron transfer dehalogenation.

MWNTs functionalized with pyrene derivatives comprising a hydroxyl group can be utilized as a polybutylene terephthalate (PBT) composites filler [95]. In [95] researchers physically functionalized MWNTs with 1-hydroxypyrene (POH), 1-pyrenebutanol (PBOH), as well as with 1-pyrenemethanol (PMOH). Then, functionalized MWNTs are used for compound fabrication with PBT with the help of melt extrusion. In the process of extrusion, a transesterification reaction appears between the PBOH hydroxyl group and PBT ester group, thus grafting PBT onto MWNTs. Because of the most stable intermediate formation as well as the minimum change in free energy at the rate-determining step for PBOH, the PBT reactivity turned out to be greater with PBOH than with PMOH or POH, which was proved via density functional theory calculations. Due to the same reason, the biggest grafted PBT amount can be observed in the structure PBOH-MWNTs. Estimated energies of interfacial adhesion between MWNTs and PBT showed that the PBOH-MWNTs structure possesses greater interfacial adhesion energy with PBT than MWNTs functionalized with POH or POMH. In addition, the structure PBT/PBOH-MWNT demonstrates the maximum fatigue life tensile property as well as tensile property between the presented composites.

Functionalized CNTs can be used as selective and highly sensitive elements of photodetectors and light-harvesting devices [96–99]. In [96] researchers developed a tunable photodetector based on SWNTs functionalized with pyrenecyclodextrin. They build a field-effect transistor (FET) based on the synthesized material in order to detect a fluorescent adamantyl-modified Ru complex (ADA-Ru). Under light irradiation of 280 nm wavelength, the obtained photodetector transfer curve shifts to a negative gate voltage by approximately 1.6 V as well as the aforementioned device conductivity decreases sharply, signifying a charge-transfer process to the SWNTs from pyrenecyclodextrins. On the other hand, the photodetector transfer curve shifts to a positive gate voltage by approximately 1.9 V and its conductivity growths slowly under 490 nm light exposure when ADA-Ru complex is introduced in the system. It also demonstrates a charge-transfer process, but at that time it goes from pyrenecyclodextrin-SWNT hybrids toward the ADA-Ru complex. As soon as the radiation exposure is turned off, the conductivity volume recovers to its initial value.

In [97] researchers developed a NIR–photodetector based on functionalized with ferricyanide SWNTs. It was demonstrated that SWNTs functionalized with quite big diameter molecules of ferricyanide show the change in emission and photo-absorption in the same CNT simultaneously. It is caused by partial electron transfer which vanishes the optical transition and localizes electrons. The ferricyanide can be reduced by hydrogen peroxide that will reverse the occurred electron transfer, which leads to a re-establishment of emission and absorption spectra. The selectivity of such device manifests itself in the form of different absorption that leads to a higher response with smaller gap species. In [98] researchers functionalized SWNTs with CdS. They developed a CdS-SWNTs film-based device with two terminals. Initially, researchers intended to develop solar cells based on functionalized CNTs but they also observed its behavior under light exposure that demonstrates its applicability in visible light photodetectors. This device shows current flow increase under light irradiation. The change of current saturates in approximately 30 seconds under light exposure. In [99] a tunable photodetector based on functionalized SWNTs was developed. SWNTs were functionalized with ZnO nanoparticles that were utilized as antennas in this structure. The improved responsivity of as-prepared device refers to the synergistic effects of the inherent SWNTs transistors sensitivity as weak as the ZnO nanoparticles photoresponsivity occurring from the zero dimensionality materials nature and the big surface-to-volume ratio.

The photodetector demonstrates significant photoswitching effects with decent reproducibility and reversibility because of photoinduced oxygen desorption as well as adsorption from the surface of nanoparticles. The photodetector under consideration possesses fine conductance tunability in the range of UV light spectrum. Such devices can be utilized in various applications like imaging, UV detection, optical communications, and memory storage.

CNTs are a very useful material for development of gas and molecule sensors [100–105]. Thus, in [100] researchers used covalently functionalized SWNTs as a sensitive element for ammonia detectors. Researchers functionalized SWNTs with poly(*m*-aminobenzene sulfonic acid) (PABS). SWNTs-PABS detector fast recovers its conductivity when ammonia is replaced with nitrogen. When NH_3 is supplied to the detector, significant changes occur in the SWNTs-PABS structure. Due to this fact, such detector can detect 5 ppm concentration of NH_3. In [101] alcohol vapor detectors based on covalently functionalized MWNTs were developed. Construction of this detector includes bundles of functionalized MWNTs with COOH groups deposited across golden electrodes on SiO_2/Si substrate with the help of AC electrophoresis. This structure allows to detect alcohol–vapor with the usage of low input power of approximately 0.01–1 μW. This detector possesses a sensitivity of 0.9 ppm alcohol–vapor concentration. Moreover, this detector can be reset by annealing the functionalized MWNTs sensitive element at a current from 100 up to 200 A in a time duration of 100–200 seconds. A chloroform vapor detector based on covalently functionalized MWNTs with COOH with poly(ethylene glycol) (PEG) was developed in [102]. Selective response of the structure can be explained via PEG chains swelling behavior wrapped around MWNTs and the interaction among PEG chains, MWNTs, as well as analytes, might be a vital factor in the sensing of gases. The change in conductivity of films occurs due to the change in intertube distance caused by polymer swelling through gas absorption. In contrast, the MWNTs-g-PEG film response to chloroform is different from the response to other solvents that are referred to hydrogen bond interaction between PEG chains and solvent molecules. MWNTs features significantly affect this interaction. Moreover, CNTs can be utilized as hydrogen gas sensors [103–105]. Hydrogen gas detector based on functionalized MWNTs with acids was investigated in [103]. Such sensor can allow detecting H_2 gas in a quantity of 0.05% at room temperature. In comparison with pristine MWNTs, MWNTs functionalized with acids possess increase in the current carrying capacity up to 35 mA from 49 μA and smaller recovery time which is 100 seconds for H_2 gas in a quantity of 0.05% compared with 190 seconds in case of pristine MWNTs. In addition, MWNTs functionalized with oxygen plasma can be used as a gas sensor for such volatile organic composites as ethanol, acetone, toluene, benzene, as well as methanol [105]. The conductivity of the functionalized MWNTs decreases with the increase of detecting vapor concentration. Such sensors possess repeatable response and reproducibility in case of benzene or toluene vapor. Methanol, acetone, and ethanol show faster response coupled with difficulties in a full baseline resistance recovery. It can be referred as the chemisorption interaction mechanism with MWNTs. In contrast, toluene and benzene are physisorbed onto MWNTs that causes fully baseline resistance recovering and slower response. It is demonstrated that MWNTs detectors are more sensitive to toluene than to benzene vapor. It can be caused by the methyl group existence in the toluene molecule that interacts better with MWNTs sidewalls than the benzene ring. This structure feature allows to develop detectors that will detect benzene selectively in the presence of xylenes or toluene with the usage of plasma-treated MWNTs functionalized with different metals or functional groups arrays chemoresistors.

1.2.4.8 Applications of CNTs

As mentioned above, the peculiar, strong and powerful optical, electronic, mechanical, as well as chemical properties, make CNTs useful in innumerable applications. In the current decade, CNTs have become a hot material for electronic, chemical, mechanical, and optical applications. Due to their excellent electronic properties, CNTs allow the fabrication of basic solid-state devices with smaller dimensions than conventional silicon-based technology. Recently developed basic electronic devices based on CNTs are diode, tunnel diode, FETs, and supercapacitors. It is also successfully used in the fabrication of devices in some other important applications:

microwave amplifier, flat panel display, radio, X-ray source, random-access memory (RAM) as well transparent conductor [106–108]. CNTs have also been used in very important communications and information technology inventions such as fiber lasers and nonlinear photonics. These inventions play a role in routing, wavelength conversion, and optical switching. It is also used in applications such as photovoltaic devices, photodetectors, and nano-light sources [109–111].

CNTs have also proven to be a promising future material in the field of mechanical applications. Worldwide researchers are working on nano electro mechanical system (NEMS) technology based on CNTs. Researchers have successfully developed mechanical devices based on high-frequency oscillators [112], rotational actuators [113], nanometer tweezers [114], and nanometer cargoes [115]. Outstanding inventions are ultra-miniature sensors for observing the working force between two molecules [116], for measuring the force produced by the magnetic resonance of a single spin [117], and for measuring the displacement created by disturbance between individual atoms. Due to their relatively high Young's modulus as well as specific weight, SWNTs have proved to be perfect for manufacturing mechanical resonators for observing mechanical movement in the quantum regime [118].

CNTs have also been part of surprising inventions in the field of electrochemistry. Several types of chemical sensors, humidity sensors, gas sensors for toxic gases, and alcohol sensors are developed so far. CNTs-based chemical sensors have the following advantages over conventional solid-state sensors.

I. Existing sensors show poor sensitivity at room temperature. Their normal operating temperature is in the range of 200–500 °C. However, CNTs-based sensors work successfully at room temperature.
II. Existing toxic gas sensors show a limited maximum sensitivity in comparison with CNTs-based sensors that have shown order of magnitude higher sensitivities.
III. The main drawback of using oxide-based gas sensors is the use of microfabrication techniques to build the sensor that puts a limit on the size and geometry of the sensor. In the case of CNTs, due to their small dimensions, no barrier is created in front of the device design and geometry.
IV. For existing technology, the sensor fabrication cost is very high compared to CNTs-based sensors.
V. The operating power of existing sensors is also very high. It is possible to operate CNTs-based sensors at very low power.
VI. Existing toxic gas sensors show very poor selectivity. They possess nonspecific gas sensing mechanism and very poor response time. These problems are resolved with CNTs-based sensor technology.

1.2.4.9 Synthesis of CNTs

In order to develop a material with specific properties, it becomes most important to be aware of all the growth techniques so that it becomes possible to synthesize material with the best quality and yield. There are three main techniques used to grow CNTs: arc discharge, laser ablation, and CVD. Amongst them, CVD is the most used technique so far.

Arc Discharge Method Arc discharge method is a widely used method for CNTs growth. It was invented by Krätschmer et al. to produce a high yield of fullerenes [119]. It was observed that in addition to fullerenes, some other allotropes of carbon were also present as by-products. Later, a microscopist Sumio Iijima reported the presence of tubular allotrope of carbon, i.e. CNTs, in the arc discharge method by-products [120]. Therefore, arc discharge was the first known method for the CNTs growth. This method was further developed and optimized for exclusive CNTs growth. Initially, the synthesis of CNTs in the soot form using this method was reported [120].

Earlier, only MWNTs were produced using this method, but at the later stage, it has been successfully tuned for the growth of SWNTs [121]. The arc discharge method involves the vaporization of graphite in an inert atmosphere using an electric arc. In specific cases, catalysts also take part in the synthesis of CNTs. At the heart of the arc discharge system, there are two graphite rods situated inside the electric arc chamber. The diameters of graphite rods are of the order of 2 mm and the separation between these rods is kept about 1 mm (see Figure 1.5a). These graphite rods act as electrodes and a high DC current of around 50–100 A is passed through them. Due to such high DC current, a very large amount of thermal energy is produced, which increases the temperature of the system up to around 4000 °C, which, in turn, evaporates the carbon of the graphite anode. These carbon vapors subsequently deposit on a cathode in the form of different carbon nanostructures such as CNTs, fullerenes, and amorphous carbon [122–125]. The electrodes composition determines the type and quality of as-synthesized CNTs among the grown carbon nanostructures [126]. While pure graphite electrodes support the synthesis of MWNTs, SWNTs can be produced using graphite electrodes mixed with metal catalysts [119, 127]. Another property of as-grown CNTs is their diameter, which can be tuned by controlling other growth parameters [123].

Laser Ablation The laser ablation method is quite similar to the arc discharge method. In both techniques, carbon is vaporized and deposited at different places. The only difference in both growth methods is the evaporation process. In the laser ablation technique, a high-intensity laser beam is used for carbon vaporization instead of high DC current. A small amount of carbon is transmitted into its plasma. CNTs with other allotropes of carbon are synthesized into a plasma plume. The plasma plume subsequently deposits on the chamber walls. This technique mainly uses CO_2 and Nd:YAG laser sources in both pulsed and continuous modes [128, 129]. This method, like the arc discharge technique, can be used to deposit both MWNTs and SWNTs. When pure carbon source is used, the growth of MWNTs takes place. On the other hand, the use of metal catalyst mixed carbon source leads to the synthesis of a mixture of MWNTs and SWNTs. Figure 1.5b shows a schematic diagram of the laser equipment.

Chemical Vapor Deposition The CVD system is basically a film depositing equipment that has been used for many years and is capable of depositing a number of nanomaterials. CVD is a method in which gaseous precursors react either in the gas phase or

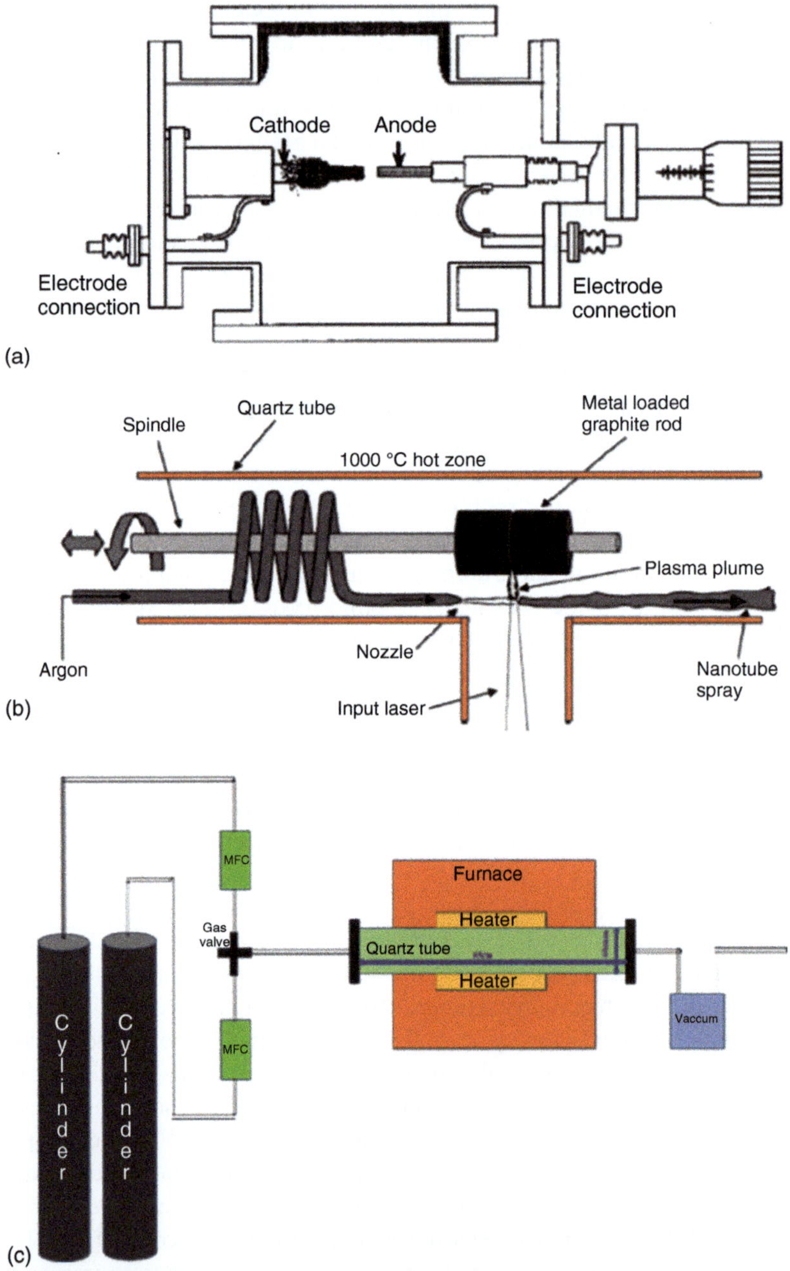

Figure 1.5 Figure (a–c) show the schematic diagram of arc-discharge, laser ablation, and CVD techniques. Source: Saito et al. [122]/with permission of AIP Publishing.

at the substrate gas interface, producing a thin film on the substrate. In the research world, there are many types of CVD systems having various types of heating arrangements as well as some extra accessories. Atypical thermal chemical vapor deposition (TCVD) setup includes a durable, high-temperature compatible tube, usually made of quartz or alumina, which is heated by a chamber furnace or heating coils (see Figure 1.5c).

All types of CVDs have the same operating concept. Users customize CVD according to requirements. Thermal and plasma enhanced CVDs are being used to fabricate CNTs. The CNTs growth by the CVD process requires three ingredients: anablation precursor gas (also called feedstock), a catalyst surface, and an energy source. Energy may be generated thermally (in TCVD) or by plasma (in plasma enhanced CVD). We preferred using TCVD for CNTs growth (since CNTs are prone to be damaged by plasma [130], which reduces their electrical performance). In TCVD, hydrocarbons are broken at high temperatures to produce atomic carbon vapors. These vapors are deposited on the metal catalyst deposited substrate via vapor–liquid–solid (V–L–S) growth process to form CNTs. The V–L–S growth process involves three steps, i.e. absorption, saturation, and structure extrusion (see Figure 1.6a). Any transition metal that has a low melting point can be used as a seed or catalyst for CNTs growth via the V–L–S growth process. These catalysts have the ability to absorb vaporized carbon [131–133]. These catalysts may be deposited on a substrate such as silicon, Al_2O_3, etc. in the form of thin film by sputtering, dip coating, or spin coating. Prior to deposition, the catalyst-coated substrate is heated. Because of heat treatment, the continuous layer breaks into lumps of nanoparticles to keep minimum surface energy.

This process is known as catalyst activation or annealing. In some reports, catalysts were injected directly into the growth chamber. Some oxide nanoparticles (SiO_2, Al_2O_3) have also been reported as catalysts for CNTs growth [2]. After putting the annealed catalyst film into the chamber, the chamber is evacuated. The precursor gas, which is usually a lower-order hydrocarbon (such as methane, acetylene, and ethylene), is then introduced into the chamber at an appropriate pressure. In some specific cases, vaporized benzene or toluene can be used as a feedstock. Occasionally, an inert carrier gas is also used as a carrier, which also serves as an etchant for excess amorphous carbon deposited on the walls of the chamber. The chamber temperature is raised to several hundred degrees Celsius. Such a high temperature induces the metal catalysts nanoparticles to get into a molten state. On the other hand, high temperature also decomposes hydrocarbon. Decomposition of hydrocarbons produces carbon vapor in the chamber, which is absorbed by the molten catalyst. As more and more carbon is absorbed by the catalyst, the concentration of carbon exceeds the solubility of the catalyst particle. At this point, the catalyst particle begins to extrude a solid formation in the form of a CNT as shown in Figure 1.6a,b. The schematic diagram of CNTs growth via V–L–S process is given in Figure 1.6. Depending on the final location of the catalyst particle, nanotubes are typically classified as either tip-grown or root-grown. Other one-dimension nanostructures can also be grown via V–L–S growth mechanism. The growth of germanium nanowires from gold nanoparticle is shown in Figure 1.6c–i, and the growth

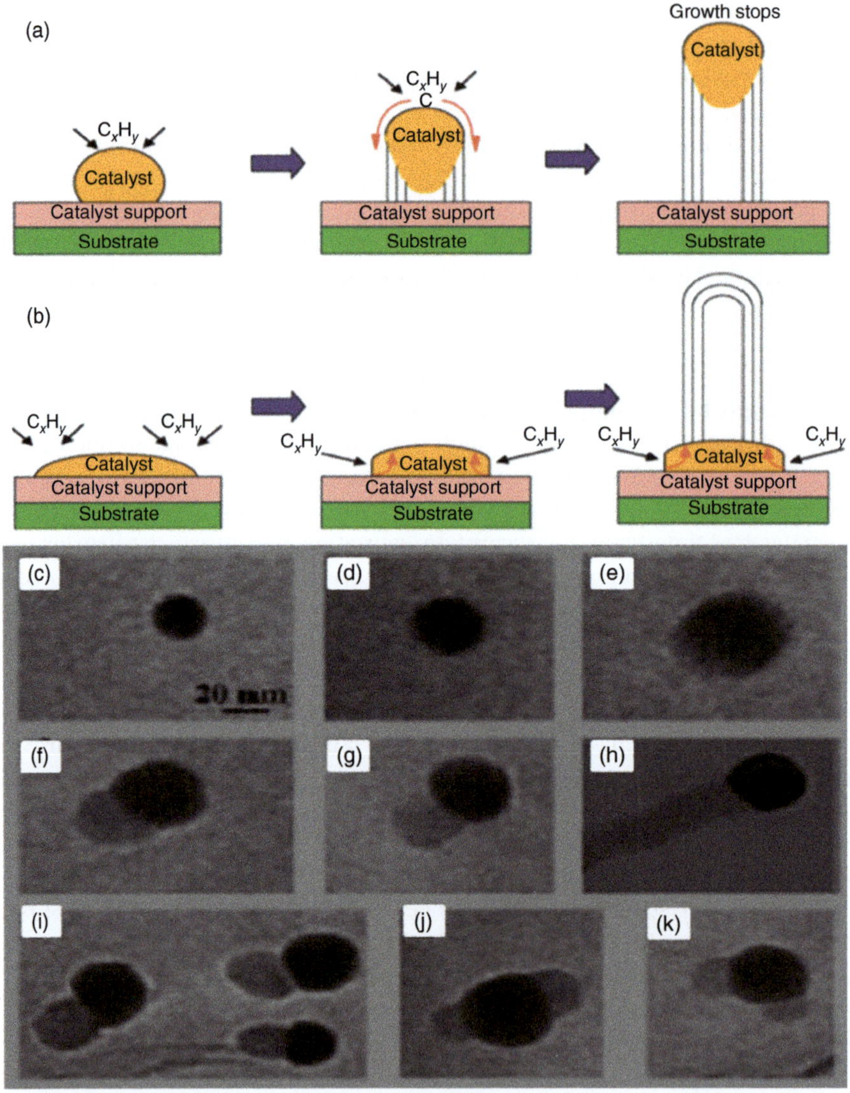

Figure 1.6 Schematic diagram of various stages of V–L–S growth mechanism. Figure (a) represent the tip-growth model while figure (b) shows the base growth model. Source: Sengupta [131]/with permission of Elsevier. Figure (c–k) shows the formation of germanium nanowires from gold nanoparticles via the vapor liquid solid growth mechanism. These transmission electron microscopy images reveal the extrusion of the germanium nanowires (e–k) from the saturated particle (c, d). Source: Wu et al. [131]/ with permission of American Chemical Society.

of multiple nanowires is shown in Figure 1.6j,k. The major benefit of using CVD for the synthesis of CNTs lies in its ability to be scaled up in size, allowing the large production facilities to produce several kilograms of CNTs per day [134].

Hydrocarbons are channeled into the heated reactor zone via mass flow controllers and eventual by-products are evacuated through a fume hood. TCVD can be used to grow MWNTs as well as SWNTs. CVD parameters such as temperature, pressure, flow of gases, etc. play a crucial role in deciding CNTs dimensions, structure as well as quality [135, 136]. They have varying degrees of interaction with growth conditions, which is further complicated by the fact that different research results often predict different behavior. All this makes it a quite challenging task to find out the optimum growth conditions for producing CNTs with desired properties. Now, after gaining knowledge about the CVD system, a detailed analysis of the various parameters used in the CVD system has become necessary. Therefore, now we are going to present a detailed literature survey on various CVD parameters.

CNT Precursors In the CVD system for growing CNTs, the most commonly used carbon feedstock compounds are methane (CH_4), ethylene (C_2H_4), acetylene (C_2H_2), benzene (C_6H_6), xylene (C_8H_{10}) as well as carbon monoxide [130, 137–141]. The CNTs growth by pyrolysis of benzene was reported at 1100 °C [142, 143], while MWNTs with an acetylene precursor had been grown at 700 °C [144]. In both cases, benzene and acetylene, iron catalysts are used for hydrocarbon decomposition. Later, cyclohexane (C_6H_{12}) and fullerene were also successfully tested as precursors for CNTs growth [145–147]. If we are talking about SWNTs, H. Dai et al. first reported about SWNTs growth using carbon monoxide as a carbon precursor and molybdenum as a catalyst while keeping the growth temperature at 1200 °C [148]. After H. Dai et al. made a milestone report on SWNTs synthesis, many other groups also became succeeding SWNTs growth with different carbon precursors, such as acetylene [149], methane [150], and benzene [151], etc. with various catalysts. A report of Maruyama et al. (2012) introduced a new carbon precursor, i.e. alcohol for SWNTs growth [152]. They grew high-quality SWNTs at a lower growth temperature on zeolite-supported Fe–Co nanoparticles. Due to its extraordinary qualities, alcohol became the highest used hydrocarbon source for SWNTs synthesis [153, 154]. It is reported that due to the carbon etching ability of alcohol, CNTs grown with alcohol precursors have a lesser amount of amorphous carbon [155]. After some time, the same group also synthesized vertically aligned SWNTs on a quartz substrate using Mo-Co catalyst with an alcohol precursor [156]. Maruyama et al. also enhanced the catalytic quality of the catalyst by using a discontinuous flow of acetylene in an ethanol-based CVD and increased the growth rate of CNTs [157]. The structural formation of the hydrocarbon source has the potential to decide the CNTs structure. Linear carbon precursors like acetylene, ethylene, and methane produced straight and hollow tubes, while benzene and cyclohexane-type cyclic hydrocarbons produced curved CNTs [147]. Generally, MWNTs synthesis is easy as compared to SWNTs. Moreover, MWNTs grow at lower temperatures as compared to SWNTs. One more observation is that MWNTs can grow with any precursor while SWNTs growth takes place on selective precursors. Common carbon precursors used for MWNTs synthesis show unstable behavior at higher temperatures. It leads to the synthesis of other carbonaceous compounds instead of CNTs. To solve that problem, in 2004 Hata et al. published a report, in which water-assisted CVD is proposed to grow impurity-less SWNTs. In water-assisted

CVD, water steam is supplied in a controlled manner as an oxidizer. It etches out amorphous carbon without damaging the grown SWNTs [158]. Another work is reported by Zhong et al., in which they proved that no water steam type oxidant is required in cold wall CVD systems. It is successful only for low reactive carbon precursors [159]. The above discussions emphatically show that hydrocarbon or carbon precursors have an important role in CNTs growth. Therefore, researchers get the best yield as well as structure of grown CNTs by proper tuning and selection of carbon precursors.

Apart from the types of hydrocarbons, the flow rate and flow duration of hydrocarbons also play an important role in CNTs growth by TCVD. To analyze the various effects of hydrocarbon flow rate and flow duration, we performed several experiments with different flow rates (10–50 sccm) and flow duration (2–16 minutes) of C_2H_2while keeping other parameters of the CVD system constant. It has been observed that initially the structural quality of CNTs enhanced with an increment of C_2H_2 flow rate and, after a certain limit of C_2H_2 flow rate the structural quality of CNTs starts to degrade. The best quality of CNTs has been grown at a 12 sccm flow rate of C_2H_2. Similarly, in case of C_2H_2 flow duration, the quality of CNTs has been enhanced when we increased flow duration from 2 to 12 minutes, but beyond this limit structural quality of CNTs started to degrade. The best quality of CNTs has been found with 12 minutes flow duration of C_2H_2 [136].

Catalyst Engineering As we discussed earlier, for CNTs growth nano-scale particles are needed to trigger the carbon precursor decomposition at a lesser temperature than its actual decomposition temperature as well as to initiate CNTs growth. In the early time of CNTs invention, transition metals (Fe, Co, Ni) were used as catalysts for CNTs growth. Due to high solubility as well as high carbon diffusion rate of transition metals, they are preferable in CNTs growth. High melting point and low equilibrium vapor pressure of transition metals also give a large range for selecting growth temperature as well as a large number of carbon precursors. It is also reported by some groups that transition metals have better adhesion with grown nanotubes compared to other metals. That quality of transition metals provides grown CNTs with large curvature and lower diameter [160]. Some reports also made evidence of solid organometallic precursors having quality of catalyzed CNTs indeed better than transition metals. Apart from Fe, Co, and Ni, some other metals such as Pd, Pt, Cu, and Au also have the potential to catalyze CNTs with different carbon feedstock [161]. Generally, a tube diameter depends very much on the catalyst particles' size [147]. Some groups manipulate the size of catalyst particles prior to growth and hence the diameter of CNTs [162]. To get pure nanotubes, it is necessary that carbon precursors used in CNTs growth decompose only on catalyst particles and avoid aerial pyrolysis. As mentioned previously, metal catalysts are also used in arc discharge as well as laser ablation techniques of growth CNTs, so it may be possible that these different techniques have the same growth mechanism.

It has been observed that metal salts such as iron nitrate, magnesium nitrate, cobalt nitrate, etc. can be also utilized as catalysts for growing CNTs. The density and orientation of CNTs can be tuned by selecting the appropriate metal salt as a catalyst. It has been observed that cobalt nitrate and magnesium nitrate are capable to produce horizontal network type CNTs with acceptable structural quality. The CNTs density is much higher in the case of magnesium nitrate as compared to cobalt nitrate. In case of iron nitrate, vertically aligned CNTs have taken place when iron nitrate was used as a catalyst. Therefore, by utilizing different metal salts, someone can tune the yield as well as orientation of CNTs. Vertically aligned CNTs can be also produced by oxygen plasma treatment of catalyst film prior to CNTs growth. Oxygen plasma treatment enhances the nucleation sites on catalyst particles. Hence, the density of CNTs sharply increases and due to crowding effect vertically aligned CNTs grow [1]. The height of CNTs is another important requirement for application development. It has been observed that the flow rate of carrier gas is affecting significantly on CNTs height/length. We performed several experiments on CNTs synthesis at different argon flow rates (0–80 sccm). Initially, without argon gas very short length (~2 μm) CNTs growth took place but the height of CNTs increased sharply (~57 μm) when argon flow turned on 20 sccm flow rate. The height of CNTs is found varying as a function of the argon flow rate. In case of structural quality and diameter of CNTs, it has been found that they are improving up to 60 sccm flow rate of argon; beyond this limit, the diameter starts to increase and structural quality starts to degrade. Carrier gas increases the distribution of hydrocarbons inside the chamber, balances the production of carbon atoms and, as per requirement, grows CNTs. The best structural quality, less diameter, and acceptable height have been found at 60 sccm argon flow rate [5]. In the general CNTs growth mechanism, catalyst nanoparticles either remain on the bottom of CNTs or they peeled-off from the substrate and come on the CNTs tip. If the catalyst remains on the bottom of CNTs then it is called base-growth and when the catalyst particles come on tip then it is called tip-growth. It has also been observed that sometimes these catalyst particles remain inside CNTs. Many researchers reported that CNTs growth with a metal catalyst is not suitable for electronics, optoelectronics, medical, and agriculture applications. The metal catalyst is very harmful for all living organisms as well as for solid state devices [163]. Although metal catalysts can be removed from CNTs by performing certain post-growth purification processes, these processes are very complex and also create defects in CNTs. To solve this issue, many research groups tried to grow CNTs without catalyst as well as without metal catalyst [163–165]. To make CNTs more useful for mentioned applications, in our group we have grown CNTs without catalyst directly on Al_2O_3 substrate as well as by using green catalyst [2, 163]. It has been reported that Al_2O_2 is capable to decompose hydrocarbons at lower temperatures, and surface roughness of Al_2O_3 substrate provides nucleation sites for CNTs growth. The as-grown CNTs were few walls CNTs having diameters in the range of 4–8 nm and very high structural quality (see Figure 1.7a,b).

We also tried unconventional catalysts for growing CNTs, i.e. green catalysts. The walnut plant extract has been utilized as a green catalyst. Green catalysts are found able to grow CNTs at lower temperatures. To grow CNTs by using metal catalyst, generally 800 °C temperature is required but in case of green catalysts, CNTs growth took place at 575 °C. When CNTs growth performed at 800 °C, then a new carbon nanostructure with rectangular cross section took place (see

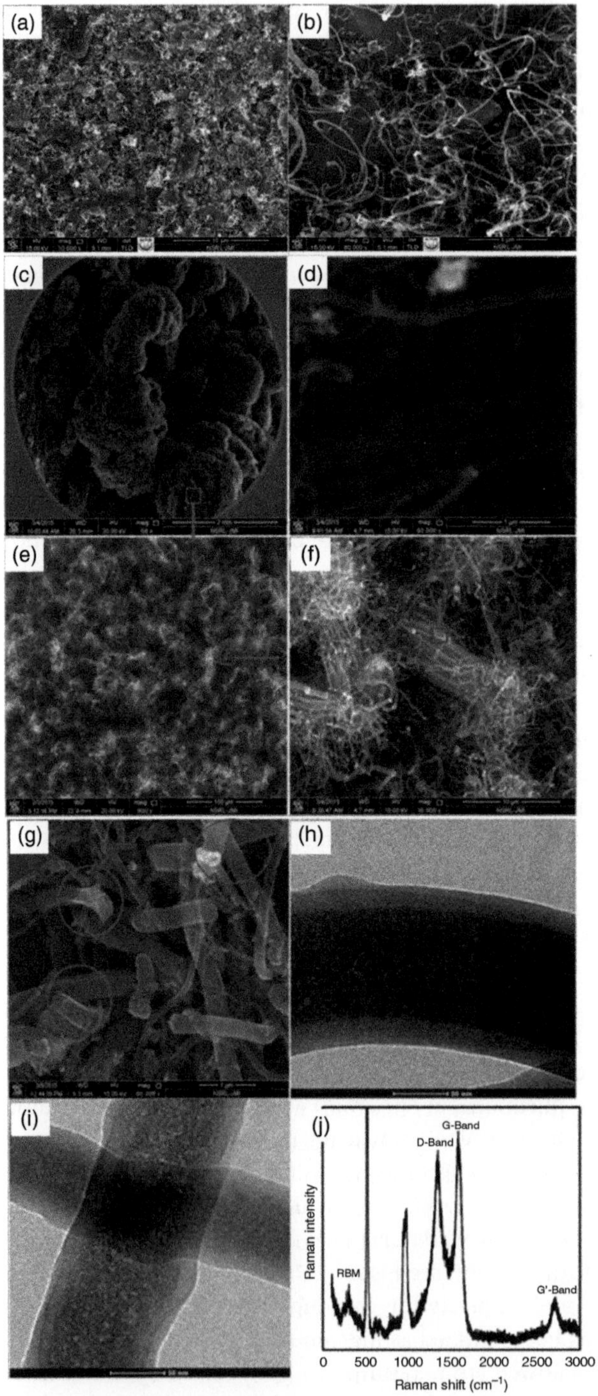

Figure 1.7 Figure (a) shows the FE-SEM image of CNTs growing on the Al_2O_3 substrate, its magnified image can be seen in figure (b). Source: Tripathi et al. [2]/with permission of Elsevier. Figure (c) shows CNTs growth by using green catalyst and its magnification image can be seen in figure (d–f). Figure (g) shows the carbon nanostructure with rectangular cross-section and figure (h, i) show TEM images of the as-grown nanostructure. Figure (j) shows its Raman analysis. Source: Tripathi et al. [163] / With permission of Springer Nature / CC BY 4.0.

Figure 1.7c–j). Apart from low temperatures growth, green catalysts are capable of growing CNTs on mass scale.

To verify the fact that only walnut plant extract can catalyzed CNTs growth or any plant extract, the CNTs growth has been performed by using *Azadirachta indica*, *Cascabela thevetia*, rose flower, and garden grass extract as catalysts. It has been observed that all mentioned plants are able to grow CNTs. Green catalyst finished the requirement of complex instruments, which required metal catalyst deposition. Since CNTs growth is conducted at low temperatures and green catalysts available freely in nature, all mentioned factors reduced the cost for CNTs producing [163]. Another important parameter of CNTs growth is the growth temperature and temperature increasing rate. The growth temperature and temperature increasing rate make a direct impact on catalyst particles and on the CNTs morphology and structural quality. We analyzed CNTs growth by performing multiple experiments by taking different growth temperatures (650–850 °C). At 650 °C growth temperature, obtained CNTs had a very poor structural quality and a large diameter. The structural quality/diameter of CNTs started to improve/decrease with rising growth temperature. The best structural quality with lesser diameter CNTs was obtained at 775 °C growth temperature. When the growth temperature further increased the structural quality/diameter started to decrease/increase. At a lower growth temperature, the catalyst film made large diameter catalyst particles. When the growth temperature increased, the diameter of catalyst particles has been found decreasing. After a certain limit of the growth temperature, the catalyst particles started to agglomerate and converted into larger size particles that is the possible reason for growth temperature dependent CNTs growth. Similar to the growth temperature, the temperature increasing rate makes a big impact on the CNTs structural quality as well as on the diameter of as-grown CNTs.

To analyze the various effects of the temperature increasing rate on CNTs growth, multiple experiments with different temperature increasing rates (from 100 to 600 °C per hour) have been performed. As explained earlier, when the catalyst film deposited sample is annealed in the presence of argon gas, the catalyst film is converted into nanosized hemispherical particles. It happened due to different thermal expansion coefficients of a metal catalyst film and a substrate. When the CVD temperature slowly increased, the catalyst film and substrate got sufficient time to balance thermal expansion rates. In that case, large size catalyst particles were produced. Opposite to this, when the sample was heated with a fast temperature increasing rate, the catalyst film has been broken into small size particles because of a large thermal expansion gradient created between the substrate and catalyst film. Also, when the sample was heated at a fast temperature increasing rate, lots of defects were created on catalyst particles and as-produced defects provided small size nucleation sites for CNTs growth. When CNTs growth experiment was conducted with 100°C per hour temperature increasing rate, large size poor structural quality CNTs were grown. The structural quality/diameter/no. of walls has been found increasing/decreasing/decreasing with respect to the temperature increasing rate increment. The best quality CNTs with a single wall have been grown at 600 °C per hour temperature increasing rate (see Figure 1.8) [166].

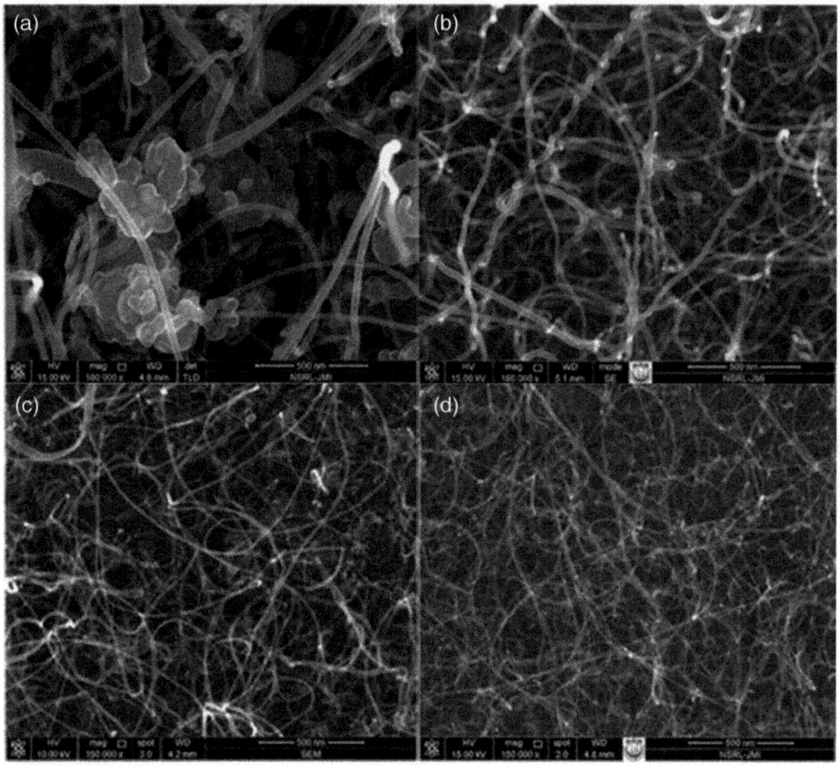

Figure 1.8 Figure shows the SEM monographs for as-grown CNTs with taking temperature increasing rate as (a) 100, (b) 300, (c) 500, and (d) 600 °C per hour, respectively. Source: Tripathi et al. [166]/with permission of Springer Nature/CC BY 4.0.

1.2.4.10 Analysis of CNTs by Raman Spectroscopy

Raman spectroscopy is one of the main tools used for molecular vibrations investigation. It is utilized to gain knowledge about chemical elements and structural forms to recognize materials by characteristic fingerprinting. It is also used to get quantitative information about the amount of material in a sample. This equipment has the potential of scanning samples in all states as well as in all conditions; for example, powder, vapors or liquids, surface layers, or bulk form of nanoparticles, in different temperature states. Prior, IR absorption was widely used as compared to Raman scattering (RS) due to complexity in Raman spectroscopy. But improvements in technology made Raman spectroscopy simple, compact and reduced the problems substantially. Mentioned improvements in Raman spectroscopy along with the capability of investigation of different type of samples without any preparation, make Raman spectroscopy a powerful investigation tool in the field of materials science. In 1928, Sir C.V. Raman reported the historical invention of inelastic light scattering. Raman scattering (RS) is defined as inelastic scattering of light. In inelastic scattering, the photon energy shifted with respect to the incident energy (see Figure 1.9a). The schematic diagram of the experiment is displayed in Figure 1.9b. In a basic concept, a laser beam is incident on materials, and the properties of scattered light

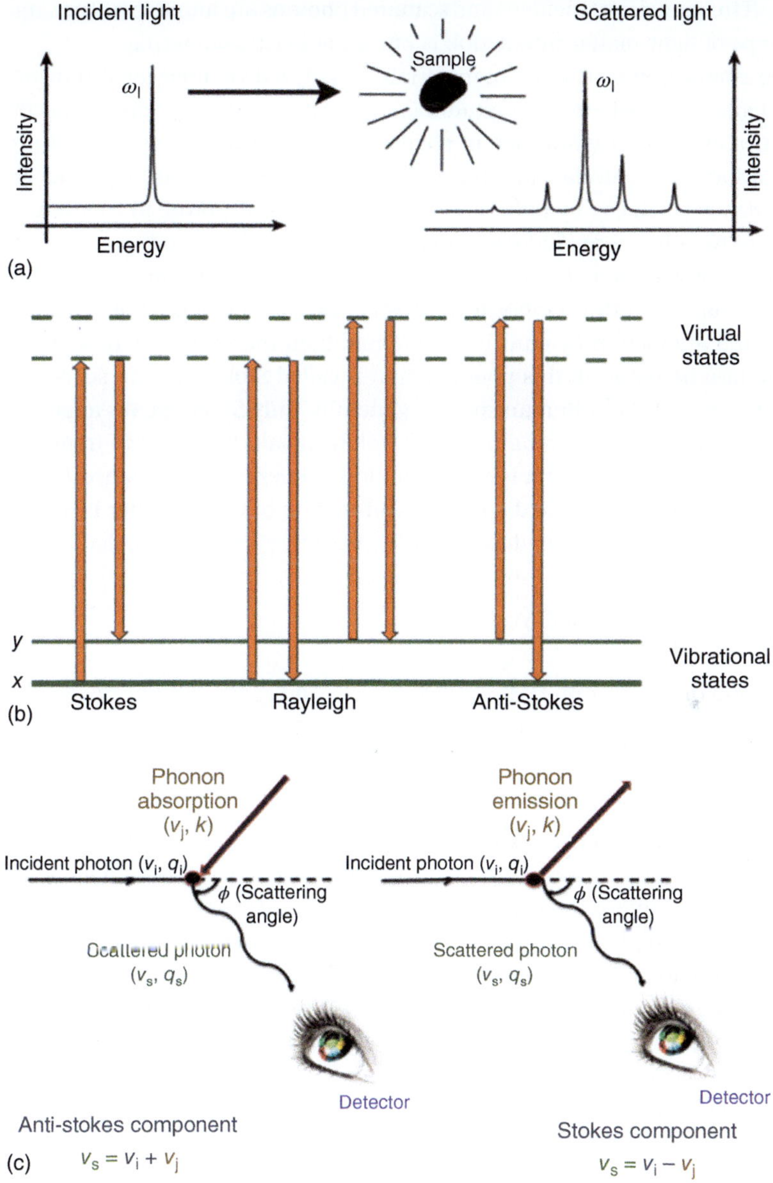

Figure 1.9 Figure (a) shows the phenomenon for Raman Scattering. Plot of figure (b) shows the sample identification and phase quantification Figure (c) shows the interaction mechanism that causes a frequency change.

from material is analyzed. Raman spectroscopy is known as a powerful and non-destructive tool that can rapidly give information about the low-frequency modes of materials along with rotational as well as vibrational modes [167–169]. Raman scattering phenomenon depends on the frequency change of photons after interaction with materials and such type of light–matter interaction is known as inelastic interaction. If frequencies of incident and scattered photons are found to be the same, then such type of light-matter interaction is known as elastic scattering.

In RS, the atomic bond structure of materials is analyzed via inelastic interaction of incident light with the help of longitudinal/transverse modes of phonon. Spectral characteristics of phonon vibrations in solids can be investigated by RS. RS provides the platform for materials identification as well as phase quantification. The shift in frequency produced by light–matter interaction is displayed in Figure 1.9c. If the incident photon energy increase after the interaction by absorbing a phonon then the energy of scattered photons will also be increased and this process is called anti-Stokes Raman scattering. Opposite to anti-Stokes, if an incident photon reduces energy after the interaction by emitting a phonon, then the frequency of scattered photons will also be reduced, this phenomenon is called Stokes Raman scattering. It has been observed that in Raman spectra, generally, only Stokes peaks appeared. It happened due to the poor strength of anti-Stokes signals. The shift in frequency is represented as $\upsilon = \upsilon_s - \upsilon_i$, and it is equivalent to the energy change, since $E = h\upsilon$. The results of Raman spectra are displayed as a function of wavenumber instead of frequency. Wavenumber is defined as $\frac{\text{no. of waves}}{\text{cm}}$. Wavenumber can be calculated by following expression:

$$\text{Wavenumber} = \frac{\text{Frequency}}{c} \tag{1.13}$$

where the value of c is around 2.99×10^8 m s^{-1}, i.e. the speed of light in vacuum. Additionally, wavenumber is inversely proportional to the wavelength since:

$$\text{Frequency} = \frac{c}{\text{Wavelength}} \tag{1.14}$$

To represent wavenumber, mostly cm^{-1} unit is used. The atomic bonds vibrations within the material are responsible for frequency shift in scattering while the phonons energy of material is comparable to the difference of energies between incident and scattered photons. The investigation of scattered frequencies provides important information about the material structure. Generally, materials have a very narrow distribution of bond energies, that is why Raman spectra associated with such materials also show narrow lines. In case of amorphous materials, the bonding energy has a broad distribution that is why Raman spectra associated with amorphous materials have broad signals, which conclude that if Raman spectra have broad signals then material has a disordered structure. In general, RS is a very weak phenomenon. In 10^6–10^8 scattered photons only one Raman scattered photon comes out. This does not make the process unachievable since with high-tech CCD detectors and lasers, very high power densities can be incident to small objects. But it triggers other unacceptable things such as sample degradation and fluorescence. Figure 1.9c shows the fundamental phenomenon, which takes place for one vibration. In that figure, virtual energy states are imaginary states. These states are created by an incident laser light on a sample. Laser sources also determine the energy of mentioned states. In general, most molecules are located at their lowest energy state. The Rayleigh has the greatest intensity among all processes, since a major portion of scattered photons is taken in that way. They do not show any shift in energy. So, this type of scattering is useless for investigation purposes. Upon illumination by a laser, the electrons present in a vibrational level get excited by a higher vibration level. During the de-excitation process, some electrons return to the initial vibrational energy level. The photons generated during this type of de-excitation have the same wavelength as the incident photons. This is called Raleigh de-excitation or Rayleigh scattering. Some electrons get de-excited to the lower vibrational level with respect to the initial vibrational level. The photons generated in this type of de-excitation have fewer wavelengths than that of incident photons. This process is called anti-stock de-excitation or anti-stock scattering. Rest of the electrons get de-excited to higher vibrational levels with respect to the initial vibrational level. The photons generated in the process have more wavelength than the wavelength of incident photons. This process is Stokes de-excitation or Stokes scattering. Intensity of anti-Stokes scattering also becomes very poor. At higher temperatures, both types of scattering increase. Raman spectroscopy is divided into types according to its way of signal collection, i.e. dispersive Raman spectroscopy and FT-Raman spectroscopy. Both types have their own advantages and can analyze specific samples [167–169].

Finally, having gone through the appropriate theory, all the details of the experimental setup, we are now ready to consider the experimental Raman spectra of CNTs. Raman Spectroscopy provides major information about CNTs structure, defects in CNTs as well as their purity. It is also used in the identification between MWNTs and SWNTs as well as other carbon materials [144]. In general, Raman spectroscopy is more fruitful for SWNTs as compared to MWNTs. For MWNTs, it gives very complex spectra. Because MWNTs behave like a group of SWNTs, the diameter ranges from very small to large [170]. The decoding of the Raman spectra for MWNTs is based on the well-established Raman spectra for SWNTs. It is reported in many articles that Raman spectroscopy has the potential to provide qualitative as well quantitative information about CNTs. An ideal Raman spectra of CNTs have the following important peaks (also see Figure 1.10a):

I. RBM
II. D-Band
III. G-Band
IV. G′-Band

Radial Breathing Mode (RBM) The radial breathing mode (RBM) (<200 cm^{-1}) is a strong line in the SWNT Raman spectrum. This line is the only strong line that has no analog in the spectra of graphite. It was revealed that it has an elementary dependence on diameter $\omega\text{RBM} = \frac{\alpha\text{RBM}}{dt}$. The RBM is strongly related to a tube diameter. It is inversely proportional to the diameter. For large diameter SWNTs, greater than 2 nm, the RBM may disappear [172]. Due to radial motion of CNT atoms, the RBM is more strongly affected with the environment of a nanotube than the other high-intensity modes. For example, other tubes can be in a bindle or on a supporting substrate. In addition to this fact, the RBM is a direct demonstration of the SWNTs one-dimensional tubular structure that makes it a valuable probe for the properties of nanotubes as well as for the majority of papers concerned

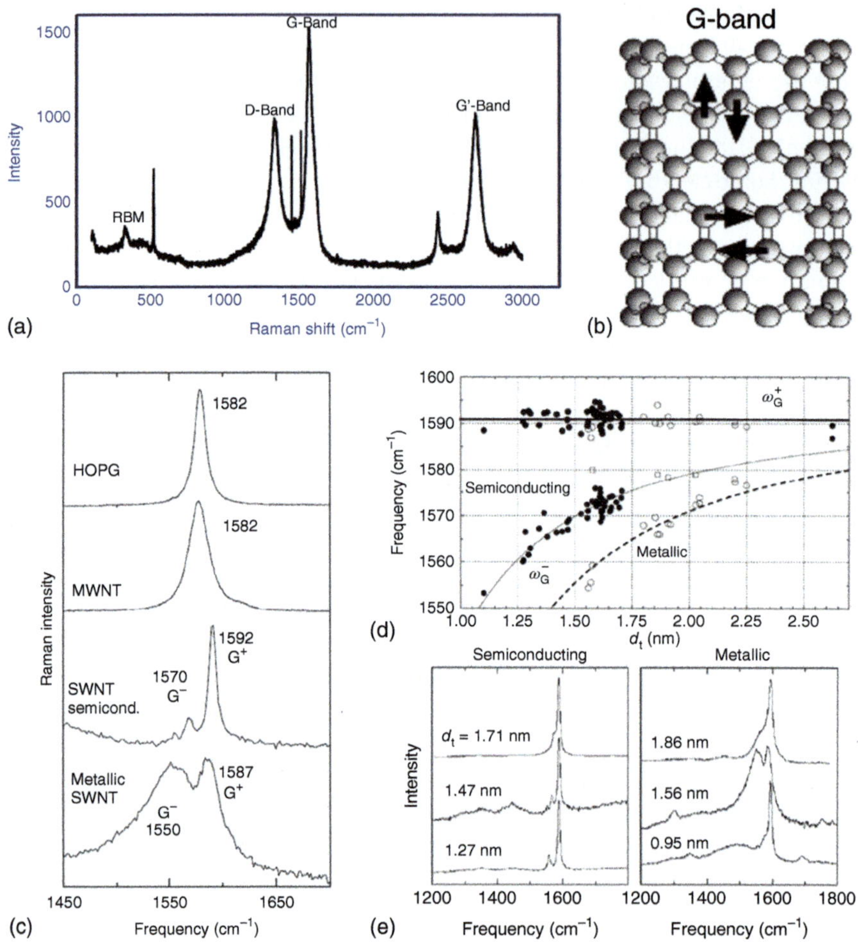

Figure 1.10 Figure (a) shows the Raman spectra of SWNTs. Figure (b) shows the schematic picture of atomic vibrations for the G band modes. Figure (c) represents the G-band peak of various graphitic nanostructures. Figure (d) shows the ω_{G+} and ω_{G-} for different types of SWNTs as a function of the tube diameter. Figure (e) shows the G-band and D-band peaks of Raman spectra of various types of SWNTs. Source: (b, c, e) Jorio et al. [167]/with permission of IOP Publishing and (d) Jorio et al. [171]/with permission of American Physical Society.

with Raman effect in SWNTs. In the case of RBM, peak broadening also has a very meaningful information. The RBM-peak can broaden due to the contribution of different effects. The nanotube sample in a bundle form results in the RBM expansion with regard to the case of an isolated tube, because the interactions between tubes affect ωRBM, and the interactions are determined by the resonant tube position in the bundle. The sample temperature enlarges broadening because of electron–phonon and phonon–phonon interactions [30]. It has been reported that the peaks shift during heating perhaps appears owing to changes in the intra- and inter-tube interactions inside the bundles as well as due to a fewer degree caused by tubes thermal expansion. Therefore, the peaks will expand in case of occurred temperature differences within the sample (e.g. because of the intensity profile of the impinging beam heating specific areas more than others or due to various couplings to the substrate). Moreover, peaks broadening can be caused by defects in the tubes as well as due to other interactions. Due to the fact that several peaks are narrower than 6 cm^{-1}, it is concluded that broadening caused by electron–phonon interaction is bigger in graphite than in nanotubes [173]. Notably, Raman studies on some samples of nanotubes bundles revealed peaks equal to or greater than 20 cm^{-1} [174]. In addition, low-temperature measurements on free tubes with a big quantity of defects demonstrate line widths up to 0.35 cm^{-1} [171]. Hence, the linewidth is crucially reliant in the nature of sample and in the parameters of measurement.

G-Band In Raman spectra, G-band, a bunch of peaks around 1582 cm^{-1} also verify the existence of CNTs. G-band has the ability to investigate a diameter-dependent study of SWNTs without using RBM. In resonance, for a particular laser light, the metallic character of CNTs is also observed. For CNTs, G-band has six peaks. Out of six, two peaks have maximum intensity and for simplicity, we consider only two peaks, i.e. G+ and G−. These peaks are produced by symmetry breaking of the tangential vibration when the graphene sheet is rolled to make a cylindrically shaped tube. G+ is originated through atomic displacements along the tube axis and G− is produced by atomic displacement along the circumferential direction (see Figure 1.10b,c). Figure 1.10c shows the line shape difference of G− for semiconducting and MWCNTs and it is broaden for metallic with respect to semiconducting CNTs (see Figure 1.10c), and its broadening is coming from free electrons in nanotubes with metallic behavior [175–177]. If charged impurity exists in SWNTs then G-band for semiconducting CNTs shows the behavior of metallic CNTs [178]. If talking about a diameter-dependent study of G− band then the frequency of G+ does affected by diameter but in the frequency of G− band shifting with variation in diameter of CNTs is observed (see Figure 1.10d,e) [179, 180]. Hence, G-band and its splitting may be used for the CNTs diameter calculation. Recently, some research groups gave the evidence about G-band formation from double resonance mechanisms [176–181]. In that case, the intensity of the G-band very much depends on CNTs defects, and the frequency of G-band depends on the laser energy. This evidence creates complications in the use of Raman spectroscopy for SWNTs. However, it can be observed that double, as well as single resonance, show the same intensity for a defective material. For defect-less materials, intensity of single resonance is double as compared to double resonance phenomenon [182, 183]. So, according to this report, Raman spectroscopy is the only effective for good quality of SWNTs.

Disorder Bands (D-Band and G$^{|}$-Band) A bunch of peaks appear in Raman spectra of CNTs around 1332 and 2700 cm^{-1}. It represents defects and disorders in the structure of CNTs. It is highlighted in both types of CNTs. Due to double resonance criteria, D-band also appears due to some specific chirality [184]. In case of isolated SWNTs, D-band is split into two parts. Split distance of D-band relates to energy of incident light. G$^{|}$-Band shows an intrinsic property of graphite. The G$^{|}$-Band is caused by two-phonon scattering around the K-point of the Brillouin-zone. $G^{|}$-mode comes in a picture only at high-defect density; it has less significant as compared to the first order mode. The mode has more contributions from regions near the K and M-points, $G^{|}$ peak is visualized in Raman spectra at 2700 and 2730 cm^{-1}. It has been reported that $G^{|}$ peak provides information about CNTs where other phonon modes failed and the metallicity of CNTs is responsible for the intensity of $G^{|}$ peak [185].

1.3 Conclusion

CNT is one of the most rigorously researched materials in the contemporary world. Owing to their extraordinary properties, CNTs have become one of the front runners in the world of contemporary research. If a cross-section area of CNTs walls is considered then the values of elastic modulus and tensile strength have been estimated as 1 TPa and 100 GPa, respectively. The mechanical strength of CNTs has been found 10 times more as compared to any other fiber. Generally, MWNTs show metallic behavior with a very high current capacity ($\sim10^9$ A cm^{-2}). Individual walls in MWNTs can act as metallic or semiconducting as per folding angle of graphene sheet from which CNT is constructed. The orientation of graphene sheet with the tube axis is defined as chirality. Room-temperature thermal conductivity of SWNTs can even reach up to 3500 W m^{-1} K^{-1} that is more than of a diamond. Because of these ultimate properties, CNTs are being used in a variety of nanosensors, biosensors, optoelectronic devices such as photodetectors, solar cells, and light-emitting diodes, in different types of high-performance electronic devices, as reinforcement in different composite systems, etc. For the preparation of devices at nanoscale, where we deal with the ultra-small regime, control of synthesis and growth of CNTs having precise specifications is very important. Having understood the importance of CNTs, we made a detailed discussion on the various allotropes of carbon, especially on CNTs. The basic structures of CNTs, defects in structure, electronics, optoelectronics, and chemical as well as mechanical properties of CNTs have been covered. A detailed discussion on the various synthesis techniques of CNTs with comparative analysis have been covered. CVD method is the most widely used method to grow MWNTs and SWNTs. There are different factors such as dynamics of hydrocarbons as well as carrier gases, growth temperature, temperature ramp, catalyst engineering, substrate engineering, etc. that affect the specifications of CNTs. We made a detailed study on all mentioned factors. For CNTs analysis, various nanomaterials characterization tools such as SEM, TEM, AFM, Raman Spectroscopy, XRD, etc. are required. A detailed study on mentioned characterization tools, their instrumentation, and utilization for CNTs has been made. The chapter will be helpful for the development of modern electron materials and applications.

Declaration of Competing Interest

Authors do not have any conflict of interest.

Companies Dealing with Chemically Modified CNTs

I. Sigma-Aldrich, USA
II. Cheaptubes.com
III. Nanoshel, UK
IV. Nanostructured & Amorphous Materials, Inc., USA

Acknowledgments

This work was supported by Russian Science Foundation (Grant No. 21-79-00272).

References

1 Tripathi, N. and Islam, S.S. (2017). A new approach for orientation controlled growth of CNTs: an in-depth analysis on the role of oxygen plasma treatment to catalyst. *Appl. Nanosci.* 7: 125–129.

2 Tripathi, N., Mishra, P., Joshi, B. et al. (2014). Catalyst free, excellent quality and narrow diameter of CNT growth on Al_2O_3 by thermal CVD technique. *Phys. E* 62: 43–47.

3 Sharma, P., Pavelyev, V., Kumar, S. et al. (2020). Analysis on the synthesis of vertically aligned carbon nanotubes: growth mechanism and techniques. *J. Mater. Sci.: Mater. Electron.* 31: 4399–4443.

4 Kumar, M. and Ando, Y. (2010). Chemical vapor deposition of carbon nanotubes: a review on growth mechanism and mass production. *J. Nanosci. Nanotechnol.* 10: 3739–3758.

5 Tripathi, N., Mishra, P., Joshi, B., and Islam, S.S. (2015). Precise control over physical characteristics of carbon nanotubes by differential variation of argon flow rate during chemical vapor deposition processing: a systematic study on growth kinetics. *Mater. Sci. Semicond. Process.* 35: 207–215.

6 Krueger, A. (2010). *Carbon Materials and Nanotechnology*. Weinheim: Wiley-VCH Verlag GmbH & Co. KGaA.

7 Meyyappan, M. (2004). *Carbon Nanotubes Science and Applications*. NASA, Ames Research Center Moffett Field, CA, CRC Press.

8 Debye, P. and Scherrer, P. (1917). Interferenzen a regellos orientierten teilchen im Röntgenlicht. *Phys. Z* 18: 291.

9 Geim, A.K. and Novoselov, K.S. (2007). The rise of graphene. *Nat. Mater.* 6: 183.

10 Bokros, J.C. (1969). *Chemistry and Physics of Carbon*, vol. 5 (ed. P.L. Walker Jr.). New York: Marcel Dekker Inc.

11 Pierson, H.O. (1993). *Handbook of Carbon, Graphite, Diamond and Fullerenes, Properties, Processing and Applications*. Park Ridge, NJ: Noyes Publications.

12 Guy, G. (1959). *Elements of Physical Metallurgy*. Reading, MA: Addison-Wesley Publishing.

13 Cullity, D. (1956). *Elements of X-ray Diffraction*. Reading, MA: Addison-Wesley Publishing.
14 Spear, K.E. (1969). Diamond: ceramic coating of the future. *J. Am. Ceram. Soc.* 72: 171–191.
15 Eggers, D.F. Jr. and Halsey, G.D. Jr. (1964). *Physical Chemistry*. New York: Wiley.
16 Gardinier, F. (1988). Physical properties of superabrasives. *Ceram. Bull.* 67: 1006.
17 Spear, K.E., Phelps, A.W., and White, W.B. (1990). Diamond polytypes and their vibrational spectra. *J. Mater. Res.* 5: 2271.
18 Dawson, J.B. (1967). *The Properties of Diamond* (ed. J.E. Field), 539–554. London: Academic Press.
19 Wong, H.S.P. and Jiakinwande, D.E. (2001). *Carbon Nanotube and Graphene Device Physics*. Cambridge University Press.
20 Pauling, L. (1960). *The Nature of Chemical Bonds*. Cornell University Press.
21 Bhushan, B. (2011). *Nanoscience and Technology (Scanning Probe Microscopy in Nanoscience and Nano Technology)*. Springer.
22 Rogers, G.L. (1950). Gabor diffraction microscopy: the hologram as a generalized zone-plate. *Nature* 166: 237.
23 Radushkevich, L.V. and Lukyanovich, V.M. (1952). O struktureugleroda, obrazujucegosjapritermiceskomrazlozeniiokisiuglerodanazeleznomkontakte. *Zurn. Fisic. Chim.* 26: 88.
24 Baker, R.J., Mengden, G.A., and Bull, J.J. (1972). Karyotypic studies of thirty-eight species of North American snakes. *Copeia* 1972: 257.
25 Oberlin, A., Endo, M., and Koyama, T. (1976). Filamentous growth of carbon through benzene decomposition. *J. Cryst. Growth* 32: 3.
26 Dupuis (2005). The catalyst in the CCVD of carbon nanotubes- a review. *Prog. Mater. Sci.* 50 (8): 929–961.
27 Aqel, A., Abou El-Nour, K.M.M., Ammar, R.A.A., and Al-Warthan, A. (2012). Carbon nanotubes, science and technology part (I) structure, synthesis and characterization. *Arabian J. Chem.* 5: 1–23.
28 Saito, R., Dresselhaus, G., and Dresselhaus, M.S. (1998). *Physical Properties of Carbon Nanotubes*. Imperial College Press.
29 McEuen, P.L. (1999). Disorder, pseudospins, and backscattering in carbon nanotubes. *Phys. Rev. Lett.* 24: 5098.
30 Dresselhaus, M.S.E. (2000). Phonons in carbon nanotubes. *Adv. Phys.* 49: 705.
31 Hone, J. (2000). Quantized phonon spectrum of single-wall carbon nanotubes. *Science* 289: 1730.
32 Tans, S.J. (1997). Individual single-wall carbon nanotubes as quantum wires. *Nature* 386: 474.
33 Frank, S. (1998). Carbon nanotube quantum resistors. *Science* 280: 1744.
34 Soh, H.T. (1999). Integrated nanotube circuits: controlled growth and ohmic contacting of single-walled carbon nanotubes. *Appl. Phys. Lett.* 75: 627.
35 Rao, A.M. (1997). Diameter-selective Raman scattering from vibrational modes in carbon nanotubes. *Science* 275: 187.

36 Bachilo, S.M. (2002). Structure-assigned optical spectra of single-walled carbon nanotubes. *Science* 298: 2361.

37 Hagen, A. and Hertel, T. (2003). Quantitative analysis of optical spectra from individual single-wall carbon nanotubes. *Nano Lett.* 3: 383.

38 Paulson, S. (1999). In situ resistance measurements of strained carbon nanotubes. *Appl. Phys. Lett.* 75: 2936.

39 Tsang, S.C. (1994). A simple chemical method of opening and filling carbon nanotubes. *Nature* 372: 159.

40 Ajayan, P.M. (1995). Carbon nanotubes as removable templates for metal-oxide nanocomposites and nanostructures. *Nature* 375: 564.

41 Dujardin, E. (1994). Capillarity and wetting of carbon nanotubes. *Science* 265: 1850.

42 Ugarte, D. (1996). Nanocapillarity and chemistry in carbon nanotubes. *Science* 274: 1897.

43 Zhao, J., Lu, J.P., Han, J., and Yang, C. (2003). Noncovalent functionalization of carbon nanotubes by aromatic organic molecules. *Appl. Phys. Lett.* 82: 3746.

44 Kruger, M. (2003). Sensitivity of single multiwalled carbon nanotubes to the environment. *New J. Phys.* 5: 138.

45 Zhao, J., Buldum, A., and Han, J. (2000). First-principles study of Li-intercalated carbon nanotube ropes. *Phys. Rev. Lett.* 85: 1706.

46 Frackowiak, E., Gautier, S., Gaucher, H. et al. (1999). Electrochemical storage of lithium in multiwalled carbon nanotubes. *Carbon* 37: 61.

47 Wu, G.T. (1999). Structure and lithium insertion properties of carbon nanotubes. *J. Electrochem. Soc.* 146: 1696.

48 Lehman, J.H., Terrones, M., Mansfield, E. et al. (2011). Evaluating the characteristics of multiwall carbon nanotubes. *Carbon* 4 (9): 2581.

49 Bao, B. (1999). Carbon nanotubes science and application. *Chem. Phys. Lett.* 307: 153.

50 Bonard, J.M., Croci, M., Klinke, C. et al. (2002). Carbon nanotube films as electron field emitters. *Carbon* 40: 1715.

51 Thrower, P.A. (1969). The study of defects in graphite by transmission electron spectroscopy. *Chem. Phys. Carbon* 5: 217.

52 Terrones, M., Terrones, G., and Terrones, H. (2002). Structure, chirality, and formation of giant icosahedral fullerenes and spherical graphitic onions. *Struct. Chem.* 13: 373.

53 Charlier, J.C., Terrones, M., Baxendale, M. et al. (2002). Enhanced electron field emission in B-doped carbon nanotubes. *Nano Lett.* 2: 1191–1195.

54 Girit, C.O., Meyer, J.C., Erni, R. et al. (2009). Graphene at the edge: stability and dynamics. *Science* 323: 1705.

55 Terrones, M., Jorio, A., Endo, M. et al. (2004). New direction in nanotube science. *Mater. Today* 7: 30.

56 Maciel, O., Delgado, J.C., Silva, E.C. et al. (2009). Synthesis, electronic structure, and Raman scattering of phosphorus-doped single-wall carbon nanotubes. *Nano Lett.* 9: 2267.

57 Maciel, O., Anderson, N., Pimenta, M.A. et al. (2008). Electron and phonon renormalization near charged defects in carbon nanotubes. *Nat. Mater.* 7: 878.

58 Sadegh, H. and Shahryari-ghoshekandi, R. (2015). Functionalization of carbon nanotubes and its application in nanomedicine: a review. *Nanomed. J.* 2: 231–248.

59 Mallakpour, S. and Soltanian, S. (2016). Surface functionalization of carbon nanotubes: fabrication and applications. *RSC Adv.* 6: 109916–109935.

60 Venkatesh, K., Arthanareeswaran, G., Chandra Bose, A. et al. (2021). Diethylenetriaminepentaacetic acid-functionalized multi-walled carbon nanotubes/titanium oxide-PVDF nanofiber membrane for effective separation of oil/water emulsion. *Sep. Purif. Technol.* 257: 117926.

61 Bouanis, F.Z., Bensifia, M., Florea, I. et al. (2021). Non-covalent functionalization of single walled carbon nanotubes with Fe-/Co-porphyrin and Co-phthalocyanine for field-effect transistor applications. *Org. Electron.* 96: 106212.

62 He, X., Kevlishvili, I., Murcek, K. et al. (2021). Photocycloaddition of enones to single-walled carbon nanotubes creates fluorescent quantum defects. *ACS Nano* 15: 4833–4844.

63 Khan, D., Ali, Z., Asif, D. et al. (2021). Incorporation of carbon nanotubes in photoactive layer of organic solar cells. *Ain Shams Eng. J.* 12: 897–900.

64 Muhammad-Ahson Aslam, M., Kuo, H.-W., Den, W. et al. (2021). Functionalized carbon nanotubes (CNTs) for water and wastewater treatment: preparation to application. *Sustainability* 13: 5717.

65 Ma, P.C., Siddiqui, N.A., Marom, G., and Kim, J.K. (2010). Dispersion and functionalization of carbon nanotubes for polymer-based nanocomposites: a review. *Composites, Part A* 41: 1345–1367.

66 Hirsch, A. (2002). Functionalization of single-walled carbon nanotubes. *Angew. Chem. Int. Ed.* 41: 1853–1859.

67 Hirsch, A. and Vostrowsky, O. (2005). Functionalization of carbon nanotubes. *Top. Curr. Chem.* 245: 193–237.

68 Kim, W., Nair, N., Lee, C.Y., and Strano, M.S. (2008). Covalent functionalization of singlewalled carbon nanotubes alters their densities allowing electronic and other types of separation. *J. Phys. Chem. C* 112: 7326–7331.

69 Niyogi, S., Hamon, M.A., Hu, H. et al. (2002). Chemistry of single-walled carbon nanotubes. *Acc. Chem. Res.* 35: 1105–1113.

70 Tasis, D., Tagmatarchis, N., Georgakilas, V., and Prato, M. (2003). Soluble carbon nanotubes. *Chem. Eur. J.* 9: 4000–4008.

71 Sinnott, S.B. (2002). Chemical functionalization of carbon nanotubes. *J. Nanosci. Nanotechnol.* 2: 113–123.

72 Mickelson, E.T., Huffman, C.B., Rinzler, A.G. et al. (1998). Fluorination of single-wall carbon nanotubes. *Chem. Phys. Lett.* 296: 188–194.

73 Kelly, K.F., Chiang, I.W., Mickelson, E.T. et al. (1993). Insight into the mechanism of sidewall functionalization of single-walled nanotubes: an STM study. *Chem. Phys. Lett.* 313: 445–450.

74 Touhara, H., Inahara, J., Mizuno, T. et al. (2002). Fluorination of cup-stacked carbon nanotubes, structure and properties. *Fluorine Chem.* 114: 181–188.
75 Stevens, J.L., Huang, A.Y., Peng, H. et al. (2003). Sidewall amino-functionalization of SWNTs through fluorination and subsequent reactions with terminal diamines. *Nano Lett.* 3: 331–336.
76 Tseng, C.-H., Wang, C.-C., and Chen, C.-Y. (2007). Functionalizing carbon nanotubes by plasma modification for the preparation of covalent-integrated epoxy composites. *Chem. Mater.* 19: 308–315.
77 Xie, L., Xu, F., Qiu, F. et al. (2007). Single-walled carbon nanotubes functionalized with high bonding density of polymer layers and enhanced mechanical properties of composites. *Macromolecules* 40: 3296–3305.
78 Shi, X., Wang, J., Jiang, B., and Yang, Y. (2013). Hindered phenol grafted carbon nanotubes for enhanced thermal oxidative stability of polyethylene. *Polymer* 54: 1167–1176.
79 Zhou, Z., Yan, X., Cook, T.R. et al. (2016). Engineering functionalization in a supramolecular polymer: hierarchical self-organization of triply orthogonal non-covalent interactions on a supramolecular coordination complex platform. *J. Am. Chem. Soc.* 138: 806–809.
80 Ji, X., Xu, Y., Zhang, W. et al. (2016). Review of functionalization, structure and properties of graphene/polymer composite fibers. *Composites, Part A* 87: 29–45.
81 Orellana, W. and Correa, J.D. (2015). Noncovalent functionalization of carbon nanotubes and graphene with tetraphenylporphyrins: stability and optical properties from ab initio calculations. *J. Mater. Sci.* 50: 898–905.
82 McCarthy, B., Coleman, J.N., Czerw, R. et al. (2001). Microscopy studies of nanotube-conjugated polymer interactions. *Synth. Met.* 121: 1225–1226.
83 Gong, X., Liu, J., Baskaran, S. et al. (2000). Surfactant-assisted processing of carbon nanotube/polymer composites. *Chem. Mater.* 12: 1049–1052.
84 Vaisman, L., Marom, G., and Wagner, H.D. (2006). Dispersions of surface-modified carbon nanotubes in water-soluble and water-insoluble polymers. *Adv. Funct. Mater.* 16: 357–363.
85 Geng, Y., Liu, M.Y., Li, J. et al. (2008). Effects of surfactant treatment on mechanical and electrical properties of CNT/epoxy nanocomposites. *Composites, Part A* 39: 1876–1883.
86 Islam, M.F., Rojas, E., Bergey, D.M. et al. (2003). High weight fraction surfactant solubilization of single-wall carbon nanotubes in water. *Nano Lett.* 3: 269–273.
87 Yu, J., Grossiord, N., Koning, C.E., and Loos, J. (2007). Controlling the dispersion of multi-wall carbon nanotubes, in aqueous surfactant solution. *Carbon* 45: 618–623.
88 Whitsitt, E.A. and Barron, A.R. (2003). Silica coated single walled carbon nanotubes. *Nano Lett.* 3: 775–778.
89 Vaisman, L., Wagner, H.D., and Marom, G. (2006). The role of surfactants in dispersion of carbon nanotubes. *Adv. Colloid Interface Sci.* 128–130: 37–46.

90 Yuan, W. and Chan-Park, M.B. (2012). Covalent cum noncovalent functionalizations of carbon nanotubes for effective reinforcement of a solution cast composite film. *ACS Appl. Mater. Interfaces* 4: 2065–2073.

91 Yola, M.L. and Atar, N. (2021). Novel voltammetric tumor necrosis factor-alpha (TNF-α) immunosensor based on gold nanoparticles involved in thiol-functionalized multi-walled carbon nanotubes and bimetallic Ni/Cu-MOFs. *Anal. Bioanal.Chem.* 413: 2481–2492.

92 Wang, S., Wang, M., Li, C. et al. (2020). A highly sensitive and stable electrochemiluminescence immunosensor for alpha-fetoprotein detection based on luminol-AgNPs@Co/Ni-MOF nanosheet microflowers. *Sens. Actuators, B* 311: 127919.

93 Pouya, Z.A., Tofighy, M.A., and Mohammadi, T. (2021). Synthesis and characterization of polytetrafluoroethylene/oleic acid-functionalized carbon nanotubes composite membrane for desalination by vacuum membrane distillation. *Desalination* 503: 114931.

94 González-Muñoz, D., Martín-Somer, A., Strobl, K. et al. (2021). Enhancing visible-light photocatalysis via endohedral functionalization of single-walled carbon nanotubes with organic dyes. *ACS Appl. Mater. Interfaces* 13: 24877–24886.

95 Choi, E.Y., Kim, M.H., Lee, S.Y. et al. (2021). Reactivity estimation of polybutylene terephthalate with noncovalently functionalized multiwalled carbon nanotubes and reactivity effect on the composite reinforcement. *Compos. Sci. Technol.* 201: 108549.

96 Zhao, Y.L., Hu, L., Grüner, G., and Stoddart, J.F. (2008). A tunable photosensor. *J. Am. Chem. Soc.* 130.

97 Barone, P.W., Baik, S., Heller, D.A., and Strano, M.S. (2005). Near-infrared optical sensors based on single-walled carbon nanotubes. *Nat. Mater.* 4: 86–92.

98 Li, X., Jia, Y., Wei, J. et al. (2010). Solar cells and light sensors based on nanoparticle-grafted carbon nanotube films. *ACS Nano* 4: 2142–2148.

99 Liu, S., Ye, J., Cao, Y. et al. (2009). Tunable hybrid photodetectors with superhigh responsivity. *Small* 5: 2371–2376.

100 Bekyarova, E., Davis, M., Burch, T. et al. (2004). Chemically functionalized single-walled carbon nanotubes as ammonia sensors. *J. Phys. Chem. B* 108: 19717–19720.

101 Sin, M.L.Y., Chow, G.C.T., Wong, G.M.K. et al. (2007). Ultralow-power alcohol vapor sensors using chemically functionalized multiwalled carbon nanotubes. *IEEE Trans. Nanotechnol.* 6: 571–577.

102 Niu, L., Luo, Y., and Li, Z. (2007). A highly selective chemical gas sensor based on functionalization of multi-walled carbon nanotubes with poly(ethylene glycol). *Sens. Actuators, B* 126: 361–367.

103 Dhall, S., Jaggi, N., and Nathawat, R. (2013). Functionalized multiwalled carbon nanotubes based hydrogen gas sensor. *Sens. Actuators, A* 201: 321–327.

104 Kong, J., Chapline, M.G., and Dai, H. (2001). Functionalized carbon nanotubes for molecular hydrogen sensors. *Adv. Mater.* 13: 1384–1386.

105 Sayago, I., Terrado, E., Lafuente, E. et al. (2005). Hydrogen sensors based on carbon nanotubes thin films. *Synth. Met.* 148: 15–19.

106 Iijima, S., Ichihashi, T., and Ando, Y. (1992). Pentagons, heptagons and negative curvature in graphite microtubule growth. *Nature* 356: 776.

107 Ryu, J.H., Kang, J.S., and Park, K.C. (2012). Carbon nanotube electron emitter for X-ray imaging. *Materials* 5: 2353.

108 Jensen, K., Weldon, J., Garcia, H., and Zettl, A. (2007). Nanotube radio. *Nano Lett.* 11: 3508.

109 Rueckes, T., Kim, K., Joselevich, E. et al. (2000). Carbon nanotube-based nonvolatile random access memory for molecular computing. *Science* 289: 5476.

110 Xi, N. and La, K. (2012). *Nano Optoelectronic Sensors and Devices: Nanophotonics from Design to Manufacturing*. Elsevier.

111 Yamashita, S., Saito, Y., and Choi, J.H. (2013). *Carbon Nanotubes and Graphene for Photonic Applications*. Woodhead Publishing.

112 Melisi, D., Nitti, M.A., Valentini, M. et al. (2014). Photodetectors based on carbon nanotubes deposited by using a spray technique on semi-insulating gallium arsenide. *J. Nanotechnol.* 5: 1999.

113 Sazonova, V., Yaish, Y., Üstünel, H. et al. (2004). A tunable carbon nanotube electromechanical oscillator. *Nature* 431: 284.

114 Fennimore, M., Yuzvinsky, T.D., Han, W.-Q. et al. (2003). Rotational actuators based on carbon nanotubes. *Nature* 424: 408.

115 Kim, P. and Lieber, C.M. (1999). Nanotube nanotweezers. *Science* 286: 2148.

116 Zhang, Q. (2012). *Carbon Nanotubes and Their Applications*, Pan Stanford Series on Carbon Based Nanomaterials, vol. 1. CRC Press.

117 Popov, A.M., Lebedeva, I.V., Knizhnik, A.A. et al. (2014). Force and magnetic field sensor based on measurement of tunnelling conductance between ends of coaxial carbon nanotubes. *Comput. Mater. Sci.* 92: 84.

118 Poot, M. and van der Zant, H.S.J. (2012). Mechanical systems in the quantum regime. *Phys. Rep.* 511: 273.

119 Saito, Y., Yoshikawa, T., Inagaki, M. et al. (1993). Growth and structure of graphitic tubules and polyhedral particles in arc-discharge. *Chem. Phys. Lett.* 204: 227.

120 Iijima, S. (1991). Helical microtubules of graphitic carbon. *Nature* 354 (6348): 56.

121 Journet, C., Maser, W.K., Bernier, P. et al. (1997). Large-scale production of single-walled carbon nanotubes by the electric-arc technique. *Nature* 388: 756.

122 Saito, Y., Nishikubo, K., Kawabata, K., and Matsumoto, T. (1996). Carbon nanocapsules and single-layered nanotubes produced with platinum-group metals (Ru, Rh, Pd, Os, Ir, Pt) by arc discharge. *J. Appl. Phys.* 80: 3062.

123 Farhat, S., de La Chapelle, M.L., Loiseau, A. et al. (2001). Diameter control of single-walled carbon nanotubes using argon–helium mixture gases. *J. Chem. Phys.* 115: 6752.

124 Tang, D., Xie, S., Liu, W. et al. (2000). Evidence for an open-ended nanotube growth model in arc discharge. *Carbon* 38: 480.

125 Ebbesen, T. and Ajayan, P. (1992). Large-scale synthesis of carbon nanotubes. *Nature* 358: 220.

126 Cadek, M., Murphy, R., McCarthy, B. et al. (2002). Optimisation of the arc-discharge production of multi-walled carbon nanotubes. *Carbon* 40: 923.

127 Journet, C. and Bernier, P. (1998). Production of carbon nanotubes. *Appl. Phys. A* 67: 1.

128 Bolshakov, A.P., Uglov, S., Saveliev, A. et al. (2002). A novel CW laser–powder method of carbon single-wall nanotubes production. *Diamond Relat. Mater.* 11: 927.

129 Kokai, F., Takahashi, K., Yudasaka, M. et al. (1999). Growth dynamics of single-wall carbon nanotubes synthesized by CO_2 laser vaporization. *J. Phys. Chem.* B103: 4346.

130 Wei, B.Q., Vajtai, R., Jung, Y. et al. (2002). Microfabrication technology: organized assembly of carbon nanotubes. *Nature* 416: 495.

131 Sengupta, J. (2018). Carbon nanotube fabrication at industrial scale: opportunities and challenges, Chapter 9. In: *Handbook of Nanomaterials for Industrial Applications* (ed. C.M. Hussain), 172–194. Elsevier.

132 Nihei, M., Kawabata, A., Kondo, D. et al. (2005). Electrical properties of carbon nanotube bundles for future via interconnects. *Jpn. J. Appl. Phys.* 44: 1626.

133 Dikonimos Makris, T., Giorgi, L., Giorgi, R. et al. (2005). CNT growth on alumina supported nickel catalyst by thermal CVD. *Diamond Relat. Mater.* 14: 815.

134 Wu, Y. and Yang, P. (2001). Direct observation of vapor–liquid–solid nanowire growth. *J. Am. Chem. Soc.* 123: 3165–3166.

135 Tripathi, N., Mishra, P., Harsh, H., and Islam, S.S. (2015). Fine-tuning control on CNT diameter distribution, length and density using thermal CVD growth at atmospheric pressure: an in-depth analysis on the role of flow rate and flow duration of acetylene (C_2H_2) gas. *Appl. Nanosci.* 5: 19.

136 Shaikjee, A. and Coville, N.J. (2012). The role of the hydrocarbon source on the growth of carbon materials. *Carbon* 50: 3376–3398.

137 Hernadi, K., Fonseca, A., Nagy, J.B. et al. (1996). Fe-catalyzed carbon nanotube formation. *Carbon* 34: 1249.

138 Kong, J., Cassell, A.M., and Dai, H. (1998). Chemical vapor deposition of methane for single-walled carbon nanotubes. *Chem. Phys. Lett.* 292: 567.

139 Fan, S., Chapline, M., Frankline, N. et al. (1999). Self-oriented regular arrays of carbon nanotubes and their field emission properties. *Science* 283: 512.

140 Sen, R., Govindaraj, A., and Rao, C.N.R. (1997). Carbon nanotubes by the metallocene route. *Chem. Phys. Lett.* 267: 276.

141 Nikolaev, P., Bronikowski, M.J., Bradley, R.K. et al. (1999). Gas-phase catalytic growth of single-walled carbon nanotubes from carbon monoxide. *Chem. Phys. Lett.* 313: 91.

142 Endo, M., Fijiwara, H., and Fukunaga, E. (1991). *18th Meeting Japanese, Carbon Society*, 34. Saitama: Japanese Carbon Society.

143 Endo, M., Takeuchi, K., Igarashi, S. et al. (1993). The production and structure of pyrolytic carbon nanotubes (PCNTs). *J. Phys. Chem. Solids* 54: 1841.

144 Jose-Yacaman, M., Miki-Yoshida, M., Rendon, L., and Santiesteban, J.G. (1993). Catalytic growth of carbon microtubules with fullerene structure. *Appl. Phys. Lett.* 62: 657.

145 Liua, Z., Cheb, R., Xu, Z., and Peng, L. (2002). Preparation of Fe-filled carbon nanotubes by catalytic decomposition of cyclohexane. *Synth. Met.* 128: 191.

146 Li, N., Chen, X., Stoica, L. et al. (2007). The catalytic synthesis of three-dimensional hierarchical carbon nanotube composites with high electrical conductivity based on electrochemical iron deposition. *Adv. Mater.* 19: 2957.

147 Morjan, R.E., Nerushev, O.A., Sveningsson, M. et al. (2004). Growth of carbon nanotubes from C_{60}. *Appl. Phys. A* 78: 253.

148 Dai, H., Rinzler, A.G., Nikolaev, P. et al. (1996). Single-wall nanotubes produced by metal-catalyzed disproportionation of carbon monoxide. *Chem. Phys. Lett.* 260: 471.

149 Satishkumar, B.C., Govindaraj, A., Sen, R., and Rao, C.N.R. (1998). Single-walled nanotubes by the pyrolysis of acetylene-organometallic mixtures. *Chem. Phys. Lett.* 293: 47.

150 Flahaut, E., Govindaraj, A., Peigney, A. et al. (1999). Synthesis of single-walled carbon nanotubes using binary (Fe, Co, Ni) alloy nanoparticles prepared in situ by the reduction of oxide solid solutions. *Chem. Phys. Lett.* 300: 236.

151 Cheng, H.M., Li, F., Sun, X. et al. (1998). Bulk morphology and diameter distribution of single-walled carbon nanotubes synthesized by catalytic decomposition of hydrocarbons. *Chem. Phys. Lett.* 289: 602.

152 Maruyama, S., Kojima, R., Miyauchi, Y. et al. (2002). Low-temperature synthesis of high-purity single-walled carbon nanotubes from alcohol. *Chem. Phys. Lett.* 360: 229.

153 Okubo, S., Sekine, T., Suzuki, S. et al. (2004). Purification of single-wall carbon nanotubes synthesized from alcohol by catalytic chemical vapor deposition. *Jpn. J. Appl. Phys.* 43: L396.

154 Nasibulin, G., Moisala, A., Jiang, H., and Kauppinen, E.I. (2006). High quality double wall carbon nanotubes with a defined diameter distribution by chemical vapor deposition from alcohol. *J. Nanopart. Res.* 8: 465.

155 Murakami, Y., Miyauchi, Y., Chiashi, S., and Maruyama, S. (2003). Direct synthesis of high-quality singlewalled carbon nanotubes on silicon and quartz substrates. *Chem. Phys. Lett.* 377: 49.

156 Murakami, Y., Chiashi, S., Miyauchi, Y. et al. (2004). Growth of vertically aligned single-walled carbon nanotube films on quartz substrates and their optical anisotropy. *Chem. Phys. Lett.* 385: 298.

157 Xiang, R., Einarsson, E., Okawa, J. et al. (2009). Acetylene-accelerated alcohol catalytic chemical vapor deposition growth of vertically aligned single-walled carbon nanotubes. *J. Phys. Chem. C* 113: 7511.

158 Hata, K., Futaba, D.N., Mizuno, K. et al. (2004). Water-assisted highly efficient synthesis of impurity-free single-walled carbon nanotubes. *Science* 306: 1362.

159 Zhong, G., Hofmann, S., Yan, F. et al. (2009). Acetylene: a key growth precursor for single-walled carbon nanotube forests. *J. Phys. Chem. C* 113: 17321.

160 Ding, F., Larsson, P., Larsson, J.A. et al. (2008). The importance of strong carbon-metal adhesion for catalytic nucleation of single-walled carbon nanotubes. *Nano Lett.* 8: 463.

161 Moisala, A., Nasibulin, A.G., and Kauppinen, E.I. (2003). The role of metal nanoparticles in the catalytic production of single-walled carbon nanotubes—a review. *J. Phys. Condens. Mater.* 15: S3011.

162 Ago, H., Komatsu, T., Ohshima, S. et al. (2000). Dispersion of metal nanoparticles for aligned carbon nanotube arrays. *Appl. Phys. Lett.* 77: 79.

163 Tripathi, N., Pavelyev, V., and Islam, S.S. (2017). Synthesis of carbon nanotubes using green plant extract as catalyst: unconventional concept and its realization. *Appl. Nanosci.* 7: 557–566.

164 Domagała, K., Borlaf, M., Traber, J. et al. (2019). Purification and functionalisation of multi-walled carbon nanotubes. *Mater. Lett.* 253: 272–275.

165 Biro, L.P., Khanh, N.Q., Vertesy, Z. et al. (2002). Catalyst traces and other impurities in chemically purified carbon nanotubes grown by CVD. *Mater. Sci. Eng., C* 19: 9–13.

166 Tripathi, N., Pavelyev, V., and Islam, S.S. (2018). Tunable growth of single-wall CNTs by monitoring temperature increasing rate. International. *Nano Lett.* 8: 1–9.

167 Jorio, A., Pimenta, M.A., Filho, A.G.S. et al. (2003). Characterizing carbon nanotube samples with resonance Raman scattering. *New J. Phys.* 5: 139.

168 Jorio, A., Saito, R., Dresselhaus, G., and Dresselhaus, M.S. (2011). *Raman Spectroscopy in Graphene Related Systems*. Wiley-VCH.

169 Zumbusch, A., Holtom, G.R., and Xie, X.S. (1999). Three-dimensional vibrational imaging by coherent anti-Stokes Raman scattering. *Phys. Rev. Lett.* 82: 4142.

170 Tan, H., Hu, C.Y., Dong, J. et al. (2001). Polarization properties, high-order Raman spectra, and frequency asymmetry between Stokes and anti-Stokes scattering of Raman modes in a graphite whisker. *Phys. Rev. B* 64: 214301.

171 Jorio, A., Filho, A.G.S., Dresselhaus, G. et al. (2002). G-band resonant Raman study of 62 isolated single-wall carbon nanotubes. *Phys. Rev. B* 65: 155412.

172 Lina, D.H.M., Manaraa, D., Reis, P.L. et al. (2014). The use of different dispersive Raman spectrometers for the analysis of uranium compounds. *Vib. Spectrosc.* 73: 102.

173 Raravikar, R., Keblinski, P., Rao, A.M. et al. (2002). Temperature dependence of radial breathing mode Raman frequency of single-walled carbon nanotubes. *Phys. Rev. B* 66: 235424.

174 Bandow, S., Asaka, S., Saito, Y. et al. (1998). Effect of growth temperature on the diameter distribution and chirality of single-wall carbon nanotubes. *Phys. Rev. Lett.* 80: 3779.

175 Sethi, R. and Barron, A.R. (2009). Characterization of single-walled carbon nanotubes by Raman spectroscopy, a report of RICE University, Report number 2.

176 Pfeiffer, R., Kuzmany, H., Kramberger, C. et al. (2003). Unusual high degree of unperturbed environment in the interior of single-wall carbon nanotubes. *Phys. Rev. Lett.* 90: 225501.

177 Pimenta, M.A., Marucci, S., Empedocles, S. et al. (1998). Raman modes of metallic carbon nanotubes. *Phys. Rev. B* 58: 6016.

178 Jiang, C., Kempa, K., Zhao, J. et al. (2002). Strong enhancement of the Breit–Wigner–Fano Raman line in carbon nanotube bundles caused by plasmon band formation. *Phys. Rev. B* 66: 161404.

179 Dresselhaus, M.S. and Dresselhaus, G. (1981). Intercalation compounds of graphite. *Adv. Phys.* 30: 139.

180 Rao, A.M. (1997). Evidence for charge transfer in doped carbon nanotube bundles from Raman scattering. *Science* 388: 257.

181 Jorio, A. (2003). Resonance Raman spectra of carbon nanotubes by cross-polarized light. *Phys. Rev. Lett.* 90: 107403.

182 Maultzsch, J., Reich, S., and Thomsen, C. (2002). Raman scattering in carbon nanotubes revisited. *Phys. Rev. B* 65: 233402.

183 Maultzsch, J., Reich, S., Schlecht, U., and Thomsen, C. (2003). High-energy phonon branches of an individual metallic carbon nanotube. *Phys. Rev. Lett.* 91: 087402.

184 Saito, S. and Zettl, A. (2008). *Carbon Nanotubes: Quantum Cylinders of Graphene*. Elsevier.

185 Jushi, R.A., Venkataraman, L., Dresselhaus, M.S., and Dresselhaus, G. (1993). Phonon modes in carbon nanotubules. *Chem. Phys. Lett.* 209: 77.

2

Surface Modification Strategies for the Carbon Nanotubes

Shiv K. Prajapati[1], Tejas Agnihotri[2], and Aakanchha Jain[2]

[1] *Institute of Pharmacy, Ram-Eesh Institute of Vocational and Technical Education, Knowledge Park I, Greater Noida, Uttar Pradesh, 201310, India*
[2] *National Institute of Pharmaceutical Education and Research (NIPER) – Ahmedabad, Department of Pharmaceutics, An Institute of National Importance, Government of India, Opposite Air force Station Palaj, Gandhinagar, Gujarat, 382355, India*

2.1 Introduction

Carbon nanotubes (CNTs) are commonly known as nanoscale hollow tubes composed of carbon atoms. CNTs are distinguished from other nanosystems by their unique arrangement of carbon atoms, sp^2 hybridized and rolled up in cylindrical form with a 1.42 Å C–C distance and an interlayer arrangement of 3.4 Å [1–3]. CNTs are newer generation materials with desirable properties such as size (1 nm to tens of nm) and aspect ratio (length-to-diameter values) typically above 10^3, good electrical, mechanical, and thermal characteristics give them endless potential in nanotechnology-related applications, including biomedical applications and targeted drug delivery [4–6]. Due to their unique attributes, CNTs have been expanded in their applications due to the following characteristics: hollow tube and vast surface area make CNTs suitable for drug delivery due to the greater area for drug loading. Due to hydrophobicity, delivery of biomolecules (DNA, RNA, proteins, immunomodulators, etc.) is possible. Because of their superior optical attributes, CNTs have demonstrated enormous application potential in a wide range of disciplines, including photodynamic treatment (PDT) and photoacoustic imaging [7, 8]. CNTs may be used in biosensors due to their excellent conductive capabilities [9–11]. Because aggregation greatly influences CNT characteristics, dispersibility of CNTs in organic and aqueous phases is a crucial issue in drug delivery. However, dispersibility may well be maximized by enabling simple procedures to be performed on the CNTs surface [12, 13]. CNTs have sparked the interest of researchers, and they are now being investigated with surface engineering to provide controlled drug delivery and better intracellular target selectivity. CNTs address disadvantages such as low permeability and nonspecificity

Chemically Modified Carbon Nanotubes for Commercial Applications, First Edition.
Edited by Jeenat Aslam, Chaudhery Mustansar Hussain, and Ruby Aslam.

at the tumor location in cancer treatment and other disorders [14, 15]. This chapter summarizes the classification of CNTs, purification, and various approaches to surface modification.

2.2 Classification of Carbon Nanotubes and Their Fabrication

CNTs are differentiated as single-walled carbon nanotubes (SWCNTs) or multiwalled carbon nanotubes (MWCNTs) based on the number of graphene sheets [16, 17]. SWCNTs are tubular structures made by wrapping a single graphene sheet into a cylinder with a diameter of around 1 nm and can be capped with fullerene cages at both ends. SWCNTs comprise sidewalls of graphene sheets composed mainly of adjacent hexagonal cells [18]. Sidewall defects typically involve pentagons and heptagons, among other polygon hierarchies. MWCNTs are considerably larger and composed of many single-walled tubes layered on top of each other. The space between adjacent shells is approximately 0.34 nm. MWCNTs differ from single wall nanotubes (SWNTs) not only in their morphology but also in their corresponding properties [19, 20]. A broad classification of CNTs is described in Figure 2.1.

2.2.1 Arc-Discharge Method

It comprises a vertically placed chamber with two graphite rods, carbon molecules, and metal catalysts such as cobalt, nickel, and iron [21]. An arcing method is used to

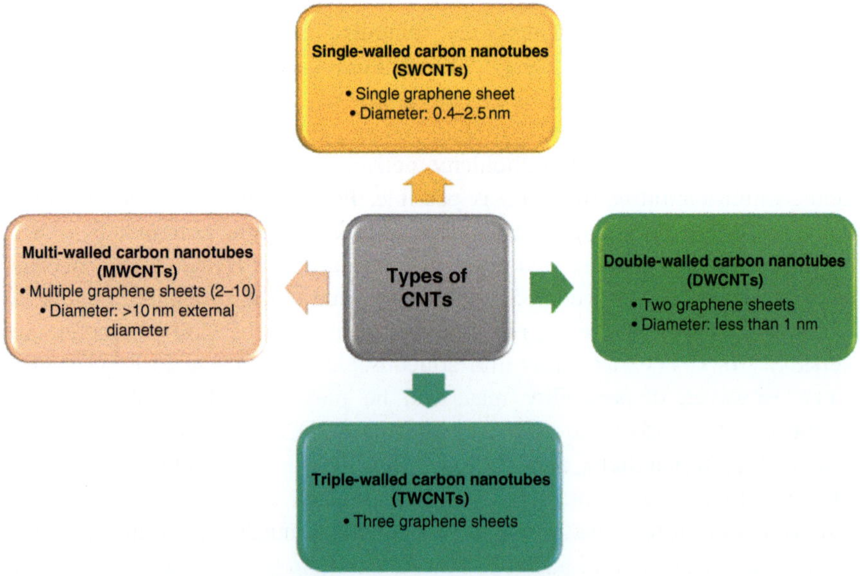

Figure 2.1 Classification of carbon nanotubes. Source: Reproduced with the permission Prajapati et al. [10]/with permission of Elsevier.

pass direct current through the chamber, and the chamber is compressed and heated to around 4000 K. The arc-discharge synthesis of CNTs can be done in two ways: one uses a variety of catalyst precursors, while the other does not. In general, MWCNTs can be produced even without a catalyst, but SWCNTs need a broad array of precursors to be synthesized, and, for arc-discharge expansion, a complicated anode composed of graphite and metal. The major downside of arc-discharge approaches is that they require higher temperatures, which often results in CNTs growing with structural defects; the alignment (i.e. chirality) of the produced nanotubes, which is critical for their function and characterization, is somewhat difficult to manage. Purification of the resulting products is also necessary due to the metallic catalyst used in the reaction [22].

2.2.2 Laser Vapor Deposition

Smalley's group invented the laser vapor deposition process for CNT manufacturing. The laser is idle for those materials having a boiling temperature such as carbon due to its higher energy density [23, 24]. Only high furnace temperatures can generate fullerenes with soccer ball architecture, emphasizing the need for annealing for nanostructures. This methodology offers various benefits, including high-quality CNT synthesis, diameter control, growth dynamics analysis, and the development of novel materials. The laser-furnace approach was combined with purification methods to generate high-quality CNTs with minimum flaws and impurities, such as amorphous carbon and catalytic metals [25, 26].

2.2.3 Chemical Vapor Deposition (CVD)

CVD is a chemical reaction that occurs when vaporized reactants react chemically to yield CNTs, which are subsequently placed on a substrate. CNTs intermediates include hydrocarbon gases like acetylene, ethylene, and methane [27–29]. Silica, silicon plates, and zeolite covered with iron particles are often used in CVD as substrates. To make SWCNTs, metal catalysts such as cobalt, iron, iron-molybdenum alloys, nickel, and molybdenum are used. Under an inert gas (argon gas) environment, the temperature is maintained at 500–900 °C. A crucible contains a catalyst-coated substrate, which is kept inside an inert gas-filled quartz tube [30]. The hydrocarbon gas is fed into a quartz tube, where it is pyrolyzed and converted into vapor carbon atoms. These carbons attach to the substrate and construct MWCNTs on the substrate due to the van der Waals attraction force [31].

2.3 Purification of CNTs

The purification protocol eliminates undesired components such as amorphous carbon, excess catalyst, and additional graphitic components from the CNTs after their synthesis. Metallic impurities, fullerenes, carbon contaminants, graphite, etc., may render with CNTs. It is very crucial to remove such impurities in case of drug delivery and biomedical applications. Impurities should be removed using efficient and moderate purification processes that do not degrade CNTs. Oxidants such as HNO_3 and $KMnO_4/H_2SO_4$ are used in this purification phase [32, 33]. Chemical and physical purification are the two most frequent types of purification techniques. Chemical techniques discriminate synthesized products depending on their reactivity, which typically incorporate undesirable defects throughout and at the ends of the CNTs, leading to significant structural and morphological damage to CNTs. Though the chemical-based method can be done in maintained reaction conditions, it renders the CNTs pure and the open tip of CNTs. Chemical purification processes include acids, annealing in an inert environment, alkali treatment, heat, and oxidation of CNTs [34]. Physical tactics such as filtration, centrifugation, and ultrasonication separate the contaminants depending on their size. These procedures are usually more complicated and less effective, but they are quite moderate and do not cause major damage to the tubes. Physical techniques are used to separate and eliminate unwanted contaminants such as nanocapsules, aggregate, and carbon [35].

2.4 Surface Modification of CNTs

CNT applications in drug delivery are still a complex area. The limited miscibility of CNTs poses a significant problem for delivery system design. CNTs are thermodynamically supposed to form aggregates and clumps due to van der Waals attraction and stacking. The dispersion of CNTs is very poor in aqueous solvents and other polymer phases due to poor interfacial contact. To resolve these concerns, functionalization techniques are used to separate nanotubes from their aggregated bundles and change their surface characteristics. CNTs that have been functionalized have excellent aqueous dispersibility, less bundling, improved cellular internalization, and a higher tendency to pair with other chemical moieties and ligands [10].

2.4.1 Methods of Functionalization

For therapeutic applications, functionalization entails attaching functional groups, polymers, nucleic acids, proteins and peptides, ligands, drug molecules, and theranostic agents to the structural design of pristine CNTs. Pristine CNTs are water-insoluble, toxic for the body, and aggregate in organs. The functionalization procedure gives CNTs great solubility and improved biocompatibility [36, 37]. As a result, functionalized CNTs are an excellent choice for encapsulating bioactive constituents for targeted therapy. Covalent functionalization and noncovalent methods are the two fundamental types of functionalization but modification with bioactive or biological molecules is one another approach for surface modification (Figure 2.2) [15, 38]. Sections 2.4.2 and 2.4.3 give details of various techniques for functionalizing of CNTs:

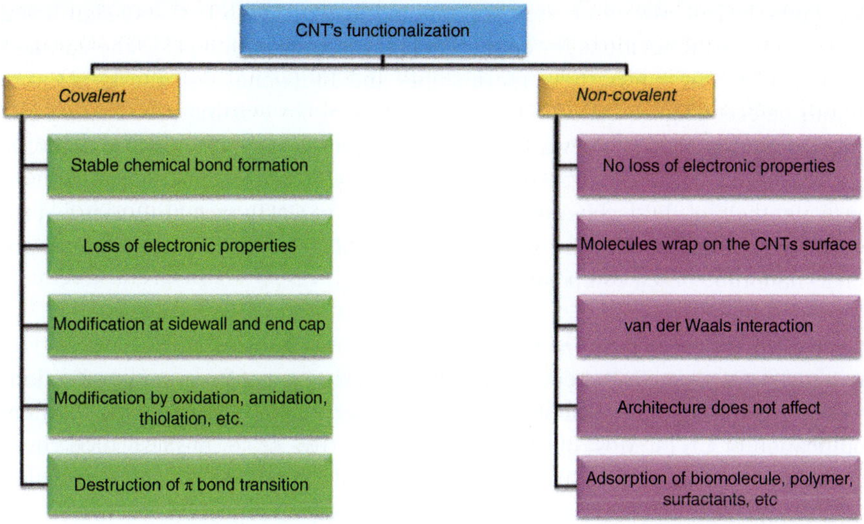

Figure 2.2 Covalent and non-covalent methods of functionalization.

2.4.2 Noncovalent Functionalization

CNTs with noncovalent functionalization can be exploited for the wrapping of distant moieties such as amphiphilic surfactants, polymers, or biomolecules. CNTs can be rendered more soluble and dispersible by using noncovalent functionalization [39]. However, because it is based on weak van der Waals forces, it has limited drug delivery potential. Noncovalent functionalization of CNTs can be done without disrupting the inherent sp^2-hybridized architecture sidewalls [40]. For the noncovalent functionalization, aromatic compounds, polymers, biopolymers, and surfactants are employed using hydrophobic interaction or π-stacking, ionic bonding, and hydrogen bonding. All these are part of supramolecular chemistry [41]. The noncovalent alterations of CNTs can accomplish a lot to retain their desirable characteristics while dramatically enhancing their solubilities. To segregate or enhance the dispersibility of CNTs, the aromatic molecules or the compounds having aromatic groups can be anchored via π–π stacking interactions on the architecture of CNTs exploiting nucleic acids, polymer, and other amphiphilic peptides [42].

2.4.3 Covalent (Chemical) Functionalization

Numerous chemical tactics have been explored toward the development of chemical bonds with CNTs such as either on the sidewalls or at the end. Amidation, halogenation (bromination, chlorination, and fluorination), hydrogenation, oxidation, thiolation, the addition of carbenes, nitrene, and radicals are all feasible chemical reactions of covalent functionalization. Covalent functionalization has limited use because it triggers sp^3 hybridization on the carbon atoms of CNTs, which restricts p-electrons from transitioning. As a result, the characteristics of

CNTs might be lost. Moreover, the covalent modification of CNTs gives a significant variety of chemical reactions for conjugation and solubilization [15]. The covalent alteration of CNTs makes them more stable and functional and can be done at sidewall defects (Figure 2.3). CNTs developed using an existing process may not be defect-free. Covalent sidewall functionalization is typically used to improve not just the dispersibility and processability of CNTs but also their reactivity. Consequently, the physicochemical characteristics of CNTs may be fine-tuned for multiple functions, paving the way for the synthesis of functional biomaterials, such as nanotube-based composites or hybrids [40].

2.4.3.1 Defect-Group Functionalization

The term "defects" in CNT chemistry implies the existence of flaws or imperfections inside the fused benzene sequence. Defects can serve as the basis for the covalent modification of CNTs. Most of the defects include five- or seven-membered rings, which are known as Stone–Wales defects [44], sp^3-hybridized defects and vacancies in the sp^2-hybridized architecture of the sidewall [45, 46], and a concurrent loss of the p-conjugates. This situation causes a local strain that increases the reactivity of adjacent C atoms. Treatment of CNTs with strong acids like HNO_3 or H_2SO_4 under extreme circumstances such as heating and/or sonication is a typical purification step. The backbone of the nanotube is damaged as a result of such a process. A variable degree of functional groups (carboxyl, carbonyl, hydroxyl groups, etc.) can be added [47, 48].

2.4.3.2 Sidewall Functionalization

Direct covalent bonding of chemical entities is much more complicated than defect-based attachment and usually necessitates the integration of highly reactive species [49]. Different types of sidewall functionalization can be done, including fluorination, cycloadditions (carbenes, nitrenes, cycloaddition, and Diels–Alder reaction), nucleophile and electrophile addition, and free radicle addition. Fluorination may be done on the sidewalls of CNTs without causing structural damage.

Fluorination Carbon atoms occupied by fluorine atoms interchange with unaltered carbon atoms during gradual fluorination. The typical approach for fluorinating CNTs involves using elemental fluorine as a catalyst in the presence of small amounts of hydrogen fluoride (HF). Fluorination has a substantial impact on solubility, electric conductivity, and optical qualities and drastically improves their chemical reactivity [50, 51].

2.4.3.3 CNTs Functionalized with Polymer

CNTs are typically functionalized with polymer molecules to increase their dispersion and build CNT-based composites for exploring distinctive features. Covalent and noncovalent attachments are the two most common methods for modifying CNTs with polymers [52]. A majority of studies have envisioned that decorating the CNTs with chemical moieties such as epoxide, –OH, and –COOH will expedite their

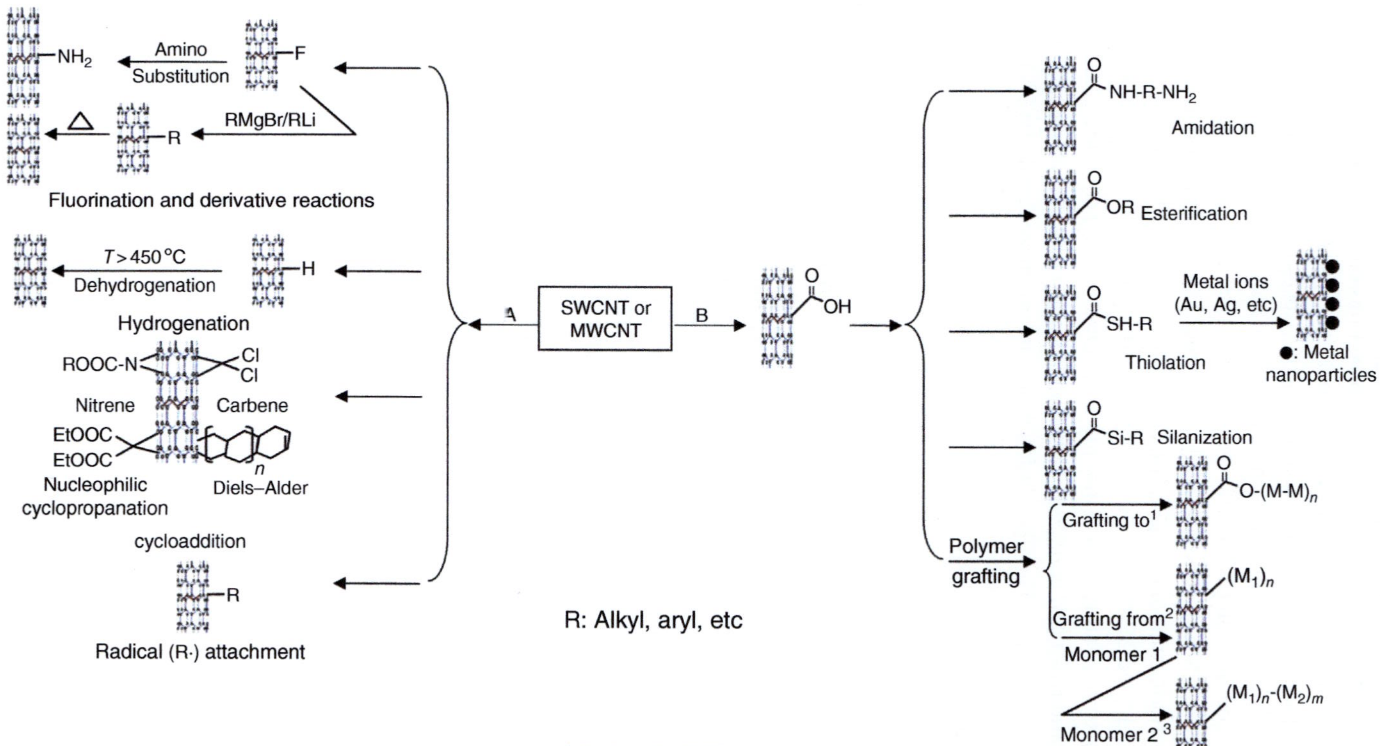

[1] Carboxylic CNTs react with reactive polymers;
[2] Functionalized CNTs act as initiators to initiate polymerization;
[3] Living polymerization, CNT-copolymer can be obtained; *n*, *m*: Degree of polymerization.

Figure 2.3 Approaches for the functionalization of CNTs via covalent reaction. (a) Defect-group. (b) Sidewall functionalization. Source: Ma et al. [43]/with permission of Elsevier.

dispersion and avoid CNT aggregation while also integrating them into polymer [53]. The covalent interaction of polymers and CNTs [54] can occur in two forms: "grafting to" [55] and "grafting from" [56]. The hydroxyl moiety on CNTs would aid in the development of a homogenous hybrid of polymer and CNT. After acid oxidation, the –OH and –COOH moiety emerge on the CNTs on defect sites [57, 58]. The "grafting to" approach involves synthesizing and attaching a polymer followed by end-group modification [59]. In the "grafting from" strategy, covalent anchoring of polymer on the CNTs occurs, followed by polymerization expansion in the existence of monomeric components [52, 60]. Prajapati et al. [61] developed hyaluronic acid (HA)-conjugated MWCNTs. In the research, the piranha treatment was done using a 3 : 1 mixture of sulfuric acid and nitric acid followed by chemical treatment (sulfuric acid and hydrogen peroxide), acylation (thionyl chloride), and finally amination (ethylene diamine). The PEGylation was done on the acylated MWCNTs and HA was conjugated on amine-functionalized MWCNTs. The developed hybrid complex showed controlled drug release, remarkable antitumor activity, and cell viability (Figure 2.4) [61].

2.4.3.4 CNTs Functionalized with Biomolecules

To explore new types of nano delivery systems or nanodevices for biological applications, the interaction between biomolecules and emerging nanomaterials like CNTs is extremely crucial. Aptamer, DNA, peptides, proteins, RNA, siRNA, and other complex molecules have mostly been functionalized on CNTs. Covalent bonding is extensively used in the coupling of proteins or peptides onto CNTs. All instances of noncovalent interactions involve the wrapping, stacking, and adsorption of DNA and peptides, as well as an electrostatic interaction that might pave the way for subsequent covalent interactions. The schematic depiction of amino acid, DNA, and peptide anchoring on CNTs is shown in Figure 2.5 [63]. The complicated interface formed by the carbon scaffold and the biological moiety displays the complexity necessary for attaining effective and robust binding while maintaining biological activity. In addition to carboxylic acids, other reactive groups on the CNTs, including amines, hydroxyls, and thiols, can be used to functionalize proteins [64]. According to Shim et al. [65], the co-adsorption of a surfactant and polyethylene glycol (PEG) on SWNTs was shown to be advantageous in avoiding nonspecific streptavidin (a protein) adsorption when delivered by SWCNTs. Optimal streptavidin coupling to SWNTs was achieved by co-functionalizing CNTs with protein-resistant polymers and biotin [65].

CNTs Conjugated with Nucleic Acid Nanotubes customized or wrapped with DNA have been explored as a core element for self-assembled nanoscale materials and are mostly employed in biosensing devices [66]. DNA may well enter the cavities of CNT. DNA movement through MWCNT channel has been tracked by fluorescence microscopy [67]. According to molecular dynamics simulations, DNA might be encased in the interior of SWCNTs in an aqueous microenvironment via an exceptionally quick interaction process if the tube size exceeds a specific threshold value. Both van der Waals [68] and hydrophobic forces [69] have been

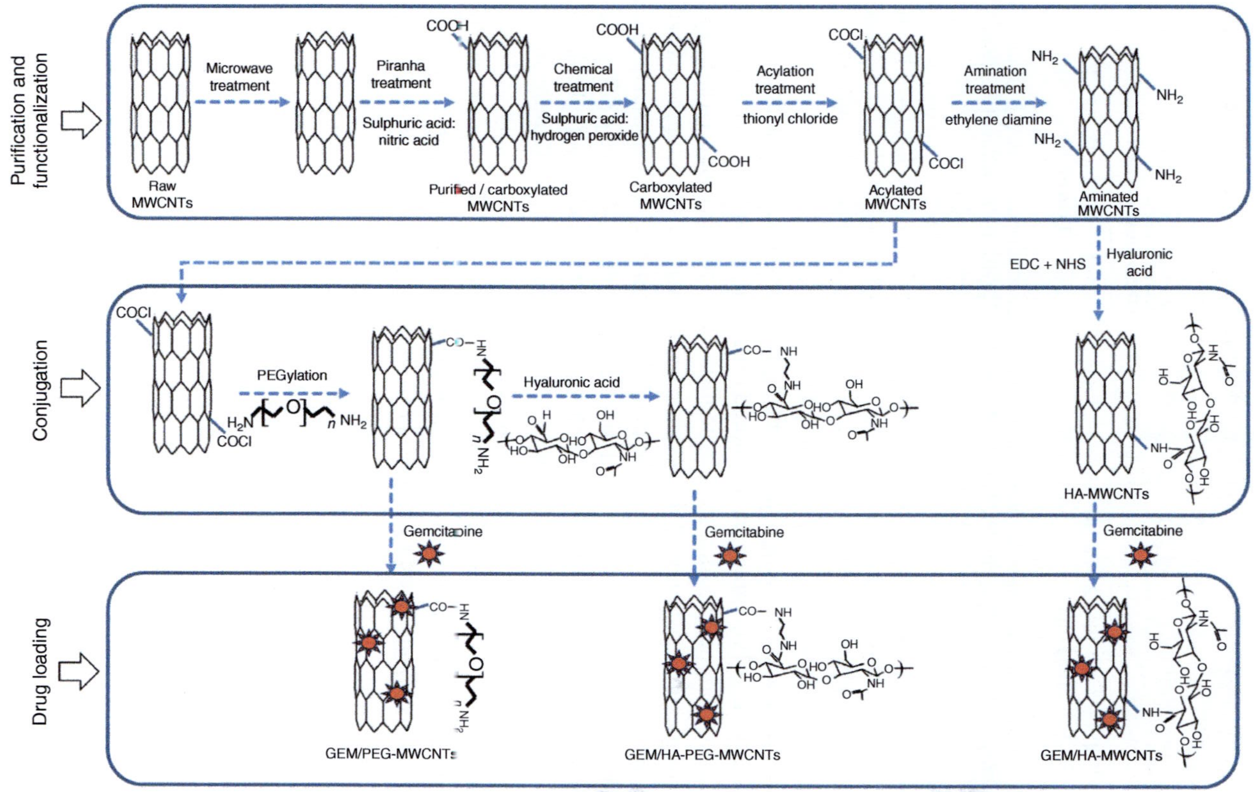

Figure 2.4 Schematic illustration of functionalization, conjugation, and drug loading in MWCNTs. Source: Prajapati et al. [61]/with permission of Elsevier.

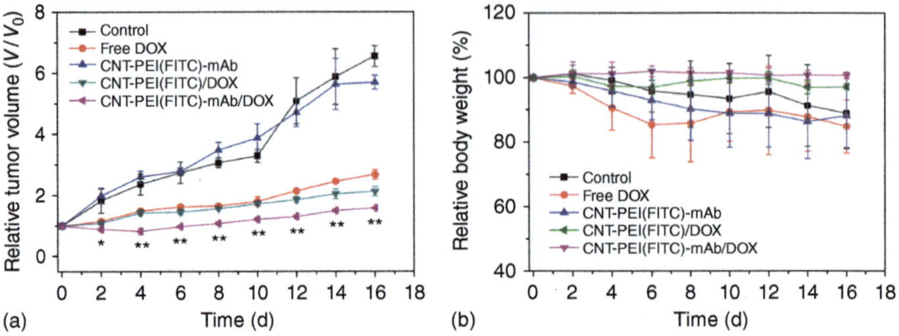

Figure 2.5 *In vivo* consequences of intervention of mentioned formulations. (a) Tumor growth curves. (b) Body weight curves. Source: Wu et al. [62]/with permission of Elsevier.

reported to be applicable in the interaction between DNA and CNTs, with the former playing a more prominent role [70]. Ozkan-Ariksoysal et al. [71] designed a DNA-wrapped MWCNTs genosensor to detect *Escherichia coli* (*E. coli*). For the first time, a DNA-bound MWCNT immobilized graphite sensor was framed, and authentic specimens were analyzed using polymerase chain reaction (PCR). The ssDNA–MWCNTs complex was anchored to the pencil graphite electrode (PGE) for immediate guanine oxidation electrochemistry. This system produced a stronger signal with greater sensitivity [71]. In this research, siRNA was complexed with SWCNTs and incubated with tumor cells to inhibit telomerase reverse transcriptase (TERT). TERT siRNA:SWCNTs+ complexes were capable of delivering siRNA and knocking down TERT expression [72]. Varkouhi et al. [73] used polyethyleneimine (PEI) and pyridinium to make cationic CNT–PEI and CNT–pyridinium for the administration of siRNA. The resulting CNTs had a silencing activity of 10–30% and cytotoxicity of 10–60%. Additionally, CNT–PEI and CNT–pyridinium did not outperform PEI and other common transfection methods, leading to reduced toxicity or vastly higher silencing efficacy [73].

CNTs Conjugated with Protein CNTs have been very well designed and synthesized with proteins since they are largely used for clinical uses such as drug delivery, DNA and gene transport, and biosensing [74]. Integrating the potential of CNTs and proteins would result in novel biomaterials for chemoprevention and other disorders. Proteins can solubilize CNTs and arrange them as per the chirality and size, detect other proteins. The surface of CNTs can be coated with proteins to make them biocompatible [75]. Oxidized CNTs were shown to absorb a variety of proteins and intercalate them onto the sidewalls of the CNTs. In addition, proteins bound to CNTs were proven to be easily carried into mammalian cells, exhibiting their biological function [76]. Covalent bonds are commonly assumed to be involved in the immobilization or conjugation of proteins on CNTs. Covalent bonds are commonly assumed to be involved in the functionalization or immobilization of proteins on CNTs. CNTs are dispersed by proteins, which may then be sorted by diameter or chirality. CNTs would then be further used to immobilize enzymes, resulting in the formation of functional materials. Proteins coupled with CNTs make them potential carriers for vaccine delivery. For instance, bioscaffolds made of protein/CNT hybrids can be used as medicinal and imaging materials. The coating of CNTs with protein renders them biocompatible. The interactions between protein and CNTs, such as lipophilic, surfactant, and π–π stacking interactions, determine the wide variety of structures, processes, and functional areas that emerge when such divergent particles interact [75]. Protein-functionalized CNTs may possess amines, carboxylic acids, hydroxyl, thiols, or other reactive groups [77]. Chymotrypsin, ferritin, fibrinogen, hemoglobin, and streptavidin are some of the proteins that have already been covalently attached to the CNTs. Human serum albumin (HSA) is a protein, and its receptors are ubiquitously existing on cancer cells, and endocytosis mediated by caveolae has the ability to internalize a substantial concentration of albumin [37]. In a study, Karajanagi et al. [77] developed bovine serum albumin (BSA) functionalized CNTs. SWNTs were dispersed in dimethylformamide to make aqueous dispersions. The nanotubes were then ultrasonically dispersed in protein solutions. The introduction of BSA significantly enhanced the solubility, while the addition of *Mucor javanicus* lipase improved the solubility even more. Even after 30 days, these CNT–protein solutions remained stable and aggregation-free. Proteins bind to frayed ends via amine moiety, π–π interactions, or hydrophobic linkages, enabling the charged or hydrophilic layer to engage with aqueous phase to make CNTs dispersible [77].

CNTs Conjugation with Antibodies The attachment of antibodies (Abs) to the CNTs produces a nano delivery vehicle that is a critical step in targeted delivery. In general, Ab-coupled CNTs for site-specific targeting are a potential nano modality in cancer therapies and diagnostics [78]. Abs are conjugated to CNTs via one of two approaches: either by direct or indirect conjugation. In the former one, the direct adsorption of Abs to CNTs occurs. However, the weak connection coupling between the CNTs and antibodies facilitates the loss of targeting ability of antibodies. The later one required an extra step, for example, the Streptavidin–biotin interaction is required to generate the antibody–biotin conjugate [79]. Venturelli et al. [80] developed anti-MUC1 Abs coupled MWCNTs and double wall carbon nanotubes (DWCNTs). Gel electrophoresis and thermogravimetric analyses indicated the presence of the Ab bound to the nanotubes. The results revealed that surface plasmon resonance molecular recognition of the antigen can determine equivalent antibody binding capabilities for both Ab-DWCNTs and Ab-MWCNTs, confirming the potential of antibody conjugation for receptor-mediated targeting of CNTs [80]. In a study, monoclonal antibody (mAb)-coupled MWCNTs were developed for site-specific targeting and imaging by Wu et al. [62]. The biocompatibility of the developed hybrid system was found to be improved as ascertained by the experimental results. The anticancer results on PC-3 tumor revealed that CNT–PEI(fluorescein isothiocyanate [FITC])–mAb potentially delivered the drug and suppressed the growth of tumor (Figure 2.5) [62].

CNTs Conjugation with Chitosan Chitosan is a well-known biomaterial for biomedical engineering and drug delivery systems because of its commendable biocompatibility, biodegradability, gel-forming ability, and nontoxic characteristics [81, 82]. Chitosan–CNTs are exploited not only as drug carriers but also broadly applied in biomedical engineering as well. Chitosan–CNTs composites or scaffolds were created as a highly conductive and porous composite to replace bone grafts for tissue engineering of the bones [83–85]. The chitosan increases the MWCNTs stability and improves their interactions with cancerous cells, allowing MWCNTs to be internalized by cells [86, 87]. Singh et al. [88] developed MWCNTs by conjugating them with chitosan-folate. The cell cytotoxicity study was performed on the A549 cell. The results suggested dose-dependent cytotoxicity of docetaxel. The folate-conjugated complex showed remarkable cytotoxicity compared to the other formulation groups. This was due to the improved cellular internalization and sustained-release effect. One more reason for the cytotoxicity of DTX–CHI–FA–CNAC conjugate is the inherent cytotoxicity of chitosan-conjugated MWCNTs. Following 24 hours of incubation with cells, the IC_{50} value of the DTX–CHI–FA–CNAC was up to 89 times lower than the outcomes of Docel™ [88]. Cirillo et al. [89] implemented a methotrexate-loaded nanohybrid system of MWCNTs by coating them with chitosan. The thermogravimetric analysis confirmed the effective coating of chitosan on MWCNTs. The developed drug-free nanosystem was highly biocompatible in both H1299 cancerous cells and MRC-5 healthy cells. The drug-loaded system greatly enhanced the anticancer efficacy on the H1299 cancer cell and lowered toxicity in MRC-5 cells. The findings of cell survival as a model for cancer and physiological conditions are described in Figure 2.6 [89].

CNTs Conjugation with Aptamer Aptamers have such significant targeting implications that they can amplify nanocarriers by allowing for localized drug delivery, higher permeability, limited cytotoxicity at the nontarget site, satisfactory

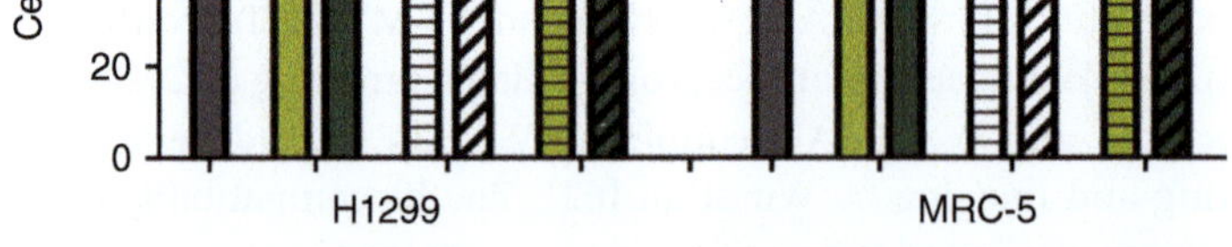

Figure 2.6 Cell viability after treatment with (i) filled bar: CS_MWCNT. (ii) Empty streaked bar: free MTX. (iii) Filled streaked bar: MTX@CS_MWCNT. (iv) Rowed bar: MTX (7.72×10^{-5} mg ml^{-1}). (v) Diagonal bar: MTX (1.51×10^{-3} mg ml^{-1}). (vi) Light green bar: CS_MWCNT (7.72×10^{-4} mg ml^{-1}). (vii) Dark green bar: 1.51×10^{-2} mg ml^{-1}. Source: Cirillo et al. [89]/MDPI/Public Domain.

therapeutic outcomes, and controlled drug retention and their release [90, 91]. Many delivery systems can promote the attributes of aptamers and shield them against nuclease degradation and rapid filtration due to their greater surface area or enhanced permeation and retention (EPR) effects [92]. The domain of aptamer-CNT-based therapies has advanced expeditiously because of the high affinity of an aptamer for binding precisely to its target [93]. Zhu et al. [94] developed aptamer-conjugated MWCNTs for PDT. The PDT works by covalently attaching a photosensitizer to an aptamer that is then embedded onto the SWNTs, permitting the photosensitizer to be triggered when it reaches the target site only by light. The binding of thrombin impairs the DNA associated, causing the DNA-aptamer to drop off the SWNTs, restoring singlet oxygen generation [94]. Gu et al. [95] developed aptamer-conjugated MWCNTs and modified them with a contrast agent. The modified MWCNTs provided more visual clarity and authenticity, along with targeting ability for the PCa cells. MWCNT–PEG–Ap ($200\,\mu g\,ml^{-1}$) rate of uptake was almost 100% and the fluorescence amplitude on average was approximately sixfold that of the concentration at $10\,g\,ml^{-1}$ (Figure 2.7a). In addition, MWCNT–PEG–Ap/Coumarin 6 showed substantially greater fluorescence intensity compared to the free Coumarin 6, suggesting that MWCNT–PEG–Ap demonstrated superior cellular uptake and was largely localized in the cytoplasm (Figure 2.7b) [95].

2.4.3.5 CNTs Functionalization with Ionic Liquid (ILs)

Without the use of solid chemicals or organic solvents, ionic liquids (ILs) are an ideal technique to improve CNT dispersibility [96]. Fukushima et al. [97] were the first to employ imidazolium ion-based ILs to enhance the dispersibility of CNTs [97].

At 100 °C, ILs are made up of an organic cation and a weakly linked anion. The interaction of ionic liquids with nanotubes causes an impact on the CNT properties, which facilitates the dispersion of CNTs in diverse mediums [98].

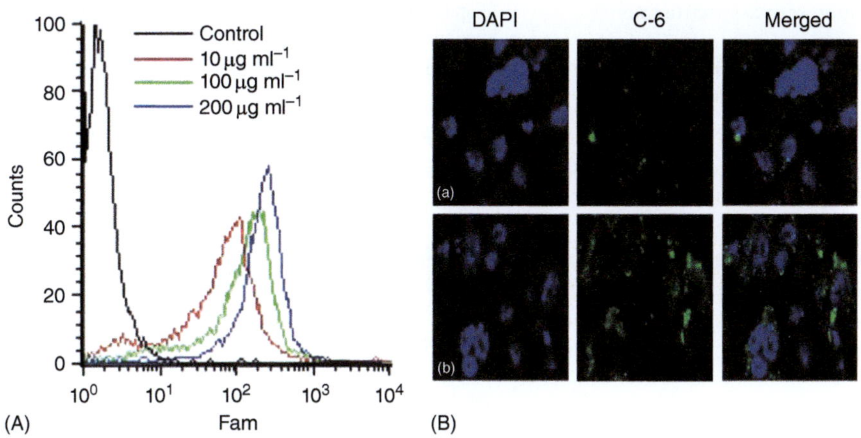

Figure 2.7 The internalization of MWCNT-PEG-Ap. (A) Flow cytometry evaluation. (B) Cellular absorption: (a) Free Coumarin 6. (b) MWCNT-PEG-Ap/Coumarin 6. Source: Gu et al. [95]/with permission of Springer Nature/CC BY 4.0.

Table 2.1 Analytical methods for CNTs and their relevance.

Characterization approach	Significance of study	Reference
Fourier transform infrared spectroscopy (FTIR)	• Determination of impurities remaining after synthesis of CNTs • Determination of functional group after surface modification • Molecular spectral properties	[101–103]
Transmission electron microscopy (TEM)	• Internal structure and diameter • Ultrastructural damage • Distribution into the cytoplasm • Number of layers and inter-wall spacing	[104–106]
UV, Vis spectroscopy	• Relative purity measurement • Dispersion ability, diameter, and length distribution	[107]
Raman spectroscopy	• Position, width, and the relative intensity of bands • Electronic structure and purity • The structural quality of CNTs	[108–110]
X-ray diffraction (XRD)	• Interlayer spacing • Crystal structure • Impurities	[111, 112]
Derivative thermogravimetry; thermal gravimetric analysis	• Thermodynamic stability is assessed • Purity of CNT	[113, 114]
Atomic force microscopy (AFM)	• Dimension and inner structure • Number of layers	[115, 116]
X-ray photoelectron spectroscopy (XPS)	• Chemical structure • Functionalization • Elemental components	[117]
Scanning electron microscopy (SEM)	• CNTs aggregation • Diameter and length	[118, 119]
Fluorescence spectroscopy	• Dispersion capacity • Size and chirality	[120, 121]
Size exclusion chromatography and capillary extrusion	• Purification and size separation	[122, 123]
Photoluminescence spectroscopy	• Electrical and optical characteristics	[13, 124]
Energy dispersive spectroscopy (EDS)	• Elemental composition	[125, 126]
Neutron diffraction	• Bond length • Probable distortion of hexagonal network • Phase specific strains	[127–129]

2.4.3.6 Plasma Activated CNTs

A new way to chemically tailor CNTs was devised, involving radio-frequency glow-discharge plasma stimulation following reactions relevant to the plasma-engendered functional moieties [99]. Schiff-base linkages have been used to immobilize amino-dextran chains onto acetaldehyde-plasma-treated oriented CNTs that were then subsequently stabilized by reduction with sodium cyanoborohydride. Another kind of chemical alteration of CNTs is plasma activation. In this method, aldehyde plasma is used to treat CNTs, and subsequently, Schiff-base bonds are formed to immobilize minodextran chains [100].

2.5 Characterization of CNTs

CNTs of various shapes, defects, and impurities are synthesized using the various processes outlined above. In biomedical applications, the characterization of CNTs is critical to avoid harmful effects and increase the efficiency of associated compositions. The physical and chemical nature of CNTs has been determined using a variety of spectroscopic approaches, diffraction techniques, thermal, and separation procedures. Table 2.1 lists various characterization methods for CNTs.

2.6 Conclusion

CNTs have made an enormous contribution to pharmaceutical research and development, including targeted drug delivery and biomedical applications. For such applications, the raw CNTs are purified via physical or chemical methods to minimize the toxic effect. The surface modification using various functional groups, polymers, biopolymers, nucleic acid, and ligands enables CNTs for the therapy of ailments, such as arthritis, cancer, diabetes, and tuberculosis, due to their good mechanical and electrical properties. CNTs are the material of choice for medical devices and biosensors. CNT technology progress is still in its early stages, primarily in terms of biomedical applications. The upgradation of CNTs from novel materials that are biologically incongruent to useful, functional, and safe devices elicit several efforts to regain various CNT properties to increase efficiency, development, and biosafety. All these are possible by developing effective surface modification methods.

References

1 Lee, S.H., Rho, W.-Y., Chang, H. et al. (2021). Carbon nanomaterials for biomedical application. In: *Nanotechnology for Bioapplications*, Advances in Experimental Medicine and Biology, vol. 1309 (ed. B.H. Jun), 257–276. Singapore: Springer, Springer.

2 Shams, R., Rizvi, Q.E.H., Dar, A.H. et al. (2021). Polysaccharides: promising constituent for the preparation of nanomaterials. In: *Polysaccharides: Properties and Applications* (ed. M.I.A. Inamuddin, R. Boddula and T. Altalhi), 441–457. Wiley Scrivener Publishing LLC.

3 Thakur, V.K. and Thakur, M.K. (2018). *Chemical Functionalization of Carbon Nanomaterials*. Warentown, NJ: CRC Press.

4 Kofoed Andersen, C., Khatri, S., Hansen, J. et al. (2021). Carbon nanotubes—potent carriers for targeted drug delivery in rheumatoid arthritis. *Pharmaceutics* 13: 453.

5 Kohls, A., Maurer Ditty, M., Dehghandehnavi, F., and Zheng, S.-Y. (2022). Vertically aligned carbon nanotubes as a unique material for biomedical applications. *ACS Appl. Mater. Interfaces* 14 (5): 6287–6306.

6 Sadeghpour, N., Vadi, M., and Bagheri, N. (2022). Engineering delivery of pantaprazole drug using multi-walled carbon nanotubes: an experimental and theoretical study. *Struct. Chem.* 33: 1–10.

7 Tang, L., Xiao, Q., Mei, Y. et al. (2021). Insights on functionalized carbon nanotubes for cancer theranostics. *J. Nanobiotechnol.* 19: 423.

8 Wang, L., Shi, J., Liu, R. et al. (2014). Photodynamic effect of functionalized single-walled carbon nanotubes: a potential sensitizer for photodynamic therapy. *Nanoscale* 6: 4642–4651.

9 Ferrier, D.C. and Honeychurch, K.C. (2021). Carbon nanotube (CNT)-based biosensors. *Biosensors* 11: 486.

10 Prajapati, S.K., Malaiya, A., Kesharwani, P. et al. (2020). Biomedical applications and toxicities of carbon nanotubes. *Drug Chem. Toxicol.* 1–16.

11 Sireesha, M., Jagadeesh Babu, V., Kranthi Kiran, A.S., and Ramakrishna, S. (2018). A review on carbon nanotubes in biosensor devices and their applications in medicine. *Nanocomposites* 4: 36–57.

12 Dubey, R., Dutta, D., Sarkar, A., and Chattopadhyay, P. (2021). Functionalized carbon nanotubes: synthesis, properties and applications in water purification, drug delivery, and material and biomedical sciences. *Nanoscale Adv.* 3: 5722–5744.

13 Rahamathulla, M., Bhosale, R.R., Osmani, R.A. et al. (2021). Carbon nanotubes: current perspectives on diverse applications in targeted drug delivery and therapies. *Materials* 14: 6707.

14 Kesharwani, P., Mishra, V., and Jain, N.K. (2015). Validating the anticancer potential of carbon nanotube-based therapeutics through cell line testing. *Drug Discovery Today* 20: 1049–1060.

15 Kumar, S., Rani, R., Dilbaghi, N. et al. (2017). Carbon nanotubes: a novel material for multifaceted applications in human healthcare. *Chem. Soc. Rev.* 46: 158–196.

16 Hasnain, M.S. and Nayak, A.K. (2019). Classification of carbon nanotubes. In: *Carbon Nanotubes for Targeted Drug Delivery*, 11–14. Singapore: Springer Singapore.

17 Varghese, R., Salvi, S., Sood, P. et al. (2022). Carbon nanotubes in COVID-19: a critical review and prospects. *Colloid Interface Sci. Commun.* 46: 100544.

18 Thakare, V.S., Das, M., Jain, A.K. et al. (2010). Carbon nanotubes in cancer theragnosis. *Nanomedicine* 5: 1277–1301.
19 Donaldson, K., Aitken, R., Tran, L. et al. (2006). Carbon nanotubes: a review of their properties in relation to pulmonary toxicology and workplace safety. *Toxicol. Sci.* 92: 5–22.
20 Katsnelson, M. I. (2016). "*graphene*" [Online]. Encyclopedia Britannica: Encyclopedia Britannica. https://www.britannica.com/science/graphene (accessed 9 February 2022).
21 Sinha, N., Ma, J., and Yeow, J.T. (2006). Carbon nanotube-based sensors. *J. Nanosci. Nanotechnol.* 6: 573–590.
22 Shoukat, R. and Khan, M.I. (2021). Carbon nanotubes: a review on properties, synthesis methods and applications in micro and nanotechnology. *Microsyst. Technol.* 27: 4183–4192.
23 Saifuddin, N., Raziah, A.Z., and Junizah, A.R. (2013). Carbon nanotubes: a review on structure and their interaction with proteins. *J. Chem.* 2013: 676815.
24 Szabó, A., Perri, C., Csató, A. et al. (2010). Synthesis methods of carbon nanotubes and related materials. *Materials* 3: 3092–3140.
25 Rathinavel, S., Priyadharshini, K., and Panda, D. (2021). A review on carbon nanotube: an overview of synthesis, properties, functionalization, characterization, and the application. *Mater. Sci. Eng., B* 268: 115095.
26 Thess, A., Lee, R., Nikolaev, P. et al. (1996). Crystalline ropes of metallic carbon nanotubes. *Science* 273: 483–487.
27 Honda, S.-I., Katayama, M., Lee, K.-Y. et al. (2003). Low temperature synthesis of aligned carbon nanotubes by inductively coupled plasma chemical vapor deposition using pure methane. *Jpn. J. Appl. Phys.* 42: L441–L443.
28 Manawi, Y.M., Ihsanullah, Samara, A. et al. (2018). A review of carbon nanomaterials' synthesis via the chemical vapor deposition (CVD) method. *Materials* 11: 822.
29 Patila, M., Chalmpes, N., Dounousi, E. et al. (2020). Use of functionalized carbon nanotubes for the development of robust nanobiocatalysts, Chapter 12. In: *Methods in Enzymology*, vol. 630 (ed. C.V. Kumar), 2–502. Academic Press.
30 Kumar, M. and Ando, Y. (2010). Chemical vapor deposition of carbon nanotubes: a review on growth mechanism and mass production. *J. Nanosci. Nanotechnol.* 10: 3739–3758.
31 Pham, V.P., Jang, H.-S., Whang, D., and Choi, J.-Y. (2017). Direct growth of graphene on rigid and flexible substrates: progress, applications, and challenges. *Chem. Soc. Rev.* 46: 6276–6300.
32 Jain, A.K., Dubey, V., Mehra, N.K. et al. (2009). Carbohydrate-conjugated multiwalled carbon nanotubes: development and characterization. *Nanomed. Nanotechnol. Biol. Med.* 5: 432–442.
33 Singh, R., Mehra, N.K., Jain, V., and Jain, N.K. (2013). Gemcitabine-loaded smart carbon nanotubes for effective targeting to cancer cells. *J. Drug Targeting* 21: 581–592.

34 Ismail, A.F., Goh, P.S., Tee, J.C. et al. (2008). A review of purification techniques for carbon nanotubes. *Nano* 03: 127–143.

35 Eatemadi, A., Daraee, H., Karimkhanloo, H. et al. (2014). Carbon nanotubes: properties, synthesis, purification, and medical applications. *Nanoscale Res. Lett.* 9: 393.

36 Kumar, R., Dhanawat, M., Kumar, S. et al. (2014). Carbon nanotubes: a potential concept for drug delivery applications. *Recent Pat. Drug Deliv. Formul.* 8: 12–26.

37 Mehra, N.K., Mishra, V., and Jain, N. (2014). A review of ligand tethered surface engineered carbon nanotubes. *Biomaterials* 35: 1267–1283.

38 Hong, C.-Y., You, Y.-Z., and Pan, C.-Y. (2006). A new approach to functionalize multi-walled carbon nanotubes by the use of functional polymers. *Polymer* 47: 4300–4309.

39 Gao, C., Guo, Z., Liu, J.-H., and Huang, X.-J. (2012). The new age of carbon nanotubes: an updated review of functionalized carbon nanotubes in electrochemical sensors. *Nanoscale* 4: 1948–1963.

40 Sadegh, H. and Shahryari-Ghoshekandi, R. (2015). Functionalization of carbon nanotubes and its application in nanomedicine: a review. *Nanomed. J.* 2: 231–248.

41 Kumar, A., Sharma, K., and Dixit, A.R. (2021). A review on the mechanical properties of polymer composites reinforced by carbon nanotubes and graphene. *Carbon Lett.* 31: 149–165.

42 Zhang, W., Zhang, Z., and Zhang, Y. (2011). The application of carbon nanotubes in target drug delivery systems for cancer therapies. *Nanoscale Res. Lett.* 6: 555.

43 Ma, P.-C., Siddiqui, N.A., Marom, G., and Kim, J.-K. (2010). Dispersion and functionalization of carbon nanotubes for polymer-based nanocomposites: a review. *Composites Part A* 41: 1345–1367.

44 Rahmandoust, M. and Ayatollahi, M.R. (2016). Carbon nanotubes. In: *Characterization of Carbon Nanotube Based Composites Under Consideration of Defects*, Advanced Structured Materials, vol. 39, 5–63. Cham: Springer, Springer International Publishing.

45 Hirsch, A. (2002). Functionalization of single-walled carbon nanotubes. *Angew. Chem. Int. Ed.* 41: 1853–1859.

46 Jeon, I.-Y., Chang, D.W., Kumar, N.A., and Baek, J.-B. (2011). Functionalization of carbon nanotubes. In: *Carbon Nanotubes-Polymer Nanocomposites*, vol. 6 (ed. S. Yellampalli), 91–110. London: IntechOpen.

47 Tessonnier, J.P., Villa, A., Majoulet, O. et al. (2009). Defect-mediated functionalization of carbon nanotubes as a route to design single-site basic heterogeneous catalysts for biomass conversion. *Angew. Chem. Int. Ed.* 48: 6543–6546.

48 Wang, C., Zhou, G., Liu, H. et al. (2006). Chemical functionalization of carbon nanotubes by carboxyl groups on stone-wales defects: a density functional theory study. *J. Phys. Chem. B* 110: 10266–10271.

49 Niyogi, S., Hamon, M., Hu, H. et al. (2002). Chemistry of single-walled carbon nanotubes. *Acc. Chem. Res.* 35: 1105–1113.

50 Adamska, M. and Narkiewicz, U. (2017). Fluorination of carbon nanotubes – a review. *J. Fluorine Chem.* 200: 179–189.

51 Tasis, D., Tagmatarchis, N., Bianco, A., and Prato, M. (2006). Chemistry of carbon nanotubes. *Chem. Rev.* 106: 1105–1136.

52 Liu, P. (2005). Modifications of carbon nanotubes with polymers. *Eur. Polym. J.* 41: 2693–2703.

53 Layek, R.K. and Nandi, A.K. (2013). A review on synthesis and properties of polymer functionalized graphene. *Polymer* 54: 5087–5103.

54 Salavagione, H.J., Martínez, G., and Ellis, G. (2011). Recent advances in the covalent modification of graphene with polymers. *Macromol. Rapid Commun.* 32: 1771–1789.

55 Homenick, C.M., Lawson, G., and Adronov, A. (2007). Polymer grafting of carbon nanotubes using living free-radical polymerization. *Polym. Rev.* 47: 265–290.

56 Díez-Pascual, A.M. (2021). Chemical functionalization of carbon nanotubes with polymers: a brief overview. *Macromolecules* 1: 64–83.

57 Czech, B., Oleszczuk, P., and Wiącek, A. (2015). Advanced oxidation (H_2O_2 and/or UV) of functionalized carbon nanotubes (CNT–OH and CNT–COOH) and its influence on the stabilization of CNTs in water and tannic acid solution. *Environ. Pollut.* 200: 161–167.

58 Nakajima, Y. and Shimada, S. (2015). Hydrosilylation reaction of olefins: recent advances and perspectives. *RSC Adv.* 5: 20603–20616.

59 Beliatis, M.J., Rozanski, L.J., Jayawardena, K.D.G.I. et al. (2015). Hybrid and nano-composite carbon sensing platforms. In: *Carbon for Sensing Devices* (ed. D. Demarchi and A. Tagliaferro), 105–132. Cham: Springer International Publishing.

60 Zhou, T., Zhu, Y., Li, X. et al. (2016). Surface functionalization of biomaterials by radical polymerization. *Prog. Mater. Sci.* 83: 191–235.

61 Prajapati, S.K., Jain, A., Shrivastava, C., and Jain, A.K. (2019b). Hyaluronic acid conjugated multi-walled carbon nanotubes for colon cancer targeting. *Int. J. Biol. Macromol.* 123: 691–703.

62 Wu, H., Shi, H., Zhang, H. et al. (2014). Prostate stem cell antigen antibody-conjugated multiwalled carbon nanotubes for targeted ultrasound imaging and drug delivery. *Biomaterials* 35: 5369–5380.

63 Maheshwari, N., Tekade, M., Soni, N. et al. (2019). Functionalized carbon nanotubes for protein, peptide, and gene delivery. In: *Biomaterials and Bionanotechnology* (ed. R.K. Tekade), 613–637. Elsevier.

64 Oliveira, S.F., Bisker, G., Bakh, N.A. et al. (2015). Protein functionalized carbon nanomaterials for biomedical applications. *Carbon* 95: 767–779.

65 Shim, M., Shi Kam, N.W., Chen, R.J. et al. (2002). Functionalization of carbon nanotubes for biocompatibility and biomolecular recognition. *Nano Lett.* 2: 285–288.

66 Villa, C.H., Mcdevitt, M.R., Escorcia, F.E. et al. (2008). Synthesis and biodistribution of oligonucleotide-functionalized, tumor-targetable carbon nanotubes. *Nano Lett.* 8: 4221–4228.

67 Ito, T., Sun, L., and Crooks, R.M. (2003). Observation of DNA transport through a single carbon nanotube channel using fluorescence microscopy. *Chem. Commun.* 1482–1483.

68 Wang, Y. and Ai, H. (2009). Theoretical insights into the interaction mechanism between proteins and SWCNTs: adsorptions of tripeptides GXG on SWCNTs. *J. Phys. Chem. B* 113: 9620–9627.

69 Zhao and Johnson, J.K. (2007). Simulation of adsorption of DNA on carbon nanotubes. *J. Am. Chem. Soc.* 129: 10438–10445.

70 Gao, H., Kong, Y., Cui, D., and Ozkan, C.S. (2003). Spontaneous insertion of DNA oligonucleotides into carbon nanotubes. *Nano Lett.* 3: 471–473.

71 Ozkan-Ariksoysal, D., Kayran, Y.U., Yilmaz, F.F. et al. (2017). DNA-wrapped multi-walled carbon nanotube modified electrochemical biosensor for the detection of *Escherichia coli* from real samples. *Talanta* 166: 27–35.

72 Zhang, Z., Yang, X., Zhang, Y. et al. (2006). Delivery of telomerase reverse transcriptase small interfering RNA in complex with positively charged single-walled carbon nanotubes suppresses tumor growth. *Clin. Cancer Res.* 12: 4933–4939.

73 Varkouhi, A.K., Foillard, S., Lammers, T. et al. (2011). SiRNA delivery with functionalized carbon nanotubes. *Int. J. Pharm.* 416: 419–425.

74 Nagaraju, K., Reddy, R., and Reddy, N. (2015). A review on protein functionalized carbon nanotubes. *J. Appl. Biomater. Funct. Mater.* 13: 301–312.

75 Calvaresi, M. and Zerbetto, F. (2013). The Devil and Holy Water: protein and carbon nanotube hybrids. *Acc. Chem. Res.* 46: 2454–2463.

76 Kam, N.W.S. and Dai, H. (2005). Carbon nanotubes as intracellular protein transporters: generality and biological functionality. *J. Am. Chem. Soc.* 127: 6021–6026.

77 Karajanagi, S.S., Yang, H., Asuri, P. et al. (2006). Protein-assisted solubilization of single-walled carbon nanotubes. *Langmuir* 22: 1392–1395.

78 Mcdevitt, M.R., Chattopadhyay, D., Kappel, B.J. et al. (2007). Tumor targeting with antibody-functionalized, radiolabeled carbon nanotubes. *J. Nucl. Med.* 48: 1180–1189.

79 Xiao, Y., Gao, X., Taratula, O. et al. (2009). Anti-HER2 IgY antibody-functionalized single-walled carbon nanotubes for detection and selective destruction of breast cancer cells. *BMC Cancer* 9: 351.

80 Venturelli, E., Fabbro, C., Chaloin, O. et al. (2011). Antibody covalent immobilization on carbon nanotubes and assessment of antigen binding. *Small* 7: 2179–2187.

81 Prajapati, S.K. and Jain, A. (2020). Polysaccharide-based interpenetrating polymeric network system for biomedical use. In: *Tailor-Made Polysaccharides in Biomedical Applications* (ed. A. Nayak, M.D.S. Hasnain and T. Aminabhavi), 133–150. Elsevier.

82 Prajapati, S.K., Jain, A., Jain, A., and Jain, S. (2019a). Biodegradable polymers and constructs: a novel approach in drug delivery. *Eur. Polym. J.* 120: 109191.

83 Jain, R., Prajapati, S.K., Jain, S., and Jain, A. (2020). Chemically modified polysaccharides in tissue engineering. In: *Tailor-Made Polysaccharides in*

Biomedical Applications (ed. A. Nayak, M.D.S. Hasnain and T. Aminabhavi), 197–224. Elsevier.

84 Venkatesan, J., Ryu, B., Sudha, P.N., and Kim, S.K. (2012). Preparation and characterization of chitosan-carbon nanotube scaffolds for bone tissue engineering. *Int. J. Biol. Macromol.* 50: 393–402.

85 Yang, C. and Liu, P. (2012). Chitosan/functionalized multiwalled carbon nanotubes multilayer hollow microspheres prepared via layer-by-layer assembly technique. *Ind. Eng. Chem. Res.* 51: 13346–13353.

86 Chen, C., Xie, X.X., Zhou, Q. et al. (2012). EGF-functionalized single-walled carbon nanotubes for targeting delivery of etoposide. *Nanotechnology* 23: 045104.

87 Iamsamai, C., Hannongbua, S., Ruktanonchai, U. et al. (2010). The effect of the degree of deacetylation of chitosan on its dispersion of carbon nanotubes. *Carbon* 48: 25–30.

88 Singh, R.P., Sharma, G., Sonali et al. (2017). Chitosan-folate decorated carbon nanotubes for site specific lung cancer delivery. *Mater. Sci. Eng. C* 77: 446–458.

89 Cirillo, G., Vittorio, O., Kunhardt, D. et al. (2019). Combining carbon nanotubes and chitosan for the vectorization of methotrexate to lung cancer cells. *Materials* 12 (8): 2889.

90 Lee, J.H., Yigit, M.V., Mazumdar, D., and Lu, Y. (2010). Molecular diagnostic and drug delivery agents based on aptamer-nanomaterial conjugates. *Adv. Drug Delivery Rev.* 62: 592–605.

91 Yang, L., Zhang, X., Ye, M. et al. (2011). Aptamer-conjugated nanomaterials and their applications. *Adv. Drug Delivery Rev.* 63: 1361–1370.

92 Serrano, C.M., Freeman, R., Godbe, J. et al. (2019). DNA-peptide amphiphile nanofibers enhance aptamer function. *ACS Appl. Biomater.* 2: 2955–2963.

93 Liu, M., Wang, L., Lo, Y. et al. (2022). Aptamer-enabled nanomaterials for therapeutics, drug targeting and imaging. *Cells* 11: 159.

94 Zhu, Z., Tang, Z., Phillips, J.A. et al. (2008). Regulation of singlet oxygen generation using single-walled carbon nanotubes. *J. Am. Chem. Soc.* 130: 10856–10857.

95 Gu, F., Hu, C., Xia, Q. et al. (2018). Aptamer-conjugated multi-walled carbon nanotubes as a new targeted ultrasound contrast agent for the diagnosis of prostate cancer. *J. Nanopart. Res.* 20: 303.

96 Polo-Luque, M., Simonet, B., and Valcárcel, M. (2013). Functionalization and dispersion of carbon nanotubes in ionic liquids. *TrAC, Trends Anal. Chem.* 47: 99–110.

97 Fukushima, T., Kosaka, A., Ishimura, Y. et al. (2003). Molecular ordering of organic molten salts triggered by single-walled carbon nanotubes. *Science* 300: 2072–2074.

98 Shojaei, T.R. and Azhari, S. (2018). Fabrication, functionalization, and dispersion of carbon nanotubes. In: *Emerging Applications of Nanoparticles and Architecture Nanostructures* (ed. A.S.H. Makhlouf and A. Barhoum), 668. Elsevier.

99 Tsuji, T., Kim, J., Sakakita, H. et al. (2021). Role of hydrogen in catalyst activation for plasma-based synthesis of carbon nanotubes. *ACS Omega* 6: 18763–18769.

100 Chen, Q., Dai, L., Gao, M. et al. (2001). Plasma activation of carbon nanotubes for chemical modification. *J. Phys. Chem. B* 105: 618–622.

101 Bardajee, G.R., Sharifi, M., Torkamani, H., and Vancaeyzeele, C. (2021). Synthesis of magnetic multi walled carbon nanotubes hydrogel nanocomposite based on poly(acrylic acid) grafted onto salep and its application in the drug delivery of tetracyceline hydrochloride. *Colloids Surf., A* 616: 126350.

102 Falank, C., Tasset, A.W., Farrell, M. et al. (2019). Development of medical-grade, discrete, multi-walled carbon nanotubes as drug delivery molecules to enhance the treatment of hematological malignancies. *Nanomed. Nanotechnol. Biol. Med.* 20: 102025.

103 Im, J.S., Bai, B.C., and Lee, Y.-S. (2010). The effect of carbon nanotubes on drug delivery in an electro-sensitive transdermal drug delivery system. *Biomaterials* 31: 1414–1419.

104 Yang, Z., Zhang, Y., Yang, Y. et al. (2010). Pharmacological and toxicological target organelles and safe use of single-walled carbon nanotubes as drug carriers in treating Alzheimer disease. *Nanomed. Nanotechnol. Biol. Med.* 6: 427–441.

105 Pantarotto, D., Singh, R., Mccarthy, D. et al. (2004). Functionalized carbon nanotubes for plasmid DNA gene delivery. *Angew. Chem. Int. Ed. Engl.* 43: 5242–5246.

106 Qin, C. and Peng, L.M. (2002). Measurement accuracy of the diameter of a carbon nanotube from TEM images. *Phys. Rev. B* 65: 155431.

107 Khalid, A.H., Abdulsattar, M.A., Hussain, Z.T., and Jabor, A.A. (2020. IOP Publishing). The effect of different multiwall carbon nanotubes concentration on morphology, optical, and electrical properties used as flexible anode. *J. Phys. Conf. Ser.* 1660: 012054.

108 Tan, J.M., Karthivashan, G., Arulselvan, P. et al. (2014). Characterization and in vitro sustained release of silibinin from pH responsive carbon nanotube-based drug delivery system. *J. Nanomater.* 2014: 1–10.

109 Jha, P.K., Jha, R.K., Rout, D. et al. (2017). Potential targetability of multi-walled carbon nanotube loaded with silver nanoparticles photosynthesized from *Ocimum tenuiflorum* (tulsi extract) in fertility diagnosis. *J. Drug Targeting* 25: 616–625.

110 Negri, V., Pacheco-Torres, J., Calle, D., and López-Larrubia, P. (2020). Carbon nanotubes in biomedicine. In: *Surface-Modified Nanobiomaterials for Electrochemical and Biomedicine Applications*, vol. 378(15), 177–217.

111 Cao, A., Xu, C., Liang, J. et al. (2001). X-ray diffraction characterization on the alignment degree of carbon nanotubes. *Chem. Phys. Lett.* 344: 13–17.

112 Salam, M.A. and Burk, R. (2017). Synthesis and characterization of multi-walled carbon nanotubes modified with octadecylamine and polyethylene glycol. *Arabian J. Chem.* 10: S921–S927.

113 Krishnia, L., Kumari, R., Kumar, V. et al. (2016). Comparative study of thermal stability of filled and un-filled multiwalled carbon nanotubes. *Adv. Mater. Lett.* 7: 230–234.

114 Ibrahim, R., Hussein, M.Z., Yusof, N.A., and Abu Bakar, F. (2019). Carbon nanotube-quicklime nanocomposites prepared using a nickel catalyst supported on calcium oxide derived from carbonate stones. *Nanomaterials* 9: 1239.

115 Wilson, N.R. and Macpherson, J.V. (2009). Carbon nanotube tips for atomic force microscopy. *Nat. Nanotechnol.* 4: 483–491.

116 Shetti, N.P., Nayak, D.S., Malode, S.J. et al. (2019). Sensors based on ruthenium-doped TiO_2 nanoparticles loaded into multi-walled carbon nanotubes for the detection of flufenamic acid and mefenamic acid. *Anal. Chim. Acta* 1051: 58–72.

117 Wang, L., Mu, R.-J., Lin, L. et al. (2019). Bioinspired aerogel based on konjac glucomannan and functionalized carbon nanotube for controlled drug release. *Int. J. Biol. Macromol.* 133: 693–701.

118 Sandoval, S., Kierkowicz, M., Pach, E. et al. (2018). Determination of the length of single-walled carbon nanotubes by scanning electron microscopy. *MethodsX* 5: 1465–1472.

119 Wongon, J., Thumsorn, S., and Srisawat, N. (2016). Poly(vinyl alcohol)/ multiwalled carbon nanotubes composite nanofiber. *Energy Procedia* 89: 313–317.

120 Zhang, T., Tang, M., Yao, Y. et al. (2019). MWCNT interactions with protein: surface-induced changes in protein adsorption and the impact of protein corona on cellular uptake and cytotoxicity. *Int. J. Nanomed.* 14: 993–1009.

121 Holt, B.D., Dahl, K.N., and Islam, M.F. (2011). Quantification of uptake and localization of bovine serum albumin-stabilized single-wall carbon nanotubes in different human cell types. *Small* 7: 2348–2355.

122 Sánchez-González, J., García-Otero, N., Moreda-Piñeiro, A., and Bermejo-Barrera, P. (2012). Multi-walled carbon nanotubes—solid phase extraction for isolating marine dissolved organic matter before characterization by size exclusion chromatography. *Microchem. J.* 102: 75–82.

123 Herrero-Latorre, C., Álvarez-Méndez, J., Barciela-García, J. et al. (2015). Characterization of carbon nanotubes and analytical methods for their determination in environmental and biological samples: a review. *Anal. Chim. Acta* 853: 77–94.

124 Iwamura, M., Akizuki, N., Miyauchi, Y. et al. (2014). Nonlinear photoluminescence spectroscopy of carbon nanotubes with localized exciton states. *ACS Nano* 8: 11254–11260.

125 Osikoya, A., Donbebe, W., Vala, R. et al. (2015). Synthesis, characterization and sorption studies of nitrogen-doped carbon nanotubes. *Digest J. Nanomater. Biostruct.* 10: 125–134.

126 De Souza, L.A., Nogueira, C.A.S., Ortega, P.F.R. et al. (2016). Inclusion complex between cisplatin and single-walled carbon nanotube: an integrated experimental and theoretical approach. *Inorg. Chim. Acta* 447: 38–44.

127 Zhang, X., Zhang, J., Liu, Z. et al. (2020). Microscopic stresses in carbon nanotube reinforced aluminum matrix composites determined by in-situ neutron diffraction. *J. Mater. Sci. Technol.* 54: 58–68.

128 Aqel, A., El-Nour, K.M.M.A., Ammar, R.A.A., and Al-Warthan, A. (2012). Carbon nanotubes, science and technology part (I) structure, synthesis and characterisation. *Arabian J. Chem.* 5: 1–23.

129 Burian, A., Koloczek, J., Dore, J.C. et al. (2004). Radial distribution function analysis of spatial atomic correlations in carbon nanotubes. *Diamond Relat. Mater.* 13: 1261–1265.

3

Latest Developments in Commercial Scale Fabrications for Chemically Modified Carbon Nanotubes

Shahidul Islam Bhat[1], Mohammad Mobin[1], Shayesta Islam[2], Mohd Irfan[3], Rabia Kouser[4], Kulsoom Koser[5], Ruby Aslam[1], and Saman Zehra[1]

[1]*Aligarh Muslim University, Corrosion Research Laboratory, Department of Applied Chemistry, Faculty of Engineering and Technology, Aligarh 202002, India*
[2]*SKUAST K, Shalimar Campus Srinagar Kashmir, Division of Environmental Sciences, 190025, India*
[3]*Ajanta Caves and Conservation Research Laboratory, Aurangabad, India*
[4]*Jamia Millia Islamia, Materials Research Laboratory, Department of Chemistry, New Delhi, India*
[5]*Bio Inorganic Research Laboratory, Department of Chemistry Jamia Millia Islamia, New Delhi 110025, India*

Abbreviations

APCVD	atmospheric-pressure chemical vapor deposition
CNTs	carbon nanotubes
CVD	chemical vapor deposition
HIPCO	high-pressure carbon monoxide decay technique
HPCVD	high-pressure chemical vapor deposition
LPCVD	low-pressure chemical vapor deposition
MWCNTs	multi walled carbon nanotubes
SWCNTs	single-walled carbon nanotubes

3.1 Introduction

Nanotubes have acquired considerable attention in diverse fields like energy storage [1], filtering [2, 3], sensors [4], and other areas. The methods developed for the fabrication of carbon nanotubes (CNTs) are laser vaporization and arc-discharge [5], utilizing plasma [6], chemical vapor deposition (CVD) [7], and floating catalyst technique [8–11]. The procedure used to produce CNTs with fewer structural defects is called catalytic decomposition of hydrocarbons. Through the arc-discharge method, such nanotubes can be formed that have slight control over chirality, but there is a requirement for postprocessing purification of nanotubes due to utilization of metallic catalysts. Laser ablation helps to produce nanotubes with a few metallic impurities and a good yield due to the affinity of the metallic stimulant to volatilize from the terminal of the tube when it is stopped. This process

Chemically Modified Carbon Nanotubes for Commercial Applications, First Edition.
Edited by Jeenat Aslam, Chaudhery Mustansar Hussain, and Ruby Aslam.

is very costly due to use of pure graphite rods and laser power; moreover, the nanotubes produced by this method are not homogeneous and carry branched structures. Diverse methods like laser ablation, arc-discharge, and plasma assist in the generation of powder nanotubes. CVD in which temperature is kept low is generally employed due to good command over a diameter of nanotube, cleanness, dimensions, adjustment, and cost-effective bulk production of CNTs [12]. The CVD method was utilized to generate excellent single-walled nanotubes (SWNTs) using a new aerogel-aided by Fe/Mo stimulant with outstanding productivity. Five times improvement in the amount of nanotubes procured per unit weight of Fe/Mo stimulant has been reported as compared to stimulant aided on Al_2O_3 powder. The enhanced catalytic activity arises due to the well-built interconnection between the aerogel and stimulant and the large support surface area [13]. The CVD technique with an underlayer generates fine films of CNTs in bulk, which are then packed in layers to form sheet-like structures. Industrial-level CNTs are generated by the drifting stimulant gas-phase pyrolysis technique by directly meandering a network of nanotubes onto a barrel in an uninterrupted manner. The floating catalyst method produces good-quality, uniformly dispersed, and aligned CNTs of 10 nm in diameter. To increase the development of nanotubes, benzene served as the carbon source, ferrocene as the stimulant, and thiophene as the sulfur-carrying additive [14]. The connection between the mean dimensions of nanotubes and thiophene concentration was investigated [15]. The morphology of the nanotubes is controlled by the content of sulfur added. Substantial amounts of good-grade and cost-effective single-walled carbon nanotubes (SWCNTs) are produced by advanced drifting stimulant approach as it has a lower temperature with controlled growth parameters. This process can be utilized to produce single-walled as well as multi-walled carbon nanotubes (MWCNTs) and the production is influenced by the growth foster, which is a sulfur-carrying additive [15]. Good current fetching potential [16, 17], mechanical hardness [18], and high heat conduction [19, 20] are some of the unique characteristics of CNT material. Due to anisotropic heat conduction, CNTs have potential fire-fighting applications. A smart outfit notion was shown in a simulation to govern heat toward an outward cool sink from the outfit, thereby reducing the body temperature [21]. Due to fire-fighting applications, CNTs are considered suitable for the preparation of individual safety apparatus for firefighters [22]. The applications of CNTs extend over a broader area due to incorporation of CNT materials into textiles [23]. Furthermore, thin CNT sheets have porous structures and electrical conductivity properties, which make them a capable nominee for energy store house for wearable electronics [24]. Also, the porous structure and the high domain of CNTs make them beneficial for filtration of air and water. The characteristics of CNTs like conduction and pharmacological can be utilized to inactivate and arrest pathogen/virus. This property can be enhanced by integrating CNTs with immunosuppressive nonmaterial, viz., silver, copper, zinc oxide, etc. [25]. The performance of CNTs depends not only on innate features of layers but also on the pattern of layers and can also be gauzed by the work done in the last five years (Scheme 3.1).

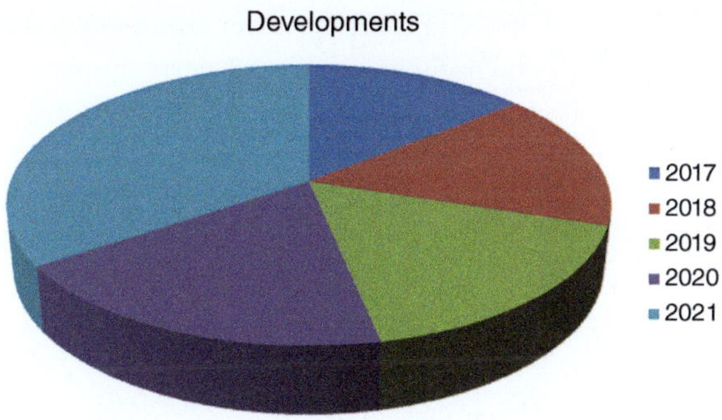

Scheme 3.1 Five year development on fabrication for chemically modified carbon nanotubes.

3.2 Industrial Scale Fabrication Strategies

The three basic methods widely used for the production of both SWCNTs and MWCNTs are chemical vapor decomposition, laser vaporization, and arc-discharge.

3.2.1 Basic Chemical Vapor Deposition (CVD) Process

The conventional methods for the synthesize of CNTs are CVD, arc-discharge, and laser ablation. The most extensively adopted method is CVD due to its manifold benefits, like cost-effectiveness, convenience synthesis, and high throughput. The position, inclination, and dimensions of CNTs for a definite purpose can be controlled by managing the factors, viz., stimulant, carbon source, gas pressure, and reaction temperature. Through arc evaporation of fullerene CNTs, they were first revealed in 1991 [26]. Guo et al. in 1995 made use of the laser ablation method to develop SWCNT [27]. On a commercial level, Ebbesen and Ajayan in 1992 synthesized for the first time CNTs by the arc-discharge method [28]. The temperature required to synthesize CNTs is very high, up to 1200 °C or more in laser and arc methods, while the position and inclination of the CNTs are hard to manage.

In the CVD process, initially only those gaseous species are kept which are needed for deposition; others are isolated from the reaction chamber. After that, the predecessor gases like hydrocarbons that approach the deposition surface are discharged into the reaction assembly and heated. After attaining ample energy, the predecessor gas molecules combine together or decay, resulting in a solid form of carbon (C), which gets quickly coated over the substratum. Through the exhaust system, the explosive secondary product hydrogen (H) is separated from the reaction assembly. In the absence of substrate, the solid form of carbon gets accumulated on the walls of the reaction compartment.

3.2.1.1 Industrial Level Fabrication of CNT Through Various CVD Methods

This is one of the excellent technique for generation of CNTs. The various types of CVD methods are catalytic chemical vapor deposition (CCVD), which is further divided into thermal [29] and plasma enhanced oxygen assisted CVD [26],

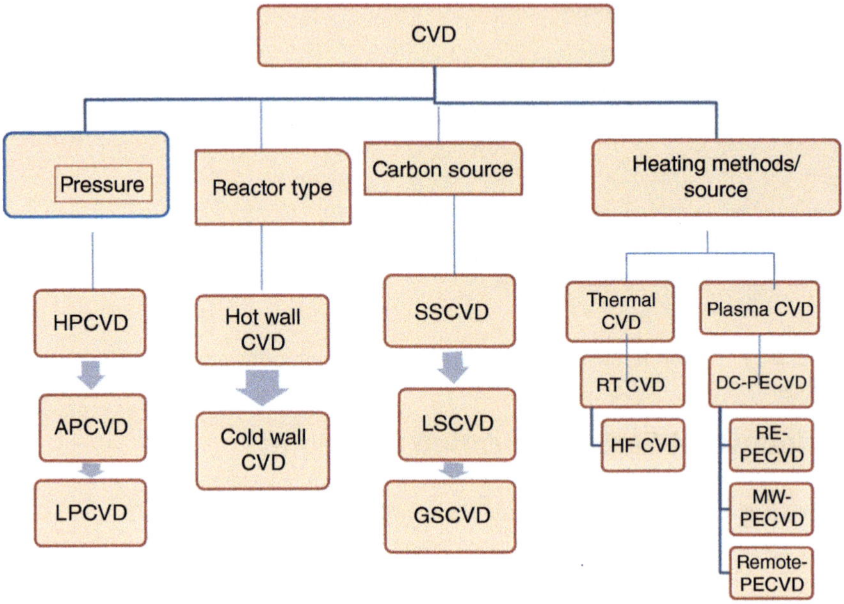

Figure 3.1 Flow chart of different CVD categories.

water-assisted CVD [30–32], microwave plasma CVD [28], radio-frequency CVD [33], or hot-filament CVD [34, 35]. Among all the methods, CVD with stimulants is presently the criterion procedure for the formation of CNTs. This process involves expansion of CNTs on variable materials and chemical disintegration of a hydrocarbon on an underlayer. The procedure of enlarging CNTs is same in CVD as well as in arc-discharge. CVD methods can be divided on the basis of pressure in the chamber, type of reactor, origin of carbon, and heating technique. A flow chart of various CVD methods is shown in Figure 3.1. In the CVD procedure, the significant factors of concern are the processes depending on the pressure of reactor. Depending on the pressure of reactor during the manufacturing of CNT, CVD is classified as (high pressure, atmospheric pressure, and low pressure) CVD.

3.2.1.2 High-Pressure Chemical Vapor Deposition

In this process, during the preparation of CNTs, the pressure on the inner side of reactor is greater than 1 atm [36] (Figure 3.2). SWCNTs were produced in bulk by Smalley's class [37] in 1999 using the high-pressure carbon monoxide decay technique (HIPCO). HIPCO involves integration of pre-treated (1000 °C), HIPCO equal to 30 atm with catalyst predecessor gas like metal carbonyl or metallocene (Royal Society of Chemistry). Under these circumstances, the predecessor molecule undergoes decomposition, resulting in clustering of metal particles over which CNTs crystallize and enlarge. 99% of CNTs formed during the process are single-walled with miniature dimensions. SWCNTs are formed in an accelerated pattern due to dismutation of CO molecules into carbon atoms resulting from high-temperature disclosure of stimulant and great pressure of CO. CoMoCAT is similar to HIPCO and was

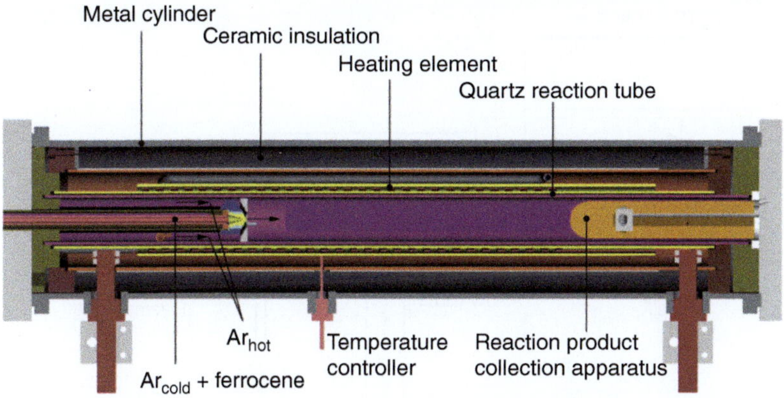

Figure 3.2 High-pressure CVD reactor cross section for SWCNT synthesis. Source: Barreiro et al. [36]/with permission of American Chemical Society.

taken on by Resasco et al. [38] for the synthesis of SWCNTs by stimulant dismutation of CO in the presence of an exceptional Co–Mo solid stimulant.

3.2.1.3 Atmospheric-Pressure Chemical Vapor Deposition (APCVD)

The process is carried out at atmospheric pressure in which underlayer is uncovered with more than one volatile predecessor that decays or reacts on the facet of the underlayer to cause accumulation (Figure 3.3). Carbon microtubules were prepared by stimulant decay of ethylene over iron particles at 700 °C in 1993 by Yacaman et al. [40]. In 2006, Wei et al. [41] used Co as a stimulant, ethylene as a carbon source, and ammonia gas to control reaction at 750 °C for preparation of aligned CNT and concluded that the flow rate ratio of NH_3 to C_2H_2 and depth of the stimulant layer predict whether the CNTs will be upright or haphazardly aligned. Majewska et al. [42] manifested a traditional gas-flow system to prepare CNTs loaded with Co nanowires without interruption at the temperature of 400 and 800 °C and at atmospheric pressure by methane degradation. Yun et al. [43] used Fe as a catalyst to prepare immensely oriented arrangement of MWCNT on layered Si underlayer. It has been observed that growth rate of CNTs can be enhanced by adding water vapor to the gas blend of hydrogen and ethylene in the reaction, resulting in the formation of larger CNTs with outstanding density. Cao et al. [44] utilized Fe/Mo/Co as a stimulant, Al_2O_3 as a substratum, and ethylene as a carbon feedstock to prepare SWCNTs. Alan M. Cassell [39] and his coworkers used Ni or Fe-coated Si underlayer at a temperature of 850 °C to synthesize loaded and empty MWCNTs by atmospheric pressure CVD of propane (C_3H_8). They reported that nanotubes have potential applications in areas like magnetic recording media due to their ferromagnetic nature caused by magnetic material filling. Boron-loaded hybrid CNTs were prepared in single-step CVD process at 950 °C under Ar and CH_4 flows of 100 and 10 sccm, respectively, by Patel et al. [45]. At high temperatures, boron-filled hybrid carbon nanotubes (BHCNTs) have outstanding mechanical properties, 31% more rigid and 233% more powerful than traditional MWCNTs. Li et al. [46] established very reliable manufacture of CNT

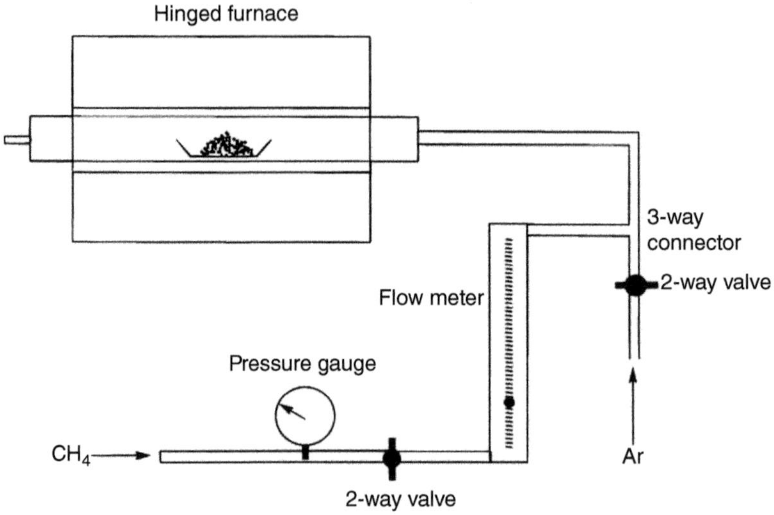

Figure 3.3 Experimental set up (schematic) for methane CVD synthesis of SWNTs. Source: Alan et al. [39]/with permission of American Chemical Society.

in bulk at 775 °C by separating the stimulant by heat treatment and hydrocarbon exposure by moving sample inside and outside of the CVD system and balancing the content of moisture and hydrocarbons between the two steps. Two-layered stimulant Fe/Al_2O_3 was accumulated by crackling on oxide-layered Si wafers and ethylene was used as the source of carbon.

3.2.1.4 Low-Pressure Chemical Vapor Deposition (LPCVD)

In this method, the pressure inside the reaction assembly is below 1 atm. Liao et al. [47] used decay of ethene at low pressure of about 10 mTorr over iron atoms aided on Si wafers at 550 °C to prepare SWCNTs. Ikuno et al. [48] used ethene at 100 Pa pressure and 800–970 °C temperature for fabricating straight chain CNT bridges in the middle of iron nanomaterials by low-pressure CVD. The CNTs obtained were constituted of packs of SWCNTs and MWCNTs. Cantoro et al. [49] prepared SWCNTs by the CVD method using concentrated ethylene at 350 °C below 1022 mbar pressure. Exploratory outcomes determined that ammonia or hydrogen exposure encourages nanostructuring and stimulation of subnanometer iron and Al/Fe/Al complex stimulant films prior to growth, leading to the formation of SWCNT at low temperatures. Chen et al. [50] reported that optimum states for the preparation of good quality MWCNTs are catalyst precursor ferrocene $Fe(C_5H_5)_2$, carbon source cyclohexane (C_6H_{12}), conveyor gas (H_2), and promoter thiophene (C_4H_4S), 15 kPa pressure, 650 °C temperature, and 60 sccm flow rate of hydrogen. SWCNT was prepared with ethylene at 0.5 mbar pressure and 800–1000 °C temperatures on Al/Fe-painted Si_3N_4 lamina with Si as underlayer, devoid of any drift in gas at any point of the whole process, together with heating and cooling of the samples by Kasumov et al. [51]. This process is different from other low-pressure techniques where the production is carried out in the gas flow for five minutes to half an hour.

In single insertion, ethylene is inserted in the sample assembly and evacuated readily. High-density oriented CNT was prepared by Wang et al. [52] on an iron-painted quartz underlayer at 900 °C at a pressure of 500 Torr.

3.3 CVD on the Basis of Reactor Wall Temperature

3.3.1 Hot-Wall Chemical Vapor Deposition (Hot-Wall CVD)

In this method, the temperature of the underlayer is similar to the reactor fence as the reactor pipe is bounded by heating components. High temperature prevents unnecessary deposition; hence, hot-wall reactor is usually preferred for heat-evolving reactions. The benefit of hot fence CVD is that the temperature is uniform, but the drawback is that high-temperature vapors are produced that react chemically with the walls of the reactor and taint the product. Zhang et al. [53] utilized hot fenced reactor to accomplish the CVD technique to produce SWCNTs by decomposing methane on very thin Ni/Al lamina-covered Si underlayer at temperature of 800 °C. Gruneis et al. [54] utilized hot-fenced CVD to manufacture extraordinary quality two-walled CNTs with a fixed diameter distribution from alcohol at a temperature of 940 °C and Co and Mo acetates served as catalysts in this process. Ayala et al. [55] utilized a catalyst consisting of two metals, viz., molybdenum and iron at a temperature of 900–1000 °C supported by ceramic for preparation of good grade nitrogen-doped single- and double-walled CNTs with definite dimensions from a concentrated C/N raw material benzylamine (C_7H_9N). An "excellent expansion" of SWCNTs was carried out by Noda et al. [56] using Fe/Al_2O_3 as a stimulant on a Si wafer with ethene as a source of carbon at temperature of 820 °C. Sugime et al. [57] evaluated most favorable stimulants and reaction circumstances using a collection of stimulants and also recognized areas having great stimulation action on the underlayer by recording the production of CNT against the orthogonal gradient thickness profiles of Co and Mo. Yamada et al. [58] studied the impact of addition of water on the growth by synthesizing CNTs from thin films of iron (1.5 nm) accumulated on 50 nm SiO_2 nanoballs kept on a transmission electron microscope (TEM) matrix at 750 °C using ethene as carbon origin.

3.3.2 Cold-Wall Chemical Vapor Deposition (Cold-Wall CVD)

This method involves heating of the substrate only by radio frequency induction or high emission light, and the reactor remains cool. This method is usually suitable for reactions involving the absorption of energy. The advantages of cold fence CVD over hot fence CVD are need for a brief operation period and short maturation term, and the avoidance of taint of the assembly fence as walls remain cool. Furthermore, there is no need to heat the whole reactor; only the sample needs to be heated. Hence, electricity and gas consumed are reduced compared to the hot fence CVD process. Cold fence CVD is best suitable for optic supervision than hot fence CVD, which can have significant benefit in the attempt to best comprehend the extension process. The temperature difference is caused as the heat is taken away by the cold surface, making the temperature uniformity factor of concern. Finnie's group [59] used this technique to prepare SWCNTs with methane as a carbon source and iron as a stimulant on silicon wafer at temperature 900 °C. Afterward, this process was optimized [60] by considering the impact of pressure, temperature, substrate conditioning, and metallization. Chiashi et al. [61] evaluated the production of outstanding-purity SWCNTs using cold-wall CVD from alcohol with Fe/Co atoms aided on zeolite by Joule heating of a silicon foot panel at temperature 850 °C. Cantoro et al. [49] prepared surface confined SWCNTs from unmixed ethene using an iron covered Si underlayer at temperatures of 350 °C by catalytic CVD. Maruyama et al. [62] investigated SWCNT extension from a Pt stimulant using a syringe injector for the ethanol gas supply in a high vacuum. In order to optimize pressure of ethanol, a temperature range from 330 to 700 °C was used and revealed that by decreasing the growth temperature, production can be decreased.

3.4 Arc-Discharge

Using this method, the first confirmed CNTs fabrication was done between graphite electrodes [26]. Production of huge amounts of fullerenes can successfully be made through this method [63]. After that, it was investigated that the addition of some stimulant elements to one of the electrodes created depositions enriched with SWCNT [5, 64, 65]. Consequently, arc-discharge has turned out to be the most extensively used method for CNT preparation. In the basic method (Figure 3.4), direct current electric discharge is established between graphite electrodes under an unreactive helium environment (500 Torr). The electrodes are classically water-cooled graphite rods apart from each other by 1 mm with diameter of cathode of 8–12 mm, which is larger than the anode with diameter 6–8 mm. To produce a current density of about 150 A cm^{-2} and current of approximately 60–100 A, a voltage of nearly 10–35 V is exerted on the electrodes to create the emissions. To

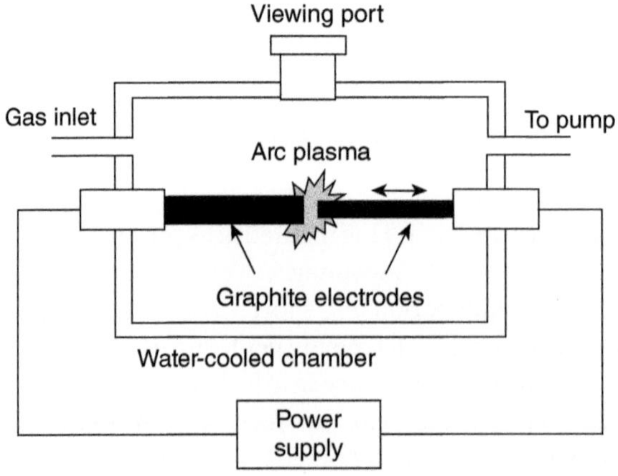

Figure 3.4 Schematic representation of an arc-discharge apparatus. Source: Kingston and Simard [66]/with permission of Taylor & Francis.

create feasible environmental conditions for the growth of nanotubes, the important factor is to achieve stable plasma discharge. The matter from the anode undergoes sublimation at elevated temperature greater than 4000 °C and gets conveyed to the negatively charged electrode and adjacent walls of the equipment so as to sustain the permanence of the emissions. The major collection of CNTs is found in the deposits of negatively charged electrodes. Based on the specific geometry and factors used in the preparation, the CNTs differ in appearance and texture. With pure graphite electrodes, a spongy black fibrous material is set down at the terminal point of the cathode, enclosed by a solid outer ring or shell. Multi-walled nanotube (MWNT) in combination with polycrystalline graphite nanoparticles gets accumulated on the inner side, while the external frame is mainly formed of meld nanotubes, nanomaterials, and amorphous carbon [28, 67]. Catalysts are inserted into the reaction by putting a powdered blend of metal and graphite into a cavity formed in the anode. It has been reported that production of SWNT is preferred by utilizing single or bi-metal blend of Co, Ni, Y, and Fe [31, 65, 68–73]. In these cases, the nanotube-enriched matter is expressed as a spongy strip or band adjoining the tough deposit and has been noticed to hold high percentage (70–90%) of SWNT. It is commonly admitted that the most proficient for SWNT formation by arc discharge is Co/Y and Ni/Y bimetal blends. In this method, the number of nanotubes produced per unit time is very high. The limitation is that the experiment reaches culmination only after 10–20 minutes due to instability in the DC arc as a result of nonuniform utilization of the anode and accumulation of matter on the cathode. 30–50% of the total carbonaceous matter obtained is accounted for nanotubes rich portion.

3.5 Laser Vaporization

This method is an expansion of technique utilized for fullerene production [74], similar to the arc discharge method. The initial demonstration has been proposed as a check for a study on metallofullerene synthesis [75]. In an experiment, a quartz tube was taken in which solid graphite was launched and kept in a temperature-regulated oven. The target launched was then vaporized with a pulsed Nd:YAG laser, and on the inner side of the equipment a carbon-based filth was accumulated (Figure 3.5).

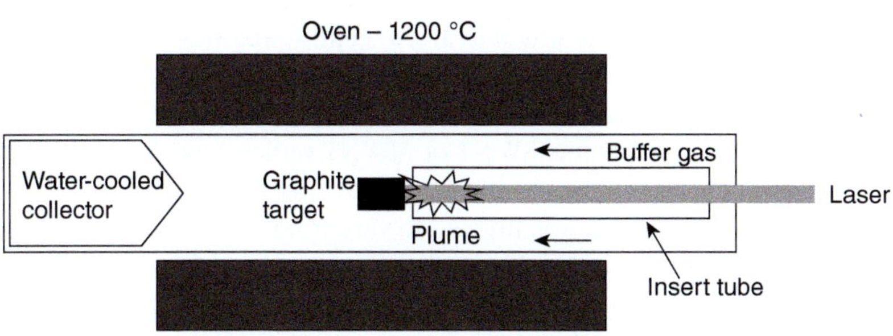

Figure 3.5 Schematic representation of a laser vaporization apparatus. Source: Kingston and Simard [66]/with permission of Taylor & Francis.

The electron microscopy of the collected soot by the researchers exposed the existence of well-graphitized MWNT. After this, Guo et al. [27] verified that similar practices can also be utilized for the manufacture of SWNT merely by contaminating the graphite target with proper transition metal stimulants. Consequently, this method has been utilized more or less entirely for preparation of SWNT. To advance the productivity of nanotubes, the original discoveries have been accompanied by an optimization period. The two most momentous improvements that have come out were adding up of a second Nd:YAG pulsed laser belated to some extent [76] as well as the thriving scaling up instrument proportions [77]. The principal trouble with the basic design was the obstruction of the quartz tube close to the target rod by the nanotube-enriched soot, which was reduced by escalating the dimensions of the equipment. By integrating both the modifications [77], the resultant soot was found to be made of 60–90% SWNT, considerably greater than that from the characteristic arc-discharge. The development in synthesis rates was also enhanced from around 80 mg per day to quantities close to 1 g/24 h. As a result of elevated clarity and high nanotube-rich soot, the double pulsed-lasers reported by Rinzler et al. [77] are the "standard" protocol for CNTs laser synthesis.

The major benefit of laser vaporization over arc-discharge is that it has additional defined control over the high-temperature situation of the synthesis atmosphere. Modification of the circumstances feasible for nanotube production is made easy by parameters like laser wavelength, emittance, rate of repetition, and temperature framework of oven. It is for this motive that laser vaporization has been time and again confirmed to be more successful at producing highly pure nanotube sublimates in comparison to arc-discharge techniques. The productivity of the laser method is usually very low compared to arc methods but can produce same amount of nanotube sublimate in one go due to the prolonged time of action. The limitation of this technique is expandability due to the reliance on potent lasers and the requirement for an elevated-temperature ambient environment in most cases [66].

3.6 Other Synthesis Methods

The utilization of concentrated radiation from the sun for the generation of CNTs has been confirmed by a group from France [78]. The powdered blend of graphite and catalyst was kept in a graphite crucible at the central end of a solar heating system proficient of providing 1600 W cm^{-2} of rays under most favorable conditions. It was noticed that the graphite and metal blend were evaporated in similar manner as in the laser-vaporization technique. SWNT or MWNT settles down in quantities of nearly 100 mg per run, depending on stimulant, gas pressure, and rate of flow. The diameter of the SWNT formed was noticed to vary from 1 to 2 nm determined by the type of stimulant utilized [79]. Various CNTs have been formed through flame-based methods. The preparation of low-combustion SWNTs at 80 Torr pressure has been reported by Diener et al. [80]. To produce a SWNT containing filth, van der Waals used an aerosol of iron nitrate passed by a gas blend of helium, CO, and hydrogen into an air/acetylene flame [81]. The production and SWNT portions were found to differ significantly with respect to the blaze constitution, carrier gas, and flow rates. Smiljanic et al. [82] reported SWNTs in which a 1300 W microwave magnetron was utilized to generate 5500 K plasma from ferrocene and ethylene. Even though the procedure was mainly not optimized, deposits constituting SWNT were created with dimensions varying from 0.9 to1.5 nm at clarity extent equivalent to those of arc-discharge synthesis. Since 1995 [83], electric breakdown of liquefied halogen salts has been utilized for nanotube synthesis but has not been comprehensively studied. In this method, halide salts with temperature higher than their melting point served as the electrolyte, and graphite rod as the negative electrode and crucible as the positive electrode. A negative electrode has been engrossed into the electrolyte and an electric current of 3–7 A is generated, accelerating the wear away of the cathode.

3.7 Applications

The diverse applications discussed so far have focused on scanty research, and there is a need to switch to applications with industrial importance. Also advancement in nanotechnology it is quite difficult to figure out if the CNTs will meet the expectations to their best. The applications of CNTs have been found to be not only limited to hydrogen storage, black body absorption, terahertz polarization, and artificial muscles [84]. Ironically, research work has also been reported in the field of cancer research using CNTs due to strong absorptions in the infrared region of the electromagnetic spectra [85].

3.7.1 Transistors

The transistors are switches with a small size, forming the microelectronic fundamental blocks. For an efficient transistor, a semiconducting material like chiral SWCNTs is used [86]. To create a fully operational integrated circuit (IC) care should be taken that a high level of chirality control and individual SWCNT deposition is achieved. Thus, such applications do not require mass production and it has been found that SWCNTs have found profound use in solar panels. Thus it serves two important functions, i.e. a flexible conductor and optically active component part [84].

3.7.2 Conductor

As a conductor, CNTs have another promising application where the conductor wires can be replaced by CNT-based wires. This application requires production of CNTs at a large scale. The advantage of use of CNT-based conductors over copper ones is quite small for low voltage at smaller scale. It has been found that the metallic SWCNTs possess superior electrical conductivity in comparison to those of copper-based ones in terms of higher current carrying capacity [87]. Researchers are trying their level best to develop alloys based on copper metal and CNTs, although from alloys the benefits of pure CNT cables can never be reared [88]. However, its use as a conductor will prove to be an efficient shielding material for use in satellites [84].

3.7.3 Composites

The prime use of CNTs will be in the field of composites that form a fairly new class of materials. The possibility of combining the already known composites with the CNTs has made this domain of research more interesting. The use of carbon fiber together with CNT composites has been considered the lightest and toughest material [89]. In the current era, aircraft, and space rockets have slowly switched to carbon fiber composites [90]. Thus, in near future we expect CNT composites to be incorporated into aerospace applications. In terms of mechanical strength, CNT-based carbon fibers are one of the toughest materials, but the mechanical properties of MWCNTs have been found to be better [91]. The applications of CNTs may be in gas might be in designing gas turbines, where blades can be made from super alloys, which can withstand drastic temperatures and pressure [92].

3.7.4 Aerogels

The aerogels have been developed from CNTs and are considered the lightest solids, with a weight even lighter than air [88]. As a result of its porous nature, the bulk gets filled with air, which makes it difficult to get afloat. The carbon-based aerogels behave like foam in comparison to the brittle silica aerogels. As a result of their porous nature, they behave as efficient thermal insulators and have found their applications as heat shields.

3.8 Future Scope

As aforementioned, various fabrications of CNTs have widened their area of applications, but still there are some unexplored areas where work must be done. The major challenge is the functionalization of CNTs without disintegrating their basic nature, especially their chiral nature. Despite having emerged from decades of antibacterial application, it is still a matter of concern. For a lot of reasons, CNTs are unable to compete/substitute with the polymer-based antibacterial species, which is evident from the fact that their toxicity toward human cells has not been addressed yet. Safety issues have not been thoroughly examined, which will pave ways for the development of CNTs in nanomedicine. The progress in CNT technology may lead to insights into various physical/biological processes that will open gateways to search for innovative compounds compatible with CNTs for facilitated use of nanotube morphology.

3.9 Conclusion

The stunning message conveyed in this review is that the process of CNT functionalization by chemical means is a steadily growing process. Currently, the researchers are focusing their attention on technological aspects that arise from the versatile nature of CNTs and the possibility of creation of novel architectures possessing unique properties. CNT production has gone cheaper over the years and a considerable improvement has been seen in the production quality so far as contamination by impurities like metal catalysts/amorphous carbon is concerned. In view of this, the current chapter focuses on the developments at commercial scale for fabrication of chemically modified CNTs. The chapter highlights the different methods of CVD like atmospheric-pressure CVD, low-pressure CVD, arc-discharge, and laser vaporization. Various applications like transistors in electronic devices, however the applications, have been extended to possible other fields like terahertz polarization, hydrogen storage, near-ideal black body absorption, and artificial muscles.

Author's Contributions

All the authors have contributed equally.

Conflict of Interest

There is no conflict of interest between the authors.

Other Sources

https://www.nanowerk.com/nanotechnology/introduction/introduction_to_nanotechnology_22.php
https://www.marketsandmarkets.com/Market-Reports/carbon-nanotubes-139.html
https://www.marketwatch.com/press-release/industrial-carbon-nanotubes-market-scope-and-overview-with-highest-growth-in-the-near-future-by-leading-key-players-2030-2022-01-09
https://www.intechopen.com/chapters/17304
https://arxiv.org/ftp/arxiv/papers/1909/1909.01242.pdf

Acknowledgments

Shahidul Islam Bhat would like to thank the UGC, New Delhi, India for financial support in the form of Dr. D.S. Kothari postdoc vide award number and grant number (No. F.4-2/2006 (BSR)/CH/20-21/0146).

References

1 Sun, L., Wang, X., Wang, Y., and Zhang, Q. (2017). Roles of carbon nanotubes in novel energy storage devices. *Carbon* 122: 462–474.

2 Zhang, R. and Wei, F. (2019). *High Efficiency Particulate Air Filters Based on Carbon Nanotubes*, 643–666. Amsterdam: Elsevier.

3 Yang, S., Zhu, Z., Wei, F., and Yang, X. (2017). Carbon nanotubes/activated carbon fiber based air filter media for simultaneous removal of particulate matter and ozone. *Build. Environ.* 125: 60–66.

4 Vashist, S.K., Zheng, D., Al-Rubeaan, K. et al. (2011). Advances in carbon nanotube based electrochemical sensors for bioanalytical applications. *Biotechnol. Adv.* 29: 169–188.

5 Iijima, S. and Ichihashi, T. (1993). Single-shell carbon nanotubes of 1-nm diameter. *Nat. Cell Biol.* 363: 603–605.

6 Lim, S.H., Luo, Z., Shen, Z., and Lin, J. (2010). Plasma-assisted synthesis of carbon nanotubes. *Nanoscale Res. Lett.* 5: 1377–1386.

7 Bronikowski, M.J. (2006). CVD growth of carbon nanotube bundle arrays. *Carbon* 44: 2822–2832.

8 Gspann, T.S., Smail, F.R., and Windle, A.H. (2014). Spinning of carbon nanotube fibres using the floating catalyst high temperature route: purity issues and the critical role of sulphur. *Faraday Discuss.* 173: 47–65.

9 Hou, G., Su, R., Wang, A. et al. (2016). The effect of a convection vortex on sock formation in the floating catalyst method for carbon nanotube synthesis. *Carbon* 102: 513–519.

10 Hoecker, C., Smail, F.R., Pick, M. et al. (2017). The dependence of CNT aerogel synthesis on sulfur-driven catalyst nucleation processes and a critical catalyst particle mass concentration. *Sci. Rep.* 7: 14519.

11 Hoecker, C., Smail, F.R., Bajada, M.A. et al. (2016). Catalyst nanoparticle growth dynamics and their influence on product morphology in a CVD process for continuous carbon nanotube synthesis. *Carbon* 96: 116–124.

12 Eatemandi, A., Daraee, H., Karimkhanloo, H. et al. (2014). Carbon nanotubes: properties, synthesis, purification, and medical applications. *Nanoscale Res. Lett.* 9: 393.

13 Su, M., Zheng, B., and Liu, J. (2000). A scalable CVD method for the synthesis of single-walled carbon nanotubes with high catalyst productivity. *Chem. Phys. Lett.* 322: 321–326.

14 Ci, L., Wei, J., Wei, B. et al. (2001). Carbon nanofibers and single-walled carbon nanotubes prepared by the floating catalyst method. *Carbon* 39: 329–335.

15 Cheng, H.-M., Li, F., Su, G. et al. (1998). Large-scale and low-cost synthesis of single-walled carbon nanotubes by the catalytic pyrolysis of hydrocarbons. *Appl. Phys. Lett.* 72: 3282–3284.

16 Jakubinek, M.B., Johnson, M.B., White, M.A. et al. (2012). Thermal and electrical conductivity of array-spun multi-walled carbon nanotube yarns. *Carbon* 50: 244–248.

17 Behabtu, N., Young, C.C., Tsentalovich, D.E. et al. (2013). Strong, light, multifunctional fibers of carbon nanotubes with ultrahigh conductivity. *Science* 339: 182–186.

18 Boncel, S., Sundaram, R., Windle, A.H., and Koziol, K. (2011). Enhancement of the mechanical properties of directly spun CNT fibers by chemical treatment. *ACS Nano* 5: 9339–9344.

19 Moisala, A., Li, Q., Kinloch, I., and Windle, A. (2006). Thermal and electrical conductivity of single- and multi-walled carbon nanotube-epoxy composites. *Compos. Sci. Technol.* 66: 1285–1288.

20 Hone, J., Llaguno, M., Biercuk, M. et al. (2002). Thermal properties of carbon nanotubes and nanotube-based materials. *Appl. Phys. A* 74: 339–343.

21 Sullivan, J., Schulz, M., Vemaganti, K. et al. (2015). Carbon nanotube fabric cooling system for firefighters and first responders: modeling and simulation. *J. Fiber Bioeng. Inf.* 8: 1–12.

22 Janas, D., Rdest, M., and Koziol, K. (2017). Flame-retardant carbon nanotube films. *Appl. Surf. Sci.* 411: 177–181.

23 Schulz, M.J., Chitranshi, M., Chauhan, D. et al. (2019). Pioneering carbon nanotube textile engineering & fashion technology. *J. Text. Eng. Fashion Technol.* 5 (2): 89–92.

24 Hu, L., Pasta, M., La Mantia, F. et al. (2010). Stretchable, porous, and conductive energy textiles. *Nano Lett.* 10: 708–714.

25 Ashley, K., Mark, S., Daniel, R.C. et al. Carbon nanotube hybrid fabric to filter and deactivate virus. US Patent 63046680, 1 July 2020.

26 Iijima, S. (1991). Helical microtubules of graphitic carbon. *Nature* 354: 56–58.

27 Guo, T., Nikolaev, P., Thess, A. et al. (1995). Catalytic growth of single-walled nanotubes by laser vaporization. *Chem. Phys. Lett.* 243: 49–54.

28 Ebbesen, T.W. and Ajayan, P.M. (1992). Large-scale synthesis of carbon nanotubes. *Nature* 358: 220–222.

29 Vander Wal, R.L., Berger, G.M., and Ticich, T.M. (2003). Carbon nanotube synthesis in a flame using laser ablation for in situ catalyst generation. *Appl. Phys. A* 77: 885–889.

30 Iijima, S., Ajayan, P.M., and Ichihashi, T. (1992). Growth model for carbon nanotubes. *Phys. Rev. Lett.* 69: 3100.

31 Journet, C., Maser, W.K., Bernier, P. et al. (1997). Large-scale production of single-walled carbon nanotubes by the electric-arc technique. *Nature* 388: 756–758.

32 He, Z.B., Maurice, J.L., Lee, C.S. et al. (2010). Nickel catalyst faceting in plasma-enhanced direct current chemical vapor deposition of carbon nanofibers. *Arabian J. Sci. Eng.* 35: 11–19.

33 Bernholc, J., Roland, C., and Yakobson, B.I. (1997). Nanotubes. *Curr. Opin. Solid State Mater. Sci.* 2: 706–715.

34 Dervishi, E., Li, Z., Xu, Y. et al. (2009). Carbon nanotubes: synthesis, properties, and applications. *Part. Sci. Technol.* 27: 107–125.

35 Ajayan, P.M., Charlier, J.C., and Rinzler, A.G. (1999). Carbon nanotubes: from macromolecules to nanotechnology. *Proc. Natl. Acad. Sci. U.S.A.* 96: 14199–14200.

36 Barreiro, A., Hampel, S., Rummeli, M.H. et al. (2006). Thermal decomposition of ferrocene as a method for production of single-walled carbon nanotubes without additional carbon sources. *J. Phys. Chem. B* 110 (42): 20973–20977.

37 Nikolaev, P., Bronikowski, M.J., Bradley, R.K. et al. (1999). Gas-phase catalytic growth of single-walled carbon nanotubes from carbon monoxide. *Chem. Phys. Lett.* 313: 91–97.

38 Resasco, D.E., Alvarez, W.E., Pompeo, F. et al. (2002). A scalable process for production of single-walled carbon nanotubes (SWNTs) by catalytic disproportionation of CO on a solid catalyst. *J. Nanopart. Res.* 4: 131–136.

39 Cassell, A.M., Raymakers, J.A., Kong, J., and Dai, H. (1999). Large scale CVD synthesis of single-walled carbon nanotubes. *J. Phys. Chem. B* 103: 6484–6492.

40 Yacaman, M.J., Miki-Yoshida, M., and Rendon, L. (1993). Catalytic growth of carbon microtubules with fullerene structure. *Appl. Phys. Lett.* 62: 202–204.

41 Wei, S., Kang, W.P., Davidson, J.L., and Huang, J.H. (2006). Aligned carbon nanotubes fabricated by thermal CVD at atmospheric pressure using Co as catalyst with NH_3 as reactive gas. *Diamond Relat. Mater.* 15: 1828–1833.

42 Majewska, J. and Michalkiewicz, B. (2013). Low temperature one-step synthesis of cobalt nanowires encapsulated in carbon. *Appl. Phys. A* 111: 1013–1016.

43 Yun, Y.H., Shanov, V., Tu, Y. et al. (2006). Growth mechanism of long aligned multiwall carbon nanotube arrays by water-assisted chemical vapour deposition. *J. Phys. Chem. B* 110 (47): 23920–23925.

44 Cao, T.T., Ngo, T.T.T., Nguyen, V.C. et al. (2011). Single-walled carbon nanotubes synthesized by chemical vapour deposition of C_2H_2 over an Al_2O_3 supported mixture of Fe, Mo, Co catalysts. *Adv. Nat. Sci.: Nanosci. Nanotechnol.* 2: 15.

45 Patel, R.B., Chou, T., Kanwal, A. et al. (2016). Boron-filled hybrid carbon nanotubes. *Sci. Rep.* 6: 30495.

46 Li, J., Bedewy, M., White, A.O. et al. (2016). Highly consistent atmospheric pressure synthesis of carbon nanotube forests by mitigation of moisture transients. *J. Phys. Chem. C* 120: 11277–11287.

47 Liao, H. and Hafner, J.H. (2004). Low-temperature single-wall carbon nanotube synthesis by thermal chemical vapour deposition. *J. Phys. Chem. B* 108: 6941–6943.

48 Ikuno, T., Katayama, M., Yamauchi, N. et al. (2004). Selective growth of straight carbon nanotubes by low pressure thermal chemical vapor deposition. *Jpn. J. Appl. Phys.* 43: 860–863.

49 Cantoro, M., Hofmann, S., Pisana, S. et al. (2006). Catalytic chemical vapour deposition of single-wall carbon nanotubes at low temperatures. *Nano Lett.* 6: 1107–1112.

50 Chen, Y., Sun, Z., Li, Y.N., and Tay, B.K. (2006). Optimization of carbon nanotube powder growth using low pressure floating catalytic chemical vapor deposition. *Mater. Chem. Phys.* 98: 256–260.

51 Kasumov, Y.A., Shailos, A., Khados, I.I. et al. (2007). CVD growth of carbon nanotubes at very low pressure of acetylene. *Appl. Phys. A* 88: 687–691.

52 Wang, C., Ryu, K., De Arco, L.G. et al. (2010). Synthesis and device applications of high-density aligned carbon nanotubes using low-pressure chemical vapor deposition and stacked multiple transfer. *Nano Res.* 3: 831–842.
53 Zhang, R.Y., Amlani, I., Baker, J. et al. (2003). Chemical vapor deposition of single-walled carbon nanotubes using ultrathin Ni/Al film as catalyst. *Nano Lett.* 3: 731–735.
54 Gruneis, A., Rummeli, M.H., Kramberger, C., and Barreiro, A. (2006). High quality double wall carbon nanotubes with a defined diameter distribution by chemical vapor deposition from alcohol. *Carbon* 44: 3177–3182.
55 Ayala, P. et al. (2007). Tailoring N-doped single and double wall carbon nanotubes from a nondiluted carbon/nitrogen feedstock. *J. Phys. Chem. C* 111: 2879–2884.
56 Noda, S., Hasegawa, K., Sugime, H. et al. (2007). Millimeter-thick single-walled carbon nanotube forests: hidden role of catalyst support. *Jpn. J. Appl. Phys.* 46: 399–401.
57 Sugime, H., Noda, S., Maruyama, S., and Yamaguchi, Y. (2009). Multiple 'optimum' conditions for CoMo catalyzed growth of vertically aligned single-walled carbon nanotube forests. *Carbon* 47: 234–241.
58 Yamada, T., Maigne, A., Yudasaka, M. et al. (2008). Revealing the secret of water-assisted carbon nanotube synthesis by microscopic observation of the interaction of water on the catalysts. *Nano Lett.* 8: 4288–4292.
59 Finnie, P., Bardwell, J., Tsandev, I. et al. (2004). Cold wall chemical vapour deposition of single walled carbon nanotubes. *J. Vac. Sci. Technol., A* 22: 747–751.
60 Finnie, P., Than, A.L., Lefebvre, J., and Austing, D.G. (2006). Optimization of methane cold wall chemical vapor deposition for the production of single walled carbon nanotubes and devices. *Carbon* 44: 3199–3206.
61 Chiashi, S., Murakami, Y., Miyauchi, Y., and Maruyama, S. (2004). Cold wall CVD generation of single-walled carbon nanotubes and in situ Raman scattering measurements of the growth stage. *Chem. Phys. Lett.* 386: 89–94.
62 Maruyama, T., Kondo, H., Ghosh, R. et al. (2016). Single-walled carbon nanotube synthesis using Pt catalysts under low ethanol pressure via cold-wall chemical vapour deposition in high vacuum. *Carbon* 96: 6–13.
63 Kratschmer, W., Lamb, L.O., Fostiropoulos, K., and Huffman, D.R. (1990). Solid C_{60}: a new form of carbon. *Nature* 347: 354–358.
64 Ajayan, P.M. and Iijima, S. (1992). Smallest carbon nanotube. *Nature* 358: 23.
65 Bethund, D.S., Kiang, C.H., de Vries, M.S. et al. (1993). Cobalt-catalysed growth of carbon nanotubes with single-atomic-layer walls. *Nature* 363: 605–607.
66 Kingston, C.T. and Simard, B. (2003). Fabrication of carbon nanotubes. *Anal. Lett.* 36 (15): 3119–3145.
67 Ebbesen, T.W. (1994). Carbon nanotubes. *Annu. Rev. Mater. Sci.* 24: 235–264.
68 Ajayan, P.M., Lambert, J.M., Bernier, P. et al. (1993). Growth morphologies during cobalt-catalyzed single-shell carbon nanotube synthesis. *Chem. Phys. Lett.* 215: 509–517.

69 Kiang, C.H., Goddard, W.A. III, Beyers, R. et al. (1994). Catalytic synthesis of single-layer carbon nanotubes with a wide range of diameters. *J. Phys. Chem.* 98: 6612–6618.

70 Seraphin, S. and Zhou, D. (1994). Single-walled carbon nanotubes produced at high yield by mixed catalysts. *Appl. Phys. Lett.* 64: 2087–2089.

71 Lambert, J.M., Ahayan, P.M., Bernier, P. et al. (1994). Improving conditions towards isolating single-shell carbon nanotubes. *Chem. Phys. Lett.* 226: 364–371.

72 Kiang, C.H., Goddard, W.A. III, Beyers, R., and Bethune, D.S. (1995). Carbon nanotubes with single-layer walls. *Carbon* 33: 903–914.

73 Lambert, J.M., Ajayan, P.M., and Bernier, P. (1995). Synthesis of single and multi-shell carbon nanotubes. *Synth. Met.* 70: 1475–1476.

74 Haufler, R.E., Chai, Y., Chibante, L.P.F. et al. (1991). Carbon arc generation of C_{60}. *Mater. Res. Soc. Symp. Proc.* 206: 627–637.

75 Guo, T., Nikolaev, P., Rinzler, A.G. et al. (1995). Self-assembly of tubular fullerenes. *J. Phys. Chem.* 99: 10694–10697.

76 Thess, A., Lee, R., Nikolaev, P. et al. (1996). Crystalline ropes of metallic carbon nanotubes. *Science* 273: 483–487.

77 Rinzler, A.G., Liu, J., Dai, H. et al. (1998). Large-scale purification of single-wall carbon nanotubes: process, product, and characterization. *Appl. Phys. A* 67: 29–37.

78 Laplaze, D., Bernier, P., Maser, W.K. et al. (1998). Carbon nanotubes: the solar approach. *Carbon* 36: 685–688.

79 Alvarez, L., Guillard, T., Sauvajol, J.L. et al. (2001). Growth mechanisms and diameter evolution of single wall carbon nanotubes. *Chem. Phys. Lett.* 342: 7–14.

80 Diener, M.D., Nichelson, N., and Alford, J.M. (2000). Synthesis of single walled carbon nanotubes in flames. *J. Phys. Chem. B* 104: 9615–9620.

81 Vander Wal, R.L. (2002). Fe-catalyzed single-walled carbon nanotube synthesis within a flame environment. *Combust. Flame* 130: 37–47.

82 Smiljanic, O., Dellero, T., Serventi, A. et al. (2001). Growth of carbon nanotubes on Ohmically heated carbon paper. *Ohm Chem. Phys. Lett.* 342 (5): 503–509.

83 Hsu, W.K., Hare, J.P., Terrones, M. et al. (1995). Condensed-phase nanotubes. *Nature* 377: 687.

84 De Volder, M.F.L., Tawfick, S.H., Baughman, R.H., and Hart, A.J. (2013). Carbon nanotubes: present and future commercial applications. *Science* 339 (6119): 535–539.

85 Inagaki, M. and Yang, Y. (2012). Carbon nanofibers prepared via electrospinning. *Adv. Mater.* 24 (19): 2547–2566.

86 Gomulya, W. and Loi, M.A. (2013). Conjugated polymer wrapped carbon nanotubes: physical properties and device applications. *Eur. Phys. J. B* 1–13.

87 Wang, X. and Fan, S. (2009). Fabrication of ultralong and electrically uniform single-walled carbon nanotubes on clean substrates. *Nano Lett.* 9 (9): 3137–3141.

88 Croitoru, C. (2020). 1000 at 1000: the lightest bakelite and beyond. *J. Mater. Sci.* 55: 15637–15642.

89 Zakaria, M.R., Akil, H.M., Kudus, M.H.A. et al. (2019). Hybrid carbon fiber-carbon nanotubes reinforced polymer composites: a review. *Composites Part B* 176: 107313.

90 Zhu, L., Li, N., and Childs, P.R.N. (2018). Light-weighting in aerospace component and system design. *Propul. Power Res.* 7 (2): 103–119.

91 Sharma, S.P. and Lakkad, S.C. (2011). Effect of CNTs growth on carbon fibers on the tensile strength of CNTs grown carbon fiber-reinforced polymer matrix composites. *Composites Part A* 42 (1): 8–15.

92 Ramachandran, K., Boopalan, V., Bear, J.C., and Subramani, R. (2022). Multi-walled carbon nanotubes (MWCNTs)-reinforced ceramic nanocomposites for aerospace applications: a review. *J. Mater. Sci.* 57: 3923–3953.

4

Economical Uses of Chemically Modified Carbon Nanotubes

Sheerin Masroor[1], Lalan Kumar[1], and Irshad Ahmad[2]

[1] *Patliputra University, A. N. College, Department of Chemistry, Boring Road, Patna, Bihar, 800013, India*
[2] *Jamia Millia Islamia University, Department of Applied Sciences and Humanities, Faculty of Engineering and Technology, New Delhi, 110025, India*

4.1 Introduction

Nanotechnology is the emerging and most advanced synthesis/manufacturing technique and has been widely explored worldwide. Nanotechnology deals with a myriad of materials produced in nanoscale through various physical and chemical methods [1]. Making structural materials within the nanoscale lays the foundation of nanotechnology. These materials are in dimensions below 100 nm. These vast groups of nanomaterials can achieve multiple functions in various fields such as mechanical, electronics, magnetic, or optical, one main group of which is nanotubes. Some nanotubes are functionalized based on boron and molybdenum; however, carbon nanotubes (CNTs) are the most acceptable and diverse group to study. CNTs have one to many concentric graphite layers, with a diameter in the range of 0.4–10 nm [2]. The great field of CNTs was earlier discovered in 1991 by Iijima via some experiments using transmission electron microscopy (TEM) and by different reports of conditions to synthesize large quantities of nanotubes [3]. These CNTs made of graphite sheets are rolled up into cylindrical shapes. Their length may vary from a few micrometers to a hundredth of a nanometer in diameter [4]. These CNTs are mixed derivatives of fullerene with carbon fibers compacted around 60 carbon atoms placed in specific muffled tubes [5]. There are two types of CNTs based on the number of carbon layers present: single-walled carbon nanotubes (SWCNTs) consisting of single-layered graphene having a diameter in the range of 0.4–2 nm and mainly occurring as hexagonal-packed bundles, and multi-walled carbon nanotubes (MWCNTs) made of two to several cylinders, each comprising graphene sheets in the diameter of 1–3 nm [6]. CNTs are synthesized via three different methods: arc discharge, laser ablation, and chemical vapor deposition. In the arc discharge method, at a high temperature (>3000 °C), carbon atoms are converted into a plasma, and SWCNTs and MWCNTs are prepared. A catalyst is not required for the synthesis of MWCNTs, whereas the synthesis of SWCNTs requires specific catalysts such as nickel, iron, cobalt, and yttrium. The chemical vapor deposition method

Chemically Modified Carbon Nanotubes for Commercial Applications, First Edition.
Edited by Jeenat Aslam, Chaudhery Mustansar Hussain, and Ruby Aslam.

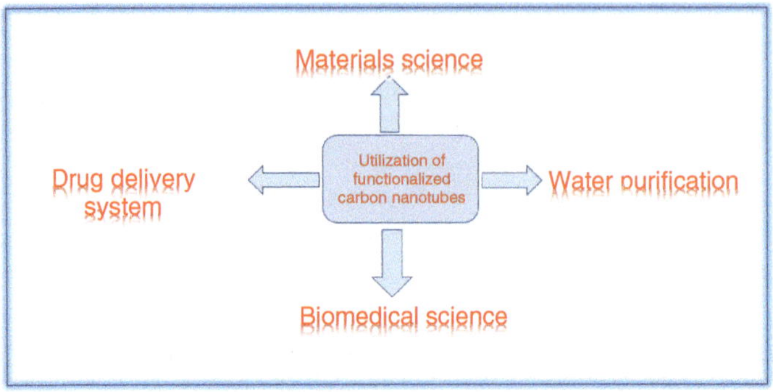

Figure 4.1 Best utilization of functionalized carbon nanotubes in different fields.

mainly involves hydrocarbon sources such as methane, ethylene. Laser ablation requires vaporization of graphite in an electrical furnace heated up to 1200 °C. The purity level of the products formed mainly depends upon the purity of graphite used and a high converting ratio. Furthermore, the macroscopic processing is also needed to obtain desired characteristic features of nanotubes, such as alignment, length. MWCNTs were discovered by Iijima using the arc discharge method. This technique is the most effective and basic method for the manufacturing of carbon fibers [7, 8]. In 2016, Khan and coworkers reported *in situ* emulsion polymerization for the synthesis of CNT composites in a polystyrene-based colloidal system to make nanostructured brushes. First, CNTs were functionalized with oleic acid, followed by salinization with (3-aminopropyl)triethoxysilane to embed the cross-linking properties [9]. The nanotubes made from carbon exhibit excellent physical and chemical properties such as extreme lightweight, specific electronic structures, high tensile strength, and high thermal and chemical stability, due to which researchers and scientists are largely attracted to work in this direction. CNTs are the most harvested nanomaterials in various applications. Their best applications include drug and drug delivery, biomolecules, and biosensors [10]. The best utilization of CNTs is illustrated in Figure 4.1.

4.2 Properties of Carbon Nanotubes

Of many properties associated with CNTs, the following are the most important that mainly deal with their applications at different levels uniquely, which discussed here briefly:

1. *Strength of nanotubes*: To evaluate this property of CNTs and consider them as one of the strongest materials, tensile strength and elastic modulus are considered. The strength comes from the covalent bonds with sp^2 hybridization, which may form in between the individual carbon atoms. The tensile strength of MWCNTs is comparable with that of high-carbon steel, which is significantly higher (63 GPa) than that of steel (1.2 GPa). CNTs also have a very high elastic modulus of around

1 TPa (Tera Pascal). Under the applied strain, CNTs undergo plastic deformation, and at a maximum strain, they tend to fracture by the release of strain energy.

2. *Electrical property*: The presence of symmetry and the unique electronic structure of graphene have a strong effect on its electrical properties. Theoretically, metallic CNTs have electrical current density thousands (1000) times greater than that of metals such as silver and copper.
3. *Kinetic property*: MWCNTs have various multiple concentric nanotubes accurately positioned within one another, exhibiting a telescoping property wherein the innermost nanotube core may slide with no friction, enveloping its outer nanotube shell, thus creating an automatically perfect linear or rational bearing. This property has already been used to create the world's nominal motor and a nano rheostat.
4. *Thermal property*: CNTs act as good thermal conductors along the tube axis. This property is called "ballistic conducting." When arranged laterally, they behave as good insulators.

4.3 Synthesis of Carbon Nanotubes

CNTs are synthesized through multiple synthesis routes, as per technology requirements and applications. The quality of synthesized CNTs means the absence of chemical or structural defects. The scale length is around 1–10 μm, along the tube axes. A huge number of papers and new innovations in the form of patents for the synthesis of CNTs have been reported at an increasing rate. There are also many challenges in the synthesis of CNTs that need to be addressed in the near future. Among them, the following are considered critical: (i) abundant production – the development of low-cost synthetic routes for the production of very high-quality CNTs including SWCNTs; (ii) inequitable production – synthetic control over the properties of nanotubes formed, (iii) establishment – control over the orientation of synthesized CNTs; (iv) procedure – the synthetic route for the growth of nanotubes.

4.4 Functionalization of Carbon Nanotubes

CNTs were reformed via chemical treatment as suggested by Montanheiro et al. [11]. The modification was adapted to generate carboxylic groups superficially through nitric acid oxidation. First, the CNTs were placed in flask-shaped glassware with an amount of about 6 mol l^{-1} of nitric acid solution. The mixture was heated and left for five hours under reflux. The suspension of CNTs was then centrifuged at 4500 rpm for about 10 minutes to eliminate the supernatant material and was washed with double distilled water multiple times until neutral pH was reached. The product formed, and it was kept in a freeze-dryer for about 48 hours; thus functionalized CNTs were obtained. To evaluate the functionalization capacity, CNTs were dispersed in water in an ultrasonic bath. The suspension was analyzed, and photographs were taken at 24 hours, 1 week, and 3 weeks to determine their stability. The common CNTs along with their functionalized counterparts were studied by Raman spectroscopy at a wavelength of 532 nm. Nanofillers could be distinguished by Fourier-transform infrared spectroscopy (FTIR). These techniques mainly confirm the functional groups on the surface of CNTs. In FTIR, the specimens were prepared on the crystals of potassium bromide (KBr). The amount of oxygen molecules in CNTs and their functionalized counterparts can be studied by X-ray photoelectron spectroscopy (XPS) using the K-alpha device having an aluminum (Al) K-alpha emission line in the array of 0–1200 eV binding energy, with an energy step size of 0.1 eV [12].

4.5 Characterization/Analysis of Functionalized Carbon Nanotubes

The CNTs synthesized through multiple routes have been characterized by the following techniques:

a. *Electrical conductivity* (σ): It is an important technique to characterize the samples formed and easily taken by electrical resistivity AC (alternating current) and impedance spectroscopy [13]. The numerical values of electrical conductivity (σ) are generally computed by taking the inverse of electrical resistivity (ρ) by Eq. (4.1). The value of ρ was obtained from Eq. (4.2):

$$\sigma = 1/\rho \tag{4.1}$$

$$\rho = (Z \times A)/T \tag{4.2}$$

   where
   Z is the real impedance;
   A is the electrical contact area; and
   T is the thickness of the samples.
   The electrical contact can be made by forming a delicate layer of gold–palladium alloy, which was conveyed using a sputter coating on both sides of the sample. All experiments were mainly performed at standard temperature and pressure at a frequency of about 10 Hz with a voltage amplitude of 0.5 V.

b. *Differential scanning calorimetry (DSC)*: The thermic properties of the functionalized carbon nanotubes samples were evaluated by DSC instrumentation in a nitrogen (N_2) atmosphere [14]. The samples were sealed in aluminum dish and warmed up to 250 °C, which were kept for three minutes at the prescribed temperature, and then suddenly cooled to 0 °C. Again, the specimens were reheated to 250 °C. The degree of crystallinity (X_c) was then determined for the second heating scan using Eq. (4.3):

$$X_c = (\Delta H_m - \Delta H_{cc})/(\Delta H_{m^\circ} \times W) \times 100$$

   where
   ΔH_m is the melting enthalpy;
   ΔH_{cc} is the cold crystallization enthalpy;

ΔH_{m° is the heat of fusion; and
W is the weight fraction.

c. *Thermogravimetric analysis (TGA)*: The sample's degradation was analyzed by TGA. Here in this experiment, samples were heated from room temperature to around 800 °C at a heating rate of 20 °C min^{-1}, under a nitrogen atmosphere.
d. *Tensile tests*: The tensile tests were carried out on thin films of functionalized CNTs in accordance with ASTM D882-12 [15]. The test was performed at a crosshead strain rate of 10 mm min^{-1} with a load cell of 5 kN. Average values were taken for five specimens. Then Young's modulus was found to be in the 0.1–0.5% range of deformation, within the elastic regime.
e. *Scanning electron microscopy with a field emission guns (SEM-FEG)*: The fracture on the surface of a tensile test film for the CNTs and its counterpart was made via gold coating using an automated sputter coater and analyzed by SEM-FEG using a TESCAM microscope operating at 5 kV.

4.6 Economy of Carbon Nanotubes

There has been worldwide concern about declining natural resources for many decades. It has been emphasized that use of fossil fuels should be reduced and replaced with sustainable sources. There is some progress in use of modified CNTs in photovoltaics-, hydrogen-, and thermoelectric energy-based economies, which further helps in the development of a nation. The greatest move is in the direction of harvesting hydrogen economy, with advances in hydrogen fuel cells that are economically feasible. Automobile industry has promoted the use of different measures to develop vehicles with hydrogen fuel cells. By taking such measures, any country can improve its economy. The key to the promotion of this type of technology is the need to improve its efficiency continuously while keeping all safety measures intact. The modified CNTs have been used to develop a variety of key areas for the hydrogen economy [16].

4.7 Economic Importance of Carbon Nanotubes

Many countries are looking into multiple alternative energy resources to address issues of environmental liabilities and energy autonomy. Some of these energy sources include hydrogen, wind energy, solar energy, and natural gas. At present, scientists are more inclined to explore hydrogen as a preferred energy source as well as how effective CNTs are in finding solutions related to safety, function, and structure in multiple industries.

4.8 Hydrogen Fuel Cells

Fuel cell is the simplest cell comprising an anode and a cathode, with a permeable membrane in between. Various types of fuel cells with different membranes are

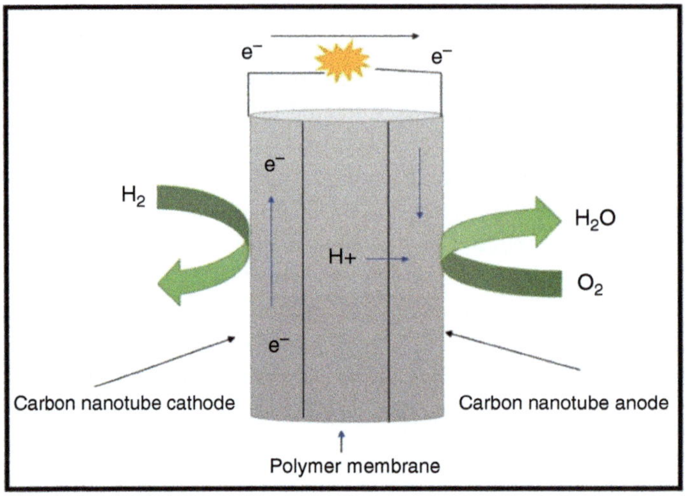

Figure 4.2 Block diagram of fuel cell based on carbon nanotubes.

methanol, polymers, phosphoric acid, molten carbonate, alkaline, and solid oxide. Hydrogen fuel cells are an example of the polymer electrolyte membrane fuel cell (PEMFC), also called proton-exchange membrane fuel cell, where the permeable membrane mainly contains a proton-conductive polymer such as perfluoro sulfonic acid, commercially known as Nafion.

The working of a fuel cell depends on the catalyst, which oxidizes hydrogen at the anode, changing it into a positively charged proton and a negatively charged electron. The electrons travel through a wire, making an electrical connection, which powers a device. The protons move through the permeable membrane to the cathode. At the cathode, the protons again combine with the free electrons and react with oxygen and form water molecules. This formed water has to be drained from the cell system. The block diagram of fuel cells based on CNTs is shown in Figure 4.2.

Despite extensive research, there are still many challenges that need to be overcome to improve the performance of fuel cells. As the recent market and technology is having high demand, so the production of fuel must be inexpensive. The construction of fuel cells is highly expensive as they are mainly made from the platinum (Pt) catalyst.

The other parameter of fuel cells that needs to be improved is operational lifetime. The synthetic polymer membrane in the cell is susceptible to degradation by different chemicals. Research is focused on to enhance the efficiency of the catalytic process and the lifetime of various components. CNTs have been used to replace the carbon powder in the cell. The excellent properties of CNTs, such as low mass density, conductive properties, and robust physical properties, make them the best material to use as electrodes in fuel cells. In addition, the assemblies of nanotubes allow them to make macrostructures, which give a high surface area and allow them to act as a better substrate for Pt catalysts and hydrogen adsorption [17]. In 2003, researchers from the University of California, Riverside, investigated the use of MWCNTs as a carbon support for platinum catalysts to maximize platinum assimilation between all the fuel cell contents.

4.9 Water Splitting

Research is inclined toward water splitting, which has provided a route for the emergence of the hydrogen economy based on clean manufacturing of hydrogen and oxygen. Also, this fact has a much more important impact if reacted with hydrogen fuel cells. Here, the waste product is water, which is formed by the reaction with oxygen. Hence, the water could fuel the process of splitting, which in turn could fuel hydrogen cells.

CNTs are excellent candidates to improve the water splitting performance in photocatalysis of titania [18]. A sulfonated tetrafluoroethylene garrisoned fluoropolymer-copolymer, named Nafion, has been used successfully in fuel cell manufacturing industries. Various scientists and researchers are engaged in continuous experiments to supplement the features of CNT-based films. The concentration of CNTs affects the functioning of membranes, such as low concentrations of CNTs have an effect on solvent permeation and mechanical stability. On the other hand, higher concentrations of CNTs enable the membrane to easily separate proton and electron conduction pathways in the membrane. Based on this, various applications have been in active use for various membranes made from different types of CNTs [19].

A membrane can be tested for its best functioning by an artificial leaf system, as demonstrated in Figure 4.3. The synthesized membrane separated the anode, which was further exposed to sunlight where water sprinklers were present, and the cathode. The anodic layer of cell contained a chromophore with an oxygen-evolving complex, whereas the cathode had a proton-reducing catalyst [19].

4.10 Dye-Sensitized Solar Cells

For economic development via chemistry, much research is focused on dye-sensitized solar cells (DSSCs). The fact behind this kind of technology is the generation of renewable energy and its use in multiple approaches to form a perfect silicon-dependent solar cell. The DSSCs are assemblies of three

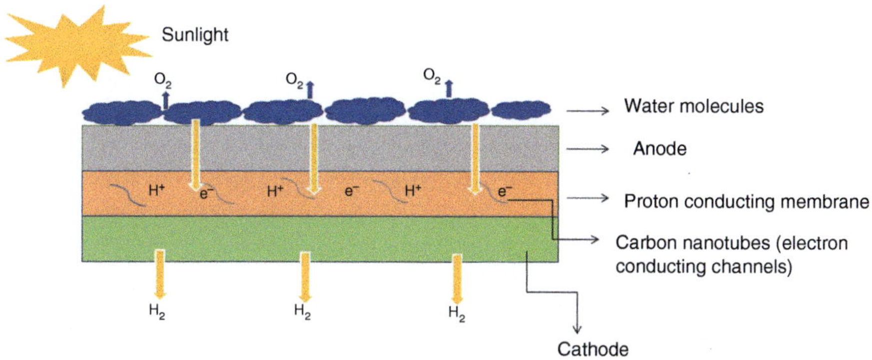

Figure 4.3 Pictorial presentation of solar water splitting.

components – an anode, a cathode, and an electrolyte. The anode used is generally nanocrystalline titania (TiO_2) and a dye attached to the particles. The cathode, also known as the counter electrode (CE), serves as a place where the catalysis occurs and specifically contains platinum (Pt) metal. The electrolyte used is iodide solution, which promotes the iodide/triiodide redox couple reaction, where after the excitation of the dye and loss of an electron, it regains one from iodide, oxidizing it to triiodide. The highest efficiency of DSSCs was reported to be 11.40%, according to the National Institute for Material Science (NIMS).

In recent years, carbon nanotubes have been used as a stand-in for the platinum-based counter electrode (CE). In 2012, Jo et al. studied attached ordered mesoporous carbon–carbon nanocomposites and demonstrated that these nanocomposites exhibit platinum-like counter electrode behavior in any DSSC [20]. The most important application of CNTs fibers is as a material that conducts electricity to support dye-impregnated TiO_2 particles [21]. This application was best used as counter and working electrodes.

4.11 Quantum Dot Solar Cells

Cadmium telluride (CdTe) is a promising and low-cost photovoltaic material. Development of quantum dot (QD) based technologies will increase the lethal effect of cadmium and cadmium-based compounds.

Considerable progress has been made in the past few years to obtain highest efficiency of 17.3% for cadmium telluride (CdTe) photovoltaic instruments. Research is currently focused on CdTe/graphene nanocomposites [22].

4.12 Silicon-Based Solar Cells

Crystalline silicon-based cells are the best choice, with a percentage efficiency of about 20.4% for any multi-crystalline structure and about 27.6% for single-based crystal cells.

There is clearly room for improvement inefficiency, which has reached a plateau in the past few decades. A versatile approach is required for the design and chemistry for photovoltaic devices. Polyaniline (PANI) and CNTs were used to build a heterojunction diode device on n-type silicon [23].

Another example to report is silicon incorporated with CNTs for use in hybrid photovoltaic devices to investigate the minimal width of SWCNT films on n-type silicon. This all enhances the percentage efficiencies from 0.4% to 2.4% to affect the number of MWCNTs on the photon to electron alteration [24, 25].

4.13 Thermoelectric Fabrics

Other recent advancements in research have shown that films made from composites of MWCNTs and polyvinylidene fluoride (PVDF) combined in a laminous structure can produce an effect of felt-like fabric [26]. This is the most important idea applied in wearable electronics, which has been predicted for many years in the research field of photovoltaics.

These fabrics are more economical to generate efficient energy-harvesting devices that can power portable devices/electronics.

4.14 Cost of Carbon Nanotubes

The production cost of CNT's is very high and mainly changes with respect to the yield, carbon content in percentage along with its purity levels. Laser ablation methods along with arc discharge largely produce on the scale of gram and hence are comparatively overpriced. At present, the chemical vapor deposition method gives high yield at lowest expense. However, the CNTs formed by this method generally contain more defects [27, 28].

In 2006, the production capacity of MWCNTs was greater than 300 t. In 2007, Bayer produced 60 t per year. In 2009, the overall production of SWCNTs was estimated to be hundreds of kilograms globally, with a production rate of $50\,g\,h^{-1}$. In 2010, Bayer built its biggest CNT facility in the world with an annual capacity of 200 t.

It is expected that for the next 10–20 years, the demand for CNTs will increase steadily, depending on the purity, cost, quality, and yield of production, which can be met soon. The production cost of CNTs has been changing considerably in the past few decades, which depends on their purity and type as well as the suppliers. The MWCNTs production cost lies between US$ 0.5 and 100 per g, while that of double walled carbon nanotubes (DWCNTs) is around US$ 10 per g. The cost of SWCNTs also varies considerably, between US$ 20 and 2000 per g [29].

4.15 Globalization of Carbon Nanotubes

The key manufacturing companies that are best profiled to synthesize CNTs are Nanocyl SA, Arkema SA, Hyperion Catalysis International, Carbon Solutions, Inc., Nanoshell LLC, SHOWA DENKO K.K., and Klean Commodities [30].

One report stated that CNTs market size will increase to reach US$ 13.65 billion by the end of 2026 globally. This improvement is witnessed by the development of sustainable solar cells, which opens a big door for the growth of economy in the coming years [31]. As per the Fortune Business Insights report, entitled "Carbon Nanotubes (CNT) Market Size, Share & Industry Analysis, By Product (Multiwalled Carbon Nanotubes (MWCNT) and Single-Walled Carbon Nanotubes (SWCNT)), By Application (Plastics & Composites, Electrical & Electronics, Energy, and Others), and Regional Forecast, 2019–2026," the global market was approximately US$ 4.56 billion in 2018 and will increase at a compound annual growth rate (CAGR) of about 15.3% during the forecast period of seven years from 2019 to 2026.

4.16 Conclusion

CNTs have been rarely used as novel composite materials. Their exploitation needs to be changed from the nano to macro level. Two decades after their discovery, their use and applications have reached into and around almost all aspects of research in a scientific manner, from basic science to medicine to engineering fields. In view of sustainable climatic change, the development of CNTs has opened all doors toward renewable energy. Major advancements happen in hydrogen fuel cells integrated with CNTs made of composite electrodes/membranes, which help decrease or eliminate the need of expensive catalysts. Assurance is another issue for the development of hydrogen-dependent CNT economy. The technology applied in the catalysis process for different fuel cells, which works in connection with CNTs, has provided remarkable sensitivity. The inbuilt solar cells with CNTs show overall higher efficiencies. Integration of CNTs with thermoelectric devices currently does not have the same level of development as other alternative energy resources, possibly because the field is largely driven toward cost saving on an industrial scale and the development of component systems for vehicles rather than consumer gadgets or devices, but research on waste heat recovery is substantial.

In summary, we are beginning to see a transition toward alternative fuel sources, with a focus on hybrid technologies present in the automotive industries, and we need to address the impact of our current economy as we transition to a more sustainable one using CNTs.

References

1 Kumar, S., Rani, R., Dilbaghi, N. et al. (2017). Carbon nanotubes: a novel material for multifaceted applications in human healthcare. *Chem. Soc. Rev.* 46: 158–196.

2 Mamalis, A.G., Vogtlander, L.O., and Markopoulos, A. (2004). Nanotechnology and nanostructured materials: trends in carbon nanotubes. *Precis. Eng.* 28 (1): 16–30.

3 Iijima, S. (1991). Helical microtubules of graphitic carbon. *Nature* 354 (6348): 56.

4 Tasis, D., Tagmatachis, N., Bianco, A., and Prato, M. (2006). Chemistry of carbon nanotubes. *Chem. Rev.* 31: 1105–1136.

5 Polizu, S., Savadogo, O., Poulin, P., and Yahia, L.H. (2006). Applications of carbon nanotubes-based biomaterials in biomedical nanotechnology. *J. Nanosci. Nanotechnol.* 6 (7): 1883–1904.

6 He, H., Pham-Huy, L.A., Dramou, P. et al. (2013). Carbon nanotubes: applications in pharmacy and medicine. *BioMed Res. Int.* 2013: 1–12.

7 Baughman, R.H., Zakhidov, A.A., and De Heer, W.A. (2002). Carbon nanotubes – the route toward applications. *Science* 297 (5582): 787–792.

8 Popov, V.N. (2004). Carbon nanotubes: properties and application. *Mater. Sci. Eng., R* 43 (3): 61–102.

9 Khan, M.U., Reddy, K.R., Snguanwongchai, T. et al. (2016). Polymer brush synthesis on surface modified carbon nanotubes via in situ emulsion polymerization. *Colloid Polym. Sci.* 294: 1599–1610.

10 Che, J., Cagin, T., and Goddard, W.A. III (2000). Thermal conductivity of carbon nanotubes. *Nanotechnology* 11 (2): 65.

11 Montanheiro, T.L.D.A., Cristóvan, F.H., Machado, J.P.B. et al. (2015). Effect of MWCNT functionalization on thermal and electrical properties of PHBV/MWCNT nanocomposites. *J. Mater. Res.* 30: 55–65.

12 Braga, N.F., Zaggo, H.M., Montagna, L.S., and Pescador, F.R. (2020). Effect of carbon nanotubes (CNT) functionalization and maleic anhydride-grafted poly(trimethylene terephthalate) (PTT-g-MA) on the preparation of antistatic packages of PTT/CNT nanocomposites. *J. Compos. Sci.* 4: 44. https://doi.org/10.3390/jcs4020044.

13 da Silva, T.F., Menezes, F., Montagna, L.S. et al. (2019). Preparation and characterization of antistatic packaging for electronic components based on poly(lactic acid)/carbon black composites. *J. Appl. Polym. Sci.* 136: 47273.

14 Szymczyk, A., Paszkiewicz, S., and Roslaniec, Z. (2013). Influence of intercalated organoclay on the phase structure and physical properties of PTT–PTMO block copolymers. *Polym. Bull.* 70: 1575–1590.

15 ASTM D882-12. (2012). *Standard Test Method for Tensile Properties of Thin Plastic Sheeting (American Society for Testing and Materials) D882-12.* West Conshohocken, PA: ASTM International.

16 Dillon, A.C., Jones, K.M., Bekkedahl, T.A. et al. (1997). Storage of hydrogen in single-walled carbon nanotubes. *Nature* 386 (6623): 377–379.

17 Oh, S.H., Kim, K., and Kim, H. (2011). Polypyrrole-modified hydrophobic carbon nanotubes as promising electrocatalyst supports in polymer electrolyte membrane fuel cells. *Int. J. Hydrogen Energy* 36 (18): 11564–11571.

18 Li, N., Ma, Y.F., Wang, B. et al. (2011). Synthesis of semiconducting SWNTs by arc discharge and their enhancement of water splitting performance with TiO_2 photocatalyst. *Carbon* 49 (15): 5132–5514.

19 Ljeri, V., Cappelletto, L., Bianco, S. et al. (2010). Nafion and carbon nanotube nanocomposites for mixed proton and electron conduction. *J. Membr. Sci.* 363 (1, 2): 265–270.

20 Jo, Y., Cheon, J.Y., Yu, J. et al. (2012). Highly interconnected ordered mesoporous carbon-carbon nanotube nanocomposites: Pt-free, highly efficient, and durable counter electrodes for dye-sensitized solar cells. *Chem. Commun.* 363 (1–2): 265–270.

21 Chen, T., Qiu, L., Cai, Z. et al. (2012). Intertwined aligned carbon nanotube fiber based dye-sensitized solar cells. *Nano Lett.* 12 (5): 2568–2572.

22 Bi, H., Huang, F.Q., Liang, J. et al. (2011). Large-scale preparation of highly conductive three dimensional graphene and its applications in CdTe solar cells. *J. Mater. Chem.* 21 (43): 17366–17370.

23 Bourdo, S.E., Saini, V., Piron, J. et al. (2012). Photovoltaic device performance of single-walled carbon nanotube and polyaniline films on n-Si: device structure analysis. *ACS Appl. Mater. Interfaces* 4 (1): 363–368.

24 Kozawa, D., Hiraoka, K., Miyauchi, Y. et al. (2012). Analysis of the photovoltaic properties of single-walled carbon nanotube/silicon heterojunction solar cells. *Appl. Phys Express* 5 (4): 363–368.

25 Castrucci, P., Del Gobbo, S., Camilli, L. et al. (2011). Photovoltaic response of carbon nanotube-silicon heterojunctions: effect of nanotube film thickness and number of walls. *J. Nanosci. Nanotechnol.* 11 (10): 9202–9207.

26 Hewitt, C.A., Kaiser, A.B., Roth, S. et al. (2012). Multilayered carbon nanotube/polymer composite based thermoelectric fabrics. *Nano Lett.* 12 (3): 1307–1310.

27 Bierdel, M., Buchholz, S., Michele, V. et al. (2007). Multilayered carbon nanotube/polymer composite based thermoelectric fabrics. *Phys. Stat. Sol.* 244: 3939–3943.

28 Bierdel, M., et al. (2007). "Industrial production of multiwalled carbon nanotubes." *Phys. Stat. Sol. (b)* 244.11: 3939–3943.

29 https://sites.google.com/site/cntcomposites/cost-and-production.

30 Source: Fortune Business Insights (08 September 2020) 12:54 ET.

31 GLOBE NEWSWIRE, Pune (08 September 2020).

Part II

Chemically Modified Carbon Nanotubes: Energy and Environment Applications

5

Chemically Modified Carbon Nanotubes in Energy Production and Storage

Manal G. Mohamed[1], Ahmed A. Mousa[2], and Azza M. Mazrouaa[1]

[1] *Egyptian Petroleum Research Institute, Petrochemical Department, Polymer Lab, Nasr City, Cairo 11727, Egypt*
[2] *Synthomer SAE, 10th of Ramadan city, 4955235, Egypt*

Abbreviations

AA	Azure A chloride
CB	polyaromatic Celestine blue
CNT	carbon nanotube
CoO	cobalt oxide
Co_3O_4	cobalt oxides
CP	*m*-cresol purple
CVD	chemical vapor deposition
D	dimension
DD	organic liquid-*n*-dodecane
DWCNTs	double-walled carbon nanotubes
EDLCs	electrical double layers capacitors
FSCs	fiber based super capacitors
FWCNTs	few-walled carbon nanotubes
GE	germanium
GO	graphene oxide
GPa	giga pascal
Li	lithium
LiBs	lithium ions batteries
MnO	manganese(II) oxide
MWCNTs	multiwalled-carbon nanotubes
NiO_2	nickel oxide
NIR	near infrared
OHA	octanohydroxamic acid
PANI	polyaniline
PC	polycarbonate
PCMs	phase change materials
PEDOT	poly(3,4-ethylenedioxythiophene)
PSD	pore size distribution

Chemically Modified Carbon Nanotubes for Commercial Applications, First Edition.
Edited by Jeenat Aslam, Chaudhery Mustansar Hussain, and Ruby Aslam.

PV	pyrocatechol
PPy	polypyrrole
PVDF	polyvinylidene difluoride
PUCNT	partially unzipped carbon nanotube
RGO	reduced graphene oxide
S	sulfur
SCs	super capacitors
SEM	scanning electron microscope
Si	silicon
Sn	tin
SSA	specific surface area
SWCNTs	single-walled carbon nanotubes
$TEABF_4$	tetra ethyl ammonium tetra fluoroborate
UV	ultraviolet
VA	vertically aligned
VN	vanadium nitride
SMWCNTs	spun multiwalled carbon nanotubes

5.1 Introduction

The performance of energy storage devices has increased substantially since the widespread use of nanomaterials and nanotechnologies used in them 20 years ago. The most common 1D nanomaterial, carbon nanotubes (CNTs), can act as both a conductive component and an electroactive substance, making them an excellent choice for energy storage systems. Due to their small weight, good electrical conductivity, great physical toughness, and superior electrochemical performance, carbon is employed in all its different compounds: black, amorphous, or solid carbon to make CNTs utilized in conventional lithium (Li)-ion batteries and supercapacitors (SCs). Due to considerable developments in nanomaterial handling technologies, CNTs can now be fine manufactured, culminating in the development of a great number of CNT-based revolutionary energy storage devices [1]. If they are correctly developed, CNTs could open the way for the production of a host of innovative materials for a range of uses in the future. When introduced to other materials, CNTs have an open structure and enriched chirality, allowing them to improve the characteristics and quality of such materials. Depending on structural and morphological factors, CNTs have been employed as an extra in energy storage devices to advance the cathode electrical conductivity constituents or as an active anode element. They have also been used directly in supercapacitors and fuel cells as the electrode material. As a result of their underlying qualities and potential uses in energy storage research, CNTs are given a lot of attention. CNTs come in a range of shapes and sizes, and they have been used in energy storage systems, including batteries, supercapacitors, and fuel cells with great success [2]. Single-walled carbon nanotubes (SWCNTs) and multiwalled carbon nanotubes (MWCNTs) have tensile strength of 60 GPa and electrical conductivity of 108 and 107 S cm^{-1}, respectively [3]. CNTs have been used in alkali metal ion batteries, fuel cells, nano-electronic equipment [4], supercapacitors [5], and hydrogen storage systems [6], in addition to field emission materials, practical uses in chemistry, biosensors, catalysts, and electronic devices. The development of improved, cost-effective, and dependable technology for electrical energy storage is the most crucial component of shifting our society from fossil fuels to clean, sustainable, and renewable sources. Supercapacitors (SCs) are a promising energy storage technology for future technologies due to their appealing properties such as long operational lifetime, short charging time, high-power density, and safety. On the other hand, SCs' low specific energy has severely limited their standalone utility [7]. Energy research and development in the twenty-first century are focusing on a variety of sources of renewable energy (e.g. wind and solar). As the efficiency of energy production rises and demand for portable gadgets, electric cars, and fixed electrical networks develops, the administrable ability to acquire, store, and transfer generated electricity in response to a practical requirement is becoming a prominent scientific and technological challenge. Due to their favorable performance features in terms of energy density and power density, advanced energy storage for high-energy and high-power systems is increasingly focused on two important types of power sources: rechargeable lithium-ion batteries (LiBs) and electrochemical capacitors (also identified as supercapacitors or ultracapacitors) [8]. Energy storage technologies based on electrochemistry in the future must have reliable robustness when operating in a variety of specific and harsh environments, such as temperatures from 40 to 160 °C [9], pressures that are extremely low and high, and for usage in electronic devices that can be flexible/stretchable when stretched to high levels of deformation (>100%) [10]. Furthermore, it has better power and energy density, and even a longer cyclic efficiency. Carbon-based nanomaterials with structural diversity include activated carbons [11], templated carbons [12], carbide-derived carbons [13], activated carbon fabrics [14], carbon fibers [15], carbon aerogels [16], CNTs [17, 18], carbon nano horns [19], carbon onions [20], and the emerging graphene [21–23], or graphene composite [24–27] (Figure 5.1). They have a vast surface area and good conductivity (up to 2000 m^2 g^{-1}), outstanding anticorrosion qualities, stability at high temperatures, customizable porous structure, composite material compatibility, and low cost. They have been widely employed and thoroughly explored as materials for electrodes. Electro conductive additives [28], appropriate substrate for active materials [29], electrocatalysts [30], and intercalation hosts [31] are all employed in energy storage systems. CNT macrofilms having self-assembling web-like networks have improved mechanical, thermal, flexible structure, and 10 000–100 000 S cm^{-1} electrical conductivity in a more stable way, making them excellent for energy storage device applications [32].

5.2 Production of Carbon Nanotubes

CNTs are manufactured in a variety of ways, resulting in a variety of CNT kinds (SWCNTs, DWCNTs, FWCNTs, and MWCNTs) with different sizes and purity. The generated CNT film has a homogeneous electrical conductivity because of the

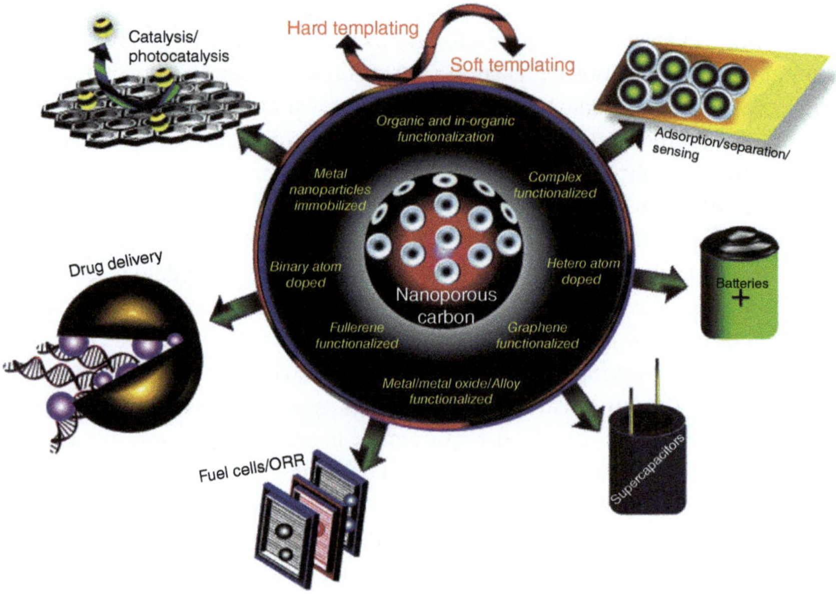

Figure 5.1 A diagram of CNT in energy production and storage [27]. Source: Paul et al. [27]/with permission of Elsevier.

huge length to diameter ratio. In recent years, sheets of aligned CNT created via dry drawing from the CNT array have piqued attention as a way to enhance the CNT film's electrical conductivity [33]. When compared to a CNT film formed by randomly scattering CNTs, the CNT film produced by dry drawing has a structure that is perfectly aligned, which means virtually all CNTs have the same alignment and are bonded end to end, allowing for a fast electron transfer channel running in a CNT-aligned direction. They could also be used to wrap active materials directly like a flexible and conductive scaffold allowing for great electrochemical performance in energy storage systems. Because of its excellent flexibility and conductivity, another essential 1D material for creating fiber energy storage devices is CNT fiber. Furthermore, active materials may be simply put on the sheet of CNT before twisting to generate electrodes of hybrid fiber, making the construction process of devices for fiber energy storage, particularly batteries, much easier [34]. CNTs are functionalized to change their electrical characteristics, introduce certain functional groups, and boost their hydrophilicity or tube adherence to a matrix. Functional groups improve molecule solubility, chemical reactivity, as well as other physical properties [35]. Nitrogen, fluorine or boron, phosphorus, sulfur are examples of dopants (heteroatoms) with higher or lower electronegativity than carbon [36, 37]. The higher electronegativity of the atoms for N and F draws electrons from carbon, causing electrons to delocalize, increasing the electrical conductivity of CNTs over pure CNTs. Energy conversion, storage, and metal-free catalysts have all been made possible by heteroatom-doped CNTs. Few-walled carbon nanotubes (FWCNTs) are a new carbon nanostructured material that consists of two to six sidewall layers having lengths of a few tens of micrometers and widths of 2–8 nm. FWCNTs are widely employed and projected to represent the next generation of nanomaterials for CNT-based reinforcement in field emission. Composites because of their exceptional electrical properties and strong mechanical strength. When SWCNTs and double-walled carbon nanotubes (DWCNTs) are grouped together, they tend to form bundles. To generate a crystal-like structure, SWCNTs are hexagonally stacked in a bundle shape [38].

5.3 History of Energy Storage Devices and Materials

The oldest fuel cells are 2200 years old and date back to the ancient Asian culture around 200 BC. They were made from a material such as asphalt and within a tube wrapped in copper foil, which was at the top end of the copper disc. A very thin iron rod was attached to the bottom cap and passed through the central tube of copper, but did not touch its walls. Another old battery is powered by acetic acid or fermented juice, and these acids allow electrons to flow from the copper tube to the iron on contact, generating a low-intensity current. Luigi Galvani (1737–1798) and Alessandro Conte di Volta (1745–1827) were named after these pioneering steps, and their names are still remembered today as "galvanic element" and "volt." Gaston Plante originally employed the lead/acid/lead dioxide (lead-acid battery) combination to store energy for telegraphy in 1859. Unfortunately, these batteries could not be used in industrial production. In the late nineteenth century, processes of electrochemical energy storage began to grow very rapidly. They invented the dynamo and electric light. Emile Alphonse Faure developed a process for coating the lead plates with a paste of lead powder and sulfuric acid. This caused the plate to have a greater storage capacity since the first charge. The next step in battery development was to increase storage capacity and electric power. C. Bush (1881), who coined the electrode surface with a larger lead but with a protective cover. This was the starting point in inventing the plate batteries that everybody knows and uses today. Another invention that has proved to be important refers to a tubular plate or an iron plate. The result of this new design is several cycles of loading and a greater use of active material. At first, the tube was made of rubber, but after the Second World War, it was made of fiberglass or polyester. Once, with the evolution of technology for electronic materials over the last 50 years, supercapacitor technology has come of age. Electronic double-layer capacitors, or ultracapacitors, were first established and patented in 1961 by SOHIO. Scientists at Ford Motor Company discovered sodium beta alumina had an unusually high ionic conductivity in 1967. In the 1970s, there was a rush of interest in producing new rechargeable batteries. Since 1990, almost all lithium batteries have featured a carbon-based anode. Goodenough announced another novel cathode in 1997, the olivine class of compounds with the formula $LiMPO_4$, where M stands for iron, nickel, cobalt, manganese, or a combination of these elements. Because of its possible low-cost, the iron compound sparked a lot of attention. It did not take long for these chemicals to show up in commercial products. Finally, later in 2012, an energy storage hub will open. This is the second battery dream period, therefore expect great technological developments [39].

5.4 Carbon Nanotubes for Energy Storage

Nanotechnology is still important in the development of energy transformation and storage technology [40]. CNTs are perfect for energy storage owing to their superior mechanical and electrical properties. The surface and pore architectures of CNTs have a significant impact on the performance of storage for gas adsorption and ion insertion. CNTs have a pore structure that is inherent to them [41]. In addition to the aggregated holes in CNTs, intrinsic pores in CNTs may play a role in storing performance. Pores are holes between MWCNTs or SWCNTs/DWCNTs bundles that are aligned/randomly packed [42]. Because of their qualities, fuel cells, solar energy storage for hydrogen, batteries of lithium-ion, and electrochemical supercapacitors are all examples of CNT-based technology for energy transformation and storage [1]. Surprisingly, CNTs are widely used as devices for conversion and creation of energy storage like solar cells [43]. Because of its unique structure, CNT has a high electrical conductivity of $105\,S\,cm^{-1}$ and a large specific surface area [44]. Due to their important advantages, CNTs have long been used as conductive additions or as part of a 3D conductive network to boost the electrochemical kinetics of active materials, enhancing electron transport, and ion diffusion for high-rate performance [45]. Furthermore, CNTs' strong mechanical strength and high aspect ratio permit them to create a stable interconnected network even when stretched, making them a desirable material for stretchy energy storage devices. Devices of stretch energy storage are critical for the advancement of electronics able to stretch, and as a result, they have gotten a lot of consideration in a lot of different sectors, including wearables and bioelectronics. Stretchable composite electrodes can be made by uniformly mixing them with elastic polymer components. In addition to strong mechanical strength, outstanding flexibility, and improved process ability, carbon materials typically perform well [46]. According to a comprehensive study provided by Fu et al. [47], CNTs have been used in flexible photovoltaic devices, portable or wearable electronics, power-generating textiles, building-integrated photovoltaic systems, electric cars, unmanned aerial vehicles, and space robots are among the technologies being developed. Mousavi et al. have raised the efficiency of power conversion to 11% through the use of the OH functionalized MWCNTs coated in rhodium as a counter electrode [48]. Because of their high-power output, rapid charge–discharge mechanism, and long cycling life, supercapacitors are the energy storage technique of the twenty-first century. Carbon-based supercapacitors have even better electrochemical cycle stability, a greater capability for charging and discharging, and a higher power density [49]. Chemical vapor phase deposition was employed to make branched CNT/carbon nanofiber nanocomposites, which were used as electrodes in supercapacitors. Hollow-structure by Zhou et al., produced NiCo hydroxide/CNT composites for high-performance supercapacitors. Incorporating CNTs increased the electrode's structural stability and resulted in a greater specific capacitance than NiCo hydroxide [50]. Aqueous supercapacitors, which include acidic solutions like H_2SO_4 [51], neutral solutions like Na_2SO_4 [52], and alkali solutions like KOH [53], and nonaqueous supercapacitors, which include organic solutions like tetraethylazanium tetrafluoroborate ($TEABF_4$) in polycarbonate (PC) [14], and ionic liquids [54]. The electrolyte determines the supercapacitor's voltage window, which is normally quite narrow, as low as 1.2 V. However by converting to an organic electrolyte, it can be raised above 3 V. Devices for storing and producing energy, for example, supercapacitors, batteries, and cells of fuel, are particularly interested in decorated CNT. Using functionally decorated CNT, advanced batteries with increased capacity [55], good cycle stability [56], and improved conductivity [57] were made. As previously demonstrated, CNTs can be coated with oxides [58], polypyrrole [59], and activated carbon [60] for applications of energy storage in supercapacitors. Fe_3O_4-adorned CNT was made possible. Fe_3O_4 is chemically precipitated in the presence of CNT. The findings indicate how the dispersants' electric charge and chelating groups affect Fe_3O_4 formation on the CNT surface. Negative electrodes with a high active mass of 40 mg cm^2 were made using decorated CNT fibrous networks. The dispersion structure was linked to the capacitive behavior. Two of the most extensively utilized energy storage technologies are the battery of lithium-ion and the supercapacitor (SC) and they continue to control the current markets for commercialized energy storage [61, 62]. Nanostructured materials have been shown to be an efficient way to address this need [63–66]. CNTs, as one of the most important nanomaterials, are frequently employed to make electrodes for new LiBs and SCs [67–69]. Shortly after its discovery, some organizations looked into the lithium storage capacity of various kinds of CNTs [70]. The specific capacity of SWCNT anodes would be higher than that of ordinary graphite anodes, which have a theoretical specific capacity of 372 mAh g^{-1} [71, 72]. Furthermore, when the LiB and SC charge/discharge current densities rise, the energy storage capacity falls because strong currents can prevent ions from entering active electrode materials completely. This difficulty could be solved by incorporating CNTs into both LiBs and SCs electrodes, resulting in an increase in their power density [73, 74]. The solid components are dispersed in a solvent to make a homogeneous slurry, which is then uniformly cast on current collectors, such as Al foil, Cu foil, and others, to form an electrochemical cell's operating electrode using a drying procedure [75]. Among the many 1D nanomaterials, CNTs are the most hopeful for manufacturing binder-free electrodes since they precisely satisfy all of the parameters described above. CNTs (particularly MWCNTs) have remarkable conductivity and quick ion transport because of their multilayered sidewalls and distinctive 1D form [76]. To achieve this purpose, CNTs have been extensively exploited in the manufacture of better binder-free electrodes for energy storage applications. To achieve this purpose, CNTs have been widely used in energy storage applications to create superior binder-free electrodes. Employing functionalized vertically aligned (VA) CNT arrays on current collectors for high-performance energy storage applications; the electrochemical behavior of many lithium metal alloy-based anodes has been improved. Jiang et al. [77] employed a VACNT array decorated with nickel nanoparticles to generate a binder-free electrode for SCs. As just a product of volume expansion during the lithiation process, LiB anode materials with thin films have been shown to suffer from aggression and pulverization. This problem can only be solved if the active materials have a thickness below a specific limit [78, 79]. The ability of CNT sheets to store more energy is a desirable goal, and some interesting ways have already been described. Yang's team [80] generated an oxidized FWCNTs thin film with a high gravimetric energy of 200 Wh kg^{-1} and a high power of 10 kW kg^{-1} by adding oxygen functional groups via acid treatment. Zhou's group [81] created functionalized CNT thin films with good power density and cycle performance. On the other hand, functionalization procedures may cause further flaws on independent CNT electrodes, reducing the functionalized CNT thin film's conductivity and limiting the energy density increase. To enhance the electrocatalytic activity of free-standing CNT thin films, researchers are beginning to mix CNT thin films with superior electroactive components to create composite materials. CNTs can operate as an electrically conductive collector as well as an electrochemical active material, as previously indicated, composite free-standing electrodes have promise for versatile energy storage applications. The large porosity of CNTs also allowed for greater interspacing, which allowed for more active material loading and easier electrolytic ion insertion [82]. The CNT/electroactive materials electrodes, on the other hand, looked to have a defect (like for metals of both oxide and sulfide, and semiconductors and others). The free-standing energy storage devices appeared to have numerous times of bending and stretching, a limited life time and poor structural reusability due to the CNTs' and the electroactive materials' adhesion being weak [83]. The mechanical characteristics of CNTs based on freestanding, in contrast to CNTs based on a flat substrate or a newly designed current collector, were principally governed by the links between CNTs and CNT/active substances [84]. Due to the substantial internal force exerted during charge/discharge operations within the electrodes, CNT-supported active materials will experience significant volume expansion associated with lithium ions penetrating and the formation of solid-electrolyte interphase layers, particularly in LiBs. Recently, it has been demonstrated that PPy/CoO nanowires array [85], polyaniline (PANI) hybrid electrodes [86], PPy/CNTs bucky paper separately [87], and some other polymer electroactive materials have improved electrochemical performances in order to ensure robust CNT adhesion. The remarkable mechanical characteristics and structural robustness of these polymer-based electrodes constitute a significant benefit. The strength and tensile modulus of the electrodes were greatly increased after polymers were injected. Furthermore, the densities of energy storage for polymer-based electrodes are comparable to those of contemporary electroactive material composite electrodes. Chen's group [71] used an *in situ* electrochemical polymerization approach to generate CNTs/PANI freestanding SCs electrodes employing a skeleton/skin strategy. With 30 seconds of conformal electrochemical deposition of PANI and a consistent diameter, the CNTs/PANI composite was successfully accomplished. Because of the thin polymer coating, the freestanding film of pure CNTs became considerably more mechanically stable, demonstrating that PANI skin has a beneficial role in CNTs-alone electrodes. Mechanically durable CNTs/PANI free-standing electrodes with 131 Wh kg^{-1} of energy and 62.5 kW kg^{-1} of power density and an extremely long lifespan of over 1000 cycles were also created, thanks to PANI's inexpensive cost and good conductivity and electrochemical performance. The development of high-capacity, self-contained electronic devices based on a single chip is a popular research issue.

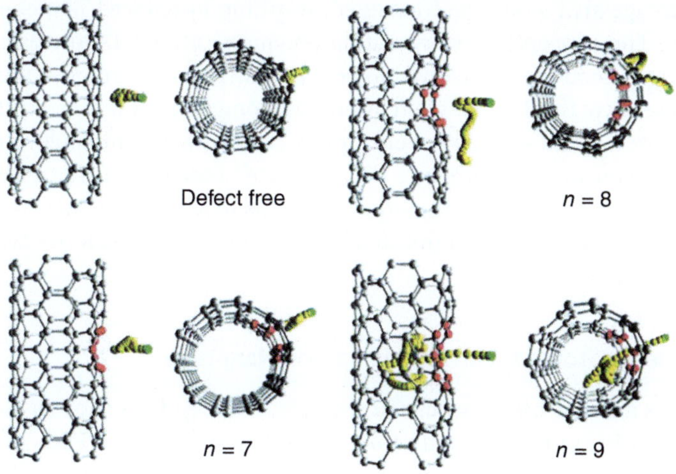

Figure 5.2 The effect of flaws on Li insertion in SWCNT (5, 5) Red balls represent faults, green balls represent Li's initial position, and yellow balls represent Li's trajectory [89]. Source: Nishidate and Hasegawa [89]/with permission of American Physical Society 2022.

Furthermore, the introduction of CNTs may improve the flexibility and energy storage devices' flexibility, which is critical for wearable applications. The usage of CNTs in the energy business is a burgeoning market that will help to ensure long-term energy production.

5.4.1 Carbon Nanotube Hybrid for Lithium-Metal Batteries

Innovative energy storage technology has revolutionized people's lives since the discovery of LiBs. Lithium (Li) metal is a promising material for future energy storage devices. As a result of dendritic Li, practical implementation is still limited. Figure 5.2 shows a Li/CNT hybrid with no dendrites for Li-metal batteries. This is accomplished by directly coating molten Li on CNTs. The favorable thermodynamic and kinetic conditions are the driving forces behind the rapid climb upwards and the infusion of molten Li into the CNT network, which is essential for producing a homogenous metallic layer in Li/CNT hybrids. Some research contributes to the fabrication of Li anodes and their high-level services for Li-metal batteries [90]. Li metal is the most attractive anode choice for post-LiBs. However, significant concerns about safety and low cycle efficiency of Li-metal batteries continue to be significant barrier to their widespread use [91, 92]. To inhibit dendrites by thermal infusion, the bulk metal of Li can be split into tiny fields to minimize the effective current density and attenuate the change in volume [93]. Molten Li, on the other hand, cannot moisten the most promising hosts, such as CNTs. Surface alteration on the host, such as a coating of Si or ZnO, can overcome this problem [94, 95]. However, these alterations are subject to additional undesirable electrochemical processes, the by-products of which may reduce the electrode conductivity. In addition, the finite electrolyte and Li storage area would be consumed, resulting in reduced Li-metal battery performance. The concept of incorporating polymers into CNTs to create composite free-standing electrodes is believed to be a potential way to improve the electrochemical efficiency of flexible and stretchy energy storage devices. Almost all innovative electroactive materials might benefit from CNT matrices. This suggests that the methods used to raise the performance of batteries of Li-ion may also be used to enhance the efficiency of different energy storage systems made of materials, like batteries of lithium–sulfur, sodium-ion, lithium-air, and more types of batteries are available [88, 96, 97] (Figure 5.2).

5.4.2 Wearable Energy Storage with Fiberic Carbon Nanotube

Wearable technologies are now often associated with electronic gadgets such as fitness and health monitoring wristbands with fiber-based supercapacitors (FSCs), it is possible to power wire or cable-type sensors or transistors, embedded into fabrics or clothes, and also act as an energy reservoir for gathering energy from fuel and solar cell conversion devices, as well as thermoelectric and piezoelectric generators [98–103]. The research on fiber-like electrode materials is primarily focused on the development of novel materials such as carbon fibers, cotton yarns, and polymer composites that can be created as a fiber or composited with other fibers/yarns [104, 105]. Another area of research is enhancing the structure of fiber by employing ways to produce fibers with high performance and with a greater area of a specific surface that can be incorporated into standard textile structures [106, 107]. CNTs may be grown as a forest to use a number of processes, then extracted and twisted into a solid-state yarn CNT for construction into fibers, yarns, and fabric structures. Whereas CNTs have a lot of potential, their practical applications are restricted because of their sluggish manufacturing rates, expensive prices, and limited functionality. CNT fibers/yarns that are continuous and large-scale have been generated using solid-state and wet-spinning techniques for use in the energy storage area. When active materials such as MnO_2, PANI, PPy, $LiMn_2O_4$, and $LiFeO_4$ are composited with CNT fibers, they have been proven to be potential electrodes in supercapacitors or lithium batteries. Consequently, creating a rising hybrid wet-spun multiwalled carbon nanotubes (SMWCNTs)/manganese oxide fiber-based supercapacitor remains a challenge [108–113]. The mesoporous flower-MnO_2 nano flakes and the CNT fiber's interior nano network structure promoted ion transport, resulting in improved electrochemical characteristics.

5.4.3 Carbon Nanotube Hybrid for Supercapacitor Energy Storage

Capacitors are small electrical energy storage devices that are made up of an insulator that separates one or more pairs of conductors. Devices which allow for greater energy density levels are called supercapacitors. In pseudocapacitors, oxides of transition metals, for example, MnO_2 and NiO_2 are often employed for redox processes, Figure 5.3. Electrical energy is stored in electrical double-layer capacitors (EDLCs) through ion adsorption at the interface between the electrode's porous surface and

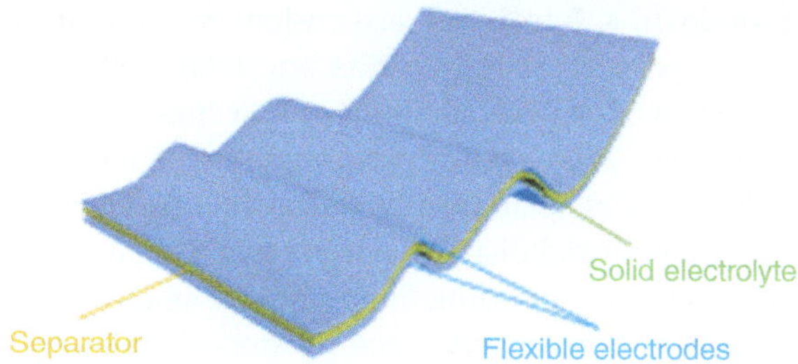

Figure 5.3 A flexible SC is depicted in this diagram. Source: Wen et al. [114]/with permission of Elsevier.

the electrolytes without undergoing any redox reactions [115]. Materials based on carbon have been utilized often for EDLC electrodes because of their high porosity, which allows for more ion adsorption [116]. EDLCs have a lower energy density and longer cycle lifetimes than pseudocapacitors, which have a higher energy density and shorter cycle lives. As a result, functional materials with pseudocapacity have been combined with EDLCs in order to improve supercapacitor efficiency. It has been demonstrated that using pseudocapacitance materials as conducting polymers increases supercapacitor energy density. Some researchers have tried to increase the energy density of CNTs by combining them with other conducting polymers or metals. To make microsupercapacitor yarns, researchers first used a conducting polymer called poly(3,4-ethylenedioxythiophene) (PEDOT) coated on CNT sheet [117]. CNT yarns have recently been shown in a number of studies to be multifunctional, such as fiber-shaped electrical sensors, actuators, supercapacitors, and energy harvesters. The usage of oxides for transition metals, for example, NiO_2, Co_3O_4, and MnO_2, has been researched to boost inherent capacitance utilizing pseudocapacitive materials [118]. Owing to the material's outstanding capacitance, low price, and environmental compatibility, MnO_2 is a popular choice for supercapacitors [119]. In the use of supercapacitors, although, MnO_2 still has a drawback in terms of low electrical conductivity and brittleness. To overcome MnO_2's weakness, which has boosted supercapacitor energy density, MnO_2 can be layered on top of carbon-based materials with high electrical conductivity [120]. Furthermore, due to the gentle manufacturing procedure, an efficient SC electrode with high mass loading and great conductivity may be made by depositing conducting polymer over independent CNTs without the need for an adhesive or current collector [121]. Through chemical reduction and wet spinning with a high concentration and low viscosity, a partially unzipped carbon nanotube/reduction graphene oxide (PUCNT/RGO) hybrid fiber with a well-ordered porosity structure and no "dead volume" was created. The hybrid fiber could be used as electrodes in flexible energy storage devices because of its high electrical conductivity, remarkable mechanical strength, and huge specific surface area. PUCNTs can operate as a solubilizer, making graphene oxide (GO) more soluble and resulting in a high-concentration spinning solution that hinders mass transfer and isolated pore creation. The PUCNTs in the PUCNT/RGO fiber can operate as a spacer, preventing graphene sheets from restacking and generating a well-defined porosity structure that enables rapid electrolyte diffusion and transport. Creating enhanced Mn_3O_4-MWCNT electrodes for supercapacitors use a unique salting-out technique that permits direct particle transfer to a 2-propanol suspension from an aqueous synthesis medium. Due to decreased particle aggregation and increased mixing of the Mn_3O_4 particles and conductive MWCNT additions, the electrodes demonstrated higher capacitance and high active mass loading performance. The method is based on the multifunctional capabilities of octanohydroxamic acid (OHA), which is used in Mn_3O_4 production as a capping and dispersing agent and also as an extractor for particle transport to the electrode processing fluid. The salting-out method has promise for the creation of better supercapacitor energy storage nanocomposites [122, 123]. Because of its higher potential energy and substantially higher power density than normal batteries, the supercapacitor has emerged as a feasible solution for meeting future energy demands among several energy storage technologies [123]. According to its energy storage technology, the EDLC is one of two types of supercapacitors. The electrostatic charge that builds up at the electrode/electrolyte junction creates a capacitor. In contrast, the pseudocapacitor relies on a reversible faradaic reaction among electro-active molecules and an electrolyte. Conductive polymers could be used as active materials because of their cheap cost, high specific capacitance, great electrical conductivity, and flexibility [109, 124, 125]. The areal capacitance of the symmetrical PANI composite CNT yarn supercapacitor, that could co-weave with conventional cotton yarns, was $38\,mF\,cm^2$, indicating that it might be used in wearable electronics [126]. Using oxidized CNT yarns after γ irradiation, which improved carboxyl group concentrations on CNT, enhanced the electrical capacitance, strength, and effectiveness of the composite of PANI/CNT yarns [127]. CNTs act as a conducting network as well as a mechanical support, enhancing the capacitor's general stability [128]. PANI has a greater theoretical capacitance than PEDOT and PPy [129], which is related to the multiredox processes. Synthesis of PANI nanowire arrays resulted in a capacitance of $950\,F\,g^{-1}$ [130]. PANI is also environmentally stable and possesses a one-of-a-kind doping mechanism [131]. However, PANI suffers from low cycle stability as a pseudocapacitive material, which may restrict its practical applicability [132]. CNT sheets retain the optimal mix of specific surface area (SSA) and controlled pore size distribution (PSD), which increases charge storage and ion accessibility (power delivery efficiency) [133]. CNTs are a form of carbon material, which has a lot of potential in energy storage applications because of their excellent electrical and mechanical characteristics [134–139]. CNT sheets and yarns, for example, have a high inherent electrical conductivity and their mechanical resilience makes them excellent wearable device integration and enables for effective charge storage and transmission. [140, 141].

5.4.4 Carbon Nanotubes/Biochar for Energy Storage

Because of its influence on social and environmental problems, the construction of an energy infrastructure that is renewable and long-term has gotten a lot of attention in today's international economy. The growth of advanced, low-cost materials is necessary for the most efficient usage of energy storage and converting processes. Carbon, after oxygen, is the most plentiful and extensively distributed element in the biosphere, and carbohydrates and other biopolymers provide sustainable

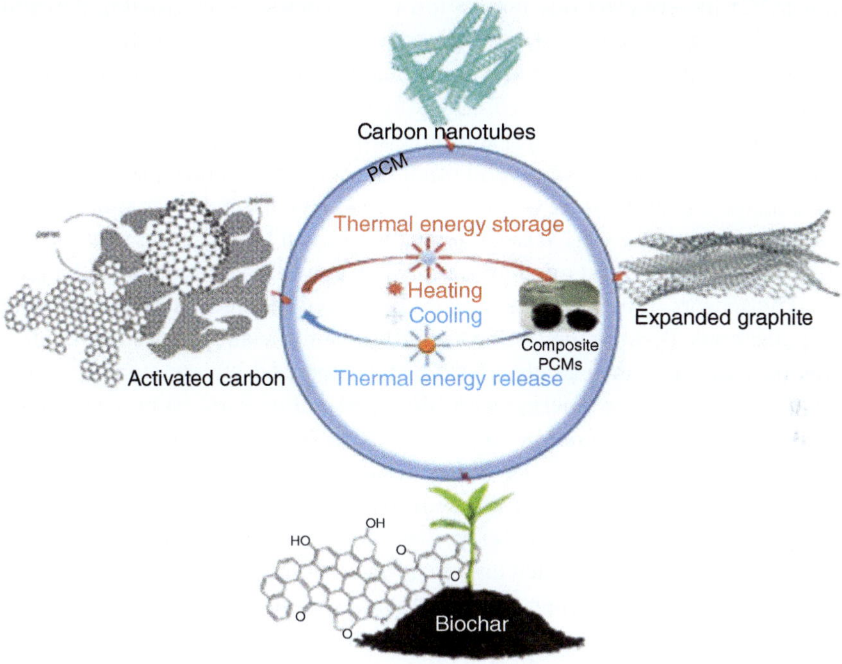

Figure 5.4 The preparation method, as well as the energy storage and release mechanisms of composite phase change materials, is depicted. Source: Dimberu et al. [143]/With permission of Elsevier.

energy [142]. In Figure 5.4, the scaffold in the composite of phase change materials (PCMs) was biochar/CNTs and the energy storage medium was organic liquid *n*-dodecane (DD). The energy storage capacity of the pure biochar supported DD rose by 26.7% after CNT integration, demonstrating that CNT integration improved the composite PCMs' energy storage capacity. Furthermore, when compared to pure biochar support, the PCM loading capacity of the biochar/CNT hybrid composite was higher by (51.3–70.2%). Chemical compatibility, form stability, and exceptional heat stability up to 600 °C without substantial weight loss were all demonstrated in the composite PCMs. After multiple cycles, the biochar/CNT/DD composite PCMs showed outstanding durability and energy storage capability, resulting in increased reversibility. The research suggests that biochar combined with functional reinforced with CNTs could be used to store thermal energy at low temperatures. As a result, shape-stabilized composite PCMs could be used to improve the energy storage efficiency of air conditioners, timber constructions, and thermal energy management systems [143].

5.5 Present and Future of Carbon Nanotubes

Energy storage batteries show an increasingly significant role in day-to-day communication and green travel, for which sophisticated lithium-ion batteries and lithium–sulfur battery electrode fabrication technologies are in growing demand. For example, high mechanical strength and good electrical conductivity and unique morphological and structural characteristics of nano-carbon materials, for example, CNT and graphene, can considerably increase the performance of energy storage batteries in various application modes. Because CNTs and graphene may be blended or composited with several other conductive structures, the synergy of multiple materials can be used to create a more ideal conductive structure. In the future, it is expected that CNT-based energy devices will help to build a new sort of personalized electronics that might be used to power wearable smart or implantable systems, biosensors, bioelectronics, and personalized biomedical devices, among other applications. Because they will constantly give sufficient power to biomedical systems on request from a single user, these customized energy storage devices are required to integrate energy-harvesting and energy-converting equipment. Pacemakers would be powered by carbon-based energy devices, implantable radio transmitters, stomach stimulators, smart gesture gloves, fitness and movement monitors, and wearable biosensors. Biomedical energy storage systems, for example, must be physically attached to the body, either by a skin patch or an implant. New novel techniques and sophisticated nanomaterials have explicitly addressed some of the primary hurdles to the general deployment of carbon-based biotechnologies in commercial applications. These innovative carbon compounds have the potential to improve the efficiency of biomedical devices and open up a host of new medicinal uses. Continuous developments in material sciences, as well as addressing and improving current issues, can assist carbon-based devices in realizing their full potential as responsible, energy-efficient industry staples.

5.6 Commercial-Scale Application of Chemically Modified CNTs for Energy Storage

The industry has shown a strong interest in the manufacture and utilization of CNTs throughout the last few decades. Many businesses have sprouted up to investigate production methods, posttreatments, and application markets. A product's economic success depends on a particular uniformity and quality can only be guaranteed by globally agreed-upon measuring standards. In 2006, a MWCNT capacity of roughly 300 t per year was announced, as well as a yearly SWCNT capacity of 7 t. The manufacturers of MWCNT were mostly distributed in China, Japan, the United States, Germany, and Korea. The availability of M/SWCNTs on the market is sufficient, providing a strong incentive to investigate CNT bulk applications. The CNT industry is quickly increasing, and a growing number of different types of CNTs are becoming available for researchers to explore and develop CNT products. With the rapid growth of mass production, the price of CNTs has dropped dramatically. Ten years ago, the price of CNTs (45 $per g) was more than the price of gold (45 $ per g). MWCNTs and SWCNTs are now priced at 0.1–25 and 50–400 $ per g, respectively. Many fine compounds have a price close to this, but carbon nanofibers (0.1–5 $ per g) cost around the same. Specialty CNTs (around

500 $ per mg) include improved semiconducting and metallic SWCNTs; CNTs with length and surface functionalization, as well as doped CNTs, which continue to be very expensive. The high cost of the CNT process and product purification, as well as the high cost of equipment, restricted scalability of manufacturing processes, and low productivity, are all contributing to the current high price of CNTs. The wide price range is due to variances in product yield caused by different CNT manufacturing techniques. It is expected that the price of CNTs will continue to fall until it reaches a level that is acceptable to end-user applications. Many delicate scientific research projects for the CNT sector have been carried out to date, but few industrial trials have been set up. Between basic scientific research and industrial progress, there is still a significant divide [144].

5.7 Companies Produced CNTs for the Application of Chemically Modified Carbon Nanotubes for Energy Storage

Several companies are starting to scale up the production of CNTs types using different techniques for industrial applications, especially in energy storage; NANOCYl (http://www.nanocyl.com) company produced NC700™, which is MWCNTs by CVD method for electronics and energy. Graph strength® MWCNTs are products developed by AREKEMA (https://www.arkema.com/global/en), which is one of the leading companies in the field of batteries. OCSiAl (https://ocsial.com/about) is a leading company and one of the largest producers of nanotubes materials modified TUBALL™ portfolio from both SWCNTs and MWCNTs for electrodes and LiBs. Cheap tubes (www.Cheaptubes.com) manufacture nanocarbon materials such CNTs and GO for a variety of applications such as nanowire, LiBs, and nano inks.

References

1 Sun, L., Wang, X., Wang, Y., and Zhang, Q. (2017). *Carbon* 122: 462–474.
2 Amin, R., Kumar, P.R., and Belharouak, I. (2020). *Carbon Nanotubes – Redefining the World of Electronics*. IntechOpen.
3 Yu, M.F., Lourie, O., Dyer, M.J. et al. (2000). *Science* 287 (5453): 637–640.
4 Collins, P.G., Zettl, A., Bando, H. et al. (1997). *Science* 278 (5335): 100–102.
5 Pan, H., Li, J., and Feng, Y. (2010). *Nanoscale Res. Lett.* 5 (3): 654–668.
6 Cheng, H.M., Yang, Q.H., and Liu, C. (2001). *Carbon* 39 (10): 1447–1454.
7 Mukhiya, T., Tiwari, A.P., Chhetri, K. et al. (2021). *Chem. Eng. J.* 420: 129679.
8 Cao, Z. and Wei, B.B. (2013). *Energy Environ. Sci.* 6 (11): 3183–3201.
9 Ue, M., Ida, K., and Mori, S. (1994). *J. Electrochem. Soc.* 141 (11): 2989.
10 Xu, S., Zhang, Y., Cho, J. et al. (2013). *Nat. Commun.* 4 (1): 1–8.
11 Frackowiak, E. and Beguin, F. (2001). *Carbon* 39 (6): 937–950.
12 Fuertes, A.B., Lota, G., Centeno, T.A., and Frackowiak, E. (2005). *Electrochim. Acta* 50 (14): 2799–2805.

13 Gogotsi, Y., Nikitin, A., Ye, H. et al. (2003). *Nonporous Nat. Mater.* 2 (9): 591–594.

14 Hung, K., Masarapu, C., Ko, T., and Wei, B. (2009). *J. Power Sources* 193 (2): 944–949.

15 Vix-Guterl, C., Frackowiak, E., Jurewicz, K. et al. (2005). *Carbon* 43 (6): 1293–1302.

16 Pekala, R.W., Farmer, J.C., Alviso, C.T. et al. (1998). *J. Non-Cryst. Solids* 225: 74–80.

17 Du, C., Yeh, J., and Pan, N. (2005). *Nanotechnology* 16 (4): 350.

18 Hwaá Lee, S., Suná Park, J., Kyungá Lim, B. et al. (2009). *Soft Matter* 5 (12): 2343–2346.

19 Yang, C.M., Kim, Y.J., Endo, M. et al. (2007). *J. Am. Chem. Soc.* 129 (1): 20–21.

20 Pech, D., Brunet, M., Durou, H. et al. (2010). *Nat. Nanotechnol.* 5 (9): 651–654.

21 Novoselov, K.S., Geim, A.K., Morozov, S.V. et al. (2005). *Nature* 438 (7065): 197–200.

22 Yoo, E., Kim, J., Hosono, E. et al. (2008). *Nano Lett.* 8 (8): 2277–2282.

23 Lee, S.H., Kim, H.W., Hwang, J.O. et al. (2010). *Angew. Chem.* 122 (52): 10282–10286.

24 Choi, D., Wang, D., Viswanathan, V.V. et al. (2010). *Electrochem. Commun.* 12 (3): 378–381.

25 Lee, S.H., Lee, D.H., Lee, W.J., and Kim, S.O. (2011). *Adv. Funct. Mater.* 21 (8): 1338–1354.

26 Lee, D.H., Kim, J.E., Han, T.H. et al. (2010). *Adv. Mater.* 22 (11): 1247–1252.

27 Paul, R., Vincent, M., Etacheri, V., and Roy, A.K. (2019). Carbon nanotubes, graphene, porous carbon, and hybrid carbon-based materials: synthesis, properties, and functionalization for efficient energy storage. In: *Carbon Based Nanomaterials for Advanced Thermal and Electrochemical Energy Storage and Conversion*, Chapter 1, 1–24. Elsevier.

28 Zaghib, K., Shim, J., Guerfi, A. et al. (2005). *Electrochem. Solid-State Lett.* 8 (4): A207.

29 Zhang, H., Cao, G., Wang, Z. et al. (2008). *Nano Lett.* 8 (9): 2664–2668.

30 Li, Y., Zhou, W., Wang, H. et al. (2012). *Nat. Nanotechnol.* 7 (6): 394–400.

31 Pollak, E., Geng, B., Jeon, K.J. et al. (2010). *Nano Lett.* 10 (9): 3386–3388.

32 Zhu, H. and Wei, B. (2008). *J. Mater. Sci. Technol.* 24 (4): 447.

33 Jiang, K., Li, Q., and Fan, S. (2002). *Nature* 419 (6909): 801.

34 Zhang, Y., Bai, W., Ren, J. et al. (2014). *J. Mater. Chem. A* 2 (29): 11054–11059.

35 Melchionna, M. and Prato, M. (2014). Functionalization of carbon nanotubes. In: *Nanocarbon-Inorganic Hybrids: Next Generation Composites for Sustainable Energy*, 43–70. Berlin/Boston: GmbH.

36 Taghizadeh-Hesary, F., Rasoulinezhad, E., and Yoshino, N. (2019). *Energy Policy* 128: 796–806.

37 Zhang, C., Chen, X., Li, Y. et al. (2018). *J. Cleaner Prod.* 195: 625–639.

38 Makgabutlane, B., Nthunya, L.N., Maubane-Nkadimeng, M.S., and Mhlanga, S.D. (2020). *J. Environ. Chem. Eng.* 8: 104736.

39 Whittingham, M.S. (2012). History, evolution, and future status of energy storage. *Proc. IEEE* 100 (Special Centennial Issue): 1518–1534.
40 Luo, Y., Li, X., Zhang, J. et al. (2014). *J. Nanomater.* 2014.
41 Ibrahim, K.S. (2013). *Carbon Lett.* 14 (3): 131–144.
42 Liu, C. and Cheng, H.M. (2005). *J. Phys. D: Appl. Phys.* 38 (14): R231.
43 Beigmoradi, R., Samimi, A., and Mohebbi-Kalhori, D. (2018). *Beilstein J. Nanotechnol.* 9 (1): 415–435.
44 Ebbesen, T.W., Lezec, H.J., Hiura, H. et al. (1996). *Nature* 382 (6586): 54–56.
45 Li, H. and Zhou, H. (2012). *Chem. Commun.* 48 (9): 1201–1217.
46 Li, L., Wang, L., Ye, T. et al. (2021). *Small* 17: 2005015.
47 Fu, X., Xu, L., Li, J. et al. (2018). *Carbon* 139: 1063–1073.
48 Mousavi, F., Shamsipur, M., Taherpour, A.A., and Pashabadi, A. (2019). *Electrochim. Acta* 308: 373–383.
49 Ojha, M. and Deepa, M. (2019). *Chem. Eng. J.* 368: 772–783.
50 Zhou, Q., Fan, T., Li, Y. et al. (2019). *J. Power Sources* 426: 111–115.
51 Vivekchand, S.R. and Rout, C.S. (2008). *J. Chem. Sci.* 120 (1): 9–13.
52 Subramanian, V., Zhu, H., and Wei, B. (2006). *J. Power Sources* 159 (1): 361–364.
53 Du, X., Wang, C., Chen, M. et al. (2009). *J. Phys. Chem. C* 113 (6): 2643–2646.
54 Conway, B.E., Birss, V., and Wojtowicz, J. (1997). *J. Power Sources* 66 (1, 2): 1–4.
55 Hong, H.P., Kim, M.S., Lee, Y.H. et al. (2013). *Thin Solid Films* 547: 68–71.
56 Wang, T., Kaempgen, M., Nopphawan, P. et al. (2010). *J. Power Sources* 195 (13): 4350–4355.
57 Ojha, M., Le Houx, J., Mukkabla, R. et al. (2019). *Electrochim. Acta* 309: 253–263.
58 Su, Y. and Zhitomirsky, I. (2015). *Appl. Energy* 153: 48–55.
59 Shi, K., Ren, M., and Zhitomirsky, I. (2014). *ACS Sustainable Chem. Eng.* 2 (5): 1289–1298.
60 Hall, P.J. and Bain, E.J. (2008). *Energy Policy* 36 (12): 4352–4355.
61 Dunn, B., Kamath, H., and Tarascon, J.M. (2011). *Science* 334 (6058): 928–935.
62 Arico, A.S., Bruce, P., Scrosati, B. et al. (2011). Nanostructured materials for advanced energy conversion and storage devices. In: *Materials for Sustainable Energy: A Collection of Peer-Reviewed Research and Review Articles*, 148–159. Nature Publishing Group.
63 Guo, Y.G., Hu, J.S., and Wan, L.J. (2008). *Adv. Mater.* 20 (15): 2878–2887.
64 Simon, P. and Gogotsi, Y. (2013). *Acc. Chem. Res.* 46 (5): 1094–1103.
65 Gao, M.R., Xu, Y.F., Jiang, J., and Yu, S.H. (2013). *Chem. Soc. Rev.* 42 (7): 2986–3017.
66 Hu, L., Choi, J.W., Yang, Y. et al. (2009). *Proc. Natl. Acad. Sci. U.S.A.* 106 (51): 21490–21494.
67 Wang, X., Lu, X., Liu, B. et al. (2014). *Adv. Mater.* 26 (28): 4763–4782.
68 Jiang, J., Li, Y., Liu, J. et al. (2012). *Adv. Mater.* 24 (38): 5166–5180.
69 Frackowiak, E., Gautier, S., Gaucher, H. et al. (1999). *Carbon* 37 (1): 61–69.
70 Ren, J., Li, L., Chen, C. et al. (2013). *Adv. Mater.* 25 (8): 1155–1159.

71 Niu, Z., Luan, P., Shao, Q. et al. (2012). *Energy Environ. Sci.* 5 (9): 8726–8733.

72 Sun, L., Wang, X., Zhang, K. et al. (2016). *Nano Energy* 22: 11–18.

73 Sun, L., Wang, X., Liu, W. et al. (2016). *J. Power Sources* 315: 1–8.

74 MacNeil, D.D., Lu, Z., Chen, Z., and Dahn, J.R. (2002). *J. Power Sources* 108 (1, 2): 8–14.

75 Wang, X.H., Guan, C., Sun, L.M. et al. (2015). *J. Mater. Chem. A* 3 (30): 15394–15398.

76 Jiang, Y., Wang, P., Zang, X. et al. (2013). *Nano Lett.* 13 (8): 3524–3530.

77 Su, X., Wu, Q., Li, J. et al. (2014). *Adv. Energy Mater.* 4 (1): 1300882.

78 Haftbaradaran, H., Xiao, X., Verbrugge, M.W., and Gao, H. (2012). *J. Power Sources* 206: 357–366.

79 Lee, S.W., Gallant, B.M., Lee, Y. et al. (2012). *Energy Environ. Sci.* 5 (1): 5437–5444.

80 Xiao, X., Li, T., Peng, Z. et al. (2014). *Nano Energy* 6: 1–9.

81 Xu, G., Zhang, Q., Zhou, W. et al. (2008). *Appl. Phys. A* 92 (3): 531–539.

82 Flahaut, E., Peigney, A., Laurent, C. et al. (2000). *Acta Mater.* 48 (14): 3803–3812.

83 Zhi, M., Xiang, C., Li, J. et al. (2013). *Nanoscale* 5 (1): 72–88.

84 Zhou, C., Zhang, Y., Li, Y., and Liu, J. (2013). *Nano Lett.* 13 (5): 2078–2085.

85 Patil, D.S., Pawar, S.A., Devan, R.S. et al. (2014). *J. Electroanal. Chem.* 724: 21–28.

86 Che, J., Chen, P., and Chan-Park, M.B. (2013). *J. Mater. Chem. A* 1 (12): 4057–4066.

87 Wang, Z.Y., Lu, Z.X., Guo, W. et al. (2021). *Adv. Mater.* 33 (4): 2006702.

88 Zhao, L., Zhao, J., Hu, Y.S. et al. (2012). *Adv. Energy Mater.* 2 (8): 962–965.

89 Nishidate, K. and Hasegawa, M. (2005). Energetics of lithium-ion adsorption on defective carbon nanotubes. *Phys. Rev. B* 71: 245418.

90 Liu, D.H., Bai, Z., Li, M. et al. (2020). *Chem. Soc. Rev.* 49 (15): 5407–5445.

91 Xiao, J. (2019). *Science* 366 (6464): 426–427.

92 Lin, D., Liu, Y., Liang, Z. et al. (2016). *Nat. Nanotechnol.* 11 (7): 626–632.

93 Liang, Z., Lin, D., Zhao, J. et al. (2016). *Proc. Natl. Acad. Sci. U.S.A.* 113 (11): 2862–2867.

94 Liu, Y., Lin, D., Liang, Z. et al. (2016). *Nat. Commun.* 7 (1): 1–9.

95 Peng, H.J., Huang, J.Q., Zhao, M.Q. et al. (2014). *Adv. Funct. Mater.* 24 (19): 2920.

96 Grande, L., Paillard, E., Hassoun, J. et al. (2015). *Adv. Mater.* 27 (5): 784–800.

97 Fang, X. and Peng, H. (2015). *Small* 11 (13): 1488–1511.

98 Lee, M.R., Eckert, R.D., Forberich, K. et al. (2009). *Science* 324 (5924): 232–235.

99 Zhang, S., Ji, C., Bian, Z. et al. (2012). *ACS Nano* 6 (8): 7191–7198.

100 Liu, D., Zhao, M., Li, Y. et al. (2012). *ACS Nano* 6 (12): 11027–11034.

101 Gao, F., Viry, L., Maugey, M. et al. (2010). *Nat. Commun.* 1 (1): 1–7.

102 Lee, M., Chen, C.Y., Wang, S. et al. (2012). *Adv. Mater.* 24 (13): 1759–1764.

103 Yadav, A., Pipe, K.P., and Shtein, M. (2008). *J. Power Sources* 175 (2): 909–913.

104 Liu, N., Ma, W., Tao, J. et al. (2013). *Adv. Mater.* 25 (35): 4925–4931.

105 Le, V.T., Kim, H., Ghosh, A. et al. (2013). *ACS Nano* 7 (7): 5940–5947.

106 Qu, G., Cheng, J., Li, X. et al. (2016). *Adv. Mater.* 28 (19): 3646–3652.
107 Yu, D., Goh, K., Zhang, Q. et al. (2014). *Adv. Mater.* 26 (39): 6790–6797.
108 Yao, F., Pham, D.T., and Lee, Y.H. (2015). *ChemSusChem* 8 (14): 2284–2311.
109 Wang, K., Meng, Q., Zhang, Y. et al. (2013). *Adv. Mater.* 25 (10): 1494–1498.
110 Choi, C., Lee, J.A., Choi, A.Y. et al. (2014). *Adv. Mater.* 26 (13): 2059–2065.
111 Cai, Z., Li, L., Ren, J. et al. (2013). *J. Mater. Chem. A* 1 (2): 258–261.
112 Chen, X., Lin, H., Deng, J. et al. (2014). *Adv. Mater.* 26 (48): 8126–8132.
113 Chen, S., Ma, W., Cheng, Y. et al. (2015). *Nano Energy* 15: 642–653.
114 Wen, J., Xu, B., Gao, Y. et al. (2021). Wearable technologies enable high-performance textile supercapacitors with flexible, breathable and wearable characteristics for future energy storage. *Energy Stor. Mater.* 37: 94–122.
115 Jang, Y., Kim, S.M., Spinks, G.M., and Kim, S.J. (2020). *Adv. Mater.* 32 (5): 1902670.
116 Wang, L. and Hu, X. (2018). *Chem. Asian J.* 13 (12): 1518–1529.
117 Lee, J.A., Shin, M.K., Kim, S.H. et al. (2012). *ACS Nano* 6 (1): 327–334.
118 Jiang, H., Li, C., Sun, T., and Ma, J. (2012). *Nanoscale* 4 (3): 807–812.
119 Lei, Z., Zhang, J., and Zhao, X.S. (2012). *J. Mater. Chem.* 22 (1): 153–160.
120 Gao, H., Xiao, F., Ching, C.B., and Duan, H. (2012). *ACS Appl. Mater. Interfaces* 4 (5): 2801–2810.
121 Wang, S., Liang, Y., Zhuo, W. et al. (2021). *Mater. Chem. Front.* 5 (3): 1324–1329.
122 Rorabeck, K. and Zhitomirsky, I. (2021). *Molecules* 26 (2): 296.
123 Zhang, L.L. and Zhao, X.S. (2009). *Chem. Soc. Rev.* 38 (9): 2520–2531.
124 Bryan, A.M., Santino, L.M., Lu, Y. et al. (2016). *Chem. Mater.* 28 (17): 5989–5998.
125 Wang, F., Zhan, X., Cheng, Z. et al. (2015). *Small* 11 (6): 749–755.
126 Su, F., Miao, M., Niu, H., and Wei, Z. (2014). *ACS Appl. Mater. Interfaces* 6 (4): 2553–2560.
127 Yan, P., Zhang, R., Jia, J. et al. (2015). *J. Power Sources* 284: 38–43.
128 Li, H., Wang, J., Chu, Q. et al. (2009). *J. Power Sources* 190 (2): 578–586.
129 Wang, K., Huang, J., and Wei, Z. (2010). *J. Phys. Chem. C* 114 (17): 8062–8067.
130 Wang, H., Wen, H., Hu, B. et al. (2017). *Sci. Rep.* 7 (1): 1–2.
131 Liu, D., Wang, X., Deng, J. et al. (2015). *Nanomaterials* 5 (2): 1034–1047.
132 Zhang, L., Yang, X., Zhang, F. et al. (2013). *J. Am. Chem. Soc.* 135 (15): 5921–5929.
133 Antiohos, D., Romano, M., Chen, J., and Razal, J.M. (2008). *Energy Policy* 36 (12): 4352–4355.
134 Hall, P.J. and Bain, E.J. (2008). Energy-storage technologies and electricity generation. *Energy Policy* 36: 4352–4355. https://doi.org/10.1016/j.enpol.2008.09.037.
135 Chung, J., Lee, K.H., Lee, J., and Ruoff, R.S. (2004). *Langmuir* 20 (8): 3011–3017.
136 Adusei, P.K., Kanakaraj, S.N., Gbordzoe, S. et al. (2019). *Electrochem. Acta* 312: 411–423.
137 Zhang, L., DeArmond, D., Alvarez, N.T. et al. (2017). *Small* 13 (10): 1603114.

138 Adusei, P.K., Hsieh, Y.Y., Kanakaraj, S.N. et al. (2018). *Science, Technology and Advanced Application of Supercapacitors*, 1–16. IntechOpen.

139 Zhou, X., Zheng, X., Xu, J. et al. (2019). *Energy Technol.* 7 (9): 1900122.

140 Lu, Z., Raad, R., Safaei, F. et al. (2019). *Front. Mater.* 6: 138.

141 Dong, L., Feng, Y., Wang, L., and Feng, W. (2018). *Chem. Soc. Rev.* 47 (19): 7339–7368.

142 Liu, W.J., Jiang, H., and Yu, H.Q. (2019). *Energy Environ. Sci.* 12 (6): 1751–1779.

143 Atinafu, D.G., Wi, S., Yun, B.Y., and Kim, S. (2021). *Energy* 216: 119294.

144 Zhang, Q., Qi Huang, J., Qian, W.Z. et al. (2013). *Small* 9 (8): 1237–1265.

6

Chemically Modified Carbon Nanotubes for Pollutants Adsorption

Suresh Ranganathan, Saravanan Rajendran, and Lorena C. Ponce

Universidad de Tarapacá, Faculty of Engineering, Department of Mechanical Engineering, Avda. General Velásquez, Arica, 1775, Chile

6.1 Introduction

The effluents from various chemical industries possess various types of harmful and/or toxic chemical substances [1, 2]. Prior to discharge into nearby ecosystems, elimination of these toxic chemicals is of high priority, as they cause critical ecological effects and severe health risks [3, 4]. For this purpose, multiple wastewater purification processes were developed and are currently used in the world [1]. The most commonly adopted water treatments are displayed in Figure 6.1. For example, precipitation can be used for the removal of metal ions from water [5], while biological methods can be used for degradation of organic substances and even removal of heavy metal ions [6, 7]. Besides, there is still a need to increase the efficiency of these methods [8]. Therefore, continuous efforts are made to improve the performance of these wastewater treatment methods. Among the important water treatments, adsorption has attracted much attention owing to its simplified procedure, relatively high removal efficiency, possibility to treat water containing least concentrated pollutants, less power consumption, and accessibility of raw materials to produce adsorbent materials [9].

Adsorption is a surface process, in which, dissolved substance (adsorbate) from a solution gets attached to a solid surface (adsorbent). Adsorption may happen either by physical force or the formation of chemical bonds. If weak physical forces such as van der Waals attraction happen between the adsorbent and adsorbate, then this process is called physisorption [10]. On the other hand, if an adsorbate is attached to the surface of the adsorbent via strong chemical bonding, then this adsorption is termed chemisorption [11]. The effectiveness of adsorption is chiefly dependent on the adsorption capacity. The adsorption capacity of an adsorbent depends on the characteristics of the adsorbent, adsorbate molecules, and medium (usually water). Many kinds of solid adsorbents have been synthesized for removal of both organic and inorganic pollutants from water. Adsorbents may be processed biomass, carbon materials, biopolymers, metal compounds, industrial wastes,

Chemically Modified Carbon Nanotubes for Commercial Applications, First Edition.
Edited by Jeenat Aslam, Chaudhery Mustansar Hussain, and Ruby Aslam.

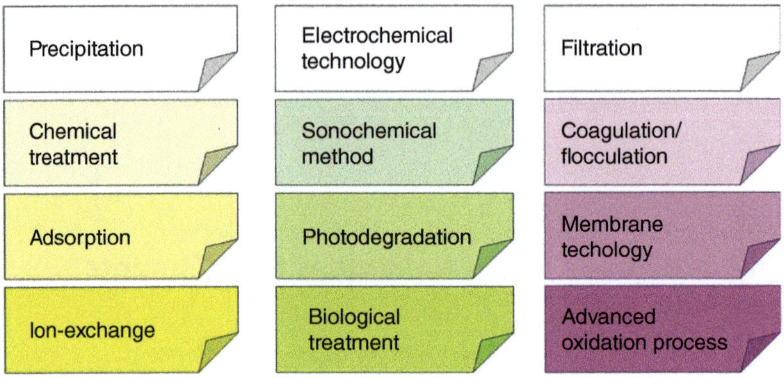

Figure 6.1 Available wastewater treatment techniques for the removal of different kinds of pollutants (physical, chemical, and biological) from water.

organic–inorganic compounds, and natural inorganic materials [12, 13]. However, each of these materials has its own confines in practical application due to the low adsorption capacity. As an alternate, researchers have proposed carbon nanotubes (CNTs) as an adsorbent material for pollutant removal from water.

CNTs are cylindrical macromolecules (radius = a few nanometers, length = up to 20 cm) that are composed of carbon atoms [14]. CNT walls are composed of a hexagonal lattice of carbon atoms, while the ends of the CNTs are capped by half of a fullerene-like structure. They show great flexibility, low density, high elastic modulus, and strength [15, 16]. Based on the number of tubes, CNTs are differentiated as single-walled carbon nanotubes (SWCNTs) and multiwalled carbon nanotubes (MWCNTs). SWCNTs possess a single-CNT cylindrical molecule (diameter = 0.4–3 nm), while MWCNTs contain two or more CNT cylinders (diameters = 3–40 nm). MWCNTs containing only two CNT cylinders are said to be double-walled CNTs (diameters = 1–4 nm). Applications of CNTs (SWCNTs and MWCNTs) are in medical diagnosis and therapy [17, 18], environmental monitoring and protection [19–21], engineering and material science [22, 23]. Studies on the adsorption behavior of CNTs toward aquatic pollutants are also enormous [21, 24]. However, reports reveal that the adsorption capacity of pure CNTs is not sufficiently high [25]. This is due to its aggregation and poor dispersion capacity in water. But these flaws could be rectified by adopting surface modification or functionalization processes.

Surface modification involves the attachment of functional groups or organic molecules or polymers to the surface of the CNTs. There are two ways of surface modification process that are commonly used in CNTs [26]. One is covalent modification, i.e. functional groups (carboxyl, amino, silane, thiol, etc.) or polymers or molecules are linked with CNTs via covalent bonds. But, the original properties of CNTs will be changed after the completion of this modification. Many scientific papers have been published on chemically modified CNTs for adsorption of organic and inorganic pollutants from water. Figure 6.2 represents the development of surface modified CNT-based adsorbents, reported in recent years [27–38]. Another way

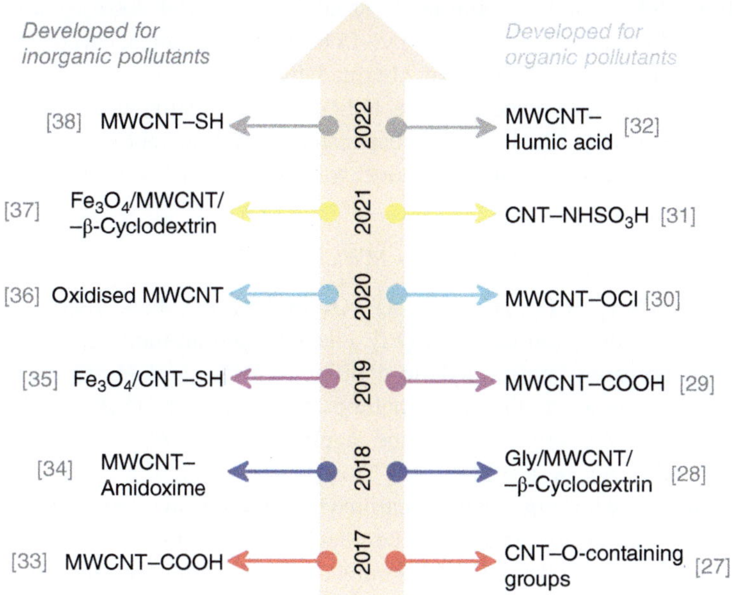

Figure 6.2 Development of chemically modified CNT-based adsorbents for the removal of organic and inorganic pollutants from water. The corresponding reference number has given within square bracket.

is a noncovalent modification. Functional moieties including surfactants, polymers, and other materials are attached to the surface of CNTs through physical forces. Noncovalent modification doesn't alter the real properties of CNTs. At the same time, this method limits the possibility of further chemical modification of CNTs. Because of surface modifications, the physical properties of CNTs like hydrophobicity/hydrophilicity, dispersibility, surface area, and charge are significantly altered. The available surface modification (covalent) methods for CNTs are discussed in Section 6.2.

6.2 Chemically Modified CNTs

Chemically modified SWCNTs and MWCNTs were prepared to increase their adsorption capacity/removal efficiency and selectivity in the adsorption of pollutants. Different surface modification methods reported for CNTs were discussed below.

(a) *Oxidation process*: In this method, functional groups/moieties (such as –COOH, –OH, –C=O, $-SO_3H$, etc.) are attached on the CNT surface via oxidation using a mixture of inorganic acids under ultrasonication. Generally, a concentrated nitric acid (HNO_3) and sulfuric acid (H_2SO_4) mixture is used [39]. In the presence of these strong acids, the caps of CNTs open and holes are created. An oxidative etching also takes place along with liberation of CO_2 gas.

The resulting CNTs attained nanometer sized lengths (100–300 nm) and received various functional groups (mainly –COOH). For instance, carboxylic acid groups modified MWCNTs were obtained by the following way [39] (Eq. (6.1)): MWCNTs were dispersed in a H_2SO_4/HNO_3 mixture using ultrasonication (symbol "((("). The resulting suspension should be kept at <100 °C for a certain period of time. Then, the product was filtered, washed with water, and dried.

$$\text{MWCNTs} + HNO_3 + H_2SO_4 + (((\rightarrow \text{MWCNTs–COOH} \tag{6.1}$$

Oxidants such as $(NH_4)_2S_2O_8$, H_2O_2, H_2SO_4, O_3, and $KMnO_4$ have also been used for surface modification of CNTs [26]. Likewise, amine modified CNTs have also been prepared using ethylene diamine, $NaNO_2$, and H_2SO_4 [40]. The advantages of these modification methods are as follows [14]: (i) The inherent properties of CNTs don't change significantly. (ii) Any impurities in raw CNT surface will be removed. (iii) This process opens the tips of the CNTs. (iv) Functional groups such as carboxyl (–COOH), hydroxyl (–OH), carbonyl (–CO–), and epoxy groups will be attached on the surface of CNTs. (v) The functional groups like carboxyl groups decrease van der Waals forces between the CNTs, thereby decreasing aggregation and making it solubility in various solvents. On the other hand, some disadvantages of this method are the creation of defects, which harshly deteriorate the structure of CNTs and strong fragmentation of the CNT [41]. The chemically modified CNTs generally showed a wide band gap and reduced electrical conductivity.

(b) *Plasma method*: The plasma method is a faster, more eco-friendly, and more efficient surface modification method for CNTs to attain surface groups. For this reaction, a plasma reactor is constructed using a cylindrical quartz tube and coil, RF power supply unit, a rotary vacuum pump, and a mass-flow controller [42]. A mixture of CNTs and maleic anhydride powder with definite proportion is kept in the plasma reactor tube and then plasma power is applied with certain pressure and time. Finally, surface modified CNTs will be formed. By using this method, CNT powder with a covalently attached O–C=O group was obtained [42].

(c) *Electrochemical method*: The electrochemical technique was also used for the surface modification of CNTs. In this method, CNT is taken as a working electrode and inserted into an electrolytic solution containing suitable reagents. By using suitable experimental arrangements, a constant potential is applied where organic radicals are produced. Those organic radicals are responsible for bringing surface modification to CNTs. Surface groups of CNTs can be tailored by selecting relevant organic reagents. For example, Jaime et al. [43] have used cyclic voltammetry (electrochemical method) for the attachment of 4-aminophenyl phosphonic acid on the surface of SWCNTs. For this purpose, they use a three-electrode system for their experiments. SWCNT/glassy carbon electrode, graphite rod, and reversible hydrogen electrode act as a working, counter and reference electrode, respectively. The background solution is a homogeneous aqueous solution (0.5 M) sulfuric acid and (1 mM) 4-aminophenyl phosphonic acid.

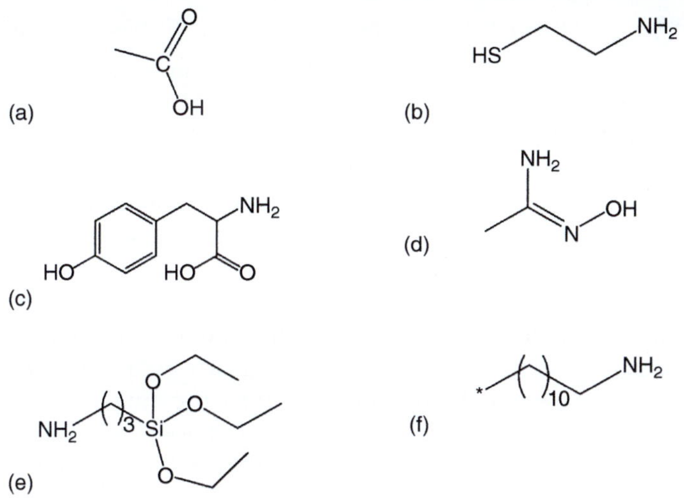

Figure 6.3 Molecular structure of some organic moieties used for surface medication in CNTs: (a) carboxylic acid, (b) Cysteamine, (c) L-tyrosine, (d) Amidoxime, (e) 3-aminopropyl triethoxysilane, and (f) Dodecylamine.

(d) *Chemical reactions*: Oxidized CNTs can be further modified by appropriate chemical reactions. Figure 6.3 shows some of the organic moieties used to modify CNTs. For instance, dodecylamine can be linked to the MWCNT–COOH surface through a dehydration method [39]. MWCNT modified with tyrosine was prepared by reacting oxidized MWCNTs and tyrosine in buffer solution (pH 7) for 48 hours at 45 °C [44]. Similarly, a number of molecules (β-cyclodextrin, cysteamine hydrochloride, amidoxime, porphyrins derivatives) can be chemically modified on the surface of CNTs. Amine-containing molecules are attached by reacting with acyl chloride (–COCl) modified CNT (prepared by using CNT–COOH and $SOCl_2$). Cycloaddition, fluorination, hydrogenation, esterification, amide or ester linkage, click coupling, and nucleophilic addition reactions have also been carried out to produce chemically modified CNTs [45].

In this chapter, research studies on the application of modified (covalent) CNTs for the adsorptive removal of pollutants from water were described. Influencing factors and possible adsorption modes have also been described in Sections 6.4 and 6.5 respectively.

6.3 Chemically Modified CNTs for Adsorptive Removal of Pollutants

Some of the research studies related to the adsorption of different kinds of water pollutants by chemically modified CNTs (SWCNTs/MWCNTs) were discussed in this section. The focused pollutants are organic dyes, pharmaceuticals, other organic pollutants (benzene and its derivatives, phenol, tetrabromobisphenol, etc.) and metal ions.

6.3.1 Organic Dyes

Synthetic organic dyes are harmful water pollutants and can cause various health-related problems in humans, animals, and plants [46, 47]. The carcinogenic nature of some synthetic dyes has also been reported [48]. Besides, synthetic dyes have the most stable structures, which make them highly resistant to biodegradation. But, the capturing of synthetic dyes from water by chemically modified CNTs was investigated. A few examples of modified CNTs-based adsorbents and their dyes removal performance have been tabulated in Table 6.1.

The studies mentioned in Table 6.1 reveal the following important conclusions: (i) Adsorption capacity of modified CNTs depends on the nature of surface functional groups. (ii) The adsorptive removal efficiency of particular surface modified CNTs is also dependent on the molecular structure of organic dye. (iii) In spite of widely optimized 25 °C, some modified CNTs adsorbents also showed good adsorption capacity at elevated temperatures, and (iv) The equilibrium adsorption time varied by varying dye for a particular adsorbent.

6.3.2 Removal of Pharmaceuticals

The persistence of pharmaceutical products, including antibiotics, in water is an emerging environmental issue, as they cause bioaccumulation and toxicity [53, 54]. Specifically, the negative effects due to the presence of tetracycline in environmental waters have been well studied [55]. A number of pharmaceuticals (Figure 6.4) are also found in water and thus their removal by using modified CNT-based adsorbents was attempted.

For example, Wei et al. [27] have performed a batch adsorption study for the removal of some selected sulfonamide compounds by using CNTs with different surface groups (–OH, –COOH, –C=O, and –COC–). The adsorption tendency of CNTs is in the following order: sulfadimethoxine > sulfamethizole > sulfamethazine > sulfamethoxazole. Based on the density functional theory (DFT), researchers have found that the oxygen-containing surface groups hinder the adsorption of targeted molecules. The carboxylic acid group has a higher inhibitory effect, while –COC– showed the lowest inhibition effect. However, these surface groups only weaken the π–π attraction only. In another investigation, the oxidized MWCNTs (MWCNT–COOH) were prepared for the capturing of pefloxacin (a fluoroquinolone-type antibiotic) and Cu^{2+} ions from water simultaneously [29]. In the presence of Cu^{2+} ions, the adsorption capacity of MWCNT–COOH was enhanced. At 0 and 5 $mg\,l^{-1}$ of Cu^{2+} ions, the pefloxacin adsorption capacity of MWCNT–COOH is 125.08 and 204.48 $mg\,g^{-1}$ respectively. But, a higher concentration of Cu^{2+} ions decreases the adsorption ability of MWCNT–COOH powder. The adsorption mechanism of pefloxacin and Cu^{2+} on the surface of MWCNT–COOH is due to the electrostatic and π–π attractions.

Recently, benzothiazole ionic liquid-MWCNTs (tablet form) were used for the adsorption of tetracycline [56]. The tetracycline removal efficiency (Langmuir isotherm) was found to be 99.76% with a single tablet. Gonzalez-Fernandez et al. [57] have targeted triclosan removal from water by using SWCNTs with high

Table 6.1 Examples of chemically modified CNT-based acsorbents and their dye removal performance.

Chemically modified CNTs[a)]-based adsorbents	Organic dye(s)	Isotherm model	Contact time (min)	Temperature (°C)	Adsorption capacity ($mg\,g^{-1}$)	Reference
$ZnCo_2O_4$/CNT–DES[b)]	Eosin Y	Langmuir	Overnight	Room temperature	414.454	[49]
MWCNT[c)]–tyrosine	Methylene blue	Langmuir	7	25	440	[44]
MWCNT–hypochlorite (–OCl)	Acid orange 7	Langmuir	—	25	47.72 ± 0.79	[30]
SWCNT[d)]–amine (–NH_2)	Methylene blue	Langmuir	90	50	13.62	[50]
CNT–poly(ionic liquids)	Congo red	Langmuir	56	Room temperature	178	[51]
Glycine/MWCNTs–β-cyclodextrin	• Methylene blue • Acid blue 113 • Methyl orange • DR1[e)]	Langmuir	• 20 • 20 • 35 • 5	Room temperature	• 90.90 • 172.41 • 96.15 • 500.0	[28]
CNT–sulfonic acid (–$NHSO_3H$)	Methylene blue	Langmuir	60	25	1584.22	[31]
MWCNT–carboxylic acid (–COOH)	Crystal violet Rhodamine B	—	4 h	55	0.88 $mmol\,g^{-1}$ 0.90 $mmol\,g^{-1}$	[52]

a) CNTs – carbon nanotubes.
b) DES – deep eutectic solvent.
c) MWCNTs – multiwalled carbon nanotubes.
d) SWCNT – single-walled carbon nanotubes.
e) DR1 – disperse red 1.

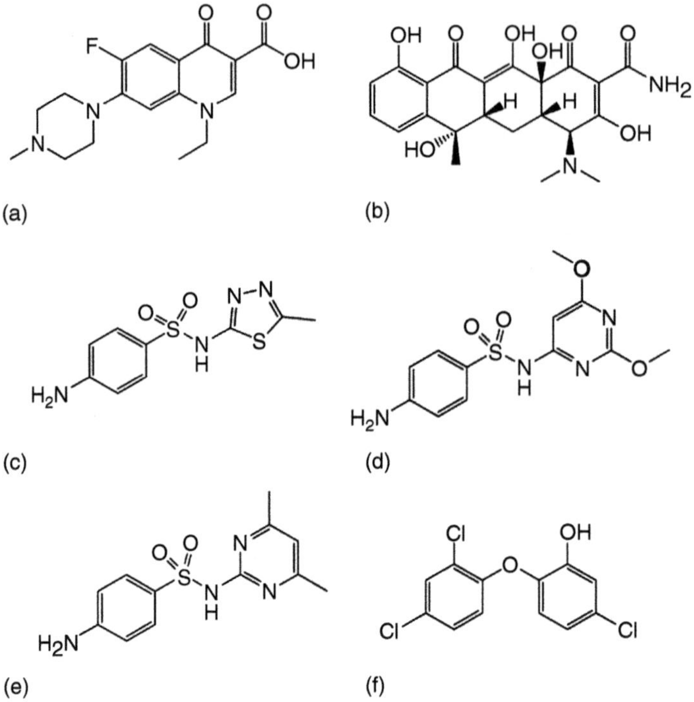

Figure 6.4 Molecular structure of pharmaceutical products, which was adsorbed by chemically modified CNT-based adsorbents. (a) Pefloxacin, (b) Tetracycline, (c) Sulfamethizole, (d) Sulfadimethoxine, (e) Sulfadimidine, and (f) Triclosan.

surface area (582 m^2 g^{-1}) and basic sites (0.1150 mmol g^{-1}). One gram of SWCNTs could bind up to 30.3 mg of triclosan (pH = 7).

6.3.3 Other Organic Pollutants

Phenol is a corrosive and nonbiodegradable organic compound. It can enter humans' bodies through the food chain and exhibit adverse effects on humans [58]. Even a low concentration of phenol persistence in the environment may cause severe effects on organisms. To remove phenol through an adsorption method, Lawal et al. [59] have synthesized deep eutectic solvent (molecular formula = $C_{22}H_{26}BrO_3P$, molar mass = 449.32 g mol^{-1}) modified CNTs. A deep eutectic solvent was synthesized by using glycerol and methyl triphenyl phosphonium bromide at 70 °C. When compared to pristine CNTs (dosage = 0.03 g), these modified CNTs (dosage = 0.03 g) have 98% phenol removal efficiency (Freundlich isotherm). The optimum pH and contact time for effective phenol removal were also identified by this study. Just recently, Lin et al. [60] have prepared β-cyclodextrin functionalized MWCNT/Fe_3O_4 composite in order to eliminate phenol from simulated oily wastewater. According to the Langmuir isotherm, the phenol adsorption capacity was estimated as 67.67 mg g^{-1} (temperature = 20 °C). In the simulated oily wastewater, this adsorbent showed 92.6% phenol removal rate.

Benzene and its derivatives (toluene and *p*-xylene) have been found as harmful water pollutants [61]. Anjum et al. [62] have prepared organic acid modified MWCNTs for the adsorption of benzene derivatives. Citric acid, tartaric acid, malic acid, and salicylic acid were used for surface modification of MWCNTs. The attached functional moieties offered higher surface wettability that enhanced adsorption efficiency. Around 99.99% removal efficiency was obtained for the adsorption of benzene and its derivatives. The effect of initial pollutant concentration, contact time, adsorbent dosage, temperature, and pH on this adsorption process has also been performed.

Tetrabromobisphenol A, a widely used flame retardant, is considered an emerging pollutant in water [63]. Hence, to remove this pollutant from water, Wu et al. [64] have synthesized α-Fe_2O_3/CNT–NH_2 sponge. Under the optimized conditions (pH = 7.0 ± 0.1 and temperature = 303 K), the highest tetrabromobisphenol A adsorption capacity of α-Fe_2O_3/CNT–NH_2 sponge was $41.35 \pm 6.09\,mg\,g^{-1}$. Importantly, the captured tetrabromobisphenol A molecules were decomposed by adding peroxymonosulfate (a source of sulfate and hydroxyl radicals). The surface groups, carbonyl, π–π^* and —$CONH_2$ bonds not only favors the adsorption of tetrabromobisphenol A but also involved in the activation of peroxymonosulfate.

Patiño et al. [65] have investigated the adsorptive removal of three important emerging contaminants by using MWCNT–COOH and MWCNT–NH_2 adsorbents. The adsorption capacity of MWCNT–NH_2 is higher than MWCNT–COOH toward 1,8-dichlorooctane, nalidixic acid, and 2-(4-methylphenoxy)ethanol in water. The studies find that this adsorption process is mainly dependent on the hydrophobicity of the pollutants and nature of modified MWCNTs.

6.3.4 Metal Ions

Industries like mining, metallurgical, tanneries, battery production, etc., produce effluents which contain a number of heavy metal ions [66]. If these effluents were discharged into the aquatic ecosystems, then heavy metal pollution would occur in those water bodies. Heavy metal ions could enter into animals and humans via the food chain and accumulate in their water ecosystems. Moreover, metal ions are not biodegradable, and cause acute health effects [67]. Therefore, research work on adsorptive removal of metal ions from water has been conducted worldwide. Some of the reported results on adsorptive removal of metal ions such as lead, nickel, cadmium, mercury, copper, zinc, uranium, and antimony from water by using modified CNT-based adsorbents were tabulated (Table 6.2).

The following observations can be drawn from Table 6.2.

(a) Like organic removal process, the nature of surface groups in CNTs also plays a vital role in the removal of metal ions from water. For example, in the adsorption of Pb^{2+} and Zn^{2+} ions, MWCNT–ttpy showed better adsorption capacity than MWCNT–COOH [68].
(b) Adsorption behavior of modified CNTs is also dependent on the nature of metal ions that need to be removed from water. For example, MWCNT–ttpy adsorbent captures Pb^{2+} ions rather than Zn^{2+} ions [68].

Table 6.2 Reported results of adsorptive removal of metal ions by using modified CNT-based adsorbents.[a)]

Adsorbents	Adsorbent dosage	Metal ions	Surface area ($m^2\ g^{-1}$)	pH	Contact time (min)	Temperature (°C)	Model of isotherm	Adsorption capacity ($mg\ g^{-1}$)	Reference
Fe_3O_4/MWCNT–β-cyclodextrin[b)]	0.06 g	Ni^{2+} (100 ml, 50 $mg\ l^{-1}$)	83	6	50	25	—	103	[37]
MWCNT–SH	0.02 g	Pb^{2+} (100 ml, 10 $mg\ l^{-1}$)	—	5	240	25		144.9	[38]
MWCNT-COOH	10 $mg\ l^{-1}$	Cu^{2+} (100 ml)	40	3	60	25	Langmuir	118.41	[33]
MWCNT–ttpy[c)]	50 mg	Pb^{2+} Zn^{2+} (20 cm^3, 100 mg dm^3)	189.2	4.5 5.5	24 h	20	Langmuir	36.23 32.60	[68]
Fe_3O_4/CNTs–SH	0.5 $g\ l^{-1}$	Hg^{2+} (50 ml, 10 $mg\ l^{-1}$)	200.21	3–11	180		Freundlich	172.4	[35]
MWCNT–ADIF[d)]	30 mg	Pb^{2+} (10 ml, 50 $mg\ l^{-1}$)	-	3	3 h	25	Langmuir	78.12	[69]
MWCNTs–amidoxime	10 mg	U^{4+} (10 ml)	—	5.0	60	25	Langmuir	67.9	[34]
FeO_x/CNTs	0.4 $g\ l^{-1}$	Sb^{3+} (50 ml, 10 $mg\ l^{-1}$)	242	6.3	12 h	20	Langmuir	172	[70]
$Fe^{(2+/3+)}$/MWCNT–Ethylenediamine	100 $mg\ l^{-1}$	As^{5+} (100 $\mu g\ l^{-1}$)	—	4	100	25	Langmuir	23.47/13.74	[71]
MnO_2/MWCNTs-o[e)]	0.05 g	Cd^{2+} (100 ml)	96	5.0	150	25	Langmuir	41.6	[72]

a) In all reports, batch adsorption experiments were used.
b) MWCNTs – Multiwalled carbon nanotubes.
c) ttpy – 40-(4-hydroxyphenyl)-2,2′:6′,2″-terpyridine.
d) ADIF – 3-amino-5*a*,10*a*-dihydroxybenzo[*b*]indeno[2,1*d*]furan-10-one.
e) MWCNTs-o – acid oxidized multiwalled carbon nanotubes.

(c) In most of the adsorption studies, the suitable experimental temperature was found to be 20 °C.
(d) Addition of metal oxide nanoparticles such as Fe_3O_4, MnO_2, etc., can further improve the adsorption capacity of modified CNTs.
(e) Metal ion adsorption process will take place through ion exchange process. For example, adsorption of Pb(II) ions on (–SH, –COOH, and –OH) modified MWCNTs happens via ion-exchange process (Eqs. (6.2)–(6.4)) [38].

$$-SH + Pb^{2+} \rightleftharpoons -SPb^{+} + H^{+} \tag{6.2}$$

$$-OH + Pb^{2+} \rightleftharpoons -OPb^{+} + H^{+} \tag{6.3}$$

$$-COOH + Pb^{2+} \rightleftharpoons -COOPb^{+} + H^{+} \tag{6.4}$$

(f) In case of Sb^{3+} ions not only adsorption but also further oxidation ($Sb^{3+} \rightarrow Sb^{5+}$) takes place. Favorably, the oxidized form of antimony (Sb^{5+}) shows less toxicity [70].
(g) Modified CNTs with higher surface area will show higher adsorption capacity [35]. However, it is not true for all adsorption processes.

6.4 Influencing Factors

The adsorption process depends on many experimental parameters (Figure 6.5). In this section, influencing parameters for adsorption of organic/metal ions by surface modified CNT-based adsorbents were discussed.

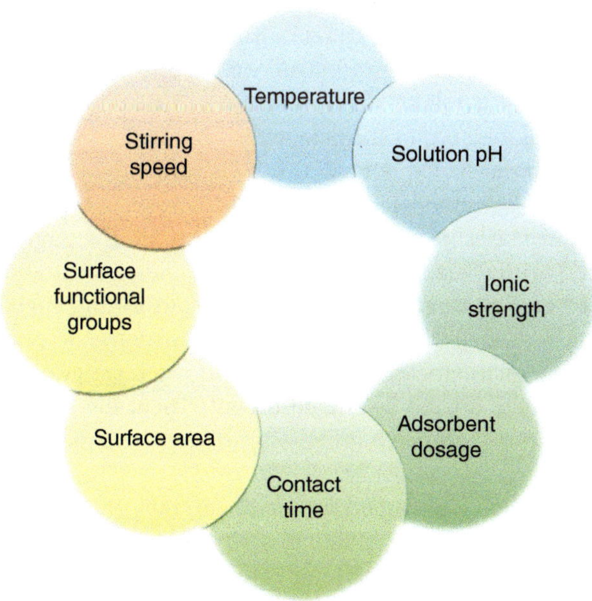

Figure 6.5 Illustration represents the parameters that affect the adsorption property of chemically modified CNT-based adsorbents.

(a) *Temperature*: As a general rule, the properties of surface modified CNT-based adsorbents are severely affected by temperature. For example, the effect of temperature (25–65 °C) on the methylene blue adsorption on CNT–tyrosine powder has been studied [44]. When the experimental temperature increases, the removal efficiency decreases. The efficiency is determined to be 99.0% at 25 °C and 94.2% at 65 °C. The optimum temperature for this adsorption process is determined as 25 °C.

(b) *The pH of the solution:* The pH value determines the ionic forms of pollutants (metal ions and charged organic molecules like methylene blue) and surface modified CNTs. For effective metal cation adsorption by modified CNTs, a low pH (acidic condition) condition is highly preferable. In a study of methylene blue adsorption by CNT–$NHSO_3H$ adsorbent [31], the maximum adsorption efficiency (90%) was found at a higher pH (= 12). In this process also, electrostatic attraction plays a major role, which was explained by the following way: under basic conditions, ionization of $-SO_3H$ groups in CNT–$NHSO_3H$ is promoted and thus the adsorbent attains negative charges ($-SO_3^-$). Whereas positively charged dye undergoes attraction toward the adsorbent. On the other hand, the surface charge density of the adsorbents is also determined by the pH value (parameter pH_{PZC}).

(c) *Ionic strength and coexisting ions:* The adsorption property of modified CNTs is influenced by the concentration of coexisting salts (ionic strength). The influence of NaCl concentrations on methylene blue adsorption by CNTs in aqueous solution was studied by Lei et al. [31]. The methylene blue adsorption capacity of CNT–$NHSO_3H$ is comparatively low in the presence of NaCl (0.01–0.2 mol l^{-1}). But, in the presence of 0.5 mol l^{-1} NaCl, the methylene blue adsorption capacity was high (477.80 mg g^{-1}). This was due to the fact that the large concentration of NaCl might affect ionization of methylene blue molecules and thus stimulate their removal efficiency.

(d) *Dosage of adsorbent:* The dosage of modified CNTs (adsorbent) has a significant influence on the adsorption of pollutants. Usually, an increase in adsorbent concentration increases its adsorption capacity toward the targeted pollutants. Saxena et al. [44] have studied the influence of the concentration of CNT–tyrosine (adsorbent) on methylene blue dye. Upto 10 mg, higher adsorptive removal efficiency (99%) was observed, and then efficiency becomes constant. Initial enhancement in efficiency is ascribed to the huge available surface area and active centers offered by the adsorbent. Then, the removal efficiency becomes constant as the maximum amount of the dye is adsorbed (or no dye for adsorption).

(e) *Contact time*: The contact time is an important factor to know the effective adsorption capacity of an adsorbent. The pollutants removal percentage will generally increase with increasing adsorption time and reach equilibrium. In the beginning of the adsorption experiment, concentration of pollutants and surface area of adsorbent will be high. Therefore, a strong driving force between adsorbent and pollutant exists at this initial time. But, adsorption equilibrium is majorly reliant on the initial concentration of pollutants. For example, the adsorption capacity of Ni^{2+} ions on Fe_3O_4/MWCNT–β-cyclodextrin has been enhanced with increasing contact time and maximized at 50 minutes [37].

(f) *Surface area:* Generally, adsorbent with higher surface area will show relatively higher adsorption capacity. However, in some reports, adsorbents with less surface area showed greater removal efficiency. For example, the surface area of MnO_2/MWCNTs-o, MWCNTs-o, and pure MWCNTs is 96, 177, and $\geq$200 $m^2\ g^{-1}$, respectively [72]. It seems that pure MWCNTs should possess higher removal efficiency. In contrast, MnO_2/MWCNTs-o with a relatively low surface area showed higher removal efficiency toward adsorption of Cd^{2+} ions. Similarly, [59] deep eutectic solvent modified CNT/$ZnCo_2O_4$ possesses less surface area (52.312 $m^2\ g^{-1}$) than that of pure CNT/$ZnCo_2O_4$ (67.609 $m^2\ g^{-1}$). However, adsorption capacity of pure CNT/$ZnCo_2O_4$ and deep eutectic solvent-CNT/$ZnCo_2O_4$ is 109.32 and 412.497, respectively.

(g) *Surface functional groups*: The surface groups on the CNTs are having much influence on the adoption tendency of modified CNTs. As already mentioned, Cd^{2+} adsorption capacity of MWCNTs-o (–OH, –COOH, and –C=O) was greater than that of pure MWCNTs [72]. Similarly, MWCNT with –SH groups showed higher Pb^{2+} adsorption capacity than that of CNTs with –COOH groups [38].

(h) *Stirring speed*: The stirring speed of the adsorption aliquots is also an important factor. Usually, higher stirring speeds improve the interaction between the pollutant and the adsorbent. Saxena et al. [44] have studied the effect of stirring speed (200–800 rpm) on the methylene blue adsorption of MWCNTs-tyrosine adsorbent. Maximum removal efficiency was observed at 400 rpm (99.0%). Further increases in stirring speed slightly decreased the removal efficiency.

6.5 Adsorption Mechanisms of Chemically Modified CNTs

The modes of adsorption mechanisms by modified CNT-based adsorbents toward organic and inorganic pollutants vary considerably. In the case of organic adsorptive removal, π–π stacking interaction, van der Waals forces, hydrophobic interactions, hydrogen bonding attraction, and electrostatic interactions will be the adsorption pathway of organics on modified CNT-based adsorbents. In certain conditions, modified CNTs, especially CNTs with hydrophilic functional groups (–COOH and –OH) wouldn't adsorb hydrophobic organic pollutants. The aromatic character of oxidized CNTs decreases as well as, showing in a weak π–π stacking attraction. Hence, the adsorption tendency of modified CNTs is mainly attributed to hydrogen bonding attraction and van der Waals attraction. On the other hand, the metal ion adsorption pathway is due to physisorption, ionic interaction, ion exchange, and surface complex formation. Each type of adsorption mechanisms is explained below:

(a) *π–π stacking interaction*: The π–π stacking or π–π interaction is a kind of noncovalent attraction, occurs between aromatic molecules through π-bonds. The presence of delocalized π-electrons (sp^2 hybridized carbon atoms) in the surface modified CNT could attract π-electrons present in organic pollutants. If the covalently attached functional organic moiety contains π-bonds, that will also contribute to π–π stacking in the organic adsorption. For example, MWCNTs–tyrosine adsorbs methylene blue dye via π–π interactions [44].

(b) *Hydrogen bonding*: The interaction of hydrogen atoms with functional groups (–OH, –COOH, $-HN_2$, etc.) containing more electronegative atoms such as nitrogen, oxygen, and fluorine is known as hydrogen bonding. Hydrogen bonding will happen between surface groups (like –COOH and –OH) of modified CNTs and organic pollutants. As an example, methylene blue, which contains nitrogen atoms, will form hydrogen bonds with hydrogen atoms in the modified CNT [44].

(c) *Electrostatic attraction*: Electrostatic force or Coulomb force may be either an attractive or repulsive force. This force acts between the charged species. In the case of modified CNTs involved in the adsorption process, solution pH decides the presence of electrostatic force. Generally, at higher pH (alkaline conditions), the carboxylic acid group of CNTs will be deprotonated and they will become a negatively charged. Then, cationic pollutants will efficiently attract CNTs via electrostatic attraction. Adsorption due to the electrostatic force of attraction between methylene blue and porphyrin derivatives functionalized CNTs was reported [40].

(d) *van der Waals force*: van der Waals force is a physical force that is normally exerted between molecules. It is further classified as weak London force and dipole–dipole interaction. van der Waals force played a role in the eosin Y adsorption by $ZnCo_2O_4$/CNT-deep eutectic solvent adsorbent [59].

(e) *Hydrophobic interaction*: Molecules that repel water are known as hydrophobic. Generally, aromatic organic pollutants are hydrophobic and they can show hydrophobic interaction with modified CNT with hydrophobic parts.

(f) *Ion-exchange process*: The exchange of ions between two functional moieties is called an ion-exchange process. Ion-exchange-based adsorption is possible between oxygen-containing surface groups of CNTs and metal ions. For instance, the adsorption of Cd^{2+} on MnO_2/MWCNTs–COOH/–OH/–C=O occurs through ion-exchange processes (Eqs. (6.5) and (6.6)) [72].

$$2(\text{MWCNT}_\pi\text{–H}^+) + \text{Cd}^{2+} \leftrightarrow (\text{MWCNT}_\pi)^2\text{–Cd}^{2+} + 2\text{H}^+ \tag{6.5}$$

$$\text{Cd}^{2+} + 2(\text{MWCNT–COOH}) \leftrightarrow \text{Cd(MWCNT–COO)}_2 + 2\text{H}^+ \tag{6.6}$$

(g) *Surface complex formation*: The formation of coordinate covalent bond between metal ions and surface groups of CNTs is known as surface complex formation. This process is a dominant driving force in the adsorption phenomenon. For example, based on the experimental evidence, Cu^{2+} ion adsorption by MWCNTs–COOH follows surface complex formation (6.7) [33].

$$\text{MWCNT–COOH} + \text{Cu}^{2+} \rightarrow \text{MWCNT–COOHCu(OH)}^+ \tag{6.7}$$

Similarly, surface complex formation was found to be the important reason for Hg^{2+} ion adsorption by Fe_3O_4/CNTs–SH [35]. This is because of the soft basic and soft acidic nature of –SH groups and Hg^{2+} ions, respectively [72].

6.6 Modified CNT-Based Materials Toward Commercialization

Apart from research articles, attempts have been made to commercialize the modified CNT adsorbents for the removal of aquatic contaminants. Mainly, patents related to nanoparticle modified CNTs adsorbents could be seen. For instance, in the patent publication of Abbas et al., metal oxide (Al_2O_3, ZnO, and iron oxide nanoparticles) modified MWCNTs have been prepared and used for adsorptive elimination of *para*-xylene and toluene from water [73]. In another patent published by Atieh et al. [74], Al_2O_3 or ZnO or Fe_2O_3 impregnated on the surface of MWCNTs has been synthesized for the capturing of some aromatic hydrocarbons (benzene, xylene, toluene, and ethyl benzene) from water. In this work, an optimum ratio between MWCNT and metal oxides was identified. They also found the optimum experimental condition to obtain the maximum adsorption capacity of targeted aromatic compounds. Recently, CNT-grafted acrylic acid or acrylamide sorbent was demonstrated as a sorbent for the phenol removal process [75]. This patent elaborates on the synthetic procedure and composition of sorbent materials. Also, optimum parameters for phenol adsorption capacity of this sorbent were also described. On the other hand, globally, there are many suppliers of chemically modified CNT-based materials. A few of them are listed below:

a. Nano-C Inc. United States, https://nano-c.com
b. NoPo Nanotechnologies Private Limited, India, https://nopo.in/web
c. Otto Chemie Pvt. Ltd., India https://www.ottokemi.com/category
d. Chengdu Organic Chemicals Co. Ltd., Chinese Academy of Sciences, China, http://www.timesnano.com/en
e. Alfa Chemistry, USA, https://graphene.alfa-chemistry.com
f. SabiNano, South Africa https://sabinano.co.za/about

6.7 Conclusion and Future Perspectives

Complete elimination of harmful and hazardous contaminants such as synthetic dyes, heavy metal ions, etc., from the water is an important need for environmental safety and protection of living beings from health risks. CNT-based adsorbents are important materials in the treatment of polluted waters. To improve pollutant adsorption capacity, chemical modification of CNTs is one of the best strategies. Simple groups (–COOH, –OH, $-NH_2$, –SH, –C=O, etc.) and several organic molecules (tyrosine, porphyrins and its derivatives, cyclodextrin, deep eutectic solvents, etc.) have been covalently attached to CNTs. Chemical modification was mainly focused on MWCNTs. Chemically modified CNTs showed greater adsorption capacity toward organic (dyes, pharmaceuticals, phenol, benzene and its derivatives, etc.) and inorganic (metal ions) water pollutants. The adsorption process depends on the surface groups, surface area, and dosage of modified CNT-based adsorbents. Experimental parameters such as pH, temperature, stirring speed, and ionic strength of electrolytes also significantly affect the adsorption capacity of modified CNT powder. Also, the adsorption tendency of modified CNTs is strongly dependent on the molecular structure of organic pollutants and the nature of metal ions. The adsorption process involves various interactions such as π–π stacking, hydrophobic, van der Waals forces, hydrogen bonding, ion exchange, Columbic force, and surface complex formation. There is also a possibility to couple modified CNT-based adsorption technology with other pollutant removal technologies for effective water remediation.

Besides, finding the optimized adsorbents for various pollutants is still a challenge. More studies are essential to abate the adsorbed pollutants consecutively. It is important to know the toxicity of modified CNTs because disposal of used CNT adsorbents may cause environmental issues. Alternatively, chemically modified CNT-based membranes, sponges, tablets, or beads could be developed as eco-friendly adsorbents.

Acknowledgments

The authors would like to thank Solar Energy Research Center, SERC-Chile (FONDAP/ANID/15110019).

References

1 Saravanan, A., Senthil Kumar, P., Hemavathy, R.V. et al. (2022). A comprehensive review on sources, analysis and toxicity of environmental pollutants and its removal methods from water environment. *Sci. Total Environ.* 812: 152456. https://doi.org/10.1016/j.scitotenv.2021.152456.

2 Sun, S., Chen, Y., Lin, Y., and An, D. (2018). Occurrence, spatial distribution, and seasonal variation of emerging trace organic pollutants in source water for Shanghai, China. *Sci. Total Environ.* 639: 1–7. https://doi.org/10.1016/j.scitotenv.2018.05.089.

3 Du, Y., Xu, X., Liu, Q. et al. (2022). Identification of organic pollutants with potential ecological and health risks in aquatic environments: Progress and challenges. *Sci. Total Environ.* 806: 150691. https://doi.org/10.1016/j.scitotenv.2021.150691.

4 Senthil Rathi, B., Senthil Kumar, P., and Vo, D.-V.N. (2021). Critical review on hazardous pollutants in water environment: occurrence, monitoring, fate, removal technologies and risk assessment. *Sci. Total Environ.* 797: 149134. https://doi.org/10.1016/j.scitotenv.2021.149134.

5 Chen, Q., Yao, Y., Li, X. et al. (2018). Comparison of heavy metal removals from aqueous solutions by chemical precipitation and characteristics of precipitates. *J. Water Process. Eng.* 26: 289–300. https://doi.org/10.1016/j.jwpe.2018.11.003.

6 Li, R., Li, L., Zhang, Z. et al. (2021). Limiting factors of heavy metals removal during anaerobic biological pretreatment of municipal solid waste landfill

leachate. *J. Hazard. Mater.* 416: 126081. https://doi.org/10.1016/j.jhazmat.2021.126081.

7 Wang, D., Sun, D., Tian, X. et al. (2021). Role of microbial communities on organic removal during petrochemical wastewater biological treatment with pure oxygen aeration. *J. Water Process. Eng.* 42: 102151. https://doi.org/10.1016/j.jwpe.2021.102151.

8 Ray, S.S., Gusain, R., and Kumar, N. (2020). Water purification using various technologies and their advantages and disadvantages, Chapter 3. In: *Carbon Nanomaterial-Based Adsorbents for Water Purification Fundamentals and Applications Micro and Nano Technologies*, 37–66, https://doi.org/10.1016/B978-0-12-821959-1.00003-9. Elsevier.

9 Novikau, R. and Lujaniene, G. (2022). Adsorption behaviour of pollutants: heavy metals, radionuclides, organic pollutants, on clays and their minerals (raw, modified and treated): a review. *J. Environ. Manage.* 309: 114685. https://doi.org/10.1016/j.jenvman.2022.114685.

10 Cychosz, K.A. and Thommes, M. (2018). Progress in the physisorption characterization of nanoporous gas storage materials. *Engineering* 4: 559–566. https://doi.org/10.1016/j.eng.2018.06.001.

11 Cuenya, B.R. and Behafarid, F. (2015). Nanocatalysis: size- and shape-dependent chemisorption and catalytic reactivity. *Surf. Sci. Rep.* 70: 135–187. https://doi.org/10.1016/j.surfrep.2015.01.001.

12 Okoro, H.K., Pandey, S., Ogunkunle, C.O. et al. (2022). Nanomaterial-based biosorbents: adsorbent for efficient removal of selected organic pollutants from industrial wastewater. *Emerging Contam.* 8: 46–58. https://doi.org/10.1016/j.emcon.2021.12.005.

13 Samuel, M.S., Selvarajan, E., Sarswat, A. et al. (2022). Nanomaterials as adsorbents for As(III) and As(V) removal from water: a review. *J. Hazard. Mater.* 424: 127572. https://doi.org/10.1016/j.jhazmat.2021.127572.

14 Balasubramanian, K. and Burghard, M. (2005). Chemically functionalized carbon nanotubes. *Small* 1 (2): 180–192. https://doi.org/10.1002/smll.200400118.

15 Wei, C., Ye, N., Xia, W. et al. (2022). An electroless deposition strategy for preparing ultrathin CNTs/Cu composite foils with excellent mechanical properties. *Diamond Relat. Mater.* 121: 108785. https://doi.org/10.1016/j.diamond.2021.108785.

16 Ding, C., Yu, K., Nodooshan, H.R.J. et al. (2021). Effect of powder microstructure on the thermal and mechanical properties of hot extruded Al-CNT composite. *J. Alloys Compd.* 891: 162059. https://doi.org/10.1016/j.jallcom.2021.162059.

17 Siahkouhi, M., Razaqpur, G., Hoult, N.A. et al. (2021). Utilization of carbon nanotubes (CNTs) in concrete for structural health monitoring (SHM) purposes: a review. *Constr. Build. Mater.* 309: 125137. https://doi.org/10.1016/j.conbuildmat.2021.125137.

18 Benos, L., Spyrou, L.A., and Sarris, I.E. (2019). Development of a new theoretical model for blood-CNTs effective thermal conductivity pertaining to hyperthermia therapy of glioblastoma multiform. *Comput. Methods Programs Biomed.* 172: 79–85. https://doi.org/10.1016/j.cmpb.2019.02.008.

19 Dhall, S., Mehta, B.R., Tyagi, A.K., and Sood, K. (2021). A review on environmental gas sensors: materials and technologies. *Sensors Int.* 2: 100116. https://doi.org/10.1016/j.sintl.2021.100116.

20 Vesali-Naseh, M., Naseh, M.R.V., and Ameri, P. (2021). Adsorption of Pb(II) ions from aqueous solutions using carbon nanotubes: a systematic review. *J. Cleaner Prod.* 291: 125917. https://doi.org/10.1016/j.jclepro.2021.125917.

21 Khan, F.S.A., Mubarak, N.M., Khalid, M. et al. (2021). A comprehensive review on micropollutants removal using carbon nanotubes-based adsorbents and membranes. *J. Environ. Chem. Eng.* 9: 106647. https://doi.org/10.1016/j.jece.2021.106647.

22 Mikhalchan, A. and Vilatela, J.J. (2019). A perspective on high-performance CNT fibres for structural composites. *Carbon* 150: 191–215. https://doi.org/10.1016/j.carbon.2019.04.113.

23 Sharma, V., Sharma, V., Goyat, M.S. et al. (2020). Recent progress in nano-oxides and CNTs based corrosion resistant superhydrophobic coatings: a critical review. *Prog. Org. Coat.* 140: 105512. https://doi.org/10.1016/j.porgcoat.2019.105512.

24 Fiyadh, S.S., AlSaadi, M.A., Jaafar, W.Z. et al. (2019). Review on heavy metal adsorption processes by carbon nanotubes. *J. Cleaner Prod.* 230: 783–793. https://doi.org/10.1016/j.jclepro.2019.05.154.

25 Salem, M.A.S., Khan, A.M., Manea, Y.K., and Wani, A.A. (2022). Nano chromium embedded in f-CNT supported CoBi-LDH nanocomposites for selective adsorption of Pb^{2+} and hazardous organic dyes. *Chemosphere* 289: 133073. https://doi.org/10.1016/j.chemosphere.2021.133073.

26 Lavagna, L., Nistico, R., Musso, S., and Pavese, M. (2021). Functionalization as a way to enhance dispersion of carbon nanotubes in matrices: a review. *Mater. Today Chem.* 20: 100477. https://doi.org/10.1016/j.mtchem.2021.100477.

27 Wei, J., Sun, W., Pan, W. et al. (2017). Comparing the effects of different oxygen-containing functional groups on sulfonamides adsorption by carbon nanotubes: experiments and theoretical calculation. *Chem. Eng. J.* 312: 167–179. https://doi.org/10.1016/j.cej.2016.11.133.

28 Mohammadi, A. and Veisia, P. (2018). High adsorption performance of β-cyclodextrin-functionalized multi-walled carbon nanotubes for the removal of organic dyes from water and industrial wastewater. *J. Environ. Chem. Eng.* 6: 4634–4643. https://doi.org/10.1016/j.jece.2018.07.002.

29 Zhou, Y., He, Y., Xiang, Y. et al. (2019). Single and simultaneous adsorption of pefloxacin and Cu(II) ions from aqueous solutions by oxidized multiwalled carbon nanotube. *Sci. Total Environ.* 646: 29–36. https://doi.org/10.1016/j.scitotenv.2018.07.267.

30 Jia, L., Liu, W., Cao, J. et al. (2020). Modified multi-walled carbon nanotubes assisted foam fractionation for effective removal of acid orange 7 from the dyestuff wastewater. *J. Environ. Manage.* 262: 110260. https://doi.org/10.1016/j.jenvman.2020.110260.

31 Lei, Y., Huang, Q., Dou, J. et al. (2021). Fast adsorptive removal of cationic organic dye by anionic group functionalized carbon nanotubes with high efficiency. *Colloids Interface Sci. Commun.* 40: 100328. https://doi.org/10.1016/j.colcom.2020.100328.

32 Dong, M., Guo, J., Wang, Y. et al. (2022). Humic acid non-covalent functionalized multi-walled carbon nanotubes composite membrane and its application for the removal of organic dyes. *J. Environ. Chem. Eng.* 10: 107320. https://doi.org/10.1016/j.jece.2022.107320.

33 Gupta, V.K., Agarwal, S., Bharti, A.K., and Sadegh, H. (2017). Adsorption mechanism of functionalized multi-walled carbon nanotubes for advanced Cu(II) removal. *J. Mol. Liq.* 230: 667–673. https://doi.org/10.1016/j.molliq.2017.01.083.

34 Wu, J., Tian, K., and Wang, J. (2018). Adsorption of uranium(VI) by amidoxime modified multiwalled carbon nanotubes. *Prog. Nucl. Energy* 106: 79–86. https://doi.org/10.1016/j.pnucene.2018.02.020.

35 Fan, L., Zhou, A., Zhong, L. et al. (2019). Selective and effective adsorption of Hg(II) from aqueous solution over wide pH range by thiol functionalized magnetic carbon nanotubes. *Chemosphere* 226: 405–412. https://doi.org/10.1016/j.chemosphere.2019.03.154.

36 Yılmaz, C.E., Aslani, M.A.A., and Aslani, C.K. (2020). Removal of thorium by modified multi-walled carbon nanotubes: optimization, thermodynamic, kinetic, and molecular dynamic viewpoint. *Prog. Nucl. Energy* 127: 103445. https://doi.org/10.1016/j.pnucene.2020.103445.

37 Lin, S., Zou, C., Liang, H. et al. (2021). The effective removal of nickel ions from aqueous solution onto magnetic multi-walled carbon nanotubes modified by β-cyclodextrin. *Colloids Surf., A* 619: 126544. https://doi.org/10.1016/j.colsurfa.2021.126544.

38 Qu, G., Zhou, J., Liang, S. et al. (2022). Thiol-functionalized multi-walled carbon nanotubes for effective removal of Pb(II) from aqueous solutions. *Mater. Chem. Phys.* 278: 125688.

39 Le, V.T., Ngo, C.L., Le, Q.T. et al. (2013). Surface modification and functionalization of carbon nanotube with some organic compounds. *Adv. Nat. Sci.: Nanosci. Nanotechnol.* 4: 035017. 5 pp, https://doi.org/10.1088/2043-6262/4/3/035017.

40 Saleh, T.A., Elsharif, A.M., and Bin-Dahman, O.A. (2021). Synthesis of amine functionalization carbon nanotube-low symmetry porphyrin derivatives conjugates toward dye and metal ions removal. *J. Mol. Liq.* 340: 117024. https://doi.org/10.1016/j.molliq.2021.117024.

41 Mishra, N.S., Kuila, A., Nawaz, A. et al. (2018). Engineered carbon nanotubes: review on the role of surface chemistry, mechanistic features, and toxicology in the adsorptive removal of aquatic pollutants. *ChemistrySelect* 3: 1040–1055. https://doi.org/10.1002/slct.201702951.

42 Steffen, T.T., Fontana, L.C., Hammer, P., and Becker, D. (2019). Carbon nanotube plasma functionalization: the role of carbon nanotube/maleic anhydride solid premix. *Appl. Surf. Sci.* 491: 405–410. https://doi.org/10.1016/j.apsusc.2019.06.176.

43 Jaime, A.F.Q., Amoros, D.C., and Morallon, E. (2020). Effect of surface oxygen groups in the electrochemical modification of multi-walled carbon nanotubes by

4-amino phenyl phosphonic acid. *Carbon* 165: 328–339. https://doi.org/10.1016/j.carbon.2020.04.062.

44 Saxena, M., Sharma, N., and Saxena, R. (2020). Highly efficient and rapid removal of a toxic dye: adsorption kinetics, isotherm, and mechanism studies on functionalized multiwalled carbon nanotubes. *Surf. Interfaces* 21: 100639. https://doi.org/10.1016/j.surfin.2020.100639.

45 Díez-Pascual, A.M. (2021). Chemical functionalization of carbon nanotubes with polymers: a brief overview. *Macromolecules* 1: 64–83. https://doi.org/10.3390/macromol1020006.

46 Tkaczyk, A., Mitrowska, K., and Posyniak, A. (2020). Synthetic organic dyes as contaminants of the aquatic environment and their implications for ecosystems: a review. *Sci. Total Environ.* 717: 137222. https://doi.org/10.1016/j.scitotenv.2020.137222.

47 Sharma, J., Sharma, S., and Soni, V. (2021). Classification and impact of synthetic textile dyes on Aquatic Flora: a review. *Reg. Stud. Mar. Sci.* 45: 101802. https://doi.org/10.1016/j.rsma.2021.101802.

48 Hanafi, M.F. and Sapawe, N. (2020). A review on the water problem associate with organic pollutants derived from phenol, methyl orange, and remazol brilliant blue dyes. *Mater. Today: Proc.* 31: A141–A150. https://doi.org/10.1016/j.matpr.2021.01.258.

49 Lawal, I.A., Lawal, M.M., Azeez, M.A., and Ndungu, P. (2019). Theoretical and experimental adsorption studies of phenol and crystal violet dye on carbon nanotube functionalized with deep eutectic solvent. *J. Mol. Liq.* 288: 110895. https://doi.org/10.1016/j.molliq.2019.110895.

50 Maazinejad, B., Mohammadnia, O., Ali, G.A.M. et al. (2020). Taguchi L9 (34) orthogonal array study based on methylene blue removal by single-walled carbon nanotubes-amine: adsorption optimization using the experimental design method, kinetics, equilibrium and thermodynamics. *J. Mol. Liq.* 298: 112001. https://doi.org/10.1016/j.molliq.2019.112001.

51 Yang, G., Huang, Q., Gan, D. et al. (2019). Biomimetic functionalization of carbon nanotubes with poly(ionic liquids) for highly efficient adsorption of organic dyes. *J. Mol. Liq.* 296: 112059. https://doi.org/10.1016/j.molliq.2019.112059.

52 Li, Z., Sellaoui, L., Franco, D. et al. (2020). Adsorption of hazardous dyes on functionalized multiwalled carbon nanotubes in single and binary systems: experimental study and physicochemical interpretation of the adsorption mechanism. *Chem. Eng. J.* 389: 124467. https://doi.org/10.1016/j.cej.2020.124467.

53 Xin, X., Huang, G., and Zhang, B. (2021). Review of aquatic toxicity of pharmaceuticals and personal care products to algae. *J. Hazard. Mater.* 410: 124619. https://doi.org/10.1016/j.jhazmat.2020.124619.

54 Maculewicz, J., Kowalska, D., Świacka, K. et al. (2022). Transformation products of pharmaceuticals in the environment: their fate, (eco)toxicity and bioaccumulation potential. *Sci. Total Environ.* 802: 149916. https://doi.org/10.1016/j.scitotenv.2021.149916.

55 Wang, L., Chen, Y., Zhao, Y. et al. (2020). Toxicity of two tetracycline antibiotics on *Stentor coeruleus* and *Stylonychia lemnae*: potential use as toxicity indicator. *Chemosphere* 255: 127011. https://doi.org/10.1016/j.chemosphere.2020.127011.

56 Chen, C., Feng, X., and Yao, S. (2021). Ionic liquid-multi walled carbon nanotubes composite tablet for continuous adsorption of tetracyclines and heavy metals. *J. Cleaner Prod.* 286: 124937. https://doi.org/10.1016/j.jclepro.2020.124937.

57 Gonzalez-Fernandez, L.A., Medellín-Castillo, N.A., Ocampo-Perez, R. et al. (2021). Equilibrium and kinetic modelling of triclosan adsorption on single-walled carbon nanotubes. *J. Environ. Chem. Eng.* 9: 106382. https://doi.org/10.1016/j.jece.2021.106382.

58 Wang, W., Xiong, P., Zhang, H. et al. (2021). Analysis, occurrence, toxicity and environmental health risks of synthetic phenolic antioxidants: a review. *Environ. Res.* 201: 111531. https://doi.org/10.1016/j.envres.2021.111531.

59 Lawal, I.A., Dolla, T.H., Pruessner, K., and Ndungu, P. (2019). Synthesis and characterization of deep eutectic solvent functionalized CNT/$ZnCo_2O_4$ nanostructure: kinetics, isotherm and regenerative studies on Eosin Y adsorption. *J. Environ. Chem. Eng.* 7: 102877. https://doi.org/10.1016/j.jece.2018.102877.

60 Lin, S., Zou, C., Cao, Y. et al. (2021). One-pot synthesis of β-cyclodextrin magnetic carbon nanotube (β-CD@MMWCNT) for effective removal of phenol from oily wastewater. *J. Environ. Chem. Eng.* 9: 106494. https://doi.org/10.1016/j.jece.2021.106494.

61 Nascimento, M.K.S., Loureiro, S., dos Reis Souza, M.R. et al. (2020). Toxicity of a mixture of monoaromatic hydrocarbons (BTX) to a tropical marine microcrustacean. *Mar. Pollut. Bull.* 156: 111272. https://doi.org/10.1016/j.marpolbul.2020.111272.

62 Anjum, H., Johari, K., Gnanasundaram, N. et al. (2019). Investigation of green functionalization of multiwall carbon nanotubes and its application in adsorption of benzene, toluene & *p*-xylene from aqueous solution. *J. Cleaner Prod.* 221: 323–338. https://doi.org/10.1016/j.jclepro.2019.02.233.

63 Alaee, M., Arias, P., Sjodin, A., and Bergman, A. (2003). An overview of commercially used brominated flame retardants, their applications, their use patterns in different countries/regions and possible modes of release. *Environ. Int.* 29: 683–689.

64 Wu, Y., Wang, Y., Lin, Z. et al. (2019). Three-dimensional α-Fe_2O_3/amino-functionalization carbon nanotube sponge for adsorption and oxidative removal of tetrabromobisphenol A. *Sep. Purif. Technol.* 211: 359–367. https://doi.org/10.1016/j.seppur.2018.10.002.

65 Patiño, Y., Díaz, E., Ordóñez, S. et al. (2015). Adsorption of emerging pollutants on functionalized multiwall carbon nanotubes. *Chemosphere* 136: 174–180. https://doi.org/10.1016/j.chemosphere.2015.04.089.

66 González-Acevedo, Z.I., García-Zarate, M.A., Núñez-Zarco, E.A., and Anda-Martín, B.I. (2018). Heavy metal sources and anthropogenic enrichment in the environment around the Cerro Prieto Geothermal Field, Mexico. *Geothermics* 72: 170–181. https://doi.org/10.1016/j.geothermics.2017.11.004.

67 Mitra, S., Chakraborty, A.J., Tareq, A.M. et al. (2022). Impact of heavy metals on the environment and human health: novel therapeutic insights to counter the toxicity. *J. King Saud Univ. Sci.* 34: 101865. https://doi.org/10.1016/j.jksus.2022.101865.

68 Oyetade, O.A., Skelton, A.A., Nyamori, V.O. et al. (2017). Experimental and DFT studies on the selective adsorption of Pb^{2+} and Zn^{2+} from aqueous solution by nitrogen-functionalized multiwalled carbon nanotubes. *Sep. Purif. Technol.* 188: 174–187. https://doi.org/10.1016/j.seppur.2017.07.022.

69 Eyni, H., Tahermansouri, H., Kiani, F., and Jahangiri, M. (2019). Kinetics, equilibrium and isotherms of Pb^{2+} adsorption from aqueous solutions on carbon nanotubes functionalized with 3-amino-5*a*,10*a*-dihydroxybenzo[*b*]indeno[2,1-*d*]furan-10-one. *New Carbon Mater.* 34 (6): 512–523. https://doi.org/10.1016/S1872-5805(19)60027-2.

70 Cheng, Z., Lyu, H., Shen, B. et al. (2022). Removal of antimonite (Sb(III)) from aqueous solution using a magnetic iron-modified carbon nanotubes (CNTs) composite: experimental observations and governing mechanisms. *Chemosphere* 288: 132581. https://doi.org/10.1016/j.chemosphere.2021.132581.

71 Velickovic, Z., Vukovic, G.D., Marinkovic, A.D. et al. (2012). Adsorption of arsenate on iron(III) oxide coated ethylenediamine functionalized multiwall carbon nanotubes. *Chem. Eng. J.* 181, 182: 174–181. https://doi.org/10.1016/j.cej.2011.11.052.

72 Luo, C., Wei, R., Guo, D. et al. (2013). Adsorption behavior of MnO_2 functionalized multi-walled carbon nanotubes for the removal of cadmium from aqueous solutions. *Chem. Eng. J.* 225: 406–415. https://doi.org/10.1016/j.cej.2013.03.128.

73 Abbas, A. and Abussaud, B.A. (2018). Method for the adsorptive removal of *para*-xylene and toluene from wastewater using modified carbon nanotubes. US Patent 10,076,741B2, filed 09 December 2015, issued 18 September 2018.

74 Atieh, M.A., Abbas, A., and Abussaud, B.A. (2018). Adsorption of aromatic hydrocarbons from water using metal oxide impregnated carbon nanotubes. US Patent 9,988,286B2, filed 01 September 2015, issued 05 June 2018.

75 Elsharif, A.M.Z. (2021). Method of producing a carbon nanotube grafted acrylic acid/acrylamide copolymer nanocomposite sorbent. US Patent Application 20210115180, filed 18 October 2019, issued 16 August 2022.

7

Chemically Modified Carbon Nanotubes in Removal of Textiles Effluents

Mohammad Ehtisham Khan[1], Akbar Mohammad[2], Wahid Ali[1], and Abdulrahman Ahmed Sharwani[2]

[1] *Department of Chemical Engineering Technology, College of Applied Industrial Technology (CAIT), Jazan University, Jazan-45971, Kingdom of Saudi Arabia*
[2] *School of Chemical Engineering, Yeungnam University, Gyeongsan-si, Gyeongbuk 38541, South Korea*

7.1 Introduction

Globalization and the industrial revolution have changed the economy and technology in a positive direction, as well as accelerated environmental degradation [1, 2]. Despite the fact that environmental pollution has reached a crisis level worldwide that might unfavorably affect public health and the environment as the years go by, a search was made for quiet evidence and causes of marine contamination [3]. However, organic pollutants have been a major contributor to air pollution for many reasons [4]. Water is likely to be contaminated , further by oxygen-demanding wastes, synthetic organic compounds, oil, and water pollutants such as organic pollutants [5, 6]. In the United States, around 20 crore tons of sewage is discharged into freshwater bodies daily and 70% of household waste is directly discharged into waterways, contaminating up to 90% of city water supplies [7]. Microscopical studies of synthetic organic compounds entering water bodies through various sources, such as industrial effluents, indicate that these compounds include pesticides, synthetic dyes, and polyaromatic compounds [5, 8]. In other words, organic compounds like pesticides, dyes, polyaromatics, and pharmaceuticals are getting into water bodies from various sources, especially industrial effluents, resulting in a foremost threat to freshwater resources of 0.1% in surface water and 0.6% in groundwater out of 2.7% [9]. In most cases, organic dyes are removed from wastewater using several different methods. For water reuse, adsorption demonstrates several advantages over other technologies, including low-initial costs, elasticity, ease of use, and inattentiveness to toxic pollutants [10–12].

A new class of carbonaceous materials, carbon nanotubes (CNTs), has drawn considerable interest as probable adsorbents for the handling of water owing to their large precise surfaces, dissimilar structures, and controllable superficial chemistry [13, 14]. The carbonaceous adsorbent CNTs are more effective than the conventional activated carbons (ACs) because of their advanced adsorption ability, shorter steadiness time, and less loss of weight in reactivation [15, 16]. There has been great

Chemically Modified Carbon Nanotubes for Commercial Applications, First Edition.
Edited by Jeenat Aslam, Chaudhery Mustansar Hussain, and Ruby Aslam.

potential for CNTs to remove many types of pollutants, including resorcinol, aniline, ethyl benzene, and phenol [17, 18].

7.2 History of Removal of Textiles Effluents

Since prehistoric times, color has been used for symbolic purposes, for visual communication and for expression as part of the evolution of humanity as a whole [19]. There is evidence that the use of color in different prehistoric environments contributed to the enhancement of artists' skills and their reasoning abilities [20]. Many experts believe that color has been linked with a symbol arrangement that has spread across several civilizations throughout history [21]. The pigments are made up of a mixture of soil, marine, spittle, or animal fat and are utilize to color dissimilar surfaces [22]. The primary pigments were possibly used to spot the membrane for ceremonial determinations or to protect it from insects or the sun [23]. According to, ochre was likely included in prehistoric humans' diets, supplying them with iron essential for development [23].

As time progressed in prehistory, the occurrence of the utilization of dissimilar stains was amplified, though it arose in a different way at different times and in different areas. In Africa and Europe, pigments were initially used during the Middle Stone Age and Middle Paleolithic periods (150 000–30 000 BCE) [24]. In South Africa, the Blombos cave displays equivalent line carvings that date back to 100 000–70 000 years ago, and pigments made from ochre clay were used [22]. These pigments were used, however, as recently as the fifteenth century in the caverns of Mona Island (western Puerto Rico), where the last paintings were found [25].

As early as 1826, Otto Unverdured prepared aniline by destructive distillation of indigo [26]. Aniline was used as a methodological process in later research into new dyes without researchers' significant it; though, according to August Wilhelm Von Hofmann, all constituents were similar [27]. The primary artificial color, Mauveine, was created by Henry Perkin in 1854 from coal tar [28]. In 1879, Perkin resolved the industrial problem by designing the foundation of new techniques for producing synthetic dyes [28]. By 1869, after coal tar was distilled for dyes, some natural dyes, including Alizarin, began to disappear as synthetic dyes became more affordable [29]. The dye industry has been diversifying since then, with more than 100 000 synthetic dyes reported [25]. Europe was the center of dye production at the beginning of the twentieth century [30]. Currently, China and India are the most important producers of fabrics in the world [31].

7.3 Chemically Modified Carbon Nanotubes

7.3.1 Chemical Properties

In contrast with graphene sheets, the curvature of CNT surfaces enhances the chemical reactivity [32, 33]. A curvature between the π and σ orbital causes orbital hybridization, resulting in the mix of the two orbitals. The amount of hybridization increases as the diameter of single-walled carbon nanotubes (SWCNTs) decreases [34]. As a result, CNT reactivity is directly correlated with an increased curvature, leading to an π-orbital mismatch. It is therefore necessary to distinguish between a nanotube's sidewall and its end caps [35]. Additionally, a smaller nanotube diameter increases reactivity [35]. The possibility of chemically modifying sidewalls or end caps has been demonstrated. It is probable to regulate the solubility of CNTs in various thinners/solvents in this way, for example. It is, however, problematic to covalently attribute molecular species to the sp^2-bonded carbon atoms on the sidewalls of nanotubes [35]. As a result, nanotubes are inert in most cases. Depending on how the CNTs are formed, morphological variations are possible in multiwalled carbon nanotubes (MWCNTs), such as chemical vapor deposition (CVD) or high voltage discharge arc technique [36, 37].

Three general groups of strategies can be used to modify the surfaces of CNTs [38].

(1) A chemical or electrochemical activation is required in order for the modifier to form a covalent bond (chemisorption) with the CNTs.
(2) The CNT surface is physically adsorbent to the modifier (physisorption).
(3) The incorporation of CNT-modified electrodes into films is one method of modification.

7.3.2 Modification Through Chemical Reduction of Diazonium Salts

In the existence of hypo-phosphorous acid as a reducer, an aromatic diazonium salt is reduced in a one-electron chemical reaction. In the reduction, nitrogen molecules are released along with a very responsive aryl radical that attaches to the surface of carbon, which forms a covalent bond with the transformer molecule [38–40].

By this method, gram quantities of modified CNTs can be manufactured in a simple and convenient manner, as well as provide high levels of functionalization, resulting in the bulk altered substrate of carbon or carbon-based material [41, 42]. In addition to its use as a thin film for fabricated chemically modified electrodes [43], Figure 7.1.

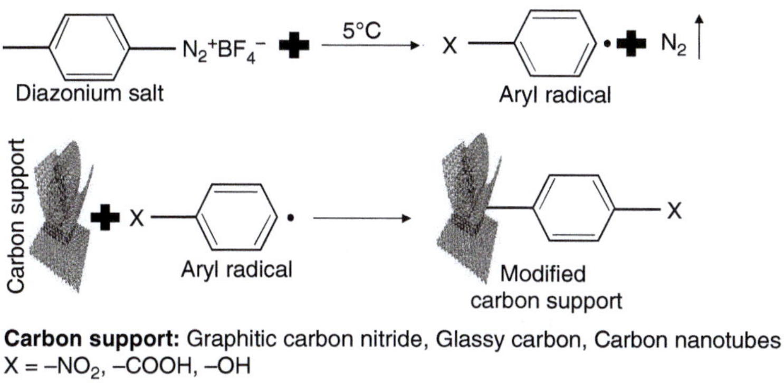

Figure 7.1 Schematic for reducing carbon substrates with hypophosphorous acid.

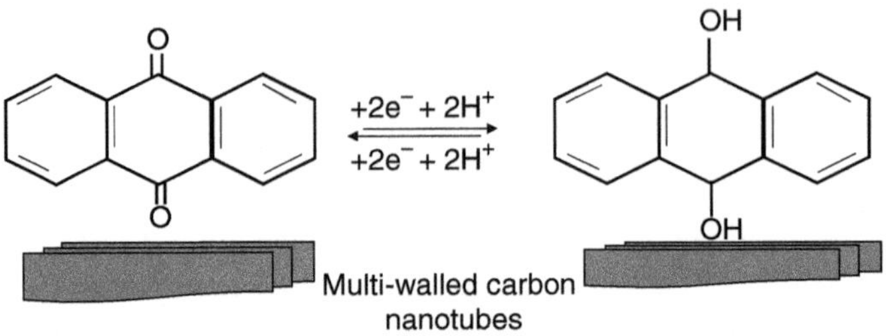

Figure 7.2 Two electrons and two protons from the anthraquinone/anthrahydroquinone redox couple of two electrons and two protons.

In general, at temperatures between 0 and 50 °C, H_3PO_2 is added to a combination of the substrate material of carbon and the diazonium salt solution with periodic moving during this procedure [44]. In order to eliminate the excess of acids, the resulting combination is filtered, cleaned with ample amounts of acetonitrile to eliminate any diazonium salt unreacted, and thoroughly diluted with water. In the end, the altered carbon is desiccated in a fume hood and kept in an airtight vessel for 12 hours [45]. By means of this approach, Heald et al. synthesized 1-anthraquinone-functionalized MWCNTs and deliberated their electrochemical behavior. According to the studies, the modified MWCNT experiences two electrons and two protons, resulting in a quasi-reversible cyclic voltammogram consistent with the quinine/hydroquinone redox couple, which is established by the pH study [41], Figure 7.2.

Masheter et al. have prolonged the derivatization technique of Heald et al. by employing various nanotube morphologies, such as h-MWCNs, b-MWCNs, and SWCNTs and projected the following alteration procedure [46]. In SWCNTs, a greater percentage of radical anthraquinonyl modifies the sidewalls than the end face of the tube, which reflects their greater reactivity when compared to MWCNTs [47]. Masheter et al. involved the anthraquinone species, which were chemically attached to the graphite powder surface after ascribing nitrophenyl groups and dropping them to aniline, then coupling them with anthraquinone-2-carboxylic acid to form CNTs modified with AQ–AN [46].

Further modifications can be made to the carboxyl groups introduced by oxidation, as they enable the formation of amide and ester bonds, which facilitate covalent coupling of molecules [48]. In this manner, nanotubes can be functionalized with a variety of molecules, such as diamines or bifunctional molecules, which function as linkers [49]. Various nanotubes, nucleic acids, enzymes, metal complexes, semiconductors, and metal nanoparticles can be considered as examples [49]. Additionally, carboxyl groups can form an anhydride at the tube end, allowing access to nanotube rings (Figure 7.3).

7.4 Dyes Removal Techniques

Industrial dyes are a significant threat to water and wastewater [50, 51]. There are different ways in which it causes problems: Despite low dye concentrations in

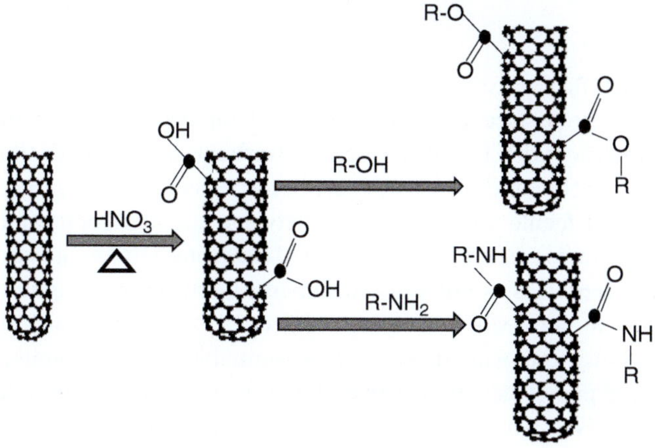

Figure 7.3 A chemical process that involves thermal oxidation of nanotubes followed by esterification or amidation of the carboxyl groups.

water, it is visible and undesirable; it avoids the penetration of sunlight into water; it discourages aquatic life; it decreases photosynthesis and impairs gas solubility in water bodies; and it causes aquatic life to die [30]. Consequently, it is primarily designed to eliminate dye constituents from wastewater. For the treatment of dye-containing wastewaters, electrochemical, chemical oxidation, flocculation, and coagulation, liquid–liquid extraction, and adsorption methods are used [52, 53]. As it is cost-effective, easy to design, easy to implement, and insensitive to poisonous elements, adsorption is an ideal method for removing organic matter from aqueous solutions [54, 55].

An extensive range of wastewater treatment methods have been industrialized for the elimination of pollutants from effluent, including membrane filtration, electrochemistry, osmosis, ion exchange, manganese dioxide oxidation, coagulation, and flocculation [56–58]. Despite their advantages, all of these techniques, except adsorption, face many complex difficulties, as well as lower efficiency, higher capital or operating expenses, excessive mud production, and high maintenance expenses that prevent them from being economically viable [59]. Nevertheless, the adsorption process stands out from the above-mentioned wastewater treatment techniques as it is easy to use, highly efficient, simple, cost-effective, and sustainable [60]. Additionally, most of the time, the process is reversible, allowing the adsorbents to be simply regenerated and recycled constantly, which makes the process of adsorption more cost-effective [61]. Moreover, this technique is versatile since there is a wide variety of adsorbents available to cater to specific needs [61]. In the recent years, CNTs have been extensively studied for their competence to eliminate environmental contaminants [62, 63]. CNTs have been widely utilized in the adsorption of colorants from aqueous media over the past few years, as described in this chapter. We reviewed and summarized the effects of several batch parameters as well as the time of contact between the adsorbate and the adsorbent, pH, initial colorant concentration, and dose of adsorbent on the adsorption procedure.

7.5 Adsorption

The adsorption of dye to CNTs, for example, is a surface procedure in which dye molecules are accrued on to the surface of the adsorbent [64]. Molecular adsorption occurs when molecules interact with molecules of an adsorbent, both chemically and physically, and chemisorption and physisorption are respective processes [65]. Physisorption is the result of feeble attractive forces between the adsorbent and adsorbate, while chemisorption is caused by the transfer of electrons between adsorbent and adsorbate species [66]. It is essential for adsorbents to possess the highest possible specific surface area, permeability, adsorption tendency, and mechanical constancy, as well as the ability to regenerate easily, be sustainable, and be selective [67, 68]. There are many types of adsorbents used today to treat wastewater. A wide range of adsorbents derived from agronomic, domestic, and industrial waste, polymers, and organic and inorganic materials are available [69, 70]. In general, however, the adsorbents attained from the overhead low-cost materials have low adsorption competence. As a consequence, more advanced, efficient, and effective adsorbents are needed for the effectual treatment of contaminated wastewater.

The nanoadsorbents can be divided into several groups according to their suitability for use in the adsorption processes [71, 72]. Examples include nanoparticles of magnetic materials, nanostructured mixed oxides, and metallic nanoparticles. In addition, CNTs, nanosheets, silicon nanoparticles, polymer nanoparticles, and nanoparticles of silicon are used as adsorbent materials for pollutant removal from wastewater [73, 74]. It is perceived that nanotechnology-decorated wastewater treatment methods are advantageous initiatives, not only in terms of one's ability to overcome the main encounters in water treatment but also to offer an efficient water supply [75]. The nanotechnology-decorated wastewater treatment methods could be helpful in a number of ways, including enhancing the efficiency of using alternative water sources [76].

7.6 Carbon-Based Nanoadsorbents

A significant amount of attention has been paid to nanoparticles made of carbon/carbon-based materials, such as graphene, fullerenes, and CNTs owing to their higher properties, including high-surface area, absorbency, thermal and mechanical stability, and elimination competence [77]. The properties of graphene, such as its specific surface area, surface alteration, high thermal constancy, mechanical strength, and ease of synthesis, have greatly drawn researchers' attention to it among the many carbon-based nanostructured materials [78, 79]. While graphene is beneficial for treating wastewater, its main disadvantage is that it tends to aggregate, which reduces its approachability, thus reducing its capacity to adsorb contaminants [80]. The CNTs are a highly effective method of removing pollutants from water. They have a greater surface area, higher absorbency, hydrophobic, hollow structure, and well-developed mesopores. In addition, environmental fates and toxic phenomena should also be considered when determining toxicity [81, 82].

In studies, CNTs were found to be noxious and the degree of noxiousness varied with their nature. There are several variables to consider: structure, degree of accumulation, aspects ratios, extent of oxidation, superficial area, constrained functional groups, and manufacture process [83].

7.7 Carbon Nanotubes

The scientist *Sumio Iijima*, a Japanese physicist who invented the arc evaporation method for CNTs [84], deserves credit. As well as arc evaporation, there have been reports of laser ablation, flame synthesis, CVD, and electrolysis for the fabrication of CNTs [85]. CNTs can be produced using approaches, such as arc evaporation, CVD, and laser ablation [86]. The solid-state carbon is utilized as an initial material in the arc evaporation and laser ablation approaches, resulting in near-perfect nanotube structures with the development of huge quantities of side products [87]. The CNTs are produced by the CVD technique using hydrocarbon gases as carbon sources and catalysts. Hydrocarbon gases are used as a seed for the growth of nucleation and to grow CNTs. Precise nanotube structures could be created on the surfaces of catalysts by placing the seeds in an arranged fashion [88]. CNTs surface functionalization is an important characteristic, and it can be performed via mechanical, physicochemical, or irradiation approaches [81]. The physicochemical process, which involves exchanging CNTs with functional groups and transformer molecules in a covalent or noncovalent manner, has been the most extensively recognized technique [89]. In general, noncovalent functionalization strategies do not significantly alter the properties of CNTs as the sp^2 structure of the hybridized orbital remains unaltered. The process of functionalizing CNTs by covalent bonding is not only difficult but also expensive and also unenviable since it alters the hybridization from sp^2 to sp^3, resulting in a probable partial damage of conjugation due to the electron acceptor and/or electron transport properties, in addition to being more precise and responsive than the noncovalent functionalization, Figure 7.4.

7.8 Carbon Nanotubes as an Adsorption of Dye Molecules

In recent years, CNTs have been extensively used to capture pollutants in water due to their high porosity, surface area, hollow structure, and strong affinity for pollutants [83]. In an effort to achieve adsorptive confiscation of Congo red, MWCNTs were exploited [83]. The effects of different batch factors on the adsorptive effectiveness toward Congo red were evaluated. These factors included adsorbate–adsorbent contact time, temperature, pH level of solution, and initial dye concentration [90]. The removal efficiency of Congo red by MWCNTs was 92% within an equilibrium time of 60 minutes [91]. According to research, adsorption capacity increases with decreasing temperature. The Langmuir isotherm and pseudosecond-order kinetic models were found to be suitable for the adsorption isotherm and kinetic data

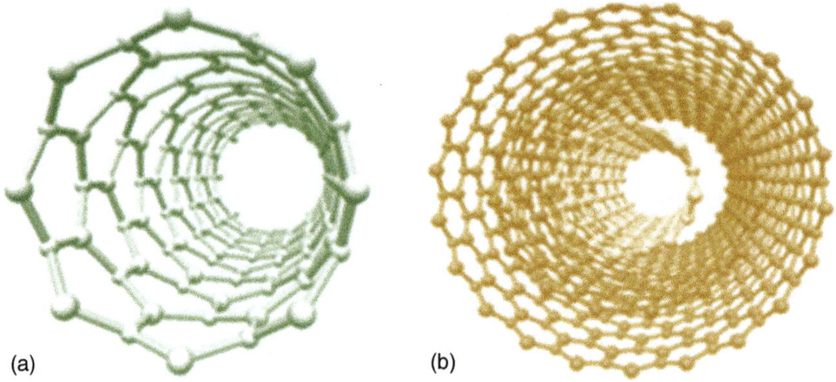

Figure 7.4 Different forms of carbon nanotubes (a) single-walled carbon nanotubes and (b) multiwalled carbon nanotubes.

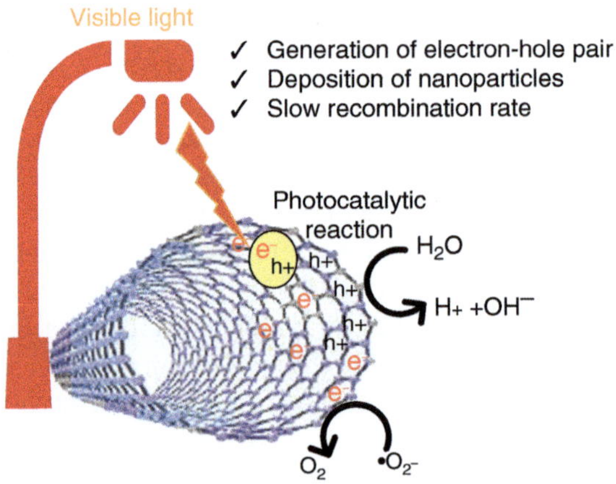

Figure 7.5 The possible mechanism of CNT in adsorption and photocatalysis.

[91, 92]. The adsorption ability of methyl orange on MWCNTs strongly depends on pH due to its effects on the surface features of the nanotubes and on the ionization and dissociation of the dye molecules. An improvement in preliminary dye absorptions from 20 to 40 $mg\,l^{-1}$ resulted in an increase in adsorption volume from 29 to 41.9 $mg\,g^{-1}$ [93].

In this study, the removal of reactive blue 4 from MWCNTs and SWCNTs made by CVD was compared [94]. The highest adsorption capacities were reported for pH 2.0 in the range of 442 and 487.6 $mg\,g^{-1}$. When compared to MWCNTs, SWCNTs have a higher adsorption capacity for reactive blue 4 due to their larger surface area and pore volume [94], Figure 7.5.

7.9 Industrial Application of Synthetic Dyes

The most commonly used dyes on an industrial scale are azo, anthraquinone, indigo, xanthene, and triarylmethane [95]. Several compounds with complex structures, such as anthraquinones, have been found to be useful in drug discovery [95]. The

Table 7.1 Applications of synthetic dyes in the service and industrial sectors.

S. no.	Sector	Dyes	Application	References
1	Textile industries	In the textile industry, synthetic dyes are widely used Azo dyes, however, are the most widely used Furthermore, textile industry dyes are classified based on their industrial applications as acid dyes, basic dyes, reactive dyes, vat dyes, disperse dyes, and direct dyes	World textile consumption is dominated by cellulosic fibers The type of dye applied depends on properties, such as affinity for the fiber, distribution, responsiveness, and constancy of the dye-fiber bond	[96, 97]
2	Printing industries	Depending on the fiber, different dyes are used. For example, disperse dyes are the only dyes that can be used in polyester fabric printing	The process of applying dyes to a restricted area of the textile, paper, or wooden that is used to apply the abstract of the proposal	[97–99]
3	Cosmetics industries	Cosmetics are primarily composed of pigments, although dyes are also used in small quantities The stains utilizes are organic in more than 80% of the cases, while azo pigments are used in more than 60% of them	There are lipsticks, blushers, eye shadows, eyeliners, and nail polishes In Europe, hair dyes make up almost 80% of cosmetic products	[100, 101]
4	Food and pharmaceutical industries	Food dyes are currently available in more than 60 different synthetic colors When using certain colors (e.g. tartrazine), they must be used according to the maximum allowable limit	Some examples of carbonated drinks are fruit drinks, energy drinks, candies, cereals, desserts, and snacks	[102, 103]
5	Leather and tanning industries	The majority of leather dyed is done with azo dyes or metal complex dyes However, dyes may differ structurally from textile dyes	Operation of dyeing leather	[104, 105]

textile and tanning industries use most synthetic dyes to color an array of products [30]. In addition, other industries, including the cosmetic and paper sectors, the food and pharmaceutical sectors, and service providers, use synthetic dyes (hospitals, universities, etc.) [30], Table 7.1.

7.10 Conclusion

There is an extensive range of applications for CNTs, which exhibit a variety of unique properties and impressive performance characteristics. Adsorption of dyes from wastewater with modified CNTs has shown remarkable improvement, and the material can be recovered and regenerated easily. It is possible to modify CNTs through oxidation, polymer grafting, the introduction of magnetic particles, or by hybridization with additional constituents. The consequences of various batch factors, such as pH values, adsorbate–adsorbent interaction time, dye concentration, temperature range, and adsorptive measure, have been investigated systematically, and the thermodynamical, isothermal, and kinetic aspects of the CNT/dye adsorption process have been explored. Dye adsorption on CNT-based composites is influenced by molecular interactions, such as π–π stacking, hydrogen bonds, aquaphobic interactions and static interactions, as well as dye molecule structure. Presently, the cost of CNTs is the primary roadblock averting them from being applied to extensive effluent handling. The cost of CNTs is likely to decrease if continuous manufacturing of nanotubes is conducted using low-priced raw materials. Additionally, there is concern about the release of CNTs into the environment and their noxiousness to plants and animals. There is, however, a limit to how much CNT can be released into the environment. As compared to other adsorbents, CNTs lose less weight during wastewater treatment, leading to a low-slung replacement of the material and a reduced release of CNTs. There still needs to be more research conducted on the extensive manufacturing of profitable and competent CNTs for effluent handling. The upcoming research is likely to lead to a breakthrough in the use of CNTs in effluent handling plants.

Important Research Groups Working on the Application of Carbon Nanotubes

1) https://sites.textiles.ncsu.edu/bradford-research-group/research
2) https://cambridgenanoaerosol.com/research
3) https://lcn.kaust.edu.sa
4) https://www.int.kit.edu/1309.php
5) www.imperial.ac.uk/nano-at-imperial/research/carbon-nanostructures

Acknowledgment

The author extends their appreciation to the Deputyship for Research & Innovation, Ministry of Education in Saudi Arabia for funding this research work through the project number (RUP-4).

References

1 Borghesi, S. and Vercelli, A. (2003). Sustainable globalisation. *Ecol. Econ.* 44: 77–89.

2 Steger, M.B. (2017). *Globalization: A Very Short Introduction*. Oxford University Press.

3 Serwecińska, L. (2020). Antimicrobials and antibiotic-resistant bacteria: a risk to the environment and to public health. *Water* 12: 3313.

4 Semple, K.T., Reid, B.J., and Fermor, T. (2001). Impact of composting strategies on the treatment of soils contaminated with organic pollutants. *Environ. Pollut.* 112: 269–283.

5 Ghangrekar, M. and Chatterjee, P. (2018). *Carbon Nanotubes for Clean Water*, 11–26. Springer.

6 Madhav, S., Ahamad, A., Singh, A.K. et al. (2020). *Sensors in Water Pollutants Monitoring: Role of Material*, 43–62. Springer.

7 Shams, M., Alam, I., and Mahbub, M.S. (2021). Plastic pollution during COVID-19: plastic waste directives and its long-term impact on the environment. *Environ. Adv.* 5: 100119.

8 Tijani, J.O., Fatoba, O.O., and Petrik, L.F. (2013). A review of pharmaceuticals and endocrine-disrupting compounds: sources, effects, removal, and detections. *Water Air Soil Pollut.* 224: 1–29.

9 Kumar, P., Bansal, V., Kim, K.-H., and Kwon, E.E. (2018). Metal-organic frameworks (MOFs) as futuristic options for wastewater treatment. *J. Ind. Eng. Chem.* 62: 130–145.

10 Crini, G. (2006). Non-conventional low-cost adsorbents for dye removal: a review. *Bioresour. Technol.* 97: 1061–1085.

11 Rafatullah, M., Sulaiman, O., Hashim, R., and Ahmad, A. (2010). Adsorption of methylene blue on low-cost adsorbents: a review. *J. Hazard. Mater.* 177: 70–80.

12 Yagub, M.T., Sen, T.K., Afroze, S., and Ang, H.M. (2014). Dye and its removal from aqueous solution by adsorption: a review. *Adv. Colloid Interface Sci.* 209: 172–184.

13 Hu, B., Wang, K., Wu, L. et al. (2010). Engineering carbon materials from the hydrothermal carbonization process of biomass. *Adv. Mater.* 22: 813–828.

14 Wang, Y., Pan, C., Chu, W. et al. (2019). Environmental remediation applications of carbon nanotubes and graphene oxide: adsorption and catalysis. *Nanomaterials* 9: 439.

15 Mohammed, M., Abdul Razak, A.A., and Hussein Al-Timimi, D.A. (2014). Modified multiwalled carbon nanotubes for treatment of some organic dyes in wastewater. *Adv. Mater. Sci. Eng.* 2014: 1–10.

16 Das, R., Abd Hamid, S.B., Ali, M.E. et al. (2014). Multifunctional carbon nanotubes in water treatment: the present, past and future. *Desalination* 354: 160–179.

17 Ren, X., Chen, C., Nagatsu, M., and Wang, X. (2011). Carbon nanotubes as adsorbents in environmental pollution management: a review. *Chem. Eng. J.* 170: 395–410.

18 Santhosh, C., Velmurugan, V., Jacob, G. et al. (2016). Role of nanomaterials in water treatment applications: a review. *Chem. Eng. J.* 306: 1116–1137.

19 Robert, E. (2017). The role of the cave in the expression of prehistoric societies. *Quat. Int.* 432: 59–65.

20 Morriss-Kay, G.M. (2010). The evolution of human artistic creativity. *J. Anat.* 216: 158–176.

21 Duarte, C.M. (2014). Red ochre and shells: clues to human evolution. *Trends Ecol. Evol.* 29: 560–565.

22 Dickerson, R. and Fortner, P. (2017). The use of geological materials in ancient and contemporary art. *Geol. Today* 33: 185–189.

23 Barnett, J.R., Miller, S., and Pearce, E. (2006). Colour and art: a brief history of pigments. *Opt. Laser Technol.* 38: 445–453.

24 Hovers, E., Ilani, S., BarYosef, O., and Vandermeersch, B. (2003). An early case of color symbolism: ochre use by modern humans in Qafzeh Cave. *Curr. Anthropol.* 44: 491–522.

25 Ardila-Leal, L.D., Poutou-Piñales, R.A., Pedroza-Rodríguez, A.M., and Quevedo-Hidalgo, B.E. (2021). A brief history of colour, the environmental impact of synthetic dyes and removal by using laccases. *Molecules* 26: 3813.

26 Johnston, W. (2008). The discovery of aniline and the origin of the term "aniline dye". *Biotech. Histochem.* 83: 83–87.

27 Abel, A. (2012). *Colour Design*, 557–587. Elsevier.

28 Perkin, W.H. (1879). *The History of Alizarin and Allied Colouring Matters, and Their Production from Coal Tar: Two Lectures Delivered Before the Chemical Section of the Society of Arts, May 8th and 22nd, 1879*. W. Trounce.

29 Clark, R.J., Cooksey, C.J., Daniels, M.A., and Withnall, R. (1993). Indigo, woad, and Tyrian Purple: important vat dyes from antiquity to the present. *Endeavour* 17: 191–199.

30 Tkaczyk, A., Mitrowska, K., and Posyniak, A. (2020). Synthetic organic dyes as contaminants of the aquatic environment and their implications for ecosystems: a review. *Sci. Total Environ.* 717: 137222.

31 Atkar, A., Pabba, M., Sekhar, S.C., and Sridhar, S. (2021). *Fundamentals of Natural Fibres and Textiles*, 741–764. Elsevier.

32 Park, S., Srivastava, D., and Cho, K. (2003). Generalized chemical reactivity of curved surfaces: carbon nanotubes. *Nano Lett.* 3: 1273–1277.

33 Wimalasiri, Y. and Zou, L. (2013). Carbon nanotube/graphene composite for enhanced capacitive deionization performance. *Carbon* 59: 464–471.

34 Sumpter, B.G., Huang, J., Meunier, V. et al. (2009). A theoretical and experimental study on manipulating the structure and properties of carbon nanotubes using substitutional dopants. *Int. J. Quantum Chem.* 109: 97–118.

35 Meyyappan, M. (2004). *Carbon Nanotubes: Science and Applications*. CRC Press.

36 Meyyappan, M., Delzeit, L., Cassell, A., and Hash, D. (2003). Carbon nanotube growth by PECVD: a review. *Plasma Sources Sci. Technol.* 12: 205.

37 Shah, K.A. and Tali, B.A. (2016). Synthesis of carbon nanotubes by catalytic chemical vapour deposition: a review on carbon sources, catalysts and substrates. *Mater. Sci. Semicond. Process.* 41: 67–82.

38 Wepasnick, K.A., Smith, B.A., Bitter, J.L., and Fairbrother, D.H. (2010). Chemical and structural characterization of carbon nanotube surfaces. *Anal. Bioanal.Chem.* 396: 1003–1014.

39 Balasubramanian, K. and Burghard, M. (2005). Chemically functionalized carbon nanotubes. *Small* 1: 180–192.

40 Bahr, J.L., Yang, J., Kosynkin, D.V. et al. (2001). Functionalization of carbon nanotubes by electrochemical reduction of aryl diazonium salts: a bucky paper electrode. *J. Am. Chem. Soc.* 123: 6536–6542.

41 Pandurangappa, M. and Raghu, G.K. (2011). *Carbon Nanotubes Applications on Electron Devices*. IntechOpen.

42 Schwenke, A.M., Hoeppener, S., and Schubert, U.S. (2015). Synthesis and modification of carbon nanomaterials utilizing microwave heating. *Adv. Mater.* 27: 4113–4141.

43 Lu, Q., Chen, J.G., and Xiao, J.Q. (2013). Nanostructured electrodes for high-performance pseudocapacitors. *Angew. Chem. Int. Ed.* 52: 1882–1889.

44 Chella, S., Venugopal, V., George, J. et al. (2016). Role of nanomaterials in water treatment applications: a review. *Chem. Eng. J.* 306: 1116–1137.

45 Zonglin, P., Fangpeng, Y., Lin, L. et al. (2019). Electrochemical microfiltration treatment of bisphenol A wastewater using coal-based carbon membrane. *Sep. Purif. Technol.* 227: 115695.

46 Masheter, A.T., Abiman, P., Wildgoose, G.G. et al. (2007). Investigating the reactive sites and the anomalously large changes in surface pK_a values of chemically modified carbon nanotubes of different morphologies. *J. Mater. Chem.* 17: 2616–2626.

47 Saifuddin, N., Raziah, A.Z., and Junizah, A.R. (2013). Carbon nanotubes: a review on structure and their interaction with proteins. *J. Chem.* 2013: 676815. https://doi.org/10.1155/2013/676815.

48 Heise, K., Delepierre, G., King, A.W.T. et al. (2021). Chemical modification of reducing end-groups in cellulose nanocrystals. *Angew. Chem. Int. Ed.* 60: 66–87. https://doi.org/10.1002/anie.202002433.

49 Laloy, J., Haguet, H., Alpan, L. et al. (2018). Impact of functional inorganic nanotubes f-INTs-WS2 on hemolysis, platelet function and coagulation. *Nano Convergence* 5: 31. https://doi.org/10.1186/s40580-018-0162-1.

50 Khan, M.E., Khan, M.M., and Cho, M.H. (2015). Biogenic synthesis of a Ag–graphene nanocomposite with efficient photocatalytic degradation, electrical conductivity and photoelectrochemical performance. *New J. Chem.* 39: 8121–8129. https://doi.org/10.1039/C5NJ01320H.

51 Khan, M.E., Khan, M.M., and Cho, M.H. (2016). CdS-graphene nanocomposite for efficient visible-light-driven photocatalytic and photoelectrochemical applications. *J. Colloid Interface Sci.* 482: 221–232. https://doi.org/10.1016/j.jcis.2016.07.070.

52 Su, C.X.-H., Low, L.W., Teng, T.T., and Wong, Y.S. (2016). Combination and hybridisation of treatments in dye wastewater treatment: a review. *J. Environ. Chem. Eng.* 4: 3618–3631. https://doi.org/10.1016/j.jece.2016.07.026.

53 Teng, T.T. and Low, L.W. (2012). *Advances in Water Treatment and Pollution Prevention* (ed. S.K. Sharma and R. Sanghi), 65–93. Dordrecht: Springer Netherlands.

54 Saleh, I.A., Zouari, N., and Al-Ghouti, M.A. (2020). Removal of pesticides from water and wastewater: chemical, physical and biological treatment approaches. *Environ. Technol. Innov.* 19: 101026. https://doi.org/10.1016/j.eti.2020.101026.

55 Ali, I., Asim, M., and Khan, T.A. (2012). Low cost adsorbents for the removal of organic pollutants from wastewater. *J. Environ. Manage.* 113: 170–183. https://doi.org/10.1016/j.jenvman.2012.08.028.

56 Yaqoob, A.A., Parveen, T., Umar, K., and Mohamad Ibrahim, M.N. (2020). Role of nanomaterials in the treatment of wastewater: a review. *Water* 12: 495.

57 Vardhan, K.H., Kumar, P.S., and Panda, R.C. (2019). A review on heavy metal pollution, toxicity and remedial measures: current trends and future perspectives. *J. Mol. Liq.* 290: 111197. https://doi.org/10.1016/j.molliq.2019.111197.

58 Lu, F. and Astruc, D. (2018). Nanomaterials for removal of toxic elements from water. *Coord. Chem. Rev.* 356: 147–164. https://doi.org/10.1016/j.ccr.2017.11.003.

59 Fu, F. and Wang, Q. (2011). Removal of heavy metal ions from wastewaters: a review. *J. Environ. Manage.* 92: 407–418. https://doi.org/10.1016/j.jenvman.2010.11.011.

60 Pirhashemi, M., Habibi-Yangjeh, A., and Rahim Pouran, S. (2018). Review on the criteria anticipated for the fabrication of highly efficient ZnO-based visible-light-driven photocatalysts. *J. Ind. Eng. Chem.* 62: 1–25. https://doi.org/10.1016/j.jiec.2018.01.012.

61 Awual, M.R. (2015). A novel facial composite adsorbent for enhanced copper(II) detection and removal from wastewater. *Chem. Eng. J.* 266: 368–375. https://doi.org/10.1016/j.cej.2014.12.094.

62 Yu, J.-G., Zhao, X.-H., Yang, H. et al. (2014). Aqueous adsorption and removal of organic contaminants by carbon nanotubes. *Sci. Total Environ.* 482, 483: 241–251. https://doi.org/10.1016/j.scitotenv.2014.02.129.

63 Peng, J., He, Y., Zhou, C. et al. (2021). The carbon nanotubes-based materials and their applications for organic pollutant removal: a critical review. *Chin. Chem. Lett.* 32: 1626–1636. https://doi.org/10.1016/j.cclet.2020.10.026.

64 Mishra, A.K., Arockiadoss, T., and Ramaprabhu, S. (2010). Study of removal of azo dye by functionalized multi walled carbon nanotubes. *Chem. Eng. J.* 162: 1026–1034. https://doi.org/10.1016/j.cej.2010.07.014.

65 Ferreira, G.M.D., Ferreira, G.M.D., Hespanhol, M.C. et al. (2017). Adsorption of red azo dyes on multi-walled carbon nanotubes and activated carbon: a thermodynamic study. *Colloids Surf., A* 529: 531–540. https://doi.org/10.1016/j.colsurfa.2017.06.021.

66 Jahandar Lashaki, M., Atkinson, J.D., Hashisho, Z. et al. (2016). The role of beaded activated carbon's surface oxygen groups on irreversible adsorption of organic vapors. *J. Hazard. Mater.* 317: 284–294. https://doi.org/10.1016/j.jhazmat.2016.05.087.

67 Liu, Y., Wang, Z.U., and Zhou, H.-C. (2012). Recent advances in carbon dioxide capture with metal-organic frameworks. *Greenhouse Gases Sci. Technol.* 2: 239–259. https://doi.org/10.1002/ghg.1296.

68 Wu, Z. and Zhao, D. (2011). Ordered mesoporous materials as adsorbents. *Chem. Commun.* 47: 3332–3338. https://doi.org/10.1039/C0CC04909C.

69 Mo, J., Yang, Q., Zhang, N. et al. (2018). A review on agro-industrial waste (AIW) derived adsorbents for water and wastewater treatment. *J. Environ. Manage.* 227: 395–405. https://doi.org/10.1016/j.jenvman.2018.08.069.

70 Gupta, V.K., Carrott, P.J.M., Ribeiro Carrott, M.M.L., and Suhas (2009). Low-cost adsorbents: growing approach to wastewater treatment—a review. *Crit. Rev. Env. Sci. Technol.* 39: 783–842. https://doi.org/10.1080/10643380801977610.

71 Gómez-Pastora, J., Bringas, E., and Ortiz, I. (2014). Recent progress and future challenges on the use of high performance magnetic nano-adsorbents in environmental applications. *Chem. Eng. J.* 256: 187–204. https://doi.org/10.1016/j.cej.2014.06.119.

72 Chowdhury, S. and Balasubramanian, R. (2014). Recent advances in the use of graphene-family nanoadsorbents for removal of toxic pollutants from wastewater. *Adv. Colloid Interface Sci.* 204: 35–56. https://doi.org/10.1016/j.cis.2013.12.005.

73 Khulbe, K.C. and Matsuura, T. (2018). Removal of heavy metals and pollutants by membrane adsorption techniques. *Appl. Water Sci.* 8: 19. https://doi.org/10.1007/s13201-018-0661-6.

74 Guerra, F.D., Attia, M.F., Whitehead, D.C., and Alexis, F. (2018). Nanotechnology for environmental remediation: materials and applications. *Molecules* 23: 1760.

75 Thirunavukkarasu, A., Nithya, R., and Sivashankar, R. (2020). A review on the role of nanomaterials in the removal of organic pollutants from wastewater. *Rev. Environ. Sci. Biotechnol.* 19: 751–778. https://doi.org/10.1007/s11157-020-09548-8.

76 Ahmed, S.N. and Haider, W. (2018). Heterogeneous photocatalysis and its potential applications in water and wastewater treatment: a review. *Nanotechnology* 29: 342001. https://doi.org/10.1088/1361-6528/aac6ea.

77 Zhang, B.-T., Zheng, X., Li, H.-F., and Lin, J.-M. (2013). Application of carbon-based nanomaterials in sample preparation: a review. *Anal. Chim. Acta* 784: 1–17. https://doi.org/10.1016/j.aca.2013.03.054.

78 Kumar, R., Joanni, E., Singh, R.K. et al. (2018). Recent advances in the synthesis and modification of carbon-based 2D materials for application in energy conversion and storage. *Prog. Energy Combust. Sci.* 67: 115–157. https://doi.org/10.1016/j.pecs.2018.03.001.

79 Mauter, M.S. and Elimelech, M. (2008). Environmental applications of carbon-based nanomaterials. *Environ. Sci. Technol.* 42: 5843–5859. https://doi.org/10.1021/es8006904.

80 Bagheri, S., TermehYousefi, A., and Do, T.-O. (2017). Photocatalytic pathway toward degradation of environmental pharmaceutical pollutants: structure, kinetics and mechanism approach. *Catal. Sci. Technol.* 7: 4548–4569. https://doi.org/10.1039/C7CY00468K.

81 Mashkoor, F., Nasar, A., and Inamuddin (2020). Carbon nanotube-based adsorbents for the removal of dyes from waters: a review. *Environ. Chem. Lett.* 18: 605–629. https://doi.org/10.1007/s10311-020-00970-6.

82 Upadhyayula, V.K.K., Deng, S., Mitchell, M.C., and Smith, G.B. (2009). Application of carbon nanotube technology for removal of contaminants in drinking water: a review. *Sci. Total Environ.* 408: 1–13. https://doi.org/10.1016/j.scitotenv.2009.09.027.

83 Gupta, V.K., Kumar, R., Nayak, A. et al. (2013). Adsorptive removal of dyes from aqueous solution onto carbon nanotubes: a review. *Adv. Colloid Interface Sci.* 193, 194: 24–34. https://doi.org/10.1016/j.cis.2013.03.003.

84 Ando, Y., Zhao, X., Hirahara, K. et al. (2000). Mass production of single-wall carbon nanotubes by the arc plasma jet method. *Chem. Phys. Lett.* 323: 580–585. https://doi.org/10.1016/S0009-2614(00)00556-X.

85 Paradise, M. and Goswami, T. (2007). Carbon nanotubes – production and industrial applications. *Mater. Des.* 28: 1477–1489. https://doi.org/10.1016/j.matdes.2006.03.008.

86 Kong, J., Cassell, A.M., and Dai, H. (1998). Chemical vapor deposition of methane for single-walled carbon nanotubes. *Chem. Phys. Lett.* 292: 567–574. https://doi.org/10.1016/S0009-2614(98)00745-3.

87 Trojanowicz, M. (2006). Analytical applications of carbon nanotubes: a review. *TrAC, Trends Anal. Chem.* 25: 480–489. https://doi.org/10.1016/j.trac.2005.11.008.

88 Jourdain, V. and Bichara, C. (2013). Current understanding of the growth of carbon nanotubes in catalytic chemical vapour deposition. *Carbon* 58: 2–39. https://doi.org/10.1016/j.carbon.2013.02.046.

89 Lin, T., Bajpai, V., Ji, T., and Dai, L. (2003). Chemistry of carbon nanotubes. *Aust. J. Chem.* 56: 635–651. https://doi.org/10.1071/CH02254.

90 Mittal, A., Thakur, V., Mittal, J., and Vardhan, H. (2014). Process development for the removal of hazardous anionic azo dye Congo red from wastewater by using hen feather as potential adsorbent. *Desalin. Water Treat.* 52: 227–237. https://doi.org/10.1080/19443994.2013.785030.

91 Zare, K., Sadegh, H., Shahryari-ghoshekandi, R. et al. (2015). Enhanced removal of toxic Congo red dye using multi walled carbon nanotubes: kinetic, equilibrium studies and its comparison with other adsorbents. *J. Mol. Liq.* 212: 266–271. https://doi.org/10.1016/j.molliq.2015.09.027.

92 Sheibani, M., Ghaedi, M., Marahel, F., and Ansari, A. (2015). Congo red removal using oxidized multiwalled carbon nanotubes: kinetic and isotherm study. *Desalin. Water Treat.* 53: 844–852. https://doi.org/10.1080/19443994.2013.867540.

93 Mao, B., Sidhureddy, B., Thiruppathi, A.R. et al. (2020). Efficient dye removal and separation based on graphene oxide nanomaterials. *New J. Chem.* 44: 4519–4528. https://doi.org/10.1039/C9NJ05895H.

94 Machado, F.M., Bergmann, C.P., Lima, E.C. et al. (2012). Adsorption of Reactive Blue 4 dye from water solutions by carbon nanotubes: experiment and theory. *Phys. Chem. Chem. Phys.* 14: 11139–11153. https://doi.org/10.1039/C2CP41475A.

95 Pereira, L. and Alves, M. (2012). *Environmental Protection Strategies for Sustainable Development*, 111–162. Springer.

96 Khattab, T.A., Abdelrahman, M.S., and Rehan, M. (2020). Textile dyeing industry: environmental impacts and remediation. *Environ. Sci. Pollut. Res.* 27: 3803–3818.

97 Holkar, C.R., Jadhav, A.J., Pinjari, D.V. et al. (2016). A critical review on textile wastewater treatments: possible approaches. *J. Environ. Manage.* 182: 351–366.

98 Singh, K., Gupta, A., and Sharma, A. (2016). Fly ash as low cost adsorbent for treatment of effluent of handmade paper industry-kinetic and modelling studies for direct black dye. *J. Cleaner Prod.* 112: 1227–1240.

99 Kiurski, J.S., Marić, B.B., Oros, I.B., and Kecić, V.S. (2017). The ecodesign practice in Serbian printing industry. *J. Cleaner Prod.* 149: 1200–1209.

100 Weisz, A., Milstein, S.R., Scher, A.L., and Hepp, N.M. (2018). Colouring agents in cosmetics: regulatory aspects and analytical methods. In: *Analysis of Cosmetic Products* (ed. A. Salvador and A. Chisvert), 123–157. Elsevier.

101 Guerra, E., Llompart, M., and Garcia-Jares, C. (2018). Analysis of dyes in cosmetics: challenges and recent developments. *Cosmetics* 5: 47.

102 Stevens, L.J., Burgess, J.R., Stochelski, M.A., and Kuczek, T. (2014). Amounts of artificial food colors in commonly consumed beverages and potential behavioral implications for consumption in children. *Clin. Pediatr.* 53: 133–140.

103 Šuleková, M., Smrčová, M., Hudák, A. et al. (2017). Organic colouring agents in the pharmaceutical industry. *Folia Vet.* 61: 32–46.

104 Bouzayani, B., Rosales, E., Pazos, M. et al. (2019). Homogeneous and heterogeneous peroxymonosulfate activation by transition metals for the degradation of industrial leather dye. *J. Cleaner Prod.* 228: 222–230.

105 Javaid, R. and Qazi, U.Y. (2019). Catalytic oxidation process for the degradation of synthetic dyes: an overview. *Int. J. Environ. Res. Public Health* 16: 2066.

8

Chemically Modified Carbon Nanotubes in Membrane Separation

R. Aiswarya and T. Kalaivani

Sri Ramasamy Memorial Institute of Science and Technology, Department of Physics and Nanotechnology, Potheri, Kattankulathur, Chengalpattu, Tamilnadu 603203, India

8.1 Introduction

Over the past decade, nanoscience and nanotechnology are the rapid emanating and advancing domain of science, which find technological significance and enormous interest for beginners in research field. The unique physical, chemical, electronic properties, tensile strength, and elastic modulus, with high tensile strength, high thermal conductivity, low thermal expansion coefficient, high electrical conductivity, and highly flexibility based on this carbon nanotube (CNT) are fascinating subject matter with a wide range of applications, such as biosensors, electrochemical-sensors, fuel cells, metal semiconductors with hetero combinations, catalysts used in organic and in-organic synthesis, random access memory units, advanced battery, field transistors, microelectrodes, and field emission equipments. Atomic force microscopy (AFM) probes used water treatment, and drug carriers in cancer treatment, and adequate adsorbents to remove organic pollutants from drinking water. CNTs' poor solubility in organic solvents makes them ideal for chemically modified CNTs utilized in membrane separation. As a result, a variety of modification procedures, including physical, chemical, and combination methods, have been employed to enhance and use them [1–4]. Great endeavor has been made to create high chemically modified CNTs in membrane separation in order to meet the urging demand and durability in contemporary applications.

Global water scarcity is increasing day by day; this is clearly explained using Figure 8.1 [6]. The absence of fresh water to satisfy daily needs is known as a water crisis. Physical and economic water shortages are the two forms of scarcity that have been identified. Physical water shortage occurs when there is insufficient water to satisfy daily demands, which includes ecosystems that need water to function properly. Physical water scarcity is common in arid locations (such as Central Asia, West Asia, and North Africa) [7]. Naive water or soil dampness, virtual water commerce, water quality, environmental flow needs, and globalization must all be included in water scarcity estimates [8]. There arouses a need for membranes for

Chemically Modified Carbon Nanotubes for Commercial Applications, First Edition.
Edited by Jeenat Aslam, Chaudhery Mustansar Hussain, and Ruby Aslam.

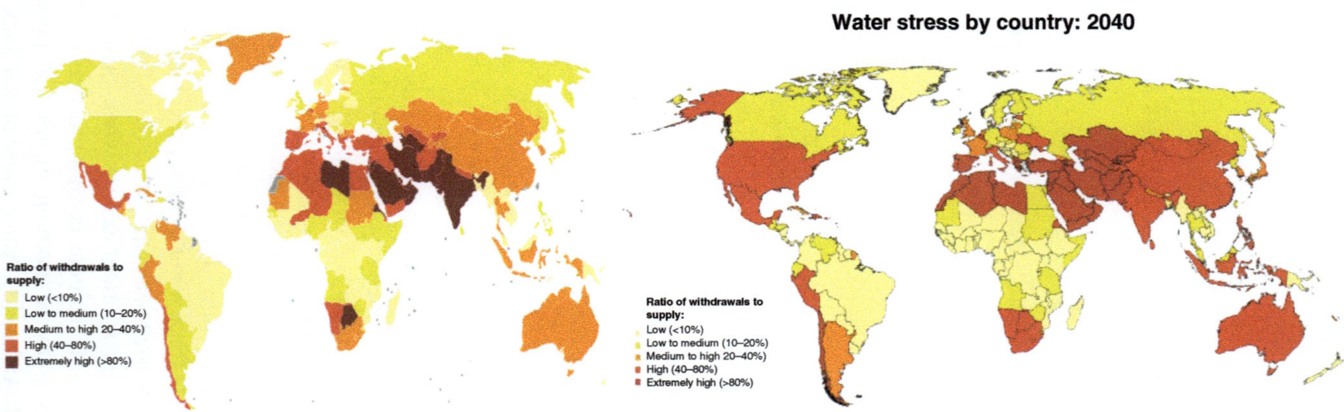

Figure 8.1 Comparison of water stress by country: 2040 and 2019 (Wikipedia source). Source: Wikipedia [5]/Wikipedia/Public Domain.

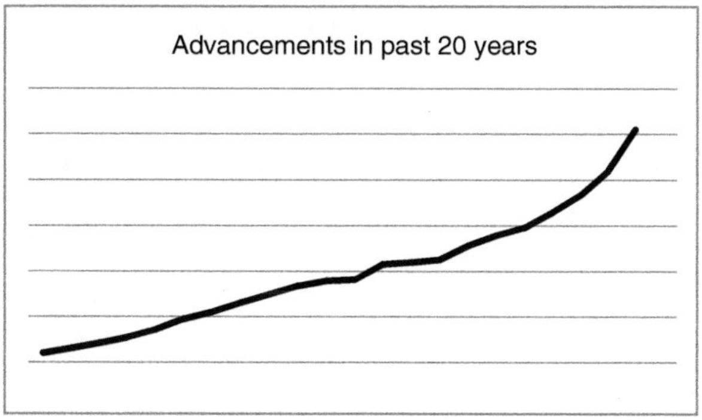

Figure 8.2 Chemically modified carbon nanotubes advancement in past 20 years.

purification as a solution to this crisis. The better choice for material is CNTs, which indeed is the cheapest, and can be easily commercialized.

There are many applications for chemically modified CNTs in membrane separation, but due to global demand, the water purification gain more importance than anyone else [9].

Many old concepts in Materials Science appear to be challenged by nanosized materials and nanoscale technology. Since the discovery of CNTs, the scholarly literature based on the chemical and physical properties of nanomaterials and chemically modified CNTs, particularly carbon-based nanomaterials (CBNs), has exploded, has the usage of CBNs in membrane applications. The rapid advancement in past 20 years is shown in the line graph, Figure 8.2 [10, 11].

8.2 Carbon Nanotubes (CNTs) Overview

CNT can be referred to as tubes with a shaky carbaslous-divider erection and diameters of fewer hundred nanometers. Radushkevich and Lukyanovich rooted out CNTs in the year 1952 [12]. CNTs were displayed to proffer extraordinary potentiality for partition implementation owing to appealing physicochemical properties, mechanical properties, electrical, and thermal effects. The tubes are made up of carbon with distances across normally estimated in nanometers. CNTs frequently allude as single-walled carbon nanotubes (SWCNTs) having breadths in the scope of a nanometer. SWCNTs are one type of allotropes of carbon, middle of the road between the fullerene enclosures and level graphene. Albeit made, thus, SWCNTs, which glorified patterns from two-layered hexagonal grid made of carbon particles moved along with one of Bravais cross-section vectors of hexagonal grid for framing an empty chamber. In this development, occasional limit circumstances are forced upon the extent of wrap-up vector to have a spiral cross-section on consistently fortified carbon atoms on the chamber surface [13].

CNTs likewise regularly allude to multi-walled carbon nanotubes (MWCNTs) comprising of settled SWCNTs pitifully bound all together by strong van der Waals connections in a huge ring-like construction. On the off chance that not indistinguishable, these cylinders are basically the same as Endo, Oberlin, and Koyama's

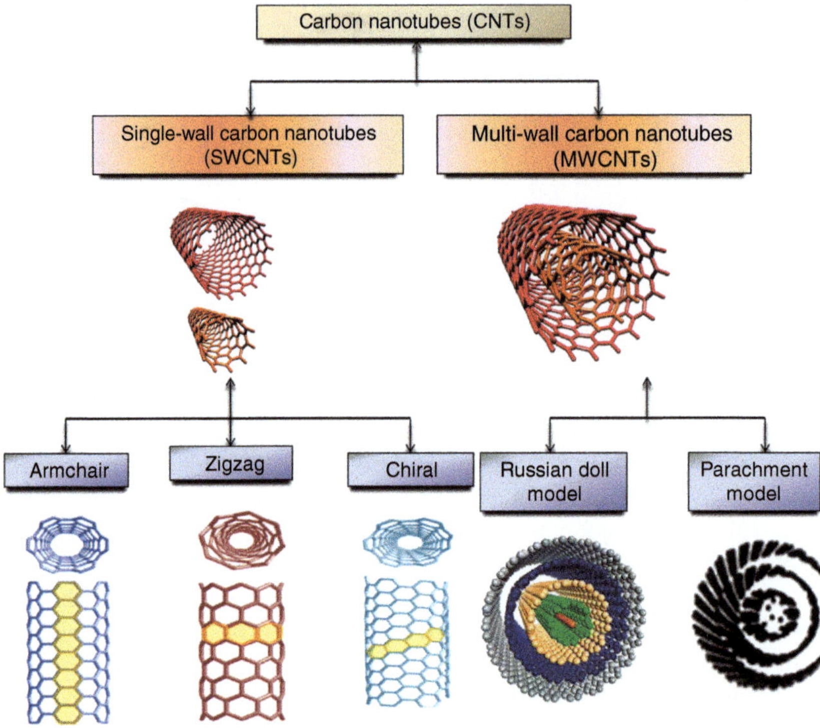

Figure 8.3 Classification of carbon nanotubes (CNTs). Source: Adapted from Saifuddin [15].

and have formed long, straight, and equally distributed carbon layers circularly organized among an empty tube [14]. MWCNT is additionally now and again used to allude to twofold and triple-walled CNTs. SWCNTs have three grouping like zigzag, chair, and armchair as shown in Figure 8.3. On a graphene cross-section, a zigzag path is defined and has a trackway that rotates 60° right and left then passes one another bonds. It's also common to describe a rocker path as one that makes two 60° left turns followed by two right turns every four stages. A closed zigzag path circumvents the cylinder on a few CNTs. The cylinders are said to be of zigzagging kind of array or are essentially snaky nanotube. The cylinder is assumed to be of the rocker type, or a rocker nanotube, if it is surrounded by a closed easy chair manner.

In recent research trends, the interest in using CNTs in membrane separation materials is growing. It is typically generated as interlaced CNTs having an open end and obtainable mesh for mesopore channels. CNT electrodes in contrast to other carbon-based electrodes, have mesopores that are connected together, hence, allowing for continuous charge distributions that use, almost completely, the surface area [16]. Due to their smaller surface areas than activated carbon, CNT electrodes are more efficient. The surface area is, therefore, more efficient. Since the electrolyte is easier to disseminate through the mesoporous network, the electron spin resonance (ESR) of the CNTs is lower than the active carbon. Furthermore, various methods of

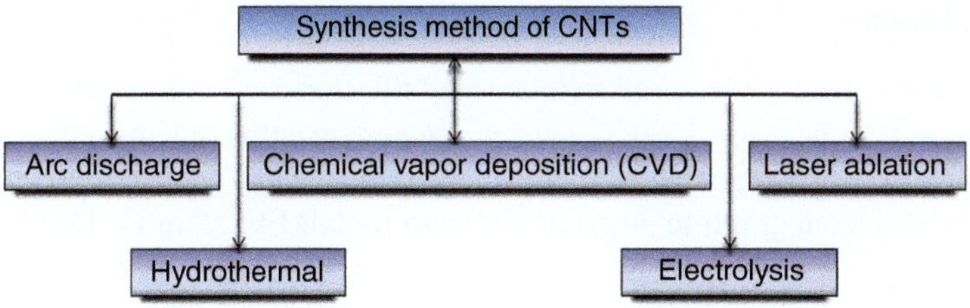

Figure 8.4 Different method of synthesis of carbon nanotubes (CNTs).

manufacturing were developed to further decrease the ESR. CNTs can be generated directly or formed into colloidal thin films in the case of current collectors and heat treatment, for example. The entangled matt structure's efficacy allows for energy densities comparable to other carbon-based materials [17].

8.3 Method of Synthesis of Carbon Nanotube (CNT)

Allotropes of carbon which has length-to-measurement correlation greater than 1 000 000 are known as CNTs. As shown in Figure 8.4, methods for delivering nanotubes in large quantities have been developed, including arc discharge, chemical vapor deposition (CVD), laser ablation, electrolysis, and hydrothermal [18].

Various designs, morphologies, and attributes are still up in the air by the technique for arrangement. Three primary methodologies can be applied for the blend of CNTs, in particular compound fume testimony (CVD), which are release and laser statement. CVD further opened up new courses for controlled blend and gadget reconciliation. Contrasted and different methods, the low-temperature CVD is viewed as achievable, conservative, and appropriate for large-scale manufacturing [4]. CNTs is planned using carbon sources, such as methane, acetylene, benzene, xylene, and carbon monoxide ethylene. Low-temperature CVD makes MWCNTs even easier to obtain from most hydrocarbons. SWCNTs are made from methane and carbon monoxide using high-temperature CVD (900–1200 °C).

8.3.1 Arc Discharge

MWCNTs were first created by Iijíma in 1991 by the curve release strategy, under an inactive medium, circular segment is produced between two carbon electrodes at high temperature (>3000 °C) to dissipate carbon atoms into plasma. Delicate stores of CNTs and other carbon stores are shaped on the cathode, electrodes are stacked with a suitable impetus, for example, Ni–Co or Co–Y to deliver SWCNTs. Notwithstanding, for the manufacture of MWCNTs, the presence of impetus is not needed [19]. The CNTs arranged by this procedure are of outstanding quality, with a slight event of deformities, and with fluctuated components of distance across and length CNTs in wastewater remediation.

8.3.2 Laser Ablation

Laser removal strategy, in graphite, target was warmed till vaporization by laser radiation in a progression of idle gas at 1200 °C, carbon species are delivered, moved by a streaming gas and afterward amassed on an authority that is water-cooled, uniform SWCNTs are delivered from graphite target doped with metals like Ni or Co [20].

8.3.3 Chemical Vapor Deposition (CVD)

Fibers and strands of carbon have been manufactured by the CVD strategy since 1960s. A gas containing carbon is permitted to stream and dissociate on a substrate within the sight of a metal impetus powder at 600 °C or higher. This low applied temperature genuinely, contrasted and different strategies, diminishes the creation costs. What's more, novel designs can be shaped through the covering of impetuses on the substrate. Be that as it may, countless imperfections were found on CNTs made by this technique [21].

8.3.4 Hydrothermal

Single-walled and multi-walled open end and shut CNTs contain the divider thickness range around a few to excess of nearly hundred carbon layer are created with a basically new strategy aqueous amalgamation utilizing required material at 700–800 °C under 60–100 MPa tension. A significant component of aqueous nanotubes is a little divider thickness, which is around 10% of the huge inward measurement of 20–800 nm. Shut CNTs is released by tight prudence of enfold exemplified water having a high-ended vacuum that can be utilized with test tubes for *in situ* tries were carried out [22].

8.3.5 Electrolysis

CNTs were made using CO_2 that had been broken down liquidified salt using the authors' initial electrolytic process. Electrolysis was carried out with the help of current and other possible controls. Electroreduction of carbon dioxide decomposed in the halide softens at an excess tension up to 15 bar was examined by cyclic voltammetry on glossy carbon (GC) terminal at a temperature of 550 °C to build up the real current and predicted reaches [23]. For elucidation of the obtained results, the electrochemical-substance electrochemical system of CO_2 electroreduction was provided. The majority of CNTs are MWCNTs, have a bended shape, and typically agglomerate into packs, according to the findings. Almost all CNTs are half-filled with electrolyte salt. MWCNTs, on the other hand, have carbon nanofibers, nanographite, and hazy carbon in the cathode. The yield of CNTs, their breadth, and design peculiarities all conspire against electrolysis systems, as well as nebulous carbon. The yield of CNTs, their breadth, and design peculiarities all conspire against electrolysis systems [24].

8.4 Fabrication Methods of CNTs

To make mixed metal oxide for nanocomposites, researchers employed a variety of solid-state processes, such as mechanical oxide mixing (i.e. ball milling process) and chemical co-precipitation. For example, by reducing $KMnO_4$ and Pb(II), mixed oxide salts form amorphous mixed oxide rainfall of Mn–Ni and Mn–Pb, and oxides nanocomposites mixers are formed [25]. In the autoclaved hydrothermal method, the mixed oxide nano-compounds containing urea were prepared by using SnO_2–Al_2O_3 as a hydrolytic agent. CNTs, nanocompogenic metal oxides were produced, via the mechanical mix of metallic oxides in the mortar, through CVD, precipitation or electrochemical deposition, directly on the CNT. CNTs are prepared through a mechanical mixture [26]. Conducting polymer nanocomposite mixed with metal oxide, CNTs, and graphene has predominantly been produced *in situ* and polymerized in solutions containing metal oxide nanoparticles (NPs), CNTs, nanosheets monomer, and conducting polymer suspension. CNT–nanocomposite (PANI), for instance, is a solution consisting of CNTs and aniline monomers. The solution consists of CNTs [27].

8.4.1 Fabrication of CNT-Reinforced Metal Matrix Composites (CNT-MMCs)

A composite of appropriate matrix and reinforcing elements can be used to develop lightweight and high-strength functional materials. Metal matrix composites (MMCs) were created in the 1950s to meet the demands of a wide range of potential industrial uses, particularly at high temperatures. The kind, amount, and shape of matrix and reinforcing elements, as well as their dispersion phase, determine the characteristics of composites. Ceramic, carbon, and fibers are used as reinforcements [28]. In reinforced aluminum, matrix composites and other composites, reinforcements such as A1,0, SiC, and TiC are common. CNT reinforcements are more appealing than ceramic materials because of their strong damping capacity, high thermal conductivity, and self-lubricating properties. In comparison to microsized ceramic reinforcements, the production of CNTs was very simple due to their nanosize. Adapting the mechanical properties of MMCs by integrating nano-reinforcements such as graphene nanofibre has been subjected to several reports in later years [29]. Owing to their superior characteristics, CNTs are the most frequently accepted reinforcement for large-scale MMC manufacture. In a scanning electron microscope (SEM) image, the CNTs appear twisted and entangled. The uniform distribution of CNTs in metal and alloy matrixes is a fundamental obstacle for successful processing [30]. The CNT structure can be damaged by applied pressures or interactions with the molten metal of the matrix material at high temperatures. These aspects should be considered while choosing a fabrication technique. The manufacture of CNT-MMCs is explained [31].

8.4.2 Microwave-Assisted Fabrication of CNTs

Single-stage gas-phase tubular microwave chemical vapor deposition (TM-CVD) using hydrogen (H) and acetylene (C·H) as precursors and a ferrocene catalyst was used to create pure (98%) CNTs under ideal conditions, with a reaction power of 900 W and a gas ratio of CHXH of 0.6. MWCNTs with sizes of 16–23 nm were found to be uniformly dispersed by TEM. The ferrocene catalyst was near the entrance, and a quartz boat was in the middle of the reaction chamber [29]. To remove any oxygen gas, argon gas was put into the system first. Before entering the microwave chamber, the CH and H flow gases were combined, and then the gas mixture was permitted to enter. The desired time period was chosen, and the CNTs were obtained in the quartz boats [32].

8.5 Functionalization of CNTs

Chemical alteration of CNTs can alter their physicochemical properties, allowing them to be tailored for a variety of purposes. This can be accomplished by performing functionalization techniques that enrich the surface of CNTs with certain functional groups connected to the carbon atoms. Several methods for functionalizing CNTs have been successfully developed [33]. The most frequent functionalization technique is oxidation, which is accomplished by refluxing CNTs with a mineral acid or a strong oxidizing agent such as $KMnO_4$. In the oxidation process, carboxylic functional groups are fixed on CNTs, and the oxidized form of CNTs can be employed for further functionalization. The most active functional groups generated on CNTs are hydroxyl (KOH) and carboxyl (–COOH). CNTs with polar groups in their surfaces spread more easily in organic solvents and solutions, making them more hydrophilic. A variety of additional methods for functionalizing CNTs have been used, including arylation and alkylation, biomolecules, esterification, thiolation, and polymer grafting [34].

8.6 Chemically Modified Derivatization of CNTs

Covalent modification is necessary in various applications to meet special needs; for example, electron mediators are required for biosensors to improve electron transit. In contrast, an electrochemical metal ion sensor requires certain functional groups with a potential affinity for a particular metal ion is modified [35]. In the 1970s, Murray et al. were the first to disclose a chemical change in a carbon substrate material. Attaching a specific molecule or entity to the substrate material conferring chemical specificity was the most common method of modification. Chemical alterations can take many different forms; they are divided into two categories as bulk manipulation and surface alterations [36]. Surface alterations involve electrochemically prompted nanoparticle deposition, metal surface modification, and polymer grafting. Some examples of the latter include hypophosphorous acid is used as a reducing agent in the chemically reduced diagnosed salts, as

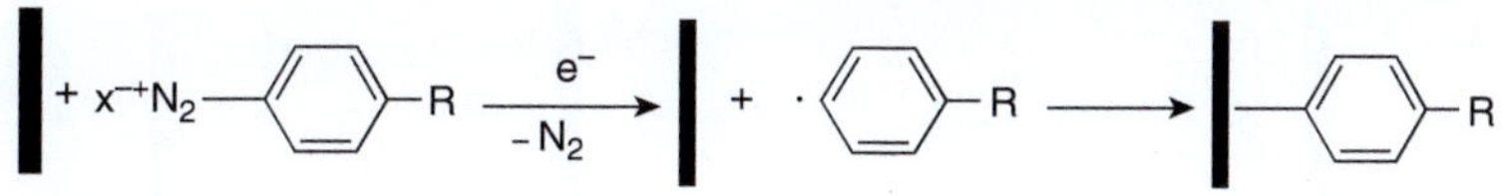

Figure 8.5 The fragrant diazonium electrolytic reduction salt is depicted in this diagram.

well as microwave-assisted modification, ball-milling modification, and thermally activated covalent modification [37].

8.6.1 Electrochemically Assisted Covalent Modification

Alison J Downard looked at electrochemically aided covalent alterations done on carbon tubes or sheets, a mod eon in chemically modified electrodes in which the substrate and the modifier molecule form a covalent bond. The reductive approach, which involves the electrochemical reductions of fragrant diazonium ions, as well as the oxidation of aromatic hydrocarbons, has both been well researched in recent reviews [38]. The reductive technique was invented by Jean Pinson and coworkers in 1990, and it was subsequently improved by a number of people, involving Downard and McCreery group. The key to this modification is the establishment of a covalent connection bridge between the carbon atoms of the substrate and the carbon atoms of the modulator molecule at the solution/electrode interface by electrochemical reduction of the matching diamine [39] (Figure 8.5).

Using methods like cyclic voltammetry or controlled potential electrolysis, electrochemical properties in nonaqueous or aqueous solutions with pH less than three may be easily changed. Variable experimental parameters such as applied voltage or current, modulator molecule length, scan rate, and concentration can all influence the thickness of the modifying molecule on the surface of CNTs [40]. According to McCreery and a group of researchers that have come up with a monolayer is formed when the modifier concentration is less than 1 mM, whereas a multilayer is formed when the modifier concentration is greater than 1 mM. The diazonium group's electron-withdrawing behavior necessitates the use of very low reduction potentials to generate radicals. Carbon–carbon coupling obtained in this process is much more rigid when compared with gold and thiol bond, and it can resist temperatures as high as 700 K without excessive fatigue. It is fairly stable throughout a wide potential range of 2.6–5.6 V, it changes according to the atom substituted and its position as para and ortho. In comparison to the alkane-thiol system, the production of those same sheets on an anode electrode is less regulated. The carbon surface captures the aryl radicals made electrochemically, a portion of the aryl radicals generated may connect to the carbon surface, while the residue penetrates solvent. According to Allongue et al., only 56% of radicals create bonds with the basal planes, whereas 84% form bonds with a glassy carbon surface by pyrolytic graphite (HOPG) [41] (Figure 8.6).

Steven et al. modified SWCNTs by electrochemical dissolution of the diamine of *p*-nitrophenol in a home-built miniaturized electrochemical cell using platinum

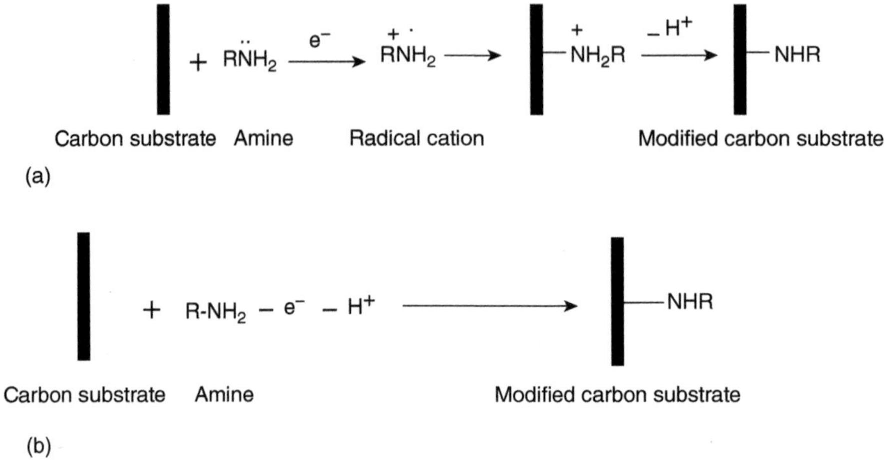

Figure 8.6 Carbon alteration by amine oxidation is depicted schematically.

wire as countering electrode and pseudo-reference electrodes. They determined that covalent bonds occur between the mesh of nanomaterials and the carbonyl group of altered phenyl groups based on experimental results [42]. CNTs had a significant degree of functionalization, with up to 5% of carbonyl group carrying a functional cluster, indicating the SWCNTs were sensitive to radical assault not only at the ends but also on the sidewalls. Another form of electrochemical modification is the oxidation of amines. Amines were functionalized on the electrode surface by oxidizing carbon substrate surfaces such as glassy carbon spheres, carbon fibers, graphitic carbon, and CNTs [43]. Several processes for alteration of carbon sheets by amines are identified here. The development of an amine ion radicals having proton deprivation before the amine ion attachment to the surface is one way, while the synthesis of an amine cation radical with proton loss prior to the amine's attachment to the surface is another [44].

8.7 Polymer Grafting

Polymers are intriguing because of their intrinsic chemical qualities, which include great stability, a large number of active sites, good adhesion, homogeneity, and electrical properties. Polymers also reduce adsorption phenomena by selectively eliminating interfering species and having high active sites number that can well before the molecules of the substrate, in addition to these properties [45]. Only certain analytes may diffuse toward them since they are semipermeable, resulting in high specificity for each analyte. Polymers and nanotubes, which possess all of these characteristics, are a particularly significant cluster of artificial matter that imitates naturally implemented by the substances. CNTs have been polymerized in three paths such as the first electrode material is soaked in a polymer solution, second is by using a chemical process that starts with an oxidizer and ends with

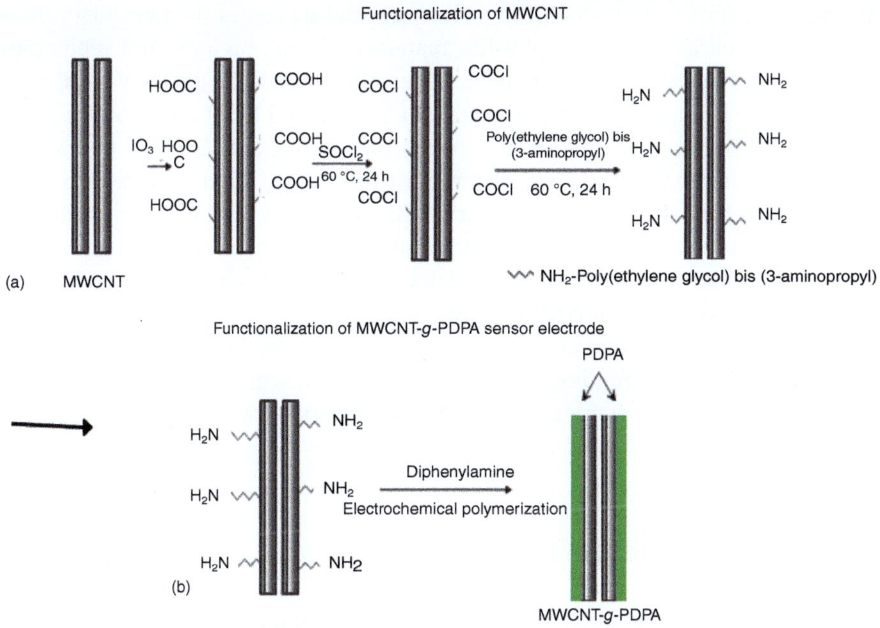

Figure 8.7 MWCNTs poly-diphenylamine sensor modification. (a) Amine functional of MWCNTs. (b) Construction of MWCNTs – PDDA Sensors.

a reductive by-products and the third is by using an electrochemical approach, which involves potential cycling or galvonostatic circumstances are both used to electro-polymerize it [45].

The idea of having significant control over the amount of the polymeric layer on the surface of CNTs has piqued interest in electrochemical polymerization on rigid electrode in recent decades [46]. Poly acrylic acid, Osmium 4,4-dimethyl 2, 2-bipyridine chloride, Polydiphenylamine (PDPA), Quinone-*o*-phenylenediamine, and the most prevalent electro conducting polymers, such as polythiophene, polyaniline or PANI, and polypyrrole or PPy and its derivative are used to modify MWCNTs in this fashion (PTh). Poly 1,2-diamino benzene sodium salt, Poly styrene sulphonic acid sodium salt and Poly 1,2-diamino benzene sodium salt (PABA) were also used to modify SWCNTs. Some altered materials were created to behave as sensing surfaces by reacting with anions, metal ions, and tiny organic molecules [47] (Figure 8.7).

8.8 Carbon Nanotubes Enhanced with Nanoparticles

Since the time of Faraday, who hypothesized that the ruby red color of gold colloid was due to its minuscule particle size, nanoparticle (NP), notably semiconductor and metal NP has been the building blocks of nanodevices [48]. In the last several decades, size-dependent characteristics have resulted in a plethora of novel applications. Sensing devices, data storage, biological interfaces, catalysis, nanoelectro-mechanical devices, nanotribology, and quantum dots are just a few domains, as well as chemistry, physics, material science, biology, and medicine, where the influence has been considerable. CNTs, include both SWCNT and MWCNT, have been extensively investigated and gained a lot of attention for their applications in various areas such as sensing, biosensing, and catalysis due to the intrinsic advantages, such as high electronic conductivity, high surface area, and outstanding chemical and electrochemical stability due to their excellent thermal, chemical, electrical, and tensile strengths [49]. Many electrochemical processes might benefit from the employment of nanoparticles to change CNTs for the building and design of nanoarchitectures, resulting in a number of applications such as nano bio-sensors, biosensors, sensors, heterocyclic for fuel cells. Catalysis for fuel cells and nanotechnology are all examples of emerging technologies. In the literature, several methods for altering CNTs with NPs have been published, each with its own set of benefits and downsides [50].

8.9 Advantages of CNTs

CNTs have a wide range of features and uses that take advantage of their unique aspect ratio, mechanical strength, electrical, and thermal conductivity [51].

- The heat conductivity of CNTs is quite high.
- The electrical conductivity of CNTs is extremely high.
- Aspect ratio of CNTs.
- CNTs have a high elongation-to-failure ratio of 18%.
- The tensile strength of CNTs is extremely great.
- CNTs are extremely flexible and may be bent to almost any shape without causing damage.
- CNTs have a low coefficient of thermal expansion.
- CNTs are strong emitters of electron fields.

8.10 Challenges in CNTs

CNTs as SPE sorbents have also made significant development in recent years, as this is one of the most essential separation science applications at the time [51]. The pesticide analysis sector continues to be the most researched, although the number of matrices explored has increased. In the realm of CNT SPME coatings, significant progress has also been made. Alternative synthetic techniques for coatings that are long-lasting and resistant have been described. Nonetheless, further research is needed to broaden their use to a larger spectrum with reagents and matrices. The insolubility, nonbiodegradability, 3–3.5-hour circulation half-life, biocompatibility, and immunogenicity of CNTs are all limitations [52]. These limitations highlight the need for adjustments in order to assess CNTs' potential as delivery vehicles. The key problem is CNT toxicity, which is detailed explained here, Table 8.1.

Table 8.1 Toxicity characteristics of several kinds of CNTs were compared in a research.

Types of CNTs	Diameter/length	Cells types	Result	References
Asbestos	Diameter (0.394 ± 1.83 μm) Length (6.22 ± 3.22 μm)	Mesothelioma (mice)	Development of extensive inflammatory and proliferative changes. The carcinogenic activity occurred.	[53, 54]
MWCNT (mixture with graphite nanofibres)	Length (5–25 μm) Diameter (10–50 nm)	Lungs (mice) Bronchial epithelial cell (human)	No significant lung inflammation or tissue damage but caused systemic immune function alterations.	[55, 56]
MWCNT	Length (0.5–6.1 μm) Diameter (10–30 nm)	Lungs (rat) Bronchial epithelial cell (human)	Release of pro-inflammatory and pro-fibrotic mediators which could lead to lung fibrosing diseases.	[57–60]
MWCNT (Nanocyl NC 7000)	Length (5–15 nm) Diameter (0.1–10 μm)	Lungs (Wistar rat)	Increase in BALF total cell count, pronounced multifocal granulomatous inflammation and lung fibrotic were the negative effects.	[61]
SWCNT	Length (0.7 μm) Diameter (0.2 μm)	Intratracheal instillation (Wistar rat)	There was no increase in total cell or neutrophil count in BALF. Well-dispersed SWCNT did not induce neutrophil inflammation in the lung.	[62]
SWCNT	Length (0.1–1 μm) Diameter (0.8–1.2 nm)	Lung (mice)	Leads to lung fibrosis effect and acute inflammatory phase reaction were also observed.	[63–65]
Rigid MWCNT	Length (5.29 μm) Diameter (50 nm)	Mesotheliomas (rat)	Mesothelial injury by thin and thick MWCNTs is responsible for the extent of inflammogenicity and carcinogenicity.	[66, 67]
Long MWCNT	Diameter (> 20 nm)	Mesothelial lining of the chest cavity (mice)	Fibroblast formation proliferation which can lead to pulmonary fibrosis. Formation of lesions known as granulomas and inflammation occurrence were included	[68, 69]

Source: Das et al. [51]/Springer Nature/CC BY 4.0.

8.11 Applications of CNTs as Membrane Separation

CNTs, either SWCNTs or MWCNTs, have been the topic of extensive and diverse study over the last two decades according to those major properties and extend the range of exertive implementation those other existing nano-materials [70]. These were dubbed the most investigated materials of the twenty-first century on multiple occasions. As a result, significant and far-reaching accomplishments have been made in a variety of domains, including material science. Previous review articles have extensively addressed and illustrated the unique features of CNTs. They have a high tensile strength that can be thermal conductivity comparable to diamond, stability, and resilience, hundred times that of stainless steel, as well as various electrical characteristics. They can be good conductors, equivalent to copper or semiconductors, and have the capacity to form various sorts of interactions with organic and inorganic analytes, depending on how the graphene sheet is arranged [71].

Low selectivity, permeability, blockage or fouling susceptibility, chemical and heat resistance, and chemical and thermal resistance are all disadvantages of polymeric combination of membranes that have historically worked for water treatment and gaseous separation. The intrinsic features of CNTs are surface area and high porosity has aroused great interest in the field of membrane construction. They have, in fact, been effective in correcting some of these faults. Furthermore, using functionalized CNTs, CNT-based membrane has been utilized for not only differentiate size of the molecules, but also to influence molecular movement. Due to its uses in gas separation and water purification, the manufacturing of CNT-based membranes has risen in recent years [72].

8.11.1 Water Treatment

A basic human right is to have access to clean and safe water. Sadly, 980 million people worldwide, especially in poor countries, lack access to safe drinking water. CNTs are the most significant nanomaterial for water filtration [73]. It can effectively remove organic, inorganic, and biological contaminants from the environment. This is because of its large surface area, strong chemical reactivity, high aspect ratio, low cost, and low energy consumption. Approximately 963 metric tonnes of CNTs have been used for energy and environmental applications in the last several years, with the number continuing to rise. Despite the fact that there is the potential for both environmental harm and human, no systematic approach for assessing the risks that exists [74] (Figure 8.8).

8.11.2 Air Filtration

When the fiber diameter is less than the gas mean free route, pressure drop and filtration efficiency are increased, making air filtration critical and difficult. Previously, free molecular flow (FMF) conditions were produced by reducing the pressure and rising the temperature to increase the gas mean free path. Under typical circumstances, CNTs of nanoscale diameters may be filtered in FMF mechanism. CNTs

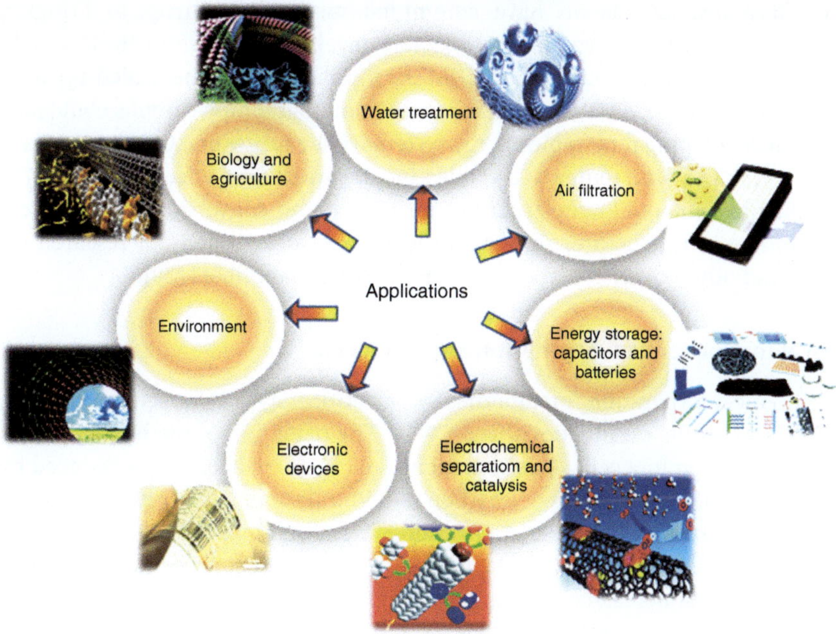

Figure 8.8 Applications of carbon nanotubes (CNTs) as membrane separation.

are used to characterize basic modeling techniques of high-efficiency particle air (HEPA) filters [75]. Conventional air filtration theory predicts a lower pressure drop in air filters operating in the FMF domain. The pressure drop of SWCNT films used to make the thinnest HEPA filters is extremely low. In dynamical filtering, nanotube air filters with a gradient nanostructure have been found to have much superior filtering performance. Agglomerated CNTs fluidized bed air sieves and stratified CNT air filters have also been presented. Finally, the problems and potential applications of CNTs in air filtration are highly intriguing [76].

8.11.3 Energy Storage: Capacitors and Batteries

CNTs have been shown to be useful in solar devices, better electro-chemical capacitor, and lithium-ion battery. Heterojunction solar cells were created by combining nanotubes with an inorganic layer. An optimized hybrid technique yielded monochromatic power conversion efficiencies of up to 10.7% with MWCNT, ZnP8 array, and pyrene MWNT. Hundreds of CNTs have also been employed to improve exciton production and photoexcited carrier transport in photovoltaic systems [77]. Photochemical solar cells were produced by Hasobe et al. employing SWCNT and readily oxidized pyridine assemblies on nano-structured SnO_2 electrodes. The organized assemblies were discovered to be photo-active, capturing a lot of light at optical wavelengths and converting 13% of photons into photo-current.

Nanotubes have been shown to act as electron donors or acceptors in a variety of new organic photovoltaic technologies. Recently, countries such as

France, Germany, and Japan have committed national resources to bringing laser vaporization-SWNT production to factory levels. As a result of these simple calculations, the United States should invest significantly in the scaled-up with laser-generated using SWNTs [78]. Other types of nanographitic carbons could also be useful in developing renewable energy systems of the future era. Graphene sheets of graphite that has been touted as a viable kind of carbon for device creation. CNTs are not the only potential nanostructures for both photo-conversion and electrical energy storage infrastructure, since there is already a huge body of research on graphene and other kinds of carbon, such as graphitic nanofibers [79].

8.11.4 Electrochemical Separation and Catalysis

Electrochemical separation membranes in electrocatalytic characteristics have high expected results on functionalized CNTs. The oxidation of methanol and the reduction of oxygen fuel cells, these reactions are crucial, which can be reduced by functionalization. The concepts of a carbon nanotubule/nanocluster membrane are the emerging materials for these applications. Metal nanoparticles are deposited, and nanotubes are formed, just inside of pores, not outside [80]. The catalyst is only inside CNTs do nanoclusters exist, and each one is unique. A carbon nanotubule reactor is a separate reactor that might be used to produce CNTs. Catalytic selectivity is improved in desired manner. Finally, membrane catalysis is a type of catalysis that occurs within a membrane, the capacity to combine reactions is tuneable [81].

8.11.5 Electronic Devices Fabrication

Dekker et al. in which single nanotube to produce a three-terminal switchable gadget. This material's mechanical and electrical qualities will surely outperform those of previous C/C composites. Electronic and mechanical properties of materials will undoubtedly be replaced with other carbon composites. The mechanical and electrical characteristics of this type of material will surely be better than other composites since the molecular model of three-dimensional network employing SWNTs structure only incorporates covalent sp^2 like carbon. Semiconducting SWNTs were coupled to metallic nano-electrodes in this room-temperature transistor [82]. However, one of nanotubes' inherent challenges is their difficulty in manipulation. In this area, another way for altering the length and, as a result, the electrical characteristics of nanotubes using STM nanostructures are revealed. CNT must have defect-free and high crystallinity in order to be employed as composite fillers. In this context, quantifying nanotube defects and ensuring uniform dispersion of CNTs in polymer array has proven problematic. Surprisingly, significant advancements in current underway lead to the following important challenges [83].

8.11.6 Environment

Depending on their unique characteristics, such as oxidative potential, electrical properties, and surface chemistry, CNTs may have a variety of outcomes in environmental matrices. Cross-linking and agglomeration states were found to impact behavior in water and soil studies. In order to assess the fate of CNTs, it would be helpful to define the kind and form of CNTs that may be released into the ambient divisions. Physiochemical circumstances in diverse natural modules such as ionic strength, conversion efficiency, heat, toxic chemical symbiosis, or ultraviolet rays are expected to change CNT properties or fully functioning adaptors, and hence their environment fate [84]. Other molecules or particles can easily adhere to the surfaces of CNTs, altering their properties. Pollutants might, therefore, be transported across environmental divisions. As a result, the question of whether and to what extent specific CNT material properties can be predicted and managed in the environment emerges.

8.11.7 Biology and Agriculture

CNTs are discharged into the environment, and preliminary research has demonstrated that they are both bioavailable and biopersistent to various organisms. As a result, the prospect of CNTs accumulating along the food chain cannot be ruled out. These findings highlight the need for more research into eco-toxicity as well as longer-term effects such as biopersistency, bioaccumulation, and detrimental on reproduction effects [85]. Once swallowed by other animals or humans, CNTs will cause inflammation, cell damage, oxidative stress, long-term pathological consequences such as fibrosis, wall thickening, granulomas, and poor cell function. In the majority of toxicology research, these effects were reported to be time and dose-dependent.

8.12 Commercial-Scale of Chemically Modified CNTs in Membrane Separation

Various forms of CNT-based membranes for water filtration have been identified in recent study. It also addresses the disadvantages of employing CNT membranes for wastewater treatment, as well as potential concerns [33, 34, 86–91]. Membrane technologists, academicians, industrialists, and environmentalists seeking for addressing aqua shortage concerns would benefit from the many features of CNT membrane-based water purification collected. KOH groups are attached to CNTs and said to enhance surface area and specific mesoporic size while also preferring oxidized functional groups anchoring on exterior part. Large pollutants like organic chemicals and dyes such as toluene, *m*-xylene and ethylbenzene are entrapped by this functionalization. Polystyrene, ZnO/MWCNT/polypyrrole-coated, MWCNTs/polysulfone support polyethersulfone matrix, Tannic Acid–FeIII functionalized MWCNTs/polysulfone support; these are some of the materials which have patents [92–94]. CNTs-based membranes, including as CNTs-based NF, RO, UF, and osmosis membranes, present many elements. The manufacture and implementation of MM, VA, and TFCs-based CNTs membrane for water treatment. Various characteristics of membranes have been thoroughly examined, including antifouling behavior, separation efficiency, reusability, and stability. This membrane has shown

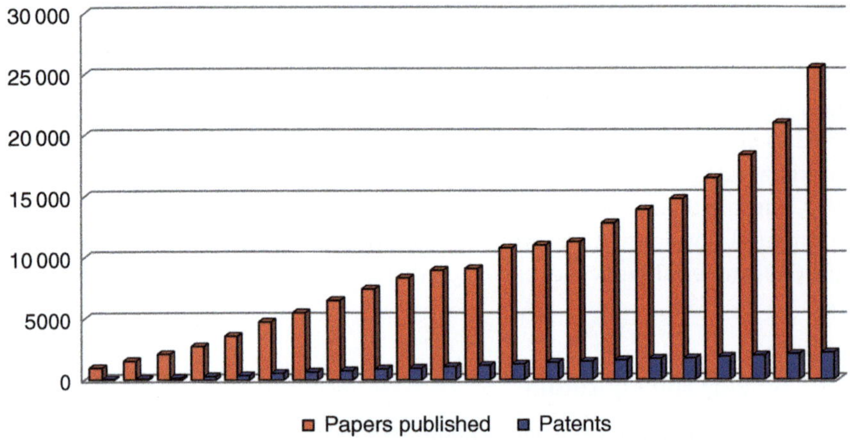

Figure 8.9 Growth of chemically modified carbon nanotubes (CNTs).

the enormous purifying capacity and might be regarded as water sieves for the next generation. The use of membranes minimizes energy, amount of pressure, and expense that are required by traditional water treatment methods. Inorganic, fouling from organic, and bio-chemical debris found in it is the fundamental obstacle that these membranes have yet to solve [95, 96].

CNT membranes that have been functionalized might offer a remedy to this problem. Usage of CNTs membrane aid in the particular filtration of microorganisms and salt. The electro-chemistry combinations with CNT sieves have produced very successful treatment approaches for a wide range of chemical and biological wastes. Electro-chemical CNT sieve has a larger surface area than ordinary CNT filters and is more specific, flexible, and stable. Rotten reduction is an add-on value for these electro-chemically active filters [34, 89, 90, 97–99] (Figure 8.9).

The recent advancements in the field of chemically modified CNTs over two decades are plotted in figures from various sources. The number of papers published and patents are increasing, this is clearly shown here. Chemically modified CNTs have piqued the interest of researchers for next-generation applications due to their superior characteristics. As a result, since 2000, the rate of research output linked to chemically modified CNTs and their application in membrane separation has increased dramatically. Rapid growth is clearly shown in the diagram [100].

8.13 Future Insights

CNTs are the most promising nanomaterials, with appealing features with a wide range of technological applications. CNTs with diverse physical properties are produced by various synthesis, purification, and postprocessing procedures, which can be used invariety of industrial spanning from nanocomposite materials to medicinal applications, electronics, and energy storage. Buckypaper membrane is macroscopic paper-like material that is randomly distributed CNTs held together by van der Waals forces and interactions. Because CNTs are expected to be widely used, it's critical to recognize their potential for harm. We noticed a wide variety of results from various toxicological research in this environmental health evaluation. Further standardization and the supply of reference materials should increase comparability [86]. During the production of BP membranes, chemical modification of CNTs is also a critical step. The chemical procedures outlined here can also be used to modify BP-CNTs chemically. During the first purification process, oxidative treatment is usually applied. To eliminate carbonaceous contaminants, which have a lower oxidation resistance than CNTs, acidic oxidation or heating the CNTs in an oxidative environment, such as air, is commonly used. This treatment makes CNTs more hydrophilic by adding carboxyl and hydroxyl groups, which improves their dispersion in polar liquids like water and the hydrophilic character of the final BP membrane.

However, the outcomes of this study point to a few key issues at this time: Because there are many types of CNTs, they cannot be regarded as a homogeneous collection of chemicals and CNTs that are bio-available to organisms in environmental compartments. Properties of CNTs indicate that they could be part of long food chain and have a long lifespan. The inherent physical and chemical characteristics of CNTs, such as CNTs coating, functionalization, agglomeration state and length are influenced by external environmental conditions during use, disposal, and production; determine their metabolism, absorption, excretion, distribution, excretion, and health effects in organisms. As a consequence, defined exposure scenarios might be useful in toxicological research. CNTs, on the other hand, induce a deadly reaction in the lungs when big enough amounts are inhaled; this is known as the CNTs reaction. The identification of possible hazards to human environment and human health is required for the effective deployment of CNTs in future applications [87].

8.14 Conclusion

Membrane separation using chemically modified CNTs Due to their prospective applications in the realms of environmental protection and physical health maintenance, applications have gotten a lots of interest in recent times. Only few groupings materials described are now important on the market. The key reason is that their pricing is currently uncompetitive. Each CNT's future is determined not just by its competitiveness, but also by society's capacity to pay for it. CNT materials have a bright future ahead of them in terms of development. Vast publications published are evidence which shows that they have a strong presence in Analytical Chemistry. These multimodal materials continue to inspire high expectations, and in the next years, new and difficult contributions will definitely emerge [33]. CNTs have shown potential as stationary phases. These applications have been hesitant to take off due to the synthesis difficulty and the demand to generate unique, stable technique for performance assessments. More articles, particularly those dealing with the assessment of authentic samples, are expected in the near future, as few have previously tackled this area [88].

Many approaches, such as random and block polymerization or grafting, have been developed to improve the characteristics of CNTs. Both the rate of CNTs and the mechanical characteristics of the final products are improved by these procedures. Another method for producing materials with various morphologies and physical properties is physical mixing [89].

Some sophisticated technologies have been used to give CNTs more value. Active packaging technologies and natural fiber reinforcements are among them. Various studies on the use of nanoclay and different fabrication perspectives for advancement in magnetic-CNT and CNTs-based membranes have recently been published, and the astounding performance in separation and ultra high water flux in magnetic-CNT, and CNTs-based membranes are all being investigated [90].

Open Source Links of Figure 8.1

https://en.wikipedia.org/wiki/Water_scarcity
https://reliefweb.int/map/world/water-stress-country-2040

References

1 Cha, S.I., Kim, K.T. et al. (2008). Mechanical and electrical properties of cross-linked carbon nanotubes. *Carbon* 46 (3): 482–488.

2 Lee, H.-J., Han, S.-W. et al. (2008). Functionalization of multi-walled carbon nanotubes with various 4-substituted benzoic acids in mild polyphosphoric acid/phosphorous pentoxide. *Carbon* 46 (14): 1850–1859.

3 Wangyang, L., Li, N. et al. (2009). The role of multiwalled carbon nanotubes in enhancing the catalytic activity of cobalt tetraaminophthalocyanine for oxidation of conjugated dyes. *Carbon* 47 (14): 3337–3345.

4 Wang, L., Zhu, D., Duan, L., and Chen, W. (2010). Adsorption of single-ringed N- and S-heterocyclic aromatics on carbon nanotubes. *Carbon* 48 (13): 3906–3915.

5 Wikipedia, the free encyclopedia. Water scarcity. https://en.wikipedia.org/wiki/Water_scarcity (accessed 20 April 2022).

6 Water Stress by Country: 2040. Source: World Resources Institute. Posted on 27 August 2015, https://reliefweb.int/map/world/water-stress-country-2040 (accessed 20 April 2022).

7 Rijsberman, F.R. (2006). Water scarcity: fact or fiction? *Agric. Water Manage.* 80 (1–3): 5–22.

8 Liu, J., Yang, H., Gosling, S.N. et al. (2017). Water scarcity assessments in the past, present, and future: review on water scarcity assessment. *Earth's Future* 5 (6): 545–559.

9 Rashed, A.O., Merenda, A., Kondo, T. et al. (2020). Carbon nanotube membranes – strategies and challenges towards scalable manufacturing and practical separation applications. *Sep. Purif. Technol.* 257: 117929.

10 Радушкевич, Л.В. (1952). Archived copy О Структуре Углерода, Образующегося При Термическом Разложении Окиси Углерода На Железном Контакте (PDF). *Журнал Физической Химии* (in Russian) 26: 88–95.

11 Monthioux, M. and Kuznetsov, V.L. (2006). Who should be given the credit for the discovery of carbon nanotubes? (PDF). *Carbon* 44 (9): 1621–1623.

12 Iijima, S. (1991). Helical microtubules of graphitic carbon. *Nature* 354 (6348): 56–58.

13 Oberlin, A., Endo, M., and Koyama, T. (1976). Filamentous growth of carbon through benzene decomposition. *J. Cryst. Growth* 32 (3): 335–349.

14 Monthioux, M. and Kuznetsov, V.L. (2006). Who should be given the credit for the discovery of carbon nanotubes? *Carbon* 44 (9): 1621–1623.

15 Saifuddin, N., Raziah, A.Z., and Junizah, A.R. (2013). Carbon nanotubes: a review on structure and their interaction with proteins. *J. Chem.* 2013: 676815.

16 Maiti, U.N., Lee, W.J., Lee, J.M. et al. (2014). 25th anniversary article: chemically modified/doped carbon nanotubes & graphene for optimized nanostructures & nanodevices. *Adv. Mater.* 26 (1): 40–67.

17 Chowdhury, Z.Z., Sagadevan, S., Johan, R.B. et al. (2018). A review on electrochemically modified carbon nanotubes (CNTs) membrane for desalination and purification of water. *Mater. Res. Exp.* 5 (10): 102001.

18 Venkata Ramana, D.K., Kumar Reddy, D.H., Kumar, B.N. et al. (2013). Adsorption of Pb(II) from aqueous solutions by chemically modified zeolite supported carbon nanotubes: equilibrium, kinetic, and thermodynamic studies. *Sep. Sci. Technol.* 48 (3): 403–412.

19 Gu, L., Chen, Q., Li, X. et al. (2019). Amino acid modified carbon nanotubes with optimal pore size for chiral separation. *Mol. Simul.* 45 (13): 1051–1057.

20 Tan, Y., Deng, W., Ge, B. et al. (2009). Biofuel cell and phenolic biosensor based on acid-resistant laccase–glutaraldehyde functionalized chitosan–multiwalled carbon nanotubes nanocomposite film. *Biosens. Bioelectron.* 24 (7): 2225–2231.

21 Pérez-López, B. and Merkoçi, A. (2012). Carbon nanotubes and graphene in analytical sciences. *Microchim. Acta* 179 (1, 2): 1–16.

22 Herrera-Herrera, A.V., González-Curbelo, M.Á., Hernández-Borges, J., and Rodríguez-Delgado, M.Á. (2012). Carbon nanotubes applications in separation science: a review. *Anal. Chim. Acta* 734: 1–30.

23 Ali, S., Rehman, S.A.U., Luan, H.-Y. et al. (2019). Challenges and opportunities in functional carbon nanotubes for membrane-based water treatment and desalination. *Sci. Total Environ.* 646: 1126–1139.

24 Tavallali, H., Malekzadeh, H., Karimi, M.A. et al. (2019). Chemically modified multiwalled carbon nanotubes as efficient and selective sorbent for separation and preconcentration of trace amount of Co(II), Cd(II), Pb(II), and Pd(II). *Arabian J. Chem.* 12 (7): 1487–1495.

25 Kim, E.-S., Hwang, G., El-Din, M.G., and Liu, Y. (2012). Development of nanosilver and multi-walled carbon nanotubes thin-film nanocomposite membrane for enhanced water treatment. *J. Membr. Sci.* 394: 37–48.

26 Burghard, M. (2005). Electronic and vibrational properties of chemically modified single-wall carbon nanotubes. *Surf. Sci. Rep.* 58 (1–4): 1–109.

27 Dumée, L., Germain, V., Sears, K. et al. (2011). Enhanced durability and hydrophobicity of carbon nanotube bucky paper membranes in membrane distillation. *J. Membr. Sci.* 376 (1, 2): 241–246.

28 Ma, L., Dong, X., Chen, M. et al. (2017). Fabrication and water treatment application of carbon nanotubes (CNTs)-based composite membranes: a review. *Membranes* 7 (1): 16.

29 Murali, R.S., Sridhar, S., Sankarshana, T., and Ravikumar, Y.V.L. (2010). Gas permeation behavior of Pebax-1657 nanocomposite membrane incorporated with multiwalled carbon nanotubes. *Ind. Eng. Chem. Res.* 49 (14): 6530–6538.

30 Sanip, S.M., Ismail, A.F., Goh, P.S. et al. (2011). Gas separation properties of functionalized carbon nanotubes mixed matrix membranes. *Sep. Purif. Technol.* 78 (2): 208–213.

31 Huang, L.-L., Shao, Q., Lu, L.-H. et al. (2006). Helicity and temperature effects on static properties of water molecules confined in modified carbon nanotubes. *Phys. Chem. Chem. Phys.* 8 (33): 3836–3844.

32 Zhao, F.-Y., Ji, Y.-L., Weng, X.-D. et al. (2016). High-flux positively charged nanocomposite nanofiltration membranes filled with poly (dopamine) modified multiwall carbon nanotubes. *ACS Appl. Mater. Interfaces* 8 (10): 6693–6700.

33 Speltini, A., Merli, D., Dondi, D. et al. (2012). Improving selectivity in gas chromatography by using chemically modified multi-walled carbon nanotubes as stationary phase. *Anal. Bioanal.Chem.* 403 (4): 1157–1165.

34 Zhao, H., Qiu, S., Wu, L. et al. (2014). Improving the performance of polyamide reverse osmosis membrane by incorporation of modified multi-walled carbon nanotubes. *J. Membr. Sci.* 450: 249–256.

35 Nejad, M.N., Asghari, M., and Afsari, M. (2016). Investigation of carbon nanotubes in mixed matrix membranes for gas separation: a review. *ChemBioEng Rev.* 3 (6): 276–298.

36 Sun, X., Luo, D., Liu, J., and Evans, D.G. (2010). Monodisperse chemically modified graphene obtained by density gradient ultracentrifugal rate separation. *ACS Nano* 4 (6): 3381–3389.

37 Khalil, M.M. and El-Aziz, G.A. (2016). Multiwall carbon nanotubes chemically modified carbon paste electrodes for determination of gentamicin sulfate in pharmaceutical preparations and biological fluids. *Mater. Sci. Eng., C* 59: 838–846.

38 Nechifor, G., Voicu, S.I., Nechifor, A.C., and Garea, S. (2009). Nanostructured hybrid membrane polysulfone-carbon nanotubes for hemodialysis. *Desalination* 241 (1–3): 342–348.

39 Yu, B., Cong, H., Li, Z. et al. (2013). Pebax-1657 nanocomposite membranes incorporated with nanoparticles/colloids/carbon nanotubes for CO_2/N_2 and CO_2/H_2 separation. *J. Appl. Polym. Sci.* 130 (4): 2867–2876.

40 Khan, M.U., Reddy, K.R., Snguanwongchai, T. et al. (2016). Polymer brush synthesis on surface modified carbon nanotubes via in situ emulsion polymerization. *Colloid. Polym. Sci.* 294 (10): 1599–1610.

41 Endo, M., Strano, M.S., and Ajayan, P.M. (2007). Potential applications of carbon nanotubes. *Carbon Nanotubes* 13–62.

42 Afzali, D., Jamshidi, R., Ghaseminezhad, S., and Afzali, Z. (2012). Preconcentration procedure trace amounts of palladium using modified multiwalled carbon nanotubes sorbent prior to flame atomic absorption spectrometry. *Arabian J. Chem.* 5 (4): 461–466.

43 Li, L., Song, C., Jiang, D., and Wang, T. (2017). Preparation and enhanced gas separation performance of carbon/carbon nanotubes (C/CNTs) hybrid membranes. *Sep. Purif. Technol.* 188: 73–80.

44 Murthy, Z.V.P. and Gaikwad, M.S. (2013). Preparation of chitosan-multiwalled carbon nanotubes blended membranes: characterization and performance in the separation of sodium and magnesium ions. *Nanoscale Microscale Thermophys. Eng.* 17 (3): 245–262.

45 Aberefa, O.A., Daramola, M.O., and Iyuke, S.E. (2019). Production and functionalization of carbon nanotubes for application in membrane synthesis for natural gas separation. *Microporous Mesoporous Mater.* 280: 26–36.

46 Brunet, L., Lyon, D.Y., Zodrow, K. et al. (2008). Properties of membranes containing semi-dispersed carbon nanotubes. *Environ. Eng. Sci.* 25 (4): 565–576.

47 Haddon, R.C., Sippel, J., Rinzler, A.G., and Papadimitrakopoulos, F. (2004). Purification and separation of carbon nanotubes. *MRS Bull.* 29 (4): 252–259.

48 Lee, H.-C., Ciou, Y.-T., and Jiang, H.-R. (2020). Separation of chemically modified carbon nanotubes by surfactant free microbubble generation. *AIP Adv.* 10 (6): 065121.

49 Burk, R. (2009). Solid phase extraction of polyhalogenated pollutants from freshwater using chemically modified multi-walled carbon nanotubes and their determination by gas chromatography. *J. Sep. Sci.* 32 (7): 1060–1068.

50 Amini, M., Jahanshahi, M., and Rahimpour, A. (2013). Synthesis of novel thin film nanocomposite (TFN) forward osmosis membranes using functionalized multi-walled carbon nanotubes. *J. Membr. Sci.* 435: 233–241.

51 Das, R., Leo, B.F., and Murphy, F. (2018). The toxic truth about carbon nanotubes in water purification: a perspective view. *Nanoscale Res. Lett.* 18 (13): 183.

52 Roe, F.J., Carter, R.L., Walters, M.A., and Harington, J.S. (1967). The pathological effects of subcutaneous injections of asbestos fibres in mice: migration of fibres to submesothelial tissues and induction of mesotheliomata. *Int. J. Cancer* 2 (6): 628–638.

53 Mitchell, L.A., Gao, J., Wal, R.V. et al. (2007). Pulmonary and systemic immune response to inhaled multiwalled carbon nanotubes. *Toxicol. Sci.* 100 (1): 203–214. https://doi.org/10.1093/toxsci/kfm196.

54 Lindberg, H.K., Falck, G.C.M., Suhonen, S. et al. (2009). Genotoxicity of nanomaterials: DNA damage and micronuclei induced by carbon nanotubes and graphite nanofibres in human bronchial epithelial cells in vitro. *Toxicol. Lett.* 186 (3): 166–173.

55 Kasai, T., Umeda, Y., Ohnishi, M. et al. (2015). Thirteen-week study of toxicity of fiber-like multi-walled carbon nanotubes with whole-body inhalation exposure in rats. *Nanotoxicology* 9 (4): 413–422. https://doi.org/10.3109/17435390.2014.933903.

56 Hussain, S., Sangtian, S., Anderson, S.M. et al. (2014). Inflammasome activation in airway epithelial cells after multi-walled carbon nanotube exposure mediates a profibrotic response in lung fibroblasts. *Part. Fibre Toxicol.* 11: 28. https://doi.org/10.1186/1743-8977-11-28.

57 Snyder-Talkington, B.N., Dong, C., Sargent, L.M. et al. (2016). mRNAs and miRNAs in whole blood associated with lung hyperplasia, fibrosis, and bronchiolo-alveolar adenoma and adenocarcinoma after multi-walled carbon nanotube inhalation exposure in mice. *J. Appl. Toxicol.* 36 (1): 161–174. https://doi.org/10.1002/jat.3157.

58 Dong, J. and Ma, Q. (2016). Myofibroblasts and lung fibrosis induced by carbon nanotube exposure. *Part. Fibre Toxicol.* 13 (1): 60.

59 Ma-Hock, L., Treumann, S., Strauss, V. et al. (2009). Inhalation toxicity of multiwall carbon nanotubes in rats exposed for 3 months. *Toxicol. Sci.* 112 (2): 468–481.

60 Doan, B.T., Seguin, J., Breton, M. et al. (2012). Functionalized single-walled carbon nanotubes containing traces of iron as new negative MRI contrast agents for in vivo imaging. *Contrast Media Mol. Imaging* 7 (2): 153–159. Yanamala, N., Kisin, E.R., Tkach, A.V., et al. (2014). Long-term effects of carbon containing engineered nanomaterials and asbestos in the lung: one year postexposure comparisons. *Am. J. Phys. Lung Cell. Mol. Phys. 306*(2): L170–L182.

61 Shvedova, A.A., Kisin, E., Murray, A.R. et al. (2008). Inhalation vs. aspiration of single-walled carbon nanotubes in C57BL/6 mice: inflammation, fibrosis, oxidative stress, and mutagenesis. American journal of physiology lung cellular and molecular. *Physiology* 295 (4): L552–L565.

62 Vietti, G., Lison, D., and van den Brule, S. (2016). Mechanisms of lung fibrosis induced by carbon nanotubes: towards an adverse outcome pathway (AOP). *Part. Fibre Toxicol.* 13: 11. https://doi.org/10.1186/s12989-016-0123-y.

63 Nagai, H., Okazaki, Y., Chew, S.H. et al. (2011). Diameter and rigidity of multiwalled carbon nanotubes are critical factors in mesothelial injury and carcinogenesis. *Proc. Natl. Acad. Sci. U.S.A.* 108 (49): E1330–E13E8.

64 Donaldson, K., Murphy, F.A., Duffin, R., and Poland, C.A. (2010). Asbestos, carbon nanotubes and the pleural mesothelium: a review of the hypothesis regarding the role of long fibre retention in the parietal pleura, inflammation and mesothelioma. *Part. Fibre Toxicol.* 7 (1): 5. https://doi.org/10.1186/1743-8977-7-5.

65 Poland, C.A., Duffin, R., Kinloch, I. et al. (2008). Carbon nanotubes introduced into the abdominal cavity of mice show asbestos-like pathogenicity in a pilot study. *Nat. Nanotechnol.* 3: 423. https://doi.org/10.1038/nnano.2008.111.

66 Wang, P., Nie, X., Wang, Y. et al. (2013). Multiwall carbon nanotubes mediate macrophage activation and promote pulmonary fibrosis through TGF-β/Smad signaling pathway. *Small* 9 (22): 3799–3811.

67 A comprehensive review on anti-fouling nanocomposite membranes for pressure driven membrane separation processes – ScienceDirect [Internet]. [cited 2022 Jan 31]. Available from: https://www.sciencedirect.com/science/article/pii/S0011916415301119?casa_token=uA5DoRefo7EAAAAA:pldZTeAE0NtE2Fd7nnrTSe85-1t8E0D8zIL_SEavV2UUJ9r6UDFrTs5zwJr4-lCbMv-3VabpKg.

68 Applications of carbon nanomaterials in chiral separation – ScienceDirect [Internet]. [cited 2022 Jan 31]. Available from: https://www.sciencedirect.com/science/article/abs/pii/S0165993620301709.

69 Carbon nanotube-based membranes: fabrication and application to desalination – ScienceDirect [Internet]. [cited 2022 Jan 31]. Available from: https://www.sciencedirect.com/science/article/abs/pii/S1226086X1200144X.

70 CNTs/TiO_2 composite membrane with adaptable wettability for on-demand oil/water separation – ScienceDirect [Internet]. [cited 2022 Jan 31]. Available from: https://www.sciencedirect.com/science/article/abs/pii/S0959652620340567.

71 Effect of raw multi-wall carbon nanotubes on morphology and separation properties of polyimide membranes: separation science and technology: Vol 45, No 16 [Internet]. [cited 2022 Jan 31]. Available from: https://www.tandfonline.com/doi/abs/10.1080/01496395.2010.484007.

72 Effect of take-up speed on polyvinylidene fluoride hollow fiber membrane in a thermally induced phase separation process – Li – 2013 – Journal of Applied Polymer Science – Wiley Online Library [Internet]. [cited 2022 Jan 31]. Available from: https://onlinelibrary.wiley.com/doi/abs/10.1002/app.37919.

73 Fabrication of carbon nanotubes incorporated double-skinned thin film nanocomposite membranes for enhanced separation performance and antifouling capability in forward osmosis process – ScienceDirect [Internet]. [cited 2022 Jan 31]. Available from: https://www.sciencedirect.com/science/article/pii/S0011916415002659?casa_token=-v8N0HBUam0AAAAA:awGWgDFBeqylCFVKACUeH6IAjQz8SJb2BtMmuRb5LPoOOHP0Q4OMw-Xlh7ojJUl0aDa31ezMGQ.

74 Gas separation properties of functionalized carbon nanotubes mixed matrix membranes – ScienceDirect [Internet]. [cited 2022 Jan 31]. Available from: https://www.sciencedirect.com/science/article/abs/pii/S1383586611000931.

75 Investigation of carbon nanotubes in mixed matrix membranes for gas separation: a review – Nejad – 2016 – ChemBioEng Reviews – Wiley Online Library [Internet]. [cited 2022 Jan 31].

76 Janus polymer/carbon nanotube hybrid membranes for oil/water separation|ACS Applied Materials & Interfaces [Internet]. [cited 2022 Jan 31]. Available from: https://pubs.acs.org/doi/abs/10.1021/am504326m.

77 Materials|Free full-text|Recent developments in carbon nanotube membranes for water purification and gas separation [Internet]. [cited 2022 Jan 31]. Available from: https://www.mdpi.com/1996-1944/3/1/127.

78 Membrane gas separation: a review/state of the art|Industrial & Engineering Chemistry Research [Internet]. [cited 2022 Jan 31]. Available from: https://pubs.acs.org/doi/abs/10.1021/ie8019032.

79 Membranes|Free full-text|Fabrication and water treatment application of carbon nanotubes (CNTs)-based composite membranes: a review [Internet]. [cited 2022 Jan 31]. Available from: https://www.mdpi.com/2077-0375/7/1/16.

80 Prospects of an electroactive carbon nanotube membrane toward environmental applications|Accounts of chemical research [Internet]. [cited 2022 Jan 31]. Available from: https://pubs.acs.org/doi/abs/10.1021/acs.accounts.0c00544.

81 Recent advances of inorganic fillers in mixed matrix membrane for gas separation – ScienceDirect [Internet]. [cited 2022 Jan 31]. Available from: https://www.sciencedirect.com/science/article/abs/pii/S1383586611004503.

82 Recent advances on electroactive CNT-based membranes for environmental applications: the perfect match of electrochemistry and membrane separation – ScienceDirect [Internet]. [cited 2022 Jan 31]. Available from: https://www.sciencedirect.com/science/article/abs/pii/S1001841720301303.

83 Synergistic effects of CNT and GO on enhancing mechanical properties and separation performance of polyelectrolyte complex membranes – ScienceDirect [Internet]. [cited 2022 Jan 31]. Available from: https://www.sciencedirect.com/science/article/pii/S0264127517300758?casa_token=4Rt6XHJXAwwAAAAA:ZZcGDglmL4SLLjADn4W3lvU8p5iP2QWQBZbvToYJ10YkkgaCyCBsGRaOLNsDkitddmwLv0Ovzw.

84 Transport and separation properties of carbon nanotube-mixed matrix membrane – ScienceDirect [Internet]. [cited 2022 Jan 31]. Available from: https://www.sciencedirect.com/science/article/abs/pii/S1383586609003815.

85 Ultra-permeable CNTs/PES membranes with a very low CNTs content and high H_2/N_2 and CH_4/N_2 selectivity for clean energy extraction applications – ScienceDirect [Internet]. [cited 2022 Jan 31]. Available from: https://www.sciencedirect.com/science/article/pii/S2238785421012618.

86 Khan, F.S.A. (2021). A comprehensive review on magnetic carbon nanotubes and carbon nanotube-based buckypaper for removal of heavy metals and dyes. *J. Hazard. Mater.* 413: 125375.

87 Deng, J.H., Zhang, X.-R., Zeng, G.M. et al. (2013). Simultaneous removal of Cd(II) and ionic dyes from aqueous solution using magnetic graphene oxide nanocomposite as an adsorbent. *Chem. Eng. J.* 226: 189–200.

88 De Volder, M.F.L., Tawfick, S.H., Baughman, R.H., and Hart, A.J. (2013). Carbon nanotubes: present and future commercial applications. *Science* 339: 535–539.

89 Yu, F., Wu, Y., Li, X., and Ma, J. (2012). Kinetic and thermodynamic studies of toluene, ethylbenzene, and m-xylene adsorption from aqueous solutions onto KOH-activated multiwalled carbon nanotubes. *J. Agric. Food. Chem.* 60: 12245–12253.

90 Zimmerman, C.M., Singh, A., and Koros, W.J. (1997). Tailoring mixed matrix composite membranes for gas separations. *J. Membr. Sci.* 137 (1): 145–154.

91 Son, M., Choi, H., Liu, L. et al. (2015). Efficacy of carbon nanotube positioning in the polyethersulfone support layer on the performance of thin-film composite membrane for desalination. *Chem. Eng. J.* 266: 376–384.

92 Song, X., Wang, L., Tang, C.Y. et al. (2015). Fabrication of carbon nanotubes incorporated double-skinned thin film nanocomposite membranes for enhanced separation performance and antifouling capability in forward osmosis process. *Desalination* 369: 1–9. Van Hooijdonk, E., Bittencourt, C., Snyders, R., and Colomer, J.F. (2013). Functionalization of vertically aligned carbon nanotubes. *Beilstein J. Nanotechnol.* 4(1): 129–152; *Membr. Technol.* 149.

93 Vatanpour, V., Madaeni, S.S., Moradian, R. et al. (2011). Fabrication and characterization of novel antifouling nanofiltration membrane prepared from oxidized multiwalled carbon nanotube/polyethersulfone nanocomposite. *J. Membr. Sci.* 375: 284–294.

94 Vecitis, C.D., Schnoor, M.H., Rahaman, M.S. et al. (2011). Electrochemical multiwalled carbon nanotube filter for viral and bacterial removal and inactivation. *Environ. Sci. Technol.* 45: 3672–3679.

95 Wu, M.B., Lv, Y., Yang, H.C. et al. (2016). Thin film composite membranes combining carbon nanotube intermediate layer and microfiltration support for high nanofiltration performances. *J. Membr. Sci.* 515: 238–244.

96 Wu, H., Sun, H., Hong, W. et al. (2017). Improvement of polyamide thin film nanocomposite membrane assisted by tannic acid–FeIII functionalized multiwall carbon nanotubes. *ACS Appl. Mater. Interfaces* 9 (37): 32255–32263.

97 Xu, R., Wang, J., Kanezashi, M. et al. (2011). Development of robust organosilica membranes for reverse osmosis. *Langmuir* 27: 13996–13999.

98 Yu, M., Funke, H.H., Falconer, J.L., and Noble, R.D. (2009). High density, vertically-aligned carbon nanotube membranes. *Nano Lett.* 9 (1): 225–229.

99 Yin, J., Zhu, G., and Deng, B. (2013). Multi-walled carbon nanotubes (MWNTs)/polysulfone (PSU) mixed matrix hollow fiber membranes for enhanced water treatment. *J. Membr. Sci.* 437: 237–248.

100 Zinadini, S., Rostami, S., Vatanpour, V., and Jalilian, E. (2017). Preparation of antibifouling polyethersulfone mixed matrix NF membrane using photocatalytic activity of ZnO/MWCNTs nanocomposite. *J. Membr. Sci.* 529: 133–141.

9

Chemically Modified Carbon Nanotubes for Water Purification System

Azza M. Mazrouaa[1], *Ahmed A. Mousa*[2], *and Manal G. Mohamed*[1]

[1]*Egyptian Petroleum Research Institute, Petrochemical Department, Polymer Lab, Nasr City, Cairo 11727, Egypt*
[2]*Synthomer SAE, 10th of Ramadan city, Al-Sharqia, 4955235, Egypt*

Abbreviations

AO	acridine orange
AR	alizarin red
AYR	alizarin yellow R
BP	bucky paper
BTB	bromothymol-blue
CMCNTs	chemically modified carbon nanotubes
CNIM	carbon nanotube immobilized membrane
CNT	carbon nanotube
CVD	chemical vapors depositions
DCMD	direct contact membrane distillation
DIF	diiodo fluorescein
DR	direct red 80
GHz	gigahertz
HACNT	horizontally aligned carbon nanotube
MB	methylene blue
MMMs	mixed matrix membranes
$MnFe_2O_4$	manganese ferrite nano particles
MNP	magnetic nanoparticles
MO	methyl orange
MPTS	mercapto propyltriethoxysilane
MWCNTs	multi-walled carbon nanotubes
NPs	nanoparticles
ox-MWCNTs	oxygenated multi-walled carbon nanotubes
PAN	1-(2-pyridylazo)-2-naphthol
PDA	polydopamine
PER	pentaerythritol
PVDF	polyvinylidene fluoride

Chemically Modified Carbon Nanotubes for Commercial Applications, First Edition.
Edited by Jeenat Aslam, Chaudhery Mustansar Hussain, and Ruby Aslam.

SDBS	sodium dodecyl benzene sulphonates
SDS	sodium dodecyl sulfate
SWCNTs	single-walled carbon nanotubes
TCE	trichloro ethylene
UV	ultra violet
VACNT	vertically aligned CNT
Y40	cationic yellow 40

9.1 Introduction

Carbon nanotubes (CNTs) have attracted the scientific community's interest due to their unusual chemical, physical, electrical, and high thermal capabilities, as well as their wide range of uses, including adsorbents for removing organic contaminants from water and wastewater [1–3]. Also, the raised-up production cost of CNTs holds back their spread along many applications. New approaches play a critical role in considerably cutting production costs. CNTs are safe to be used for water treatment applications [4]. In comparison to conventional water treatment facilities for the removal of different water contaminants, CNTs have emerged as a profitable, effective, and environmental friendly alternative. Hence, many physical/chemical modification approaches were tested to improve their solubility [5]. CNTs have a variety of physicochemical properties that can be used in environmental applications (e.g. adsorbents, membranes/filters, catalysts, oil spill sponges, and sensors) (Figure 9.1) [6]. This chapter delves into the most recent advancements of electrochemically active CNT membranes in solving the present major obstacles related with water treatment and desalination. There's also a discussion of present concerns and predictions for the future.

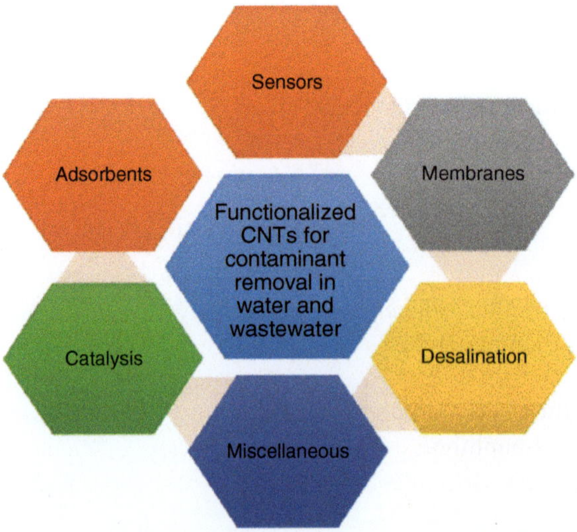

Figure 9.1 Diverse classes of environmental applications of functionalized CNTs. Source: Sarkar et al. [6]/with permission of Elsevier.

9.2 History of Water Purification Methods

The majority of prehistoric civilizations arose and flourished beside rivers and lakes, where water for drinking and agriculture was abundant. The desire for distilled water has stimulated the improvement of water purifying systems since ancient times. These procedures did not eliminate disease-causing microorganisms but they did provide the groundwork for modern purification techniques (e.g. filtration and decontamination). Minoans on the island of Crete, humans in the Indus Valley, Mesopotamians, Egyptians, and Chinese civilizations all used these technologies. Aeration basins were used to purify water in antiquity. The Hippocratic purifying method, which involved pouring boiling water through a conical fabric filter bag to capture particulates that caused poor taste and odor. The Greeks and Romans tried a variety of techniques to improve the quality of their water. However, because there was a dearth of information about water-borne bacteria, most of these procedures were likely unsuccessful, and quality requirements were rarely reached. Filtration for water purification was "commonly known" 165 years ago. Jersey City, New Jersey, will be the first city in the United States to start disinfecting community drinking water on a regular basis in 1908. Efforts were made over the last two centuries to provide communities with adequate supplies of high-quality water [7]. Adsorption, the most popular, has been shown to be a better strategy for removing a number of pollutants from water and wastewater, including organic and inorganic contaminants. Traditional adsorbents were found to have low treatment efficiency due to their small surface area, limited active sites, lack of selectivity, and slow adsorption kinetics. Due to their increased surface area combined with a larger number of active sites, variable pore size, rapid kinetics, and enhanced surface chemistry, recent improvements in nano-adsorbents such as CNT have addressed these drawbacks of conventional adsorbents [8].

9.3 Carbon Nanotubes CNTs Types

A CNT is a rolled-up graphene sheet that is cylindrically produced. It's divided into single and multiple wall tubes based on the thickness of its walls. The first study of multi-walled carbon nanotubes (MWCNTs) in 1991, and single-wall carbon nanotubes (SWCNTs) in 1993. The characteristics of MWCNTs and SWCNTs are quite comparable. Armchair, chiral, and zigzag are three different types of SWCNTs based on how they are wrapped into a cylindrical shape. The structure of a SWCNT is defined by a pair of indices (n, m) that specify the chiral vector and have a direct impact on nanotube electrical characteristics. MWCNTs, on the other hand, may be constructed in two structural models: A CNT with a smaller radial nanotube inside it is mentioned to as a Russian doll model. In the parchment idea, a single graphene sheet is wrapped around itself many times [9].

9.4 Vital of Modification of CNTs

Up-to-date studies confirms requirement to dispersion process modification to possess changes on the surface of the CNTs, which consequently affects its functionalization and its obtainable mechanical aspect. As a result, the efficient groups on the surface of the CNTs regulate the sort of interactions that take place between the tubes. Murray et al. [10] were the first to report CMCNTs in the 1970s [11]. The capability of CNTs for desalination and/or elimination of waterborne contaminants is improved by functionalization, which entails chemical/physical treatment of virgin with several types of functional groups [12].

9.5 Surface Modified CNTs for Water Purification

Surface modified is one of the most effective and extensively utilized CNT modification techniques. This kind lowers contact and interfacial resistances by increasing CNT metallicity or softening the interface between both the matrix and the CNT surfaces [13]. New composites made by integrating various forms of CNTs into various types of polymers might have excellent physical, electrical, and thermal characteristics. Enhancing the mechanical characteristics of CNT polymer nanocomposites requires optimal dispersion and adhesion of interfacial amongst both the polymer and CNTs [14]. CNTs are insoluble in most organic and aqueous solutions by definition, and their chemical compatibility with the polymer matrix is poor. As a result, one of the primary studies focuses in recent years has been surface modification of CNTs to increase compatibility and solubility [15, 16]. It is also well known that CNTs do not have enough binding sites. Surface functionalization of CNT is commonly accomplished by wrapping CNT with polymer, surfactants, or grafting dendron's onto CNT, and so on, to add more binding sites and surface attaching groups [17]. Metals can be functionalized on nanotubes in this form of modification. There are a variety of inventive ways to deposit metal nanoparticles onto CNT substrates, each of which contributes to varying degrees of distribution of particle size control along the CNT [18]. By altering the nucleation potential, deposition duration, and percentage of noble metal complexes, electrodeposition can be utilized to substitute nanotubes with nanoparticles (NPs). Upon loading NPs onto CNTs by electrochemical deposition, noble metal salts such as $H[AuCl_4]$, $H_2[PtCl_4]$, and $(NH_4)_2[PdCl_4]$ are frequently reduced, resulting in high adhesion with the tubes [19]. Pt and Pd NPs were electrochemically deposited onto the surface of CNTs at varied potentials and timings, and the results showed that the electrolyte diffusion and the number of potential pulses can readily be modified to alter the particle size of Pt NPs [20]. After that, the methodology for electrodepositing Pt and Ag on high-density SWCNT was improved, and total dependence on NP dispersion in the type of nanowires or individual nanoparticles was discovered. Arai et al. [21] reported that Cu and Ni nanoparticles were electrodeposited on CNTs. Similar changes were employed by Kim et al. to deposit RuO_2 or NiO by applying potential cycling [22]. Tsai et al. investigate electrochemical deposition in ethylene glycol with sulfuric acid to act as a supporting electrolyte to create Pt (Pt–Ru) NPs/CNTs monohybrids. Xiao et al. developed a new method for synthesizing bimetallic Au–Pt NPs on the surface of MWCNTs in ionic liquids using ultrasonic energy and the standard electrodeposition methods [23].

The fundamental problem of electrochemical techniques for depositing nanoparticles of metal onto CNTs is that bulk manufacturing, or creating gramme amounts of improved CNTs, is extremely difficult. We take a look at a few other ways for making metal (and other) nanoparticles on CNTs that don't entail electrochemical deposition but are nevertheless used in electrochemical applications. Pd, Pt, Au, and Ag nanoparticles, as well as Cu nanoparticles, were deposited from their metal salts using thermal decomposition procedures [24]. Lastly, one of the most imaginative approaches for metal reduction to make nanoparticles on CNTs is reducing Ag, Pd, and Pt–Ru alloy on poly(vinylpyrrolidone) to produce a suspension of colloidal or on SWCNTs [25]. Electroless deposition techniques depend on a chemical, rather than an electrochemical, reduction process, in which the reaction is driven by a chemical species having a lower redox potential than the metal species that is being reduced. CNT surface adhesion to numerous contaminants in water is adjusted by purification and functionalization. In Figure 9.2, we summarized various CNT

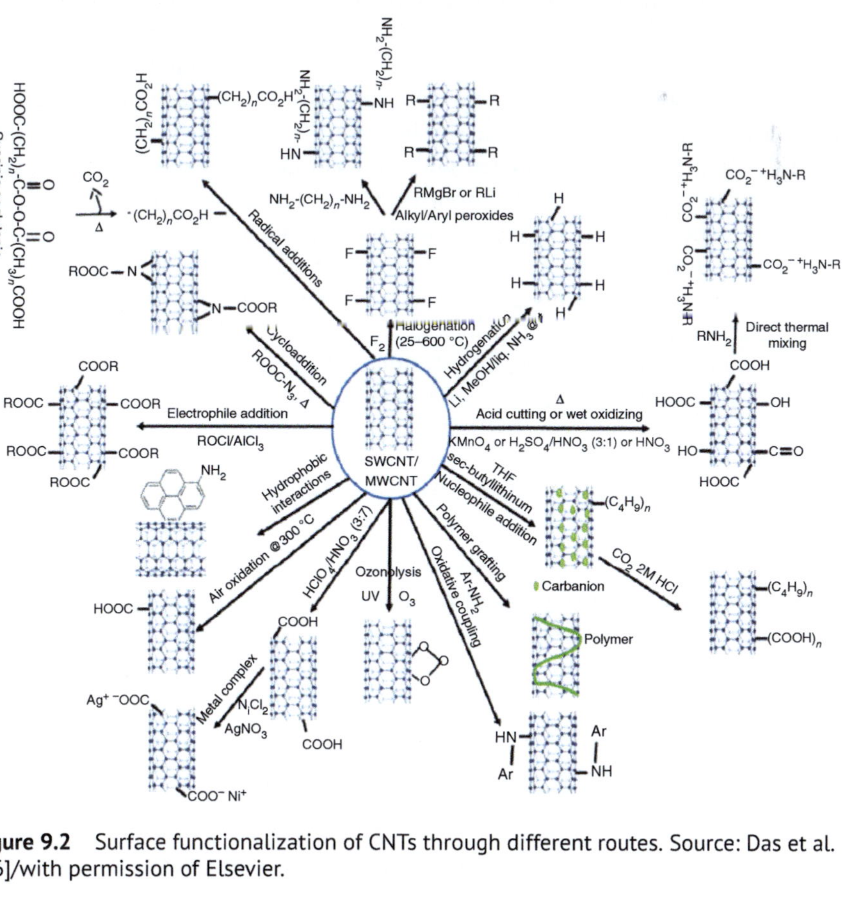

Figure 9.2 Surface functionalization of CNTs through different routes. Source: Das et al. [26]/with permission of Elsevier.

functionalization techniques. This demonstrates that different functionality has been introduced to CNT surfaces via various approaches. Such treatments improve CNT's attraction for wastewater contaminants, which are then trapped on the surface of the nanotube [26].

9.6 Polymer/CNTs Grafting for Water Purification

The addition of CNTs to polymers can improve their characteristics. If CNTs are spread appropriately, their unique characteristics can be imparted to the polymer. At first, the resulting composite can be significantly stronger and lighter than typical carbon fiber-reinforced polymer composites when added to a single polymer or polymer blend matrix [27, 28]. To enhance the properties, appropriate CNT dispersion, strong interfacial adhesion between polymer and CNTs and their solubilities in a mixture of polar and nonpolar solutions are required. Surfactants, ionic liquids, biopolymers, and polymer wrapping upon that surface of CNTs can all be used to accomplish noncovalent functionalization [29, 30]. Surfactants such as sodium dodecyl sulfate (SDS) [31–33] and sodium dodecylbenzene sulfonate (SDBS) [34, 35] interact with CNTs, can significantly reduce the aggregation rate in aqueous solutions of CNTs. Non-surfactant changes have also been proven to be effective in improving tube dispersion and debundling. The noncovalent surface functionalization of CNTs by chitosan, nucleic acids, and other biomacromolecules has been characterized as successful functionalization method [36]. When CNTs are chemically bonded with polymer chains, leading to strong bond with enhanced properties. On the surface of CNTs, oxidation can produce carbonyl, carboxyl, ketone, hydroxyl, and epoxy groups, which increase their interaction with polymers. Additions to the CNT double bonds (radical, carbanion, cycloaddition processes, etc.) are commonly utilized [37], as long as their bending renders them susceptible to a range of addition reactions. While "grafting from" technique, in which monomers having active sites (initiators) capable of starting functionality onto MWCNTs or SWCNTs are polymerized.

9.7 Bulk Modified CNTs for Water Purification

SWCNTs were designed and synthesized by electrochemical reduction of aryl diazonium salts using a bucky-paper electrode. Researcher claim that aryl diazonium salt derivatization is not restricted to electrochemically generated reactions. That is, both *in situ* diazonium synthesis using an alkyl nitrite and direct functionalization of SWCNTs with aryl diazonium tetrafluoroborate salts in solution are viable options. A feasible method is thermally induced functionalization of SWCNTs with diazonium compounds, that are produced *in situ* by the action of derivative of isoamyl nitrite on aniline [38]. Tour and coworker [39] later showed that adding more surfactants resulted in a significant increase in CNT functionalization due to increased native CNT dispersion. When CNTs are chemically modified with (bi)phenyl diazonium salts, grafting 4′-substituted phenyls and biphenyls on them is efficient and defined [40]. There are a broad variety of functional moieties that can be added to the nanotubes by using this approach, such as bifunctional molecules (e.g. diamines). Addition reactions, on the other hand, allow functional groups to be directly attached to the tubes'-conjugated carbon framework. It is preferred that small-diameter tubes be used because of their enhanced curvature, which increases the chemical reactivity, as opposed to larger ones because of their lower thermal activation potential [41]. Functional moieties are covalently attached to the sidewalls of nanotubes increasing their solubility in a way similar to functionalizing carboxyl groups on the nanotubes outside walls. Organic-modified nanotubes can be effectively purified because of their high solubility in water [42]. The nanotubes can still be covalently crosslinked by using a bifunctional agent as in an α,ω-diamine with a relatively long carbon chain [43]. In MWCNTs and few-layer graphene systems, thermal treatment also enhances preexisting vacancy defects, resulting in better mechanical behavior (than defect-free systems). During heat treatment, tube curvature has an impact on defect dynamics (rebonding and self-healing). Vazquez et al., on the other hand, tell that the enormously high measurement in the solvent-free method associated with irradiation of microwave provides functionalized nanotubes through just one hour of reaction, opening the way for large-scale functionalization manufacture have used new types of reactions that don't really work under traditional reflux heating [44]. For example, while addressing the chemistry of CNTs, Brunetti and coworkers have presented a process with solvent-free associated with irradiation of microwave. Because CNTs don't contain any solvents, they can take advantage of the high absorption of microwave that is characteristic of pure nanotubes. Acidification of CNTs using microwaves was the initial step toward synthesizing CNT-based hybrid materials. When it comes to grafting MWCNTs into polyurethane composites. By applying microwave techniques to functionalize SWCNTs, the Wallace group developed a Pt/CNTs-based nanocomposite electrocatalyst for proton-exchange membrane fuel cells [45]. As a result, ball milling has gained a lot of attention as a viable means to alter CNTs, specifically, to modify lengths and open closed ends without producing major structural modifications. [46]. Using a ball mill to break and open the CNTs' closed ends is a suitable option. For the production of iron matrix-MWCNT reinforced composite, high-intensity ball milling is used.

9.8 Important of Carbon Nanotubes for Water Purification

Nanomaterials, such as CNTs, were shown to be useful in cleaning water that has been contaminated by trace chemicals. As a result, they may be employed as stand-alone therapy procedures or integrated into current treatment systems [47]. Wastes outcome from industry contains contaminates of various organic/inorganic substances. Among these wastes also dyes and heavy metals pollutants are included. These pollutants especially are of a severe danger to living organisms and human drinking as well. Accordingly, comes the great importance of removal of such substances (dyes/heavy metals) to minimize the toxicity of the water and its harmful effect on the surrounding environment. Many studies were conducted to investigate the removal of these toxins using novel materials and technologies [48]. According to scientific references, heavy metals can be defined as metals of a density exceeding 5 g cm^{-3} such as lead Pb^{+2}, cadmium Cd^{+2}, mercury Hg^{+2}, arsenic As^{+5}, and chromium Cr^{+6}. Unsafe wastewater purification from unprocessed households and industrial wastes delivers a huge danger to the marine system [49]. The chemical composition of –Fe_2O_3, or Fe_2O_3 onto covalently modified MWCNTs is generally used to make magnetic nanoparticles embedded with MWCNTs. There are, however, a variety of additional methods for producing them [50]. Magnetic CNTs can be used as adsorbents in the process of wastewater treatment owing to their outstanding properties such as ratio of surface volume, large surface area, and appropriate separation procedures. The CNT membrane has a simple form named "bucky paper" (BP) is a self-standing film of entangled CNTs that offers a highly chemical/physical stability and flexible morphology. Bucky paper (BP) membranes, in addition to magnetic CNTs, are advantageous owing to their exclusive strength, high porosity, and adsorption capabilities. Nevertheless, BP membranes are mostly used to remove salt from the aqueous phase, with only a small amount of literature demonstrating their effectiveness in the heavy metals and dye removal [51]. Modified MWCNTs to make a CNT-based BP membrane, which they then implanted with polyvinyl chloride to remove various heavy metal ions from water sources was considered by Ali et al. [52]. In general, CNT membranes can be used for a wide range of wastewater purification or desalination applications, with no need for pretreatment or posttreatment [53]. Electrochemical CNT membranes, vertically-aligned carbon nanotube membranes (VACNT), mixed-matrix membranes (MMMs), and horizontally-aligned carbon nanotube membranes (HACNT) are among the major types of f-CNT membranes that have been produced or are under development, as indicated in Figure 9.3.

Zn^{+2} was removed from synthetic water and wastewater industrial effluent by 98% and 70%, respectively, according to the findings. Lau et al. [54] modified CNT-based BP and coupled it with polyvinyl alcohol to remove methylene blue (MB) from aqueous phase. Methylene blue was removed 99.7% of the time in the research. For humic treatment of water, Jafari et al. [55] created an antifouling CNT-based BP. The adsorption technique was used to remove 65% of the humic acid in the study. Patel et al. [56] studied a membrane-based CNTs, which were successfully used in eliminating Cd^{+2} from wastewater. According to a recent thorough analysis, the usage of surface-modified CNTs has shown considerable promise in the heavy metal removal from wastewater. The carboxylic (COOH) and hydroxyl (OH) functional groups on the surface of these CNTs are commonly imparted via oxidative techniques. Organic dyes, soluble organic materials, aromatic chemicals, phenolic derivatives, medicines, and antibiotics are only a few examples [57, 58]. Arabi et al. used a combination of ZnO and MWCNTs to remove the Congo red dye. A maximal adsorptive capacity of 249.51 mg g^{-1} was recorded under optimum conditions [59]. NH_2-MWCNTs have a good adsorption capability for the removal of

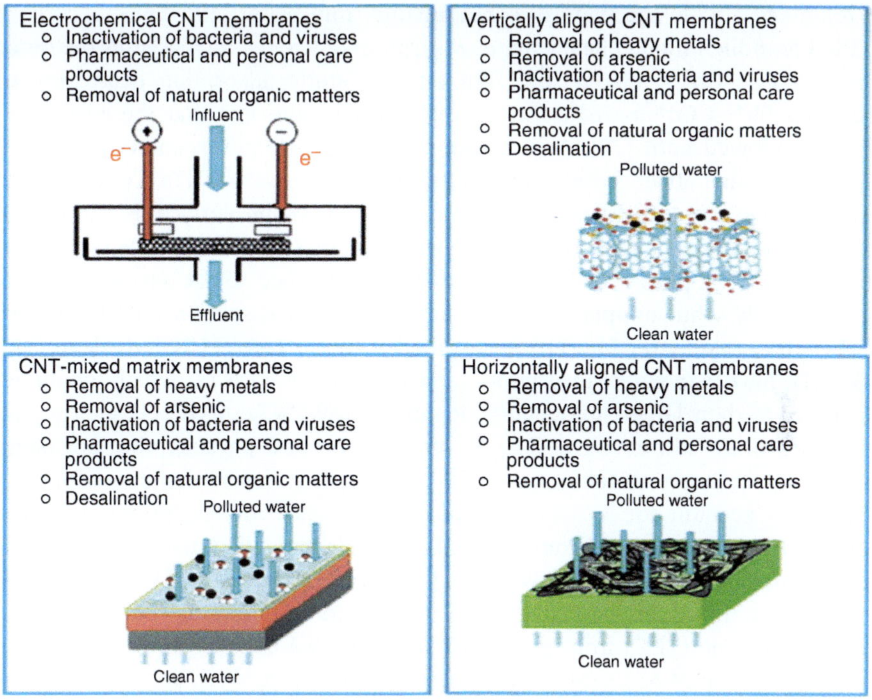

Figure 9.3 Mechanism of water passing through the four types of carbon nanotubes membranes. Source: Ali et al. [53]/with permission of Elsevier.

methylene blue dye (178.5 mg g^{-1}) [60]. CNTs and their hybrids were used to study the adsorption of some analgesic raw materials [61, 62]. Adsorption of 71–90 mg g^{-1} can be reached for these commonly used analgesics. It was strongly observed that the hydrophobicity of the organic wastes is a significant factor regulating the adsorption removal process from wastewater. CNTs' adsorption affinity is directly proportional to hydrophobicity of organic waste molecules [63]. To reduce the interactions between hydrophobic foulants and membrane surfaces, oxidized CNTs increase membranes' hydrophilic nature [64]. Additionally, CNTs' hydrophobic hollow structure makes it possible for polar water molecules to move more quickly and without friction [65]. Ragunath et al. discussed using direct contact membrane distillation to enhance the immobilization of CNTs on membrane surfaces for saltwater desalination (DCMD). The carbon nanotube immobilized membrane (CNIM) demonstrated a 76% higher water vapor flow of 51.4 l m^{-2} h^{-1} than the untreated support membrane [66]. Gupta et al. discovered that utilizing sweep gas membrane distillation, the CNIM membranes were effective at extracting organic solvents from aqueous mixtures [67]. Membrane composites with improved properties for vacuum membrane distillation were created by mixing polyvinylidene fluoride (PVDF) membranes with varying amounts of MWCNTs [68]. CNTs can be used as novel techniques or in conjunction with current conventional treatment procedures in a variety of water treatment procedures. They are extremely effective in removing a wide spectrum of contaminants from water [69]. Thus, according to the Freundlich model, heavy metal adsorption efficiency increased in the order $Mn^{+2} > Zn^{+2} > Co^{+2} > Pb^{+2} > Cu^{+2}$, but another study adsorption decreasing in the order $Pb^{+2} > Cd^{+2} > Co^{+2} > Zn^{+2} > Cu^{+2}$, despite the fact that the adsorption process followed both [70]. As a result, the radius, as well as there are several more factors that affect the adsorption affinity and process of CNTs. Contaminants may have become caught in pores or adsorbed onto the surface as a result of various interactions. Nonplanar, conjugated molecules such as Orange G, Xylenol Orange, and 1-(2-pyridylazo)-2-naphthol (PAN) adsorbed to the surface of CNTs less efficiently, while nonplanar, nonconjugated molecules such as diidofluorescein (DIF) and bromothymol blue (BTB) adsorbed to the surface of CNTs the least [71]. SWCNTs have a greater adsorption capacity than MWCNTs, which might be attributed to their larger surface area. Relative to SWCNTs, MWCNTs are roughly 100 times more cost-effective. [72]. Chemisorption and physisorption are being used to adsorb heavy metal ions onto CNTs. Yang et al. found that chemisorption procedures account for 75.3% of the adsorption, via complex formation, while physisorption processes, predominantly through electrostatic interactions, account for the remaining 24.7% of the adsorption [50]. In an intriguing study [73], pyrolysis was used to make an Al_2O_3/MWCNT hybrid for eliminating Cd^{+2} ions and trichloroethylene (TCE) from groundwater. Figure 9.4 depicts the mechanism of functionalization of Al_2O_3/MWCNTs for cadmium ion Cd^{+2} ion and TCE removal from groundwater.

Among the hybrid MWCNTs and Cd^{+2}, a range of forces, contain strong physical adsorption, precipitate of the surface, complexation of the inner sphere, attractions of cationic metal, electrostatic interactions, and complexation of the surface, were contributed to the high adsorption of Cd^{+2}. Surface carbon functionalization of CNTs by oxidation, grafting, and physical adsorption of molecules can also improve their adsorption capacity and selectivity. CNTs with oxygen, which can produce a negatively CNTs surface, can be created or deleted by oxidation [74] or heating [76]. Also, oxygen can contribute with an electron pair to remove heavy metal ions from effluent of wastewater [76]. Yu et al. used pentaerythritol (PER) to modify oxidized MWCNTs to adsorb alizarin yellow R and alizarin red S from a polluted aqueous environment [77]. They claimed that the ox-MWCNT-high PER's adsorption capacity was due in part to effective hydrogen bonding among both the hydroxyl groups on its surface and the dyes, as well as other interactions as electrostatic and interactions of p–p. Pentaerythritol was grafted onto MWCNTs and then subjected to an oxidation treatment, releasing these hydroxyl functional groups. MWCNTs can be magnetically functionalized using a variety of methods. The active involvement of oxygen containing in Fe_3O_4 to adsorb heavy metal ion Cr^{+3}, given additional active sites for CNT/Fe_3O_4's adsorption capacity higher than that of non-functionalized CNT [78]. By utilizing a simple microwave approach, Vadahanambi et al. developed graphene-CNT-Fe nanohybrids that may be used to eliminate arsenic from water. At $6\,mg\,g^{-1}$, the hybrids of magnetic graphene with Fe_2O_3 and CNT had a larger adsorption capacity than graphene with Fe_2O_3 alone because they had more surface-active sites in common with the individual

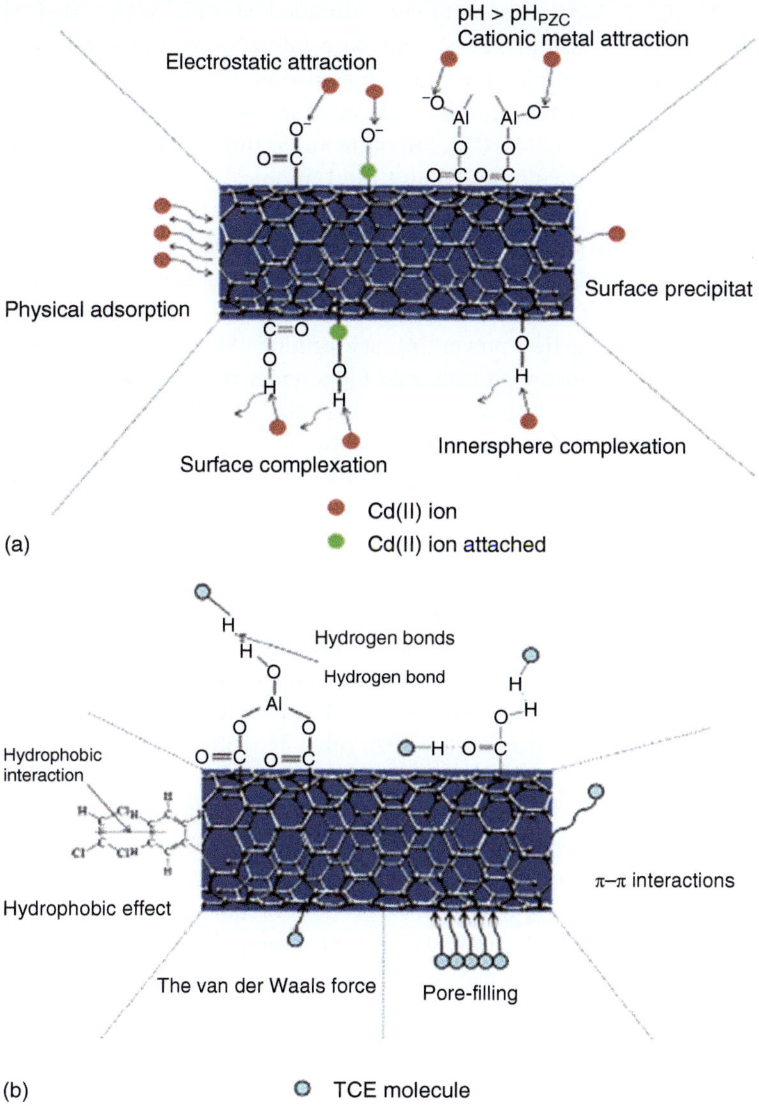

Figure 9.4 A mechanism of functionalization of Al_2O_3/MWCNTs to remove Cd^{+2} and TCE. Source: Liang et al. [73]/with permission of Elsevier.

components of those nanohybrids (2.5 mg g^{-1}) [79]. Modified MWCNTs were created by reacting 1,6-hexanediamine integrated magnetic Fe_3O_4 NPs (MNPs) with the –COOH groups of MWCNT utilizing simple solvo-thermal chemistry, according to the study by Lu and coworkers. These MNP/MWCNTs were used to remove Cr(VI) ions from polluted water after they were manufactured on the spot. At pH 2.0, MNP/MWCNTs had the greatest adsorption efficiency for Cr(VI) [80]. As a result, Cai et al. developed an mercapto propyltriethoxysilane (MPTS)/CNT/Fe_3O_4 nanomaterial with a higher adsorption capacity by functionalizing CNTs with

MPTS [81]. MPTS/CNT/Fe_3O_4 has a larger estimated specific active surface area than CNT/Fe_3O_4. Because of Lewis' acid–base interactions, thiol groups in MPTS have a considerable affinity for heavy metal ions. [82]. As a result, the MPTS/CNT/Fe_3O_4 nanomaterial proved successful in removing heavy metal ions from wastewater. $MnFe_2O_4$/MWCNTs manufactured through a hydrothermal method can be used to remove cationic yellow 40 and anionic dyes such as direct red 16 from water [83]. Chemical structure and origin, as well as adsorbate mix, influence biopolymer adsorption performance [84–88]. As a consequence, carbon-based nanomaterials are embedded into biopolymers. To enhance mechanical characteristics, chemical stability, and hydrophobicity, the functionality of biopolymer is to stabilize the carbon-based barrier, preventing aggregation [89, 90]. Chitosan is a nontoxic, widely available mucopolysaccharide biopolymer that has been utilized for a variety of purposes, including adsorption. Because it is less expensive than activated carbon, other adsorbents with amino and hydroxyl functional, chitosan is an excellent nano adsorbent for removing ions of heavy metal, dyes, organochloride insecticides, and other organic contaminants from water [91]. Water pollutants have also been removed by adsorption using MWCNT-chitosan nanocomposites [92–94]. Minteer et al. used chitosan to enhance MWCNTs to rapidly and efficiently eliminate the Cr^{+6} from wastewater [95]. Chitosan's amine groups and MWCNTs' carboxylic acid groups were protonated at low pH, resulting in anionic Cr^{+6} groups adsorbing electrostatically. The Langmuir isotherm model [96] accurately describes the adsorption process, that follows pseudo-second-order kinetics. Carbon-based nanomaterials can also be used to functionalize other biopolymers such as polyacrylamide, gum, cellulose, polydopamine, polyvinyl alcohol, polyaniline, and L-cysteine. At a pH of 5.8 and 120 minutes of contact time, Lignin-grafted CNT nanocomposites monstrated good adsorption capability (235.0 $mg\,l^{-1}$) for Pb^{+2} [97]. Duan et al. presented a one-step method for creating a polydopamine (PDA) functionalized graphene hydrogel for the adsorption of organic dyes, heavy metal ions, and aromatic contaminants from water [98]. Water pollutant adsorption has also been explored using gelatin functionalized carbon compounds. Gelatin functions as an effective adsorbent for water filtration by removing metal ions and organic contaminants because of the existence of hydroxyl, amino, and carboxyl active groups [99]. Ma et al. investigate adsorption MB and MO dyes from water upon applying MWCNT/starch/iron oxide composite and MWCNT/iron oxide. Because MWCNT/iron oxide is water insoluble, the dyes had no reaction with it. Considering this, the starch in the MWCNT/starch/iron oxide nanocomposite increased the nanocomposite's hydrophilicity, leading to improved water dispersion and adsorbate–adsorbent interactions. The iron oxide helped in the separation of magnetic of the adsorbent once the adsorption was completed. After the adsorption process is completed, separation of the adsorbent from the aqueous media is aided by functionalization with nanoparticles of magnetic. Magnetic beads made of gelatin, CNTs, and iron oxide nanoparticles were created to remove cationic methylene blue (MB) dye and anionic direct red dyes from wastewater [100]. To assess their potential in applications of adsorption, nanocomposites of biopolymer/carbon-based were compared to a variety of different commercial and noncommercial adsorbents

[101]. Nanocomposite adsorbents based on biopolymers and carbon were used to remove a wide range of pollutants from aqueous solutions. Biopolymers have poor adsorption efficiency, but their modification with nanomaterials depends on carbon resulting in increased capacities of adsorption, better physicochemical qualities, and increased surface functions and surface areas.

9.9 Conclusions and Future Research Directions

Academics, business, general public, national, and international governments, are concerned about pollution of water. Industrial wastewater treatment prior to disposal and reutilization is a high priority. It is difficult to develop a universal approach for removing all sorts of contaminants from wastewater. Adsorption has become well-established in industry, achieving good results in the removal of many types of pollution from effluent. Many factors influence the adsorption process, includes the ionic nature of adsorbate, the adsorbent types used, and the process factors. It is largely analyzed utilizing kinetic models, isotherms, and characteristics of thermodynamics, and happens predominantly through physisorption and chemisorption mechanisms, as well as ion-exchange mechanisms. The removal of hazardous inorganic and organic pollutants has recently been researched using a range of adsorbents. The importance of CNTs, their chemical modification, and functionalization for usage in water filtration were discussed in this chapter. CNTs nanocomposite, which combines inorganic and organic elements, improves pollution removal even further. Although polymers are inefficient adsorbents, their composites/hydrogels containing CNTs are excellent in removing a range of pollutants from wastewater. Nanocomposites of magnetic/CNTs offer an added benefit in separation of magnetic and reutilization of adsorbent. Desalination and membrane technologies have found CNTs to be advantageous materials. CNTs are predicted to lead to further intriguing discoveries, as well as possible low-cost industrial adsorbents for water treatment, although research is still ongoing. Researchers and scientists may be able to develop an inexpensive, convenient, and lightweight CNT modification pathway in the near future to create innovative, efficient, and stable nanocomposites based on CNT. Additionally, the development of innovative wastewater treatment technologies and reaction processes for the application of composite materials based on CNT via easier synthesis, functionalization, and controlled processing of high-quality CNT-based materials for use in purifying water from diverse pollutants.

9.10 Commercial Application of Chemically Modified CNTs in Water Purification

Flier's quality water builds commercial water purification systems for industrial, commercial, institutional, and high-end residential applications using only the best quality components and cutting-edge technology (www.fliersqualitywater

.com). Any water treatment system should include filtration. We design filtration systems that are capable of handling any type of water. Total suspended solids (TSS), iron, and manganese filtration is a common technique. Filtration media include sand, multimedia (which is made up of layers of different media), greensand, greensand +, and birm (www.hillwater.com) water conditioning by Robert B. Hill Company. CNTs have a worldwide commercial interest that is reflected in a production capacity that now surpasses several thousand tons per year. Bulk CNT powders are being used in a wide range of commercial items, including rechargeable batteries, automotive parts, and sporting goods, as well as boat hulls and water filters. CNTs can now be used in thin-film electronics and large-area coatings thanks to advancements in CNT production, purification, and chemical modification. CNT yarns and sheets have already shown promise in applications such as supercapacitors, actuators, and lightweight electromagnetic shields [102], despite the fact that they do not yet provide compelling mechanical strength or electrical or thermal conductivities for many applications [103].

9.11 Companies Produced CNTs for the Application of Chemically Modified Carbon Nanotubes for Water Purification System

Water treatments, recovery, and purifications have become a vital issue in modern life. Modern technology based on nanocomposite, membranes, and filtrations became a cornerstone in this method. Companies such as ENVIROGEN Group (www.envirogengroup.com) developed various filtrations products for water treatment in different industries. A series of multi-applied CNTs modified and available in the market by nanografi (www.nanografi.com), TFT (www.tftfluid.com) modified CRM/FCM/PCR ceramic membranes for water purification and solid–liquid separation. Global companies such as Alfa Laval (www.alfalaval.co.uk), DowDuPont (www.dow-dupont.com), GEA Group (https://www.gea.com/en/index.jsp), are involved in the development of membranes that are made of nanomaterials to eliminate pollutants during the treatment.

References

1 Cha, S.I., Kim, K.T., Lee, K.H. et al. (2008). *Carbon* 46 (3): 482–488.
2 Al-Abri, M., Al-Ghafri, B., Bora, T. et al. (2019). *npj Clean Water* 2: 2.
3 Wang, L., Zhu, D., Duan, L., and Chen, W. (2010). *Carbon* 48 (13): 3906–3915.
4 Dubey, R., Dutta, D., Sarkar, A., and Chattopadhyay, P. (2021). *Nanoscale Adv.* 3 (20): 5722–5744.
5 Lee, K.M., Wong, C.P., Tan, T.L., and Lai, C.W. (2018). *Mater. Sci. Eng., B* 236: 61–69.
6 Sarkar, B., Mandal, S., Tsang, Y.F. et al. (2018). *Sci. Total Environ.* 612: 561–581.
7 Angelakis, A.N., Vuorinen, H.S., Nikolaidis, C. et al. (2021). *Water* 13 (6): 752.

8 Aslam, M.A., Kuo, H.W., Den, W. et al. (2021). *Sustainability* 13: 5717.

9 Eatemadi, A., Daraee, H., Karimkhanloo, H. et al. (2014). *Nanoscale Res. Lett.* 9 (1): 1–3.

10 Murray, R.W., Ewing, A.G., and Durst, R.A. (1987). *Anal. Chem.* 59 (5): 379A–390A.

11 Pandurangappa, M. and Raghu, G.K. (2011). *Carbon Nanotubes Applications on Electron Devices*. Intechopen.

12 Vander Wal, R.L., Berger, G.M., and Ticich, T.M. (2003). *Appl. Phys. A* 77 (7): 885–889.

13 Atif, M., Afzaal, I., Naseer, H. et al. (2020). *ECS J. Solid-State Sci. Technol.* 9 (4): 041009.

14 Lau, K.T. (2003). *Chem. Phys. Lett.* 370 (3, 4): 399–405.

15 Hong, C.Y., You, Y.Z., and Pan, C.Y. (2006). *Polymer* 47 (12): 4300–4309.

16 Th. Le, V., Ngo, C.L., Le, Q.T. et al. (2013). *Adv. Nat. Sci.: Nanosci. Nanotechnol.* 4 (3): 035017.

17 Shi, L., Liang, R.P., and Qiu, J.D. (2012). *J. Mater. Chem.* 22 (33): 17196–17203.

18 Wildgoose, G.G., Banks, C.E., and Compton, R.G. (2006). *Small* 2 (2): 182–193.

19 He, Z., Chen, J., Liu, D. et al. (2004). *Diamond Relat. Mater.* 13 (10): 1764–1770.

20 Chen, X., Li, N., Eckhard, K. et al. (2007). *Electrochem. Commun.* 9 (6): 1348–1354.

21 Arai, S. and Endo, M. (2003). *Electrochem. Commun.* 5 (9): 797–799.

22 Kim, I.H., Kim, J.H., Lee, Y.H., and Kim, K.B. (2005). *J. Electrochem. Soc.* 152 (11): A2170.

23 Xiao, F., Mo, Z., Zhao, F., and Zeng, B. (2008). *Electrochem. Commun.* 10 (11): 1740–1743.

24 Chen, P., Wu, X., Lin, J., and Tan, K.L. (1999). *J. Phys. Chem. B* 103 (22): 4559–4561.

25 Oh, S.D., So, B.K., Choi, S.H. et al. (2005). *Mater. Lett.* 59 (10): 1121–1124.

26 Das, R., Abd Hamid, S.B., Ali, M.E. et al. (2014). *Desalination* 354: 160–179.

27 Baughman, R.H., Zakhidov, A.A., and De Heer, W.A. (2002). *Science* 297 (5582): 787–792.

28 Spitalsky, Z., Tasis, D., Papagelis, K., and Galiotis, C. (2010). *Prog. Polym. Sci.* 35 (3): 357–401.

29 Sahoo, N.G., Rana, S., Cho, J.W. et al. (2010). *Prog. Polym. Sci.* 35 (7): 837–867.

30 Jon, C.S., Meng, L.Y., and Li, D. (2019). *TrAC, Trends Anal. Chem.* 120: 115641.

31 Poulin, P., Vigolo, B., and Launois, P. (2002). *Carbon* 40 (10): 1741–1749.

32 Jiang, L., Gao, L., and Sun, J. (2003). *J. Colloid Interface Sci.* 260 (1): 89–94.

33 Yurekli, K., Mitchell, C.A., and Krishnamoorti, R. (2004). *J. Am. Chem. Soc.* 126 (32): 9902–9903.

34 Tan, Y. and Resasco, D.E. (2005). *J. Phys. Chem. B* 109 (30): 14454–14460.

35 Sáfar, G.A., Ribeiro, H.B., Malard, L.M. et al. (2008). *Chem. Phys. Lett.* 462 (1–3): 109–111.

36 Guo, Z., Sadler, P.J., and Tsang, S.C. (1998). *Adv. Mater.* 10 (9): 701–703.

37 Díez-Pascual, A.M. (2021). *Macromolecules* 1 (2): 64–83.

38 Bahr, J.L. and Tour, J.M. (2001). *Chem. Mater.* 13 (11): 3823–3824.

39 Dyke, C.A. and Tour, J.M. (2003). *Nano Lett.* 3 (9): 1215–1218.

40 Golosova, A.A., Papadakis, C.M., and Jordan, R. (2011). *MRS Online Proceedings Library (OPL)*, 1362. Cambridge University Press.

41 Bahr, J.L. and Tour, J.M. (2002). *J. Mater. Chem.* 12 (7): 1952–1958.

42 Georgakilas, V., Voulgaris, D., Vazquez, E. et al. (2002). *J. Am. Chem. Soc.* 124 (48): 14318–14319.

43 Stevens, J.L., Huang, A.Y., Peng, H. et al. (2003). *Nano Lett.* 3 (3): 331–336.

44 Vazquez, E. and Prato, M. (2009). *ACS Nano* 3 (12): 3819–3824.

45 Zhang, W., Chen, J., Swiegers, G.F. et al. (2010). *Nanoscale* 2 (2): 282–286.

46 Ahn, J.H., Shin, H.S., Kim, Y.J., and Chung, H. (2007). *J. Alloys Compd.* 434: 428–432.

47 Malik, S. (2018). *Polyhedron* 152: 90–93.

48 Wadhawan, S., Jain, A., Nayyar, J., and Mehta, S.K. (2020). *J. Water Process Eng.* 33: 101038.

49 Le, A.T., Pung, S.Y., Sreekantan, S., and Matsuda, A. (2019). *Heliyon* 5 (4): 01440.

50 Huang, Z.N., Wang, X.L., and Yang, D.S. (2015). *Water Sci. Eng.* 8 (3): 226–232.

51 Khan, F.S., Mubarak, N.M., Tan, Y.H. et al. (2021). *J. Hazard. Mater.* 413: 125375.

52 Ali, S., Rehman, S.A., Shah, I.A. et al. (2019). *J. Hazard. Mater.* 365: 64–73.

53 Ali, S., Ur Rehman, S.A., Luan, H.Y. et al. (2019). *Sci. Total Environ.* 646: 1126–1139.

54 Lau, Y.J., Karri, R.R., Mubarak, N.M. et al. (2020). *Environ. Sci. Pollut. Res.* 27 (32): 40121–40134.

55 Jafari, A., Nasseri, S., Nabizadeh, R. et al. (2017). *Glob. NEST J.* 19: 217–224.

56 Patel, F., Khraisheh, M., Atieh, M.A., and Laoui, T. (2017). *Materials* 10 (10): 1144.

57 Lico, D., Vuono, D., Siciliano, C. et al. (2019). *J. Environ. Manage.* 237: 636–643.

58 Zhang, Z., Zeng, C., and Peng, B. (2019). *Food Control* 102: 1–10.

59 Arabi, S.M., Lalehloo, R.S., Olyai, M.R. et al. (2019). *Physica E* 106: 150–155.

60 Ahamad, T., Naushad, M., Eldesoky, G.E. et al. (2019). *J. Mol. Liq.* 282: 154–161.

61 Mphahlele, K., Onyango, M.S., and Mhlanga, S.D. (2015). *J. Environ. Chem. Eng.* 3 (4): 2619–2630.

62 Fard, M.A. and Barkdoll, B. (2018). *J. Mol. Liq.* 249: 193–202.

63 Awad, A.M., Jalab, R., Benamor, A. et al. (2020). *J. Mol. Liq.* 301: 112335.

64 Ihsanullah (2019). *Sep. Purif. Technol.* 209: 307–337.

65 Ma, L., Dong, X., Chen, M. et al. (2017). *Membranes* 7 (1): 16.

66 Ragunath, S., Roy, S., and Mitra, S. (2018). *Sep. Purif. Technol.* 194: 249–255.

67 Gupta, O., Roy, S., and Mitra, S. (2018). *J. Membr. Sci.* 568: 134–140.

68 Zhou, R., Rana, D., Matsuura, T., and Lan, C.Q. (2019). *Sep. Purif. Technol.* 217: 154–163.

69 Makgabutlane, B., Nthunya, L.N., Maubane-Nkadimeng, M.S., and Mhlanga, S.D. (2021). *J. Environ. Chem. Eng.* 9 (1): 104736.
70 Tofighy, M.A. and Mohammadi, T. (2011). *J. Hazard. Mater.* 185 (1): 140–147.
71 Liu, C.H., Li, J.J., Zhang, H.L. et al. (2008). *Colloids Surf., A* 313: 9–12.
72 Gupta, V.K., Kumar, R., Nayak, A. et al. (2013). *Adv. Colloid Interface Sci.* 193: 24–34.
73 Liang, J., Liu, J., Yuan, X. et al. (2015). *Chem. Eng. J.* 273: 101–110.
74 Maleki, A., Hamesadeghi, U., Daraei, H. et al. (2017). *Chem. Eng. J.* 313: 826–835.
75 Gotovac, S., Song, L., Kanoh, H., and Kaneko, K. (2007). *Colloids Surf., A* 300 (1, 2): 117–121.
76 Peng, X., Li, Y., Luan, Z. et al. (2003). *Chem. Phys. Lett.* 376 (1, 2): 154–158.
77 Li, Y.H., Wang, S., Wei, J. et al. (2002). *Chem. Phys. Lett.* 357 (3, 4): 263–266.
78 Yang, J.Y., Jiang, X.Y., Jiao, F.P., and Yu, J.G. (2018). *Appl. Surf. Sci.* 436: 198–206.
79 Gupta, V.K., Agarwal, S., and Saleh, T.A. (2011). *Water Res.* 45 (6): 2207–2212.
80 Vadahanambi, S., Lee, S.H., Kim, W.J., and Oh, I.K. (2013). *Environ. Sci. Technol.* 47 (18): 10510–10517.
81 Lu, W., Li, J., Sheng, Y. et al. (2017). *J. Colloid Interface Sci.* 505: 1134–1146.
82 Zhang, C., Sui, J., Li, J. et al. (2012). *Chem. Eng. J.* 210: 45–52.
83 Vieira, E.F., Simoni, J.D., and Airoldi, C. (1997). *J. Mater. Chem.* 7 (11): 2249–2252.
84 Kafshgari, L.A., Ghorbani, M., and Azizi, A. (2017). *Appl. Surf. Sci.* 419: 70–83.
85 Mittal, H. and Ray, S.S. (2016). *Int. J. Biol. Macromol.* 88: 66–80.
86 Mittal, H., Kumar, V., and Ray, S.S. (2016). *Int. J. Biol. Macromol.* 89: 1–1.
87 Kera, N.H., Bhaumik, M., Pillay, K. et al. (2018). *Mater. Today Commun.* 15: 153–164.
88 Yang, Y., Wei, X., Sun, P., and Wan, J. (2010). *Molecules* 15 (4): 2872–2885.
89 Kera, N.H., Bhaumik, M., Ballav, N. et al. (2016). *J. Colloid Interface Sci.* 476: 144–157.
90 Botlhoko, O.J., Ramontja, J., and Ray, S.S. (2018). *Polymer* 139: 188–200.
91 Botlhoko, O.J., Makwakwa, D., Ray, S.S., and Ramontja, J. (2019). *J. Appl. Polym. Sci.* 136 (17): 47387.
92 Mansour, N.A., Ahmed, S.M., and Mazrouaa, A.M. (2020). *Mediterr. J. Chem.* 9 (6): 447–455.
93 Khakpour, R. and Tahermansouri, H. (2018). *Int. J. Biol. Macromol.* 109: 598–610.
94 Wasim, M., Sagar, S., Sabir, A. et al. (2017). *Carbohydr. Polym.* 174: 474–483.
95 Mallakpour, S. and Ezhieh, A.N. (2017). *J. Polym. Environ.* 25 (3): 875–883.
96 Huang, Y., Lee, X., Macazo, F.C. et al. (2018). *Chem. Eng. J.* 339: 259–267.
97 Gusain, R., Kumar, N., and Ray, S.S. (2020). *Coord. Chem. Rev.* 405: 213111.

98 Li, Z., Chen, J., and Ge, Y. (2017). *Chem. Eng. J.* 308: 809–817.

99 Gao, H., Sun, Y., Zhou, J. et al. (2013). *ACS Appl. Mater. Interfaces* 5 (2): 425–432.

100 Hayeeye, F., Sattar, M., Chinpa, W., and Sirichote, O. (2017). *Colloids Surf., A* 513: 259–266.

101 Samandari, S.S., Yekta, H.J., and Mohseni, M. (2017). *Chem. Eng. J.* 308: 1133–1144.

102 Gupta, V.K., Agarwal, S., Sadegh, H. et al. (2017). *J. Mol. Liq.* 237: 466–472.

103 De Volder, M.F.L., Tawfick, S.H., Baughman, R.H., and Hart, A.J. (2013). *Science* 339: 535–539.

Part III

Chemically Modified Carbon Nanotubes: Electronic and Electrical Applications

10

Chemically Modified Carbon Nanotubes for Electronics and Photonic Applications

Swetapadma Praharaj and Dibyaranjan Rout

Department of Physics, School of Applied Sciences, KIIT Deemed to be University, Bhubaneswar, Odisha, 754024, India

10.1 Introduction

The miniaturization of electronic devices has set a benchmark owing to modern societal needs and rising standards of living. The growing demands of the contemporary population have directly or indirectly boosted the technology for the shrinking of the primary components of any device. For instance, the emergence of lithographic techniques and thin film deposition has been instrumental in greatly reducing the size of silicon transistors, which is an essential entity in this information age. This not only results in increased density and switching speed, but also lowers the cost. Unfortunately, transistors and other silicon-based electronic devices are approaching fundamental roadblocks, indicating that we can no more exploit Si devices if further miniaturization is required. In this regard, researchers and industrialists are looking for alternative operating principles or prospective materials, which can sustain scaling to a smaller size yet with greater performance. Among the various options available, carbon nanotube (CNT)-based electronics and optoelectronics are of particular interest due to their unique set of electrical, thermal, mechanical, transport, and photoluminescence properties. Structurally, CNTs can be illustrated as rolled graphene sheets with the ends crowned with buckyball hemispheres [1]. The precise structure depends solely on the angles and curvatures in which the graphene sheets are rolled. Such orientations are described in terms of chiral vectors, which categorize CNTs into armchair, zigzag, and chiral forms. Moreover, the electronic properties of the CNTs are guided by their chiral properties. While the armchair form is usually metallic, zigzag, or chiral arrangement exhibits metallic or semiconducting behavior. On the other hand, while single-walled carbon nanotubes (SWCNTs) demonstrate both metallic and semiconducting behavior depending on geometrical sensitivity, multiwalled carbon nanotubes (MWCNTs) are generally conductors. The majority of the electrical properties of CNTs stem from graphene upon imposing some additional conditions [2]. The one-dimensional nature of the tubes gives them the flexibility of

Chemically Modified Carbon Nanotubes for Commercial Applications, First Edition.
Edited by Jeenat Aslam, Chaudhery Mustansar Hussain, and Ruby Aslam.

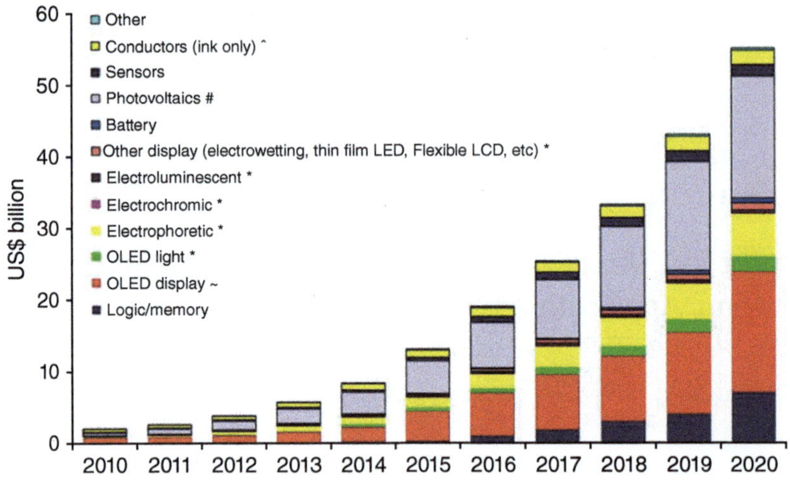

Figure 10.1 Market forecast for various CNT-based electronic and photonic products from 2010 to2020. ^Excludes ESD/RF shielding, *Front plane only, ~OLED modules (includes Thin Film Transistor TFT drive circuit and OLED front plane), #Includes all thin film photovoltaics but excludes CdTe and conventional crystalline silicon types. Source: IDTechEx Ltd.

electron conduction without scattering. The absence of scattering during electron conduction is said to be ballistic transport, which facilitates conduction without dissipation of energy [3]. Owing to extraordinary electrical response (carrier mobility $\sim 10^5\ cm^2\ V^{-1}\ s^{-1}$ at room temperature and maximum supported electrical current density $> 10^9\ A\ cm^{-2}$), they have been explored to the extent of diodes, transistors, MOSFETs (metal-oxide field-effect transistors), memory devices, photo-induced switches, and many more [4–7]. Figure 10.1 depicts a market forecast of CNT-based electronic and photonic devices in the last few years (2010–2020) in US dollars.

Despite outstanding electrical and electronic properties, the practical application of CNT in devices is far from realized. This is largely because the controlled fabrication of a specific variety of CNT is difficult to accomplish. Besides, the absence of a reliable method to synthesize extended ensembles of aligned CNTs is another bottleneck that needs to be overcome to achieve the desired location of the tubes along with a systematic connection with their neighbors. To address these issues, chemical modification or functionalization by linking specific molecules or groups to the nanotubes opens up new avenues for tuning the electronic properties. Apart from that, chemical functionalization facilitates nanotube assembly into many complicated architectures necessary for integrated device operation [8]. In this chapter, we will discuss the different types of chemical modification of CNTs and their prospective applications in electronic and photonic devices.

10.2 Chemical Modifications of CNTs

CNTs in their pristine form are not completely sufficient to cater to the specific requirements of diverse applications. For example, the chemical inertness and low surface energy of the CNTs are some of the prime roadblocks for low-cost

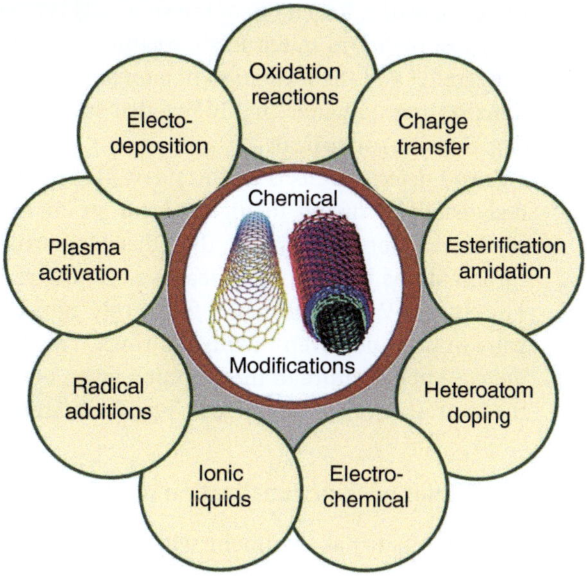

Figure 10.2 Schematic of different chemical modification techniques of CNTs.

solution processing [9]. Chemical modification/doping is identified as one of the alternatives to overcome the hurdles in processing techniques for optimizing the electrical and electronic properties of CNTs. Moreover, this is necessary for the next generation of CNT-based photonic and electronic devices demanding electrical heterojunctions with tunable carrier (p/n-types) types and Fermi energy levels. This is quite a similar strategy to that used in typical Si-based devices to modify the carrier properties by heteroatom doping. There have been many chemical modifications and doping blueprints introduced by researchers. The prime modification techniques are displayed in Figure 10.2 out of which a few are discussed below. While some of the techniques utilize the particular polymeric or other organic molecules with the aid of π–π stacking/interactions (involving very feeble rearrangement of chemical structure), others are featured by permanent alterations of local architecture and significantly affects the physical and electronic properties.

Modification of CNTs by foreign atoms or ions alters its (local) bonding environment, resulting in sp^2 to sp^3 hybrid structures useful for bringing in novel properties [10]. In addition to that, CNTs show high chemical reactivity because of their curvature. Real SWCNTs have three different areas of varying reactivity, which serve as modification sites for inorganic material deposition, chemical grafting, and ion absorption.

(i) *Tube caps*: The strong reactivity of the tube caps originates from their prominent curvature relief, ensuring significant relief of strain during the conversion of trigonal to tetragonal carbon. This is very similar to the caps in C_{60}, which is an assemblage of C=C interconnected by C–C lead to regioselectivity in addition reactions. But it is usually observed that the standard purification techniques open up the caps at the tube ends and hence the solvent remains inaccessible in most of the SWCNTs [11]. However, initial steps have been taken toward the chemical modification of caps in MWCNTs [12].

(ii) *Intact sidewalls*: Reactivity of the sidewalls is considerably less as compared to the caps, owing to much less curvature. This provides a lower threshold for stress relief and the formation of tetragonal carbon. As a result, sidewall functionalization demands highly reactive substitutions such as strong oxidizers (e.g. fluorine) or radicals.

(iii) *Sidewall defects*: Defects in the sidewalls are expected to demonstrate amplified reactivity owing to higher local strain than intact sidewalls. As soon as the reaction is initiated, the adjacent partial double bonds between the carbon atoms become more reactive and the reactions unfurl along the tube. The Stone–Wales defect is a frequently noticed defect that consists of two pairs of five and seven-membered rings. This kind of defect results in a local increase in curvature of the graphitic layer, the strongest curvature existing at the interface of five-membered rings, facilitating addition reactions [13].

10.2.1 Oxidative Functionalization of CNTs

It is a known fact that treatment with a strong acid promotes the formation of oxygen functional groups like hydroxyl, carbonyl, and carboxyl groups at the CNT surface [14]. The existence of defects such as Stone–Wales defects and vacancies at the surface of CNTs with enhanced reactivities acts as the principal oxidation sites. It has been reported earlier in the literature that reflux/sonication with HNO_3 decorates the edges of CNT because of the generated defect density. Besides, H_2SO_4–H_2O_2 and H_2SO_4–HNO_3 mixtures are also strong oxidizing agents, creating oxygen functional groups at the sidewalls of CNT extensively [15]. These species are highly reactive because of the generation of NO^{2+} or OH^+ ions. It is observed that while oxidation, sp^2 hybrid bonding is transformed into sp^3 bond at the modified sites and local scattering occurs which reduces the carrier mobility [16]. At the same time, there is some experimental as well as theoretical evidence for the increase of carrier concentration by carboxyl groups via doping [17]. However, any such changes lead to serious damage to the CNT surface and worsen the electrical conductivity [18]. To minimize oxidative deterioration, a few mild oxidation methods such as plasma and light-mediated oxidation and electrochemical oxidation methods have become popular [19–21]. In a recent report, functionalization was achieved by cold plasma discharge driven by high-voltage nanosecond pulses, which provides a controllable, rapid, and least destructive method as compared to acid treatment [22]. Figure 10.3 shows a comparison in terms of oxidation for pristine, acid treated, and cold plasma-treated MWCNTs. Furthermore, another work based on oxidation in the presence of UV light in the presence of a photosensitizer has also proved to be more useful than acid treatment with considerably less structural change [23].

10.2.2 Polymer/Ionic Liquid Modification of Oxidized CNTs

Esterification and amidation are some of the important techniques for further functionalization of oxidized CNTs using polymers, organic molecules, or ionic

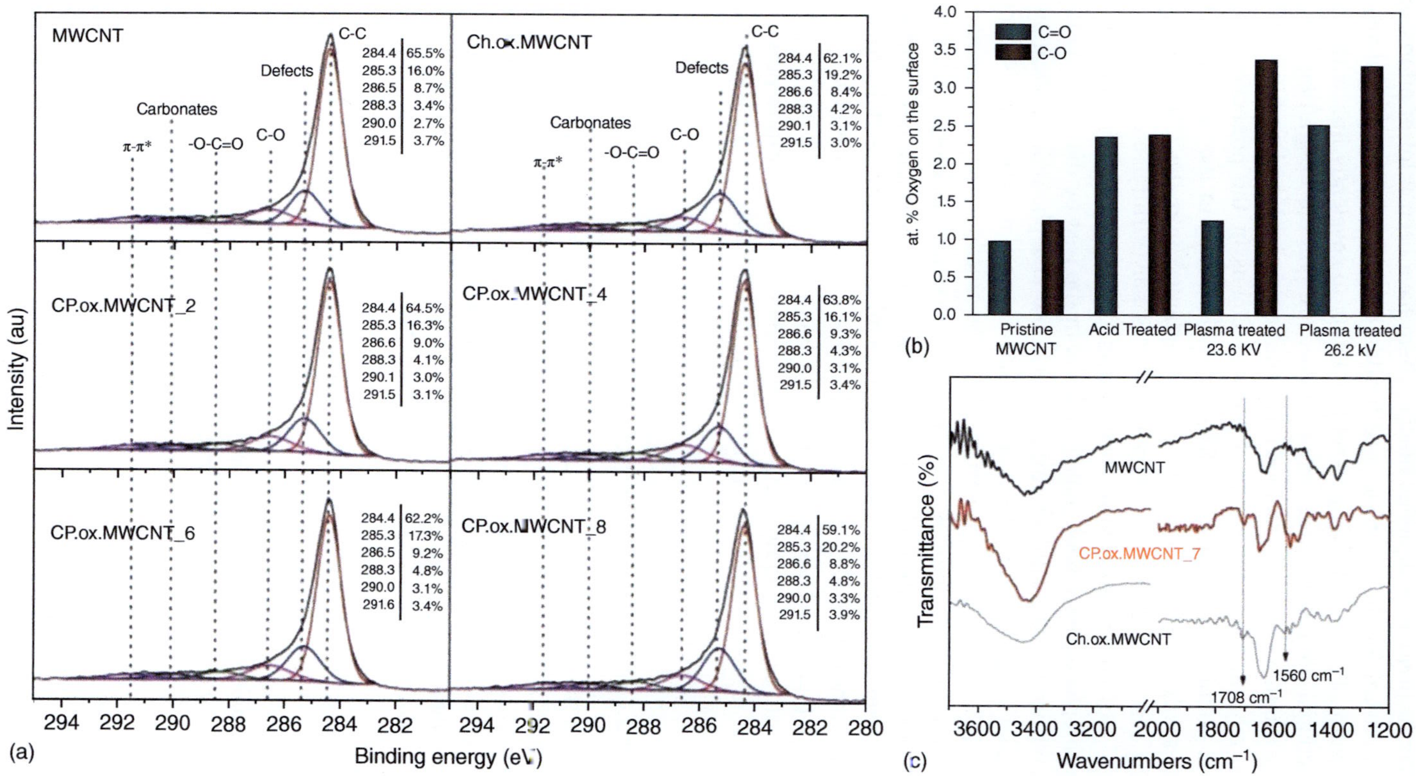

Figure 10.3 (a) XPS (deconvoluted) spectra of pristine, chemically modified, and cold plasma oxidized MWCNTs showing the presence of graphitic peak along with defects, carbon–oxygen groups; (b) Comparison of atomic percentage of oxygen attached to the surface in pristine, acid treated, and plasma-treated MWCNTs; (c) FTIR spectra of pristine, chemically modified, and cold plasma oxidized MWCNTs giving evidence of functionalization. Source: Daletou and Aggelopoulos [22]/with permission of Elsevier

liquids (ILs). These methods are principally utilized for promoting the dispersion of CNTs in organic or aqueous media. There have been a lot of efforts in recent years toward for the chemical modification of oxidized CNTs by polymer functionalization. Polymer chains possess the ability to covalently link with the oxygen functional groups by three major approaches: grafting to, grafting from, and grafting through. The later methods, i.e. grafting through and grafting from approaches, rely on the propagation of polymer chains in the presence of an initiator and double bond moiety modified CNT. However, in the "grafting to" method, presynthesized polymer chains are implanted into the CNT surface via coupling reactions. Different polymers such as poly(furfuryl methacrylate), poly(*N*-vinyl carbazole), poly[(vinyl benzyl)trimethylammonium]chloride, and poly(3-octylthiophene) have been added to CNTs in the past by the "grafting to" approach and are found useful in various electronic and photonic applications [24–27]. Similarly, grafting from the approach has also been reported for CNTs via esterification [28, 29]. ILs are also instrumental in the functionalization of oxidized CNTs. These are especially ionic salts in a liquid state comprising inorganic anions and polymeric/organic cations. The covalent linking of ILs at the CNT surface offers exotic properties such as broad electrochemical windows and increased conductivity. Zhang et al. [30] employed amidation to functionalize CNTs with imidazolium IL. In their report, the wettability of the tubes was controlled by changing the ionic part of the ionic liquid. Such a kind of chemical modification provides surface charges to assist better dispersibility in different solvents via electrostatic repulsion.

10.2.3 Direct Covalent Modification of CNT

As discussed in the Section 10.2.1., irreparable damage occurs at the graphitic surface during the oxidation of CNTs. Hence, direct covalent modification appears to be a more worthwhile method without ant oxidation or pretreatment [31]. Free radical reaction and dienophile addition are two major reaction schemes listed under direct covalent functionalization. Highly reactive aryl radicals are generated by a charge transfer mechanism from the graphitic surface to aryl diazonium salts. These radicals immediately ambush the π-conjugated bonds on the CNT surface forming sp^3 hybridized C—C bonds between the aryl moieties and the tube surface [32]. In this regard, Yang et al. [33] employed several water-soluble azo initiators to chemically modify multiwalled CNTs by a cost-effective and one step-free radical reaction. Transmission electron microscopy (TEM) images evidenced much higher dispersibility of the MWCNTs functionalized by this method rather than pristine ones in polar solvents. The proposed technique (Figure 10.4a) was promising for the development of CNT-containing high-performance materials. TEM images in Figure 10.4b–e demonstrate the direct effect of functionalization with azo-initiators on MWCNTs, leading to shorter length nanotubes, which are best in terms of dispersion in a solvent. Liu et al. [34] also proposed an aryl-free radical addition (produced by air oxidation of 4-methoxyphenylhydrazine) to SWCNTs. This mechanism is emphasized to be different from the diazonium functionalization of SWCNTs,

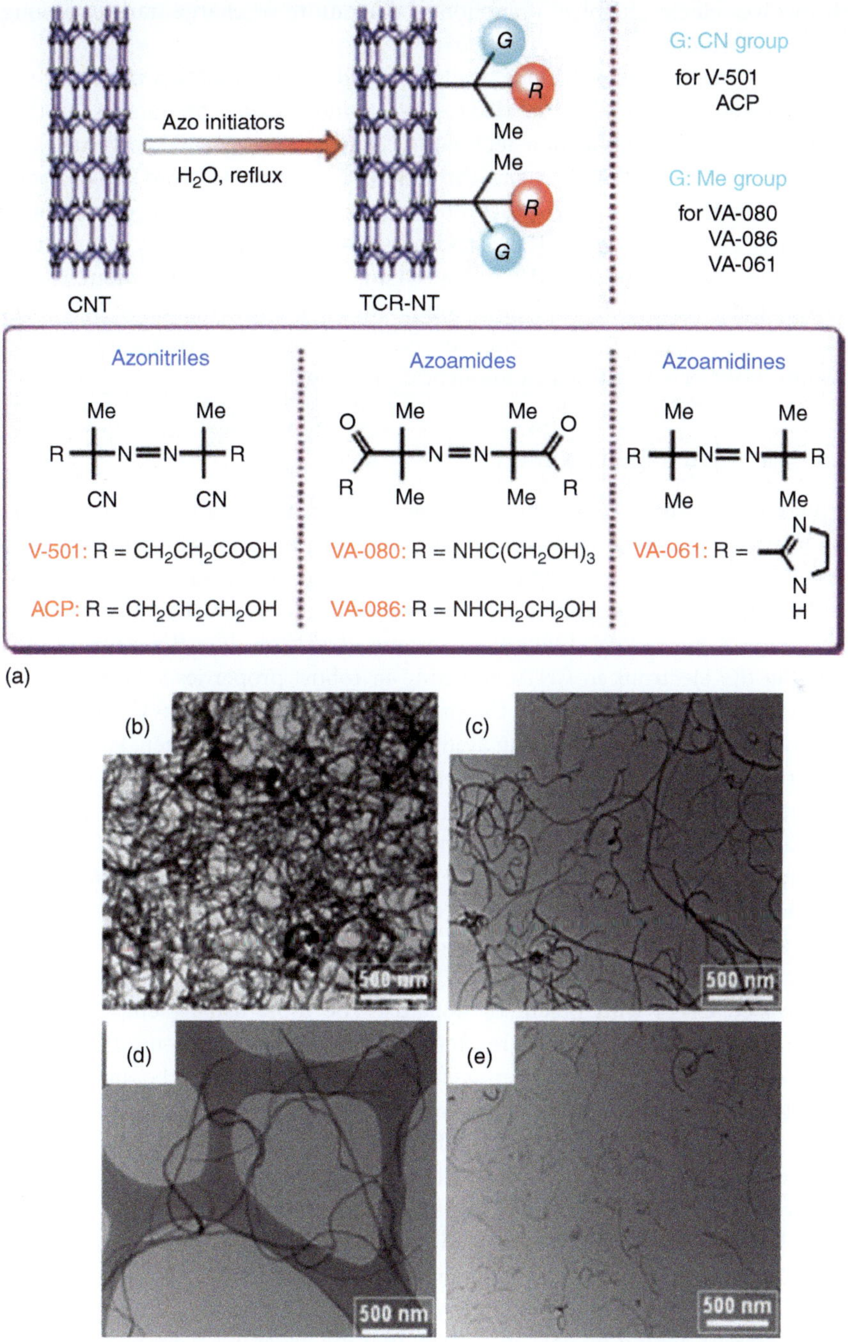

Figure 10.4 (a) Chemical addition of functional groups by one step free radical inclusion of water-soluble azo initiators onto MWCNTs; (b–e) TEM images of pristine MWCNT, 2,20-azobis{2-methyl-*N*-[1,1-bis(hydroxymethyl)-2-hydroxyethyl] propionamide} modified MWCNT, 2,20-Azobis[2-(2-imidazolin-2-yl) propane] modified MWCNT, and 4,40-Azobis (4-cyanopentanoic acid) modified MWCNT, respectively. Source: Yang et al. [33]/with permission of Elsevier.

which involves electrochemical reduction of diazonium or charge transfer among CNTs and diazonium.

Apart from free radical additions, the Diels–Alder (DA) reaction involving sp^2 carbon in the reactions and giving rise to C—C bonds as final products is another promising chemical modification technique. Here, alkyne or alkene behaves as dienophiles which react with diene through DA reaction to form [4+2] cyclic products. Though the amount of theoretical work on the surface modification of CNTs by DA reaction is more, the proportion of experimental work is less. This is due to the less reactivity of the CNTs. Chang et al. [35] used a diene (furan derivative) and dienophiles (maleimide compound) to the functionalized MWCNTs via Diels–Alder reaction. This mechanism may be useful in incorporating polymer chains into MWCNTs to fabricate nanohybrid materials.

10.2.4 Heteroatom Doping of CNTs

Controlled optimization of graphitic carbon in the nanotubes is highly essential for modulating their electronic structure for various applications. In particular, the applications in electronic and opto-electronic demand optimum conductivity and meticulous tuning of energy bandgap as well as work function. The incorporation of heteroatoms possessing different electronic structures has the capability of rearranging the electrons in CNTs, resulting in robust properties. The possibility of substitution depends greatly on the proximity of heteroatoms to C atoms. The most favorable ones are neighboring atoms including boron, nitrogen, silicon, and phosphorus. They can form sp^2 or sp^3 hybridized bonds within the CNT lattice [36]. Boron and nitrogen are the most commonly used substitutions owing to their small mismatch in atomic size as compared to others with significant size differences. Two strategies are mostly instrumental in heteroatom doping: direct synthesis and posttreatment. Direct synthesis methods entail the inclusion of precursor dopants through synthetic processes such as arc discharge, chemical vapor deposition, and laser ablation. On the other hand, posttreatment involves the treatment of presynthesized CNTs with precursor dopants. The most usually employed precursors for nitrogen substitution are N_2 and NH_3, while B_2O_3 is used for boron doping. Recently, Hong et al. [37] demonstrated the synergetic effect of microscopic densification employing boron and nitrogen dopants into the hexagonal CNT lattice. A significant improvement in the electrical conductivity was achieved ($\sim$5896 $Sm^2\ kg^{-1}$) through this process (Figure 10.5). Results obtained in this study prompt a sizeable step toward the use of CNTs as conductive materials. Takassa et al. [38] investigated the structure and electronic properties of nitrogen-doped ultrasmall (3,3) SWCNTs through theoretical calculation applying the density functional theory (DFT) model. They focused on the formation of energy bands in three different doping models and found that all structures were electronically stable. The first-model portrays (3,3) small carbon nitride nanotubes (CNNTs), while in the subsequent models (second and third), two and three linear chains of nitrogen atoms replace the carbon chain in (3,3) CNNT, respectively. As per the calculations, (3,3) SWCNT transforms its nature from semiconductor to Weyl

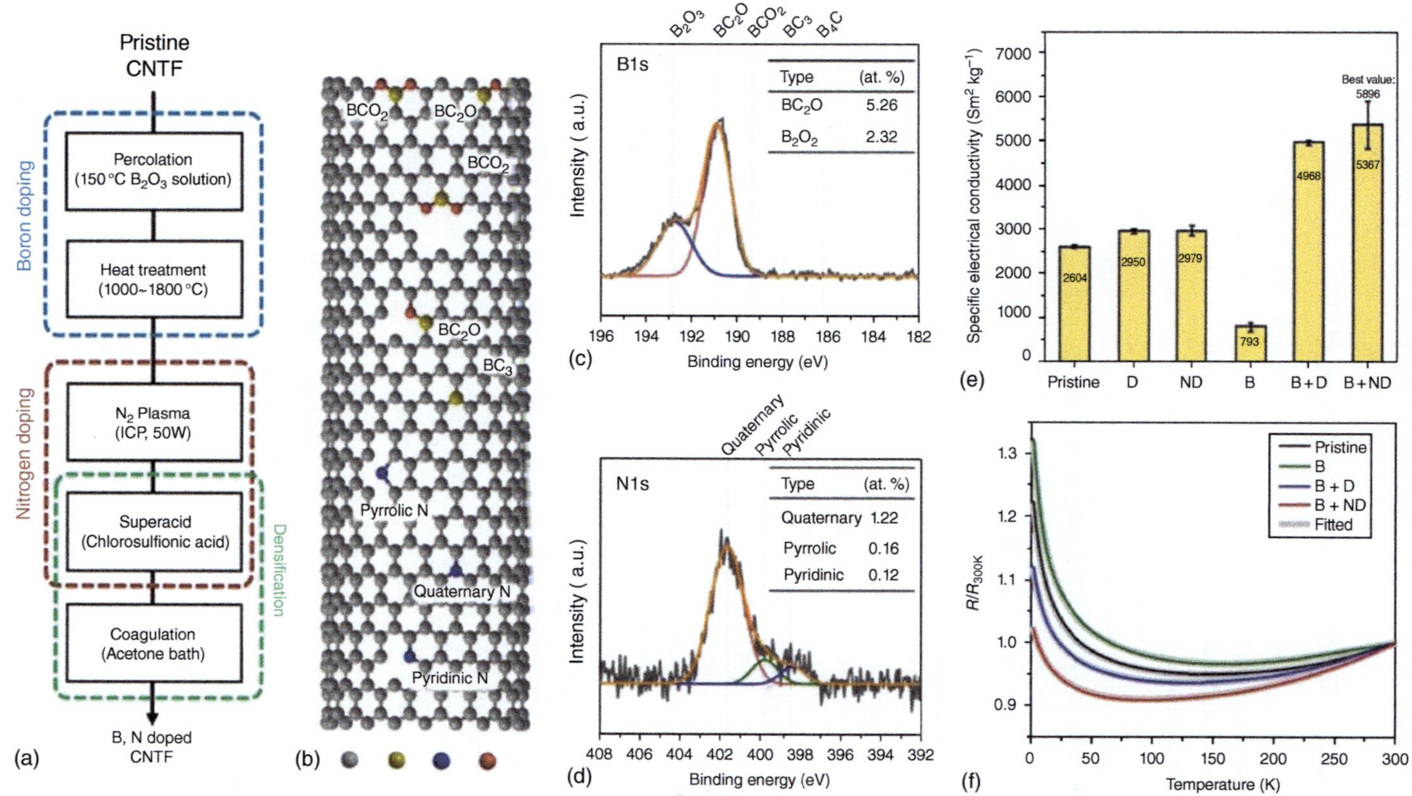

Figure 10.5 (a) Illustration showing B, N doping, and dens fication of CNTs; (b) B and N bonding on the sidewalls of CNTs; (c) and (d) XPS spectra of B and N codoped CNTs by thermal doping at 1200 °C followed by nitrogen plasma showing B 1s and N 1s, respectively; (e) and (f) Comparative plots of specific electrical conductivity and temperature dependence of electrical resistance, respectively, for pristine CNT, densified CNT, N-doped, and B-doped CNT, B, N codoped CNT, and densified B and N-doped CNT. Source: Hong et al. [37]/with permission of Elsevier.

semimetal and metal conductor in the second and first models, respectively. Such observations open up possibilities for the use of this material in small nanodevices.

10.2.5 Charge Transfer/Noncovalent Doping of CNTs

The heterogeneous atom substitution in CNTs leads to defect formation in the graphitic structure that may significantly damage the material properties like charge mobility and mechanical properties. Noncovalent doping or charge transfer is an advanced technique that involves the transfer of charge from physisorbed species such as polymers, metal/metal oxides, and organic/organo-metallic molecules to CNTs. Bouanis et al. [39] recently reported the noncovalent chemical modification of SWCNTs by Co-phthalocyanine, Co-porphyrin, and Fe-porphyrin, molecules. Experimental characterizations evidence robust electronic interactions between the metal-centered molecules and the nanotubes. These functionalized CNTs were employed for the fabrication of the field effect transistor (FET). On performing electrical measurements, a good agreement was obtained with the theoretical ones, indicating the n-type doping (electron transfer) effect. The performance of the chemically modified species was much higher than pristine CNTs, and the results are promising for future CNT-based gas sensing devices. Prato and coworkers [40] studied electron donor/acceptor functionalized polythiophene integrated with SWCNTs through electrostatic and van der Waals interactions into indium tin oxide (ITO) electrodes. The polymer, which is a chromophore, acts as the electron donor while CNTs play the role of electron acceptor. The resulting electrode provided promising photoconversion efficiencies. These chemically modified CNTs are being extensively used for various electronic and photonic applications and are listed schematically in Figure 10.6.

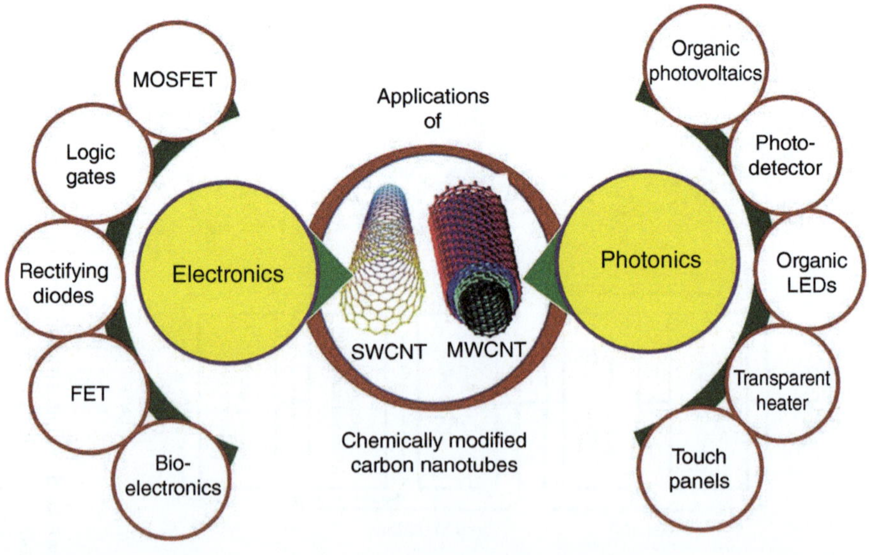

Figure 10.6 Schematic diagram showing different electronic and photonic applications of CNTs.

10.3 Chemically Modified CNTs in Electronics

Robust charge mobility and low capacitance are the salient features of CNTs that are essential for their application in basic electronic devices. Semiconductor nanotubes in particular are receiving much attention these days because of their performance in FETs. In spite of that, large-scale integration of CNTs in electronics is far from realization due to many challenges. One of the major problems is the selective positioning of CNTs on a substrate. Other limitations are precision, specificity, and interconnection, which impede their inclusion in complex circuits and large-scale devices.

10.3.1 Transistors

With a suitable atomic and electronic structure, CNTs offer unique advantages as a FET channel. Nanometer-sized diameters are instrumental in enhancing the efficiency of the gate to control the potential of the channel under the circumstances when the gate is configured to wrap around the CNT. Semiconducting CNTs possess an electrical bandgap that is inversely proportional to the diameter of the tube, which can be tailored as per the requirements of the application. CNTs in FET channels are connected to the metal electrodes such as gold/aluminum/titanium/palladium. These metal electrodes have a work function that is different than the CNTs, causing charge transfer at the interface. This creates a Schottky barrier at both ends of the FET channel. When one of the two barriers is more than the other, their sum represents the bandgap and the FET acts as a unipolar device. A high gated FET having CNT (diameter ~ 1.7 nm) and a high dielectric constant insulator can achieve a transconductance of 30 μS. However, the theoretical limits reach 155 μS which demands that a lot of improvement can be done by functionalizing the CNTs [41]. Shim et al. [42] introduced the concept of n-doping by functionalization of SWCNT sidewalls with amine-rich polymers. They prepared air-stable n-channel FETs on individual CNTs without resorting to any sealing in a vacuum or inert atmosphere. Polyethyleneimine (PEI) functionalized single-walled nanotubes demonstrated a transconductance of $4 \times 10^{-9}\,A\,V^{-1}$ in the I–V_g linear regime. With a drain-source bias voltage of 10 mV, the electron mobility was estimated to be $6 \times 10^2\,cm^2\,V^{-1}\,s^{-1}$. This is much higher than heavily doped n-Si and p-type tube FETs. Surface-functionalized CNTs increased field-effect mobility to $0.04\,cm^2\,V^{-1}\,s^{-1}$ when blended in different amounts with regular poly(3-hexylthiophene) (P3HT). This phenomenal increase in field-effect mobility as compared to pristine P3HT is due to the very high electrical conductivity of the CNTs, which act as conducting bridges between the P3HT crystals in the film. Besides, such a mechanism reduces the hole injection barrier potential with a rise in CNT concentration, thereby raising the efficiency of carrier injection [43]. Croce et al. [44] in 2011 fabricated a high-performance thrombin biosensor based on thrombin aptamer functionalized SWCNT gated silicon-based MOSFET. Figure 10.7 exhibits the schematic representation of the fabrication and implementation of a liquid-gated SWCNT MOSFET and its drain current (I_D) vs. drain voltage (V_D) characteristic in the absence of thrombin. This forms the basis of the operation of

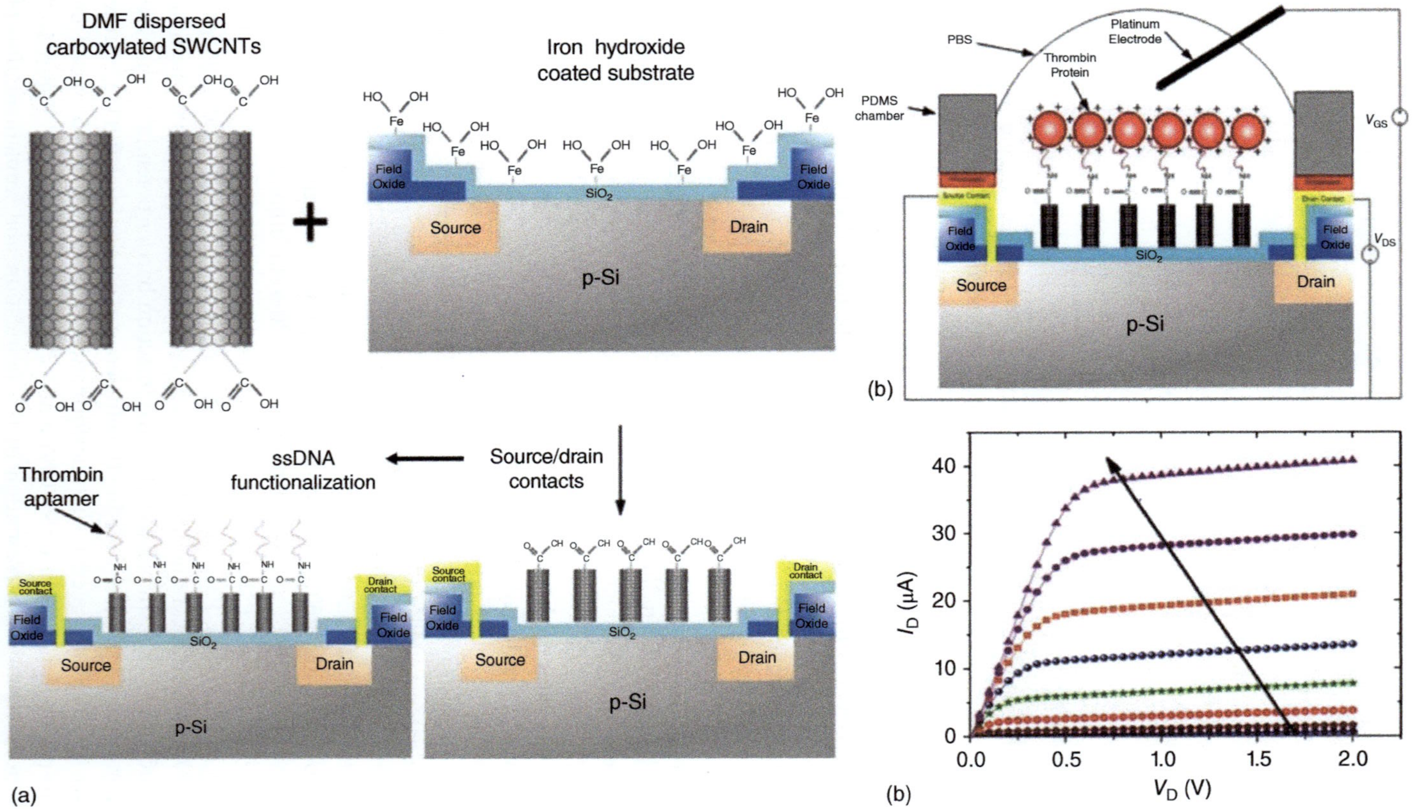

Figure 10.7 (a) Representation of fabrication and implementation of liquid-gated SWCNT MOSFET-based thrombin sensor; (b) Cross sectional view of liquid-gated MOSFET; (c) Drain current (I_D) vs. drain voltage (V_D) characteristics of the device in the absence of thrombin. Source: Croce et al. [44]/with permission of Elsevier.

MOSFET. The drain current increases linearly with the concentration thrombin, resulting in thrombin sensitivities of up to 33 μA (μM mm^2)$^{-1}$. The authors of this report also demonstrated a theoretical model to discuss the operating principle of a SWCNT gated Si MOSFET-based biosensor. It depicted that channel electron density increased with the inclusion of thrombin in addition to the flat band and voltage modification owing to the work function of the CNTs. This kind of advanced technology may prove to be very useful for scaling down electronic devices, and one can expect the dream of lab-on-a-chip to come true.

10.3.2 Rectifying Diodes

Diodes are considered the primary building blocks of a wide variety of electronic and photonic systems. The widely utilized potential of the diodes to regulate the unidirectional flow of current stemming from a structural junction permits the passage of current in one direction only. This rectifying action has tremendous commercial application in voltage regulation and surge protection; near-field communication; and wireless power transmission. Significant efforts have been made by researchers in the last few decades to overcome the challenges offered by pristine CNTs. Chemical modification is one of the promising and easy techniques to modulate the drawbacks of CNTs. For the very first time, Wei et al. [45] tried symmetric functionalization of SWCNT with an electron donor and an electron acceptor at two ends. This was helpful in constructing an exceptionally dipolar molecule that could yield interesting electronic properties, including current rectification. The asymmetric Co_2R–SWCNT–Co_2^- structure could attain a rectification ratio of 10^3, while on the other hand, the authors observed that rectification is suppressed for symmetric nanotubes. In the same year, Valentini et al. [46] also investigated the amino group functionalization of SWCNTs by sidewall fluorination using plasma treatment. Through amino bonds, the amino termination leads to hybrid nanostructures of single-walled nanotubes and alkoxy-silane (3′-(aminopropyl)tri-ethoxy silane, APTES) molecules. Under a certain applied electric field, functionalized SWCNTs are assembled into an ordered rectifying diode. Similarly, some computational studies discuss the importance of peptide linkages connecting CNTs in improving the electronic properties of the system. The junctions made out of two distinct types of CNT units (metallic and semiconducting) demonstrate rectifying current–voltage characteristics [47]. Thermal diodes and ionic diodes are two other forms of rectification diodes whose performance would be improvised inevitably by using chemically modified SWCNTs/MWCNTs [48, 49].

10.3.3 Bioelectronics

Nano-bioelectronics is an emerging interdisciplinary field that combines biology and electronics with nanomaterials. However, for the practical realization of fundamental biology and healthcare, many challenges need to be overcome to address the sensitivity and biocompatibility of shrinking transducers. In this regard, the role of chemically modified CNTs cannot be ignored. They exhibit electrical conductivity in a range from metallic to semiconducting depending on the orientation of the carbon wrap. Apart from that, biocompatibility and flexibility allow for conformal integration within dynamic tissue environments. Broadly, CNTs have been used in self-healable electronic skin, stretchable transistor-based bioelectronics, biosensors, etc. Some of the examples are discussed here. Constructing SWCNT-based transducer/sensory devices require controlled strategies to assemble biomolecules on them. This aspect was taken care of in a report by Lee et al. [50] in 2012. In their study, they functionalized SWCNT with 1,5-diaminonaphthalene via π-stacking for reliable attachment of human olfactory receptor 2AG1(hOR2AG1). This chemically modified CNT was used to construct the FET-based bioelectronic nose. The fabricated device exhibited very high sensitivity down to concentrations as low as 10^{-15} M toward various odorants. Similar work was also performed by Son et al. [51] to detect *Salmonella* infection in food, which is one of the major causes of food poisoning, diarrhea, abdominal pain, and nausea. They have functionalized CNT-based FET with a peptide derived from *Drosophila* odorant-binding protein (OBP) and developed a bioelectronic nose for instant detection of salmonella contamination in ham. This biosensor exhibited much higher sensitivity and a faster response than the traditional sensory nose. Another interesting work was reported by Song et al. [52] in which they developed a human tongue-like bioelectronic sensor from SWCNT and nanovesicles containing human taste receptors (heterodimeric class C G-protein-coupled receptors [GPCRs]). The sweet taste of various sweeteners with broad selectivity could be achieved by using the SWCNT–FET platform. This can prove to be very useful in the food and beverage industry to replace time-consuming and labor-intensive cell-based functional assays and human sensory trails generally employed for screening taste receptors and tastants.

10.4 Chemically Modified CNTs in Photonics

CNTs can act as direct bandgap-type semiconducting materials with unique optical properties. The fascinating optical properties of these materials are mainly regulated by the strongly bound 1D-excitons, whose transition energies are inversely proportional to the diameter of the nanotubes. These excitons are usually triggered through electron–hole recombination, light absorption, or internal impact excitation by hot carriers flowing through the CNTs. In addition to that, the high-energy excitons in CNTs produce sizable photocurrents and voltages that may be utilized for the spectroscopic recognition of CNTs and determination of device parameters, including Schottky barriers and band bending. The versatile optical properties of CNTs bring diverse applications in the field of photonics and optoelectronics, including organic light-emitting diodes (OLEDs), organic photovoltaics (OPV), lasers, electrically pumped solid-state nanoscale light sources; biological applications such as photosensitizers, fluorescent lamps, and sensors. Here we discuss a few photonic applications of chemically modified CNTs to overcome the previously discussed hazards.

10.4.1 Organic Photovoltaics (OPV)

Solar cells, or photovoltaic cells, are propitious sources of renewable energy conversion, which involves the production of electricity directly from the sun's energy. A crucial counterpart of such a kind of application is the optically transparent electrode, which can couple light into the devices. In this regard, our main focus will be on chemically modified CNTs in OPV due to their lower technological cost, low weight, and compatibility with flexible substrates. Wei et al. [53] in 2006 introduced an electrochemical method to functionalize huge amounts of CNT with polyaniline (PANI) in ionic liquids (1-butyl-3-methyl-imidazolium hexafluorophosphate containing 1 M trifluoroacetic acid). This technique offers the feasibility of producing donor–acceptor PANI-SWCNT dyads. The dyads formed can facilitate electron transfer between the donor and acceptor, and hence proving their candidature for highly efficient OPV. Preferential hole transport in conjugated polymer-wrapped CNT networks was noticed for the first time by Dabera et al. [54]. Regulated synthesis methodology can be effective in fabricating high to low-density hybrid CNT networks with low-density nanohybrids as efficient hole transport layers for OPV. The fabricated photovoltaic devices exhibited a high photoconversion efficiency of 7.6% with a device area of 50 mm^2. This is much higher than some of the best OPVs with 20 mm^2 of device area. Recently, work was performed to apply MWCNT/poly(3-hexylthiophene) and CNT-graft-poly(3-dodecylthiophene) in the active layers of the electrode to study the effect of nanostructures on polymer solar cells. They exhibited a high photoconversion efficiency of 4.18% with significantly high current density (short-circuited) and voltage (open-circuited) [55].

10.4.2 Organic Light-Emitting Diodes (OLEDs)

The most recent and promising electronic display technology as well as solid-state lighting devices involve OLEDs, which is a specific type of LED. It consists of an emissive electroluminescent layer of an organic compound placed between the two electrodes and emits light as a response to the electric current between the electrodes. Using CNTs as transparent electrodes may be a wise choice to realize superior flexibility in OLEDs. OLEDs work with opposite light-electric current coupling as in OPV. CNTs as OLEDs face challenges of achieving low R_s and surface roughness, hydrophilicity, proper work function, and passivation/encapsulation along with the usual drawbacks. Chemical modification can be one of the feasible solutions to solve those problems. To address the issues related to surface roughness, Ou et al. [56] fabricated a high-performance OLED using chemically modified CNT transparent and conductive anodes. The modification involves IMRE's proprietary PEDOT:PSS top coating (PS^C), soaking with nitric acid, and encapsulation with polymer. An optimum efficiency of ~10 cd A^{-1}on par with the ITO OLED device was achieved. The performance level of this fabricated device is almost 30–450 time superior to other OLEDs using pristine CNT. The authors suggested more efforts to reduce the work function for use as cathode materials. In another work, high luminance OLEDs employing acid-functionalized MWCNTs were reported as efficient hole injector electrodes. Even at a low voltage of 10 V, the luminance reaches 50 000 cd m^{-2}, with a current efficiency of more than 21 cd A^{-1} along with an external quantum efficiency of 2% [57]. A very different kind of work was performed recently by Wang et al. [58] by introducing natural polyphenol gallic acid as a noncovalent modifier to CNT, which was then amalgamated with graphene and PET by a simple spray process. The fabricated device demonstrated a relatively flat surface roughness of 12.43 nm, better bending resistance, strong adhesion, and a significant luminance of 6554.1 cd m^{-2} at 15 V, and a maximum current efficiency of 4.03 cd A^{-1}, which is very promising for OLED device application (Figure 10.8).

10.4.3 Touch Panels

With the recent popularization of smart electronic devices, starting from cell phones to tablets, laptops, digital cameras, videogame consoles, etc., touch panels are gaining much importance in contemporary society. Technically, touch panels refer to a potent human–device interface with the aid of which users are able to control the devices with single or multiple touches on the panel (without using a mouse, keyboard, or other intermediate accessories). Manufacturers from the industry have recognized touch panels as being one of the highly desirable components of modern optoelectronic devices. The mechanism by which the touch panel works is given as follows: once pressure is applied to the touch side electrode by finger or stylus pen, the electrodes are bridged together and the resulting current flow between them is measured. Simultaneously, a signal processing controller identifies the touch in x- and y-directions respectively, and may be z- also. Eventually, the computer processes the events by receiving feedback. In contrast to organic photovoltaics and OLEDs, touch panels do not demand alterations in work function, surface roughness, or hydrophilicity, but most importantly, their performance depends on the electrical conductivity and transparency of electrodes. A good amount of work has been conducted on touch panels involving CNT-based transparent conductive films. A few of them have been commercially undertaken by Foxconn Electronics (Hon Hai Precision Industry, China); Unidym Inc.; Product Press Releases to improve the durability, sensitivity, and resolution of the touch panels [59], but they may not be sufficient. Hence, the researchers are further stepping toward the chemical modification of the CNTs to achieve better goals. Li et al. [60] demonstrated the processing and characterization of transparent and highly conducting films using Pd-functionalized few-walled CNTs. The contact resistance between the metallic surface and semiconducting CNT, which is mainly governed by the Schottky barrier, was significantly reduced in the functionalized CNTs as compared to the pristine ones. Again, extremely low sheet resistance 274 Ω/sq. of poly(ethylene terephthalate) substrates coated with Pd-functionalized few-walled CNTs was achieved, along with a transmittance of 81.65% at 550 nm. Recently, Moon and coworkers constructed an electrically actuatable platform using carboxyl group functionalized CNT mats with dielectric elastomer. The fabricated composite exhibited the highest transmittance of 93.2% with significant electrical conductivity, proving its candidature for tactile touch panels [61]. In another work, composites

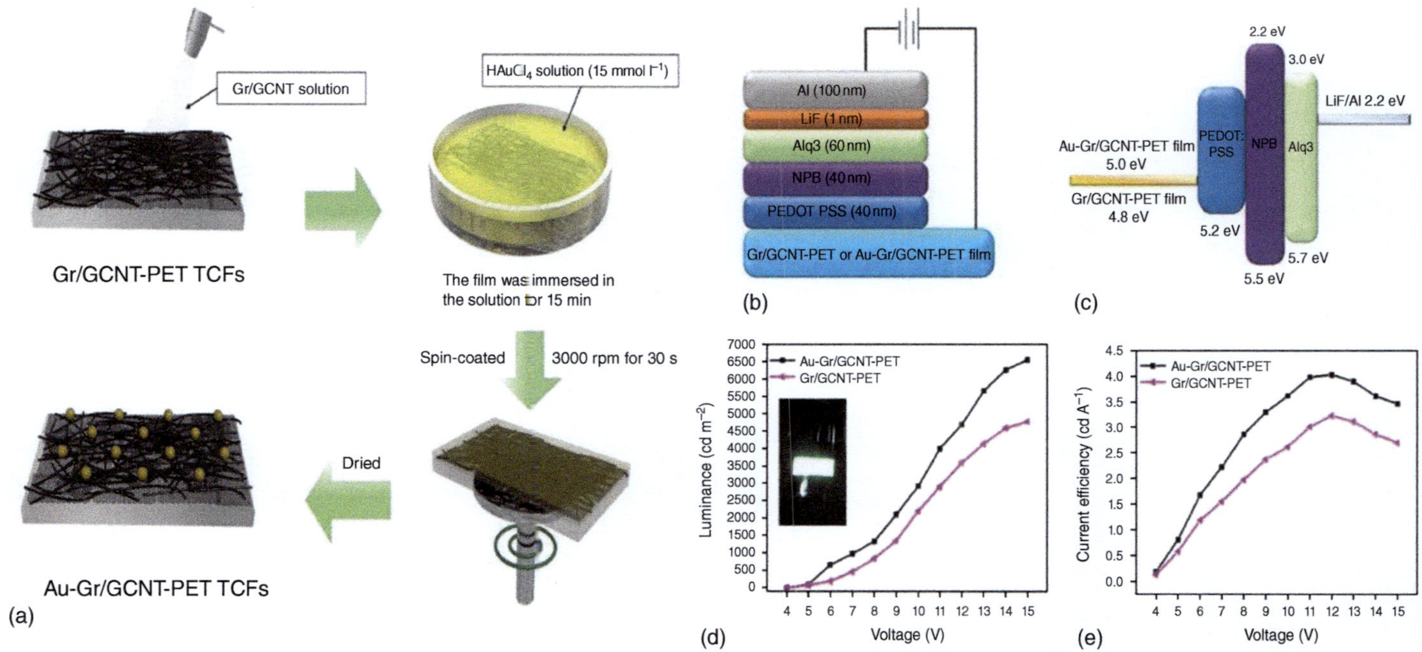

Figure 10.8 (a) Schematic figure showing fabrication of Au-Gr/GCNT-PET (Au-graphene gallic acid-modified CNT-polyethylene terephthalate) transparent conductive films; (b) Construction of green organic LEDs on Gr/GCNT-PET or Au-Gr/GCNT-PET anode; (c) Energy level diagram of organic LED device; (d) and (e) Luminance vs. voltage and current efficiency vs. voltage, respectively for the organic LEDs. Source: Wang et al. [58]/with permission of Elsevier.

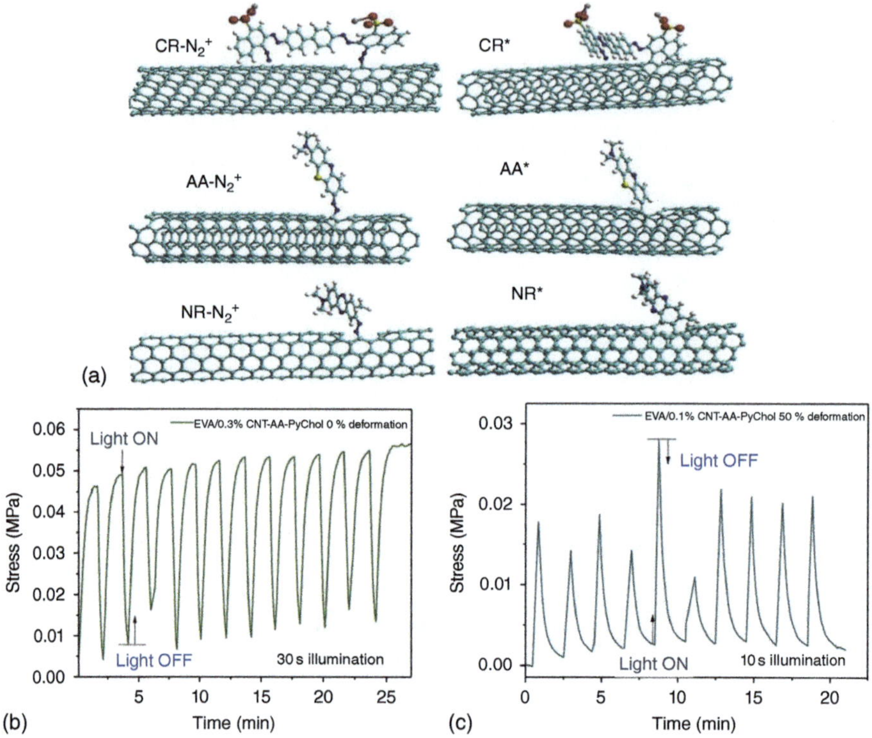

Figure 10.9 (a) Dye (three diazonium compounds: AA-N$^+$, NR-N_2^+, and CR-N_2^+) grafted structures of (5,5) defect containing CNTs. Carbon atoms are in cyan, nitrogen in blue, oxygen in red sodium in gray, sulfur in yellow, and hydrogen in white; (b) Actuation stress (in MPa) over time for unstretched EVA nanocomposite consisting of 0.3%CNT-AA-Pychol during 30 seconds of illumination; (c) Actuation stress (in MPa) over time for stretched EVA nanocomposite consisting of 0.1% CNT-AA-Pychol during 10 seconds of illumination using red LED with 58 mW power. ON/OFF light sequences are shown by arrows. Source: Bensghaïer et al. [62]/with permission of Elsevier.

were fabricated using dye-diazonium grafted MWCNTs with an ethylene-vinyl acetate (EVA) matrix. Figure 10.9 demonstrates the structure of different diazonium dye grafted CNTs and the photoactivation response of EVA/CNT composites. These nanocomposites were employed for opto-thermal actuators [62].

From the above discussions, it is clear that single-walled versions of CNT are more suitable for electronic and photonic applications owing to the strange metal-like conductivity in an organic material. However, the cost, chemical complexity, reproducibility, dispersibility, and purity are currently some of the major concerns which hinder the commercialization of CNTs. Research groups all over the globe are trying to resolve these issues through prospective chemical modification techniques. In this regard, Tour et al. [63] patented a technique to functionalize CNTs in an acidic medium. They could successfully separate individual nanotubes via facilitation of sidewall functionalization. This methodology was effortlessly scalable and affordable for industrial-scale production of CNTs.

Furthermore, Starkovich et al. [64] claimed that that addition of dopants in regulated amounts (bromine, iodine, chloroauric acid, nitric acid, potassium tetrabromoaurate, and hydroiodic acid) could increase the conductivity of the CNTs to a desired value. Another patent reports the application of a one-sided functionalized CNT fabricating FET [65]. Apart from that, some of the major companies and research laboratories involved in manufacturing and marketing chemically modified CNTs in nanoelectronics and photovoltaics are Aneeve Nanotechnologies LLC, USA, Brewer Science, USA, Ekios, USA, HRL Laboratories, USA, IBM, USA, Nanocomb Technologies, Inc. (NCTI), USA, Samsung Electronics, Korea, University of Wisconsin–Madison, USA, and Xintek Nanotechnology Innovations, USA. Besides, Prof. Takao Someya and his group from the University of Tokyo established a method to fabricate long nanotubes on an industrial scale by combining CNTs with an ionic liquid. This invention will be commercialized by Dai Nippon Printing in the coming years to construct stretchable mesh of printed transistors. Similarly, University of California (UCLA) researchers are working on CNT-based flexible solar cells; researchers from the University of Nottingham, UK, in collaboration with Nantero Inc., USA, are working on nanotube-based non-volatile memory; and scientists at DuPont, USA, in collaboration with Cornell University, USA, are deeply involved in commercialising CNT transistors.

10.5 Summary and Future Scope

The fascinating properties of semiconducting and metallic CNTs are known to be the basis of many small, fast, and flexible molecular devices. However, controlled synthesis of specific tube lengths, dispersibility in solutions, chemical inertness, and low surface energy remain the major bottlenecks in using pristine CNTs for electronics and photonics applications. In this regard, chemically modified or functionalized CNTs can be suggested as a feasible solution. Transistors with chemically modified CNTs are already in use as sensitive and selective biological and chemical sensors. Similarly, the photonic characteristics also see a major improvement with functionalized nanotubes rather than the pristine ones. Keeping a note of all these factors, we have provided an overview of some of the prospective chemical modifications of CNTs along with their applications in electronics and photonics. The evolution from CNTs to functionalized CNTs is evidence of improvement in the existing technology.

According to a survey report by Grand View Research, Inc. (https://grandviewresearch.podbean.com), global CNT market is anticipated to reach US$ 3.42 billion by 2022. Looking into the overwhelming use of chemically modified CNTs in various fields, the governments of India, China, the USA, and Japan have increased the funding aimed at the promotion of nanotechnology at a domestic level, which is expected to create new avenues for CNTs over the coming years. The hunger for improving every aspect of our smart life relies on consumer products such as televisions, mobile phones, watches, and what not!! The new e-books by Amazon Kindle are a splendid example as compared to the iconic i-Pods in their very first year of merchandize. CNTs can help with a variety of targeted accomplishments. One such technology is printed electronics, which has taken the center stage as far as modified CNTs are concerned. This is particularly important for producing flexible or stretchable devices. However, both the instability of CNT ink and the use of stabilizing agents (causes conductivity issues) are the areas needing urgent attention. There is a lot of scope for the researchers to drill in this direction. Similarly, sensory devices such as artificial retina, electronic ear, E-nose, E-skin requiring thousands of CNT transistors, photo receptors, etc., are some of the areas to be explored which can bring a boomerang in IoT (Internet of Things) and artificial intelligence. We hope the current chapter will definitely provide a rich literature review on the recent electronic and photonic applications of chemically modified CNTs to the readers and provide a roadmap for further improvements.

References

1 Novoselov, K.S., Geim, A.K., Morozov, S.V. et al. (2005). Two-dimensional gas of massless Dirac fermions in graphene. *Nature* 438 (7065): 197–200.

2 Jorio, A. and Dresselhaus, M.S. (ed.) (2008). *Carbon Nanotubes: Advanced Topics in the Synthesis, Structure, Properties and Applications*, vol. 111. Berlin: Springer.

3 Karousis, N., Tagmatarchis, N., and Tasis, D. (2010). Current progress on the chemical modification of carbon nanotubes. *Chem. Rev.* 110 (9): 5366–5397.

4 Wang, S., Zhang, Z., Ding, L. et al. (2008). A doping-free carbon nanotube CMOS inverter-based bipolar diode and ambipolar transistor. *Adv. Mater.* 20 (17): 3258–3262.

5 Franklin, A.D. (2013). The road to carbon nanotube transistors. *Nature* 498 (7455): 443–444.

6 Koswatta, S.O., Neophytou, N., Kienle, D. et al. (2006). Dependence of DC characteristics of CNT MOSFETs on band structure models. *IEEE Trans. Nanotechnol.* 5 (4): 368–372.

7 Shivananju, B.N., Asokan, S., and Misra, A. (2015). Study on effect of optical wavelength on photo-induced strain sensitivity in carbon nanotubes using fiber Bragg grating. *J. Phys. D: Appl. Phys.* 48 (27): 275502.

8 Keren, K., Berman, R.S., Buchstab, E. et al. (2003). DNA-templated carbon nanotube field-effect transistor. *Science* 302 (5649): 1380–1382.

9 Chen, J., Hamon, M.A., Hu, H. et al. (1998). Solution properties of single-walled carbon nanotubes. *Science* 282 (5386): 95–98.

10 Ding, B., Guo, D., Wang, Y. et al. (2018). Functionalized graphene nanosheets decorated on carbon nanotubes networks for high-performance supercapacitors. *J. Power Sources* 398: 113–119.

11 Monthioux, M., Smith, B.W., Burteaux, B. et al. (2001). Sensitivity of single-wall carbon nanotubes to chemical processing: an electron microscopy investigation. *Carbon* 39 (8): 1251–1272.

12 Basiuk, E.V., Monroy-Peláez, M., Puente-Lee, I., and Basiuk, V.A. (2004). Direct solvent-free amination of closed-cap carbon nanotubes: a link to fullerene chemistry. *Nano Lett.* 4 (5): 863–866.

13 Zhao, M., Xia, Y., Lewis, J.P., and Mei, L. (2004). Chemical reactivity of single-walled carbon nanotubes to amidogen from density functional calculations. *J. Phys. Chem. B* 108 (28): 9599–9603.

14 Tchoul, M.N., Ford, W.T., Lolli, G. et al. (2007). Effect of mild nitric acid oxidation on dispersibility, size, and structure of single-walled carbon nanotubes. *Chem. Mater.* 19 (23): 5765–5772.

15 Zhang, J., Zou, H., Qing, Q. et al. (2003). Effect of chemical oxidation on the structure of single-walled carbon nanotubes. *J. Phys. Chem. B* 107 (16): 3712–3718.

16 Park, H., Zhao, J., and Lu, J.P. (2006). Effects of sidewall functionalization on conducting properties of single-wall carbon nanotubes. *Nano Lett.* 6 (5): 916–919.

17 Wang, C., Zhou, G., Wu, J. et al. (2006). Effects of vacancy-carboxyl pair functionalization on electronic properties of carbon nanotubes. *Appl. Phys. Lett.* 89 (17): 173130.

18 Kovtyukhova, N.I., Mallouk, T.E., Pan, L., and Dickey, E.C. (2003). Individual single-walled nanotubes and hydrogels made by oxidative exfoliation of carbon nanotube ropes. *J. Am. Chem. Soc.* 125 (32): 9761–9769.

19 Alvarez, N.T., Kittrell, C., Schmidt, H.K. et al. (2008). Selective photochemical functionalization of surfactant-dispersed single-wall carbon nanotubes in water. *J. Am. Chem. Soc.* 130 (43): 14227–14233.

20 Zhao, B., Zhang, L., Wang, X., and Yang, J. (2012). Surface functionalization of vertically-aligned carbon nanotube forests by radio-frequency Ar/O_2 plasma. *Carbon* 50 (8): 2710–2716.

21 Ye, J.S., Liu, X., Cui, H.F. et al. (2005). Electrochemical oxidation of multi-walled carbon nanotubes and its application to electrochemical double-layer capacitors. *Electrochem. Commun.* 7 (3): 249–255.

22 Daletou, M.K. and Aggelopoulos, C.A. (2021). Highly-energy efficient oxidation of MWCNT with nanosecond pulsed dielectric barrier discharge plasma. *Appl. Surf. Sci.* 563: 150139.

23 Savage, T., Bhattacharya, S., Sadanadan, B. et al. (2003). Photoinduced oxidation of carbon nanotubes. *J. Phys.: Condens. Matter* 15 (35): 5915.

24 Pramanik, N.B. and Singha, N.K. (2015). Direct functionalization of multi-walled carbon nanotubes (MWCNTs) via grafting of poly(furfuryl methacrylate) using Diels–Alder "click chemistry" and its thermoreversibility. *RSC Adv.* 5 (114): 94321–94327.

25 Wu, H.X., Qiu, X.Q., Cai, R.F., and Qian, S.X. (2007). Poly(*N*-vinyl carbazole)-grafted multiwalled carbon nanotubes: synthesis via direct free radical reaction and optical limiting properties. *Appl. Surf. Sci.* 253 (11): 5122–5128.

26 Rahman, G.E.E., Troeger, A., Sgobba, V. et al. (2008). Improving photocurrent generation: supramolecularly and covalently functionalized single-wall carbon nanotubes-polymer/porphyrin donor–acceptor nanohybrids. *Chem. Eur. J.* 14 (29): 8837.

27 Stefopoulos, A.A., Chochos, C.L., Prato, M. et al. (2008). Novel hybrid materials consisting of regioregular poly(3-octylthiophene)s covalently attached to single-wall carbon nanotubes. *Chem. Eur. J.* 14 (28): 8715–8724.

28 Zhou, C., Wang, S., Zhuang, Q., and Han, Z. (2008). Enhanced conductivity in polybenzoxazoles doped with carboxylated multi-walled carbon nanotubes. *Carbon* 46 (9): 1232–1240.

29 Chen, X., Tao, F., Wang, J. et al. (2009). Concise route to styryl-modified multi-walled carbon nanotubes for polystyrene matrix and enhanced mechanical properties and thermal stability of composite. *Mater. Sci. Eng., A* 499 (1, 2): 469–475.

30 Zhang, Y., Shen, Y., Yuan, J. et al. (2006). Design and synthesis of multifunctional materials based on an ionic-liquid backbone. *Angew. Chem.* 118 (35): 5999–6002.

31 Tasis, D., Tagmatarchis, N., Bianco, A., and Prato, M. (2006). Chemistry of carbon nanotubes. *Chem. Rev.* 106 (3): 1105–1136.

32 Bahr, J.L., Yang, J., Kosynkin, D.V. et al. (2001). Functionalization of carbon nanotubes by electrochemical reduction of aryl diazonium salts: a bucky paper electrode. *J. Am. Chem. Soc.* 123 (27): 6536–6542.

33 Yang, Y., Qiu, S., Xie, X. et al. (2010). A facile, green, and tunable method to functionalize carbon nanotubes with water-soluble azo initiators by one-step free radical addition. *Appl. Surf. Sci.* 256 (10): 3286–3292.

34 Liu, J., Zubiri, M.R.I., Dossot, M. et al. (2006). Sidewall functionalization of single-wall carbon nanotubes (SWNTs) through aryl free radical addition. *Chem. Phys. Lett.* 430 (1–3): 93–96.

35 Chang, C.M. and Liu, Y.L. (2009). Functionalization of multi-walled carbon nanotubes with furan and maleimide compounds through Diels–Alder cycloaddition. *Carbon* 47 (13): 3041–3049.

36 Zhang, Y., Zhang, J., and Su, D.S. (2014). Substitutional doping of carbon nanotubes with heteroatoms and their chemical applications. *ChemSusChem* 7 (5): 1240–1250.

37 Hong, S., Nam, J., Park, S. et al. (2021). Carbon nanotube fibers with high specific electrical conductivity: synergistic effect of heteroatom doping and densification. *Carbon* 184: 207–213.

38 Takassa, R., Farkad, O., Ibnouelghazi, E.A., and Abouelaoualim, D. (2021). Structural and electronic properties of N-doping ultra-small diameter (3,3) armchair SWCNT by using PBE and TB-mBJ potentials. *Appl. Surf. Sci.* 563: 150283.

39 Bouanis, F.Z., Bensifia, M., Florea, I. et al. (2021). Non-covalent functionalization of single walled carbon nanotubes with Fe-/Co-porphyrin and Co-phthalocyanine for field-effect transistor applications. *Org. Electron.* 96: 106212.

40 Rahman, G.A., Guldi, D.M., Cagnoli, R. et al. (2005). Combining single wall carbon nanotubes and photoactive polymers for photoconversion. *J. Am. Chem. Soc.* 127 (28): 10051–10057.

41 Javey, A., Guo, J., Wang, Q. et al. (2003). Ballistic carbon nanotube field-effect transistors. *Nature* 424 (6949): 654–657.

42 Shim, M., Javey, A., Shi Kam, N.W., and Dai, H. (2001). Polymer functionalization for air-stable n-type carbon nanotube field-effect transistors. *J. Am. Chem. Soc.* 123 (46): 11512–11513.

43 Park, Y.D., Lim, J.A., Jang, Y. et al. (2008). Enhancement of the field-effect mobility of poly(3-hexylthiophene)/functionalized carbon nanotube hybrid transistors. *Org. Electron.* 9 (3): 317–322.

44 Croce, R.A. Jr., Vaddiraju, S., Chan, P.Y. et al. (2011). Label-free protein detection based on vertically aligned carbon nanotube gated field-effect transistors. *Sens. Actuators, B* 160 (1): 154–160.

45 Wei, Z., Kondratenko, M., Dao, L.H., and Perepichka, D.F. (2006). Rectifying diodes from asymmetrically functionalized single-wall carbon nanotubes. *J. Am. Chem. Soc.* 128 (10): 3134–3135.

46 Valentini, L., Macan, J., Armentano, I. et al. (2006). Modification of fluorinated single-walled carbon nanotubes with aminosilane molecules. *Carbon* 44 (11): 2196–2201.

47 Khazaei, M., Lee, S.U., Pichierri, F., and Kawazoe, Y. (2007). Computational design of a rectifying diode made by interconnecting carbon nanotubes with peptide linkages. *J. Phys. Chem. C* 111 (33): 12175–12180.

48 Pal, S. and Puri, I.K. (2014). Thermal rectification in a polymer-functionalized single-wall carbon nanotube. *Nanotechnology* 25 (34): 345401.

49 Peng, R., Pan, Y., Li, Z. et al. (2020). Ionotronics based on horizontally aligned carbon nanotubes. *Adv. Funct. Mater.* 30 (38): 2003177.

50 Lee, S.H., Jin, H.J., Song, H.S. et al. (2012). Bioelectronic nose with high sensitivity and selectivity using chemically functionalized carbon nanotube combined with human olfactory receptor. *J. Biotechnol.* 157 (4): 467–472.

51 Son, M., Kim, D., Kang, J. et al. (2016). Bioelectronic nose using odorant binding protein-derived peptide and carbon nanotube field-effect transistor for the assessment of *Salmonella* contamination in food. *Anal. Chem.* 88 (23): 11283–11287.

52 Song, H.S., Jin, H.J., Ahn, S.R. et al. (2014). Bioelectronic tongue using heterodimeric human taste receptor for the discrimination of sweeteners with human-like performance. *ACS Nano* 8 (10): 9781–9789.

53 Wei, D., Kvarnström, C., Lindfors, T., and Ivaska, A. (2007). Electrochemical functionalization of single walled carbon nanotubes with polyaniline in ionic liquids. *Electrochem. Commun.* 9 (2): 206–210.

54 Dabera, G.D.M., Jayawardena, K.I., Prabhath, M.R. et al. (2013). Hybrid carbon nanotube networks as efficient hole extraction layers for organic photovoltaics. *ACS Nano* 7 (1): 556–565.

55 Hadi, A., Hekmatshoar, M.H., Abbasi, F., and Agbolaghi, S. (2021). Nanostructures of chemically modified multi-walled carbon nanotubes and poly(3-hexylthiophene) to improve photophysic/photovoltaic features. *Carbon Lett.* 31 (1): 107–115.

56 Ou, E.C., Hu, L., Raymond, G.C.R. et al. (2009). Surface-modified nanotube anodes for high performance organic light-emitting diode. *ACS Nano* 3 (8): 2258–2264.

57 Shi, S. and Silva, S.R.P. (2012). High luminance organic light-emitting diodes with efficient multi-walled carbon nanotube hole injectors. *Carbon* 50 (11): 4163–4170.

58 Wang, T., Jing, L.C., Bao, Z. et al. (2021). Strong adhesion and high optoelectronic performance hybrid graphene/carbon nanotubes transparent conductive films for green-light OLED devices. *Surf. Interfaces* 24: 101137.

59 Du, J., Pei, S., Ma, L., and Cheng, H.M. (2014). 25th-anniversary article: carbon nanotube-and graphene-based transparent conductive films for optoelectronic devices. *Adv. Mater.* 26 (13): 1958–1991.

60 Li, Y.A., Tai, N.H., Chen, S.K., and Tsai, T.Y. (2011). Enhancing the electrical conductivity of carbon-nanotube-based transparent conductive films using functionalized few-walled carbon nanotubes decorated with palladium nanoparticles as fillers. *ACS Nano* 5 (8): 6500–6506.

61 Kang, S.Y., Moon, J., and Ahn, S.D. (2020). Compliant characteristics of carbon nanotube electrodes for electromechanical applications. *Thin Solid Films* 706: 138015.

62 Bensghaïer, A., Forro, K., Seydou, M. et al. (2018). Dye diazonium-modified multiwalled carbon nanotubes: light harvesters for elastomeric optothermal actuators. *Vacuum* 155: 178–184.

63 Tour, J., Hudson, J., Dyke, C., and Stephenson, J., William Marsh Rice University. (2007). Functionalization of carbon nanotubes in acidic media. US Patent Application 10/593,918, filed 24 March 2005 and issued 06 December 2007.

64 Starkovich, J.A., Silverman, E.M., and Peng, H.H., Northrop Grumman Systems Corp. (2014). Carbon nanotube conductor with enhanced electrical conductivity. US Patent 8,808,792, filed 17 January 2012 and issued 19August 2014.

65 Huang, H. and Sony Corp. (2009). Method for making field-effect transistor using carbon nanotube. US Patent 7,601,322, filed 20 July 2005 and issued 13 October 2009.

11

Chemically Modified Carbon Nanotubes for Electrochemical Sensors

Ajahar Khan, Khalid A. Alamry, and Raed H. Althomali

King Abdulaziz University, Faculty of Science, Department of Chemistry, Abdulla Sulaiman St No. 1, Jeddah 21589, Saudi Arabia

11.1 Introduction

An electrochemical sensor is a system that electronically determines the different quantities and recorded the measurements into electrical signals elsewhere [1, 2]. They are highly encouraging analytical tools owing to their large degree of stability, sensitivity, and selectivity [2, 3]. The electrochemical sensors are very practical in order to investigate the trace amount of different analytes in samples for example dissolved gases, solid, and fluid materials. Electrochemical sensors are regularly employed in environmental analysis, industrial safety, and process measuring engineering, health-care diagnostics, and pharmaceutical engineering [4, 5]. In recent years, electroactive/conducting polymers and carbon nanotubes (CNTs) have focused much attention in the bioanalytical field because of their significant function in improving the electrocatalytic activity and sensitivity of the consequent sensor systems [6]. CNTs are one of the most promising materials in comparison to other carbon-based nanomaterials (graphene, fullerenes, and carbon nanohorns) and are widely utilized in the construction of electrochemical sensor because of their chemical compatibility and dimensional with several biomolecules, which allows them to catalyze reactions [7] and to stimulate electron-transfer reactions among electrode substrates and biomolecules [8, 9]. CNTs attributed sp^2 carbon units with a diameter in the range of nanometers to many microns in one dimension. Two types of CNTs, single-walled carbon nanotubes (SWCNTs), multi-walled carbon nanotubes (MWCNTs) could be constructed through chemical vapor deposition, laser vaporization, and electrical arc discharge techniques. CNTs perform as either metals or semiconductors, determine by the degree of helicity and diameter [10]. Modification of CNTs is carried out on the nonhexagonal positions tips, and sidewall defects to achieve their improved solubilization and dispersion and for further processing and applications [11]. To improve their sensitivity, selectivity, and catalytic activity, and expand their features and applications, several physical and chemical routes have been followed to functionalize the walls of CNTs.

Chemically Modified Carbon Nanotubes for Commercial Applications, First Edition.
Edited by Jeenat Aslam, Chaudhery Mustansar Hussain, and Ruby Aslam.

Traditional chemical approaches such as covalent functionalization through oxidation and direct chemical modification of the sidewalls via incorporation and noncovalent functionalization with polymers or surfactants [12]. Other chemical approaches are also being determined such as the grafting of CNTs using monomers or polymers [11, 13]. Till now, different types of hybrid CNTs-based materials composited with organic and inorganic compounds have been studied [14, 15]. These functionalized-CNTs are favorable for the modification of several electrodes because of their better electrochemical and chemical stabilities in both nonaqueous and aqueous mediums and for the electron transfer reactions provide high electronic conductivity [16]. Moreover, the fabrication of proficient electrochemical sensors through the functionalized-CNTs is highly encouraging because they advance electron-transfer reactions in various different sizes of biomolecules [17, 18] and demonstrated very appreciable application in sensors, optics, catalysis, and electronics, etc. [19, 20]. Figure 11.1 illustrates the fabrication process of CNT-based electrochemical sensor and its performance for the detection of H_2O_2. CNTs-based sensors are also reported to develop novel electroanalytical methods to quantify essential compounds and hazardous contaminants in food samples [22–24]. The utilization of CNTs-based sensors was pioneered for biological molecules as Britto et al. reported an electrode based on CNT that was employed for the determination of reversible oxidation of dopamine (DA) [25]. However, commercial applications of CNT-based technologies are still rare and have not been widely accepted industrial scale because of their high cost and the complex interactions of CNTs with other biotic and abiotic environmental components. Usually, the price of commercially available CNTs falls in the ranges of 25 to 300 US$ per gram of SWCNTs and 0.6 to 25 US$ per gram of MWCNTs depending on the tube diameter of the products and purity level, as well as on the type of functionalization carried out during their fabrication [26]. Furthermore, to think about the commercialization of CNT-based sensors, the quality of commercially available CNT is likely to pose a major concern because the supply of CNT from the same supplier differs in features (purification protocols, surface defects, baseline impurities, etc.) from batch to batch [27]. However, the construction of sensors on a microfabrication level is expected to meet the standards of identical characteristics. Therefore, if the sensor elements are deposited with CNTs that differ in characteristics and quality, then the experimental outcomes from test systems would not generate the desired data from the sensors employed in the corresponding field. In spite of these limitations, the modularity through which nanotubes can be functionalized is empowering for the speedy advancement of sensor techniques, therefore it is obvious that we will observe a motion of commercial electrochemical CNT sensors that rival and possibly exceed the majority of other chemical sensor types.

The present book chapter summarizes the utilization of CNTs-based electrochemical sensors in the field of environmental monitoring, pharmaceutical applications, biomolecules sensing, and determination of different analytes and hazardous food contaminants in real samples. The chapter starts with the introduction, the properties and functionalization of CNTs' followed CNTs as electrochemical sensors, and their applications. Finally, in summary, and outlook some of the future perspectives

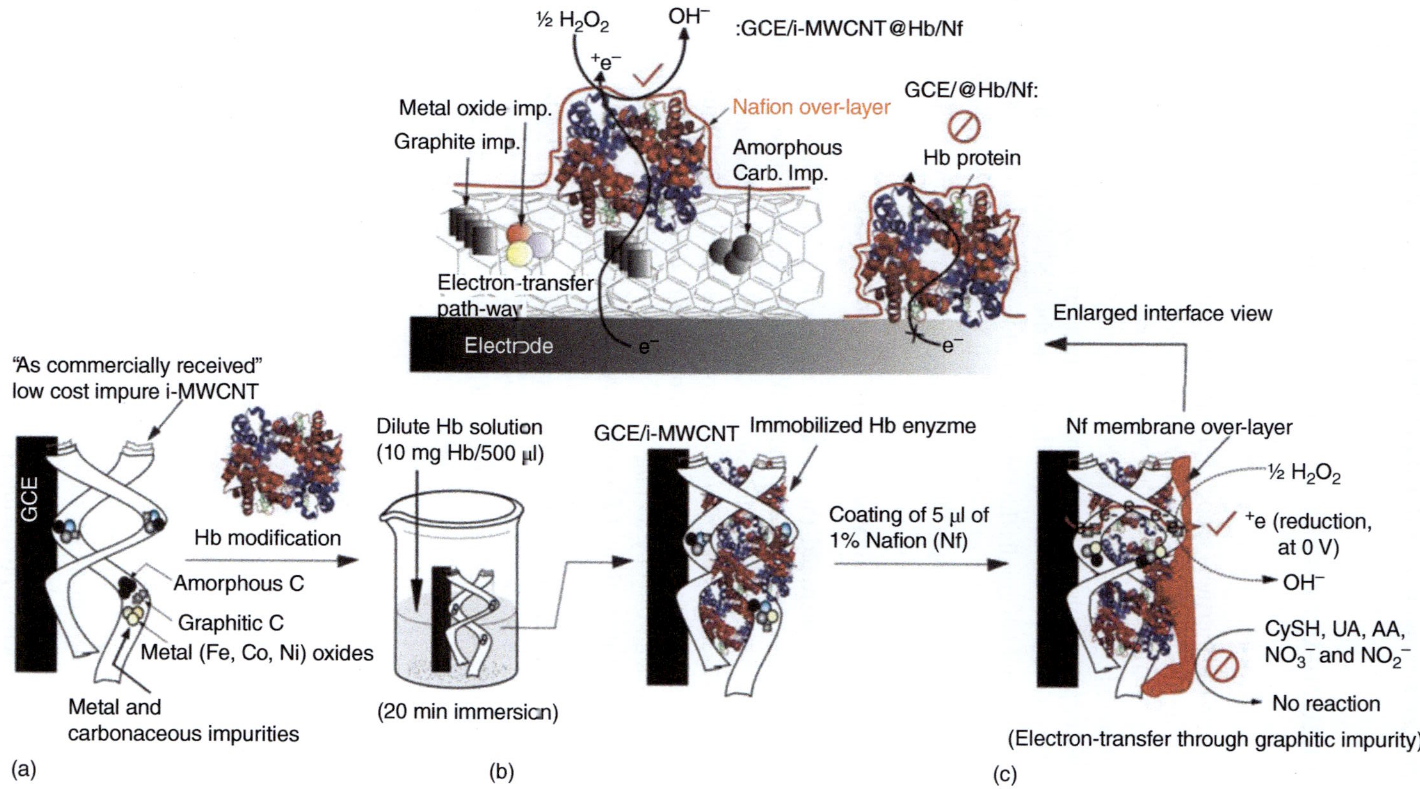

Figure 11.1 Graphical illustration for the construction of MWCNT-electrochemical sensor and its electrocatalytic performance toward selective H_2O_2 reduction. (a) GCE/i-MWCNT, (b) GCE/i-MWCNT@Hb, and (c) GCE/i-MWCNT@Hb/Nf. Source: Senthil Kumar et al. [21]/with permission of American Chemical Society.

in the field are also described by adopting specific types of nanotubes, including miniaturizing sensors and fine-tuning the characteristics of CNT sensors.

11.2 Functionalization of Carbon Nanotubes Toward Sensors

Due to the chemical inertness associated with the stiffness and strong π–π interactions CNTs are found insoluble in common water and solvents [28]. They agglomerate together because of van der Waals interactions resulting in complexity to form dispersion [29]. The major concern is not only on how to prepare CNTs to dissolve in water and ordinary solvents but also to avert re-agglomeration by separating individual CNTs. To resolve these limitations the CNTs' functionalization has been accomplished which entails the chemical or physical connection of functional groups and suitable molecules to the walls or ends of CNTs. The functionalization of CNTs makes them easily dispersible, less toxic, and biocompatible [30].

11.2.1 Covalent Functionalization of CNTs Toward Sensing

Literature survey demonstrated the covalent connections of molecular moieties on the CNTs; however, it evinces to be tricky for active species to interact by means of wholly sp^2-bonded C atoms at the CNTs' sidewalls. Basically, covalent modification of CNTs mostly carried out from organic species treated with –COOH of oxidation-reacted CNTs. Covalent functionalization can be achieved by two approaches; (i) through directly adding active species or functional groups to the CNTs' sidewalls (ii) interaction of functionalized functional groups at the ends of CNTs. The most frequent approach employed to functionalize CNTs is the oxidation process which ensures the insertion of carboxyl groups on the sidewall of CNTs. The carboxyl groups produce amide and ester bonds which allow molecular species to couple covalently. The CNTs are refluxed by treating with acid followed by exposure to a mixture of H_2SO_4:HNO_3 (3 : 1) at higher power sonification for up to six hours [31] (Figure 11.2a). It has been exhibited that functional groups associated with nanotubes via covalent bond amend the H-bond via lowering van der Waals forces thus promoting the solubility and segregation of individual CNTs [32]. Further, an amino acid such as cysteine, which possessed a higher affinity toward some heavy metals used to covalently modify carbonylated CNTs [33]. It has been found that heavy metals ions can be encapsulated over electrochemical electrodes modified by the cysteine-modified CNTs. Electrochemical and photochemical functionalization approaches have also been evaluated as other types of covalent modification. In photochemical functionalization, photoirradiation has been utilized to generate active nitrenes type species in the sidewall reactions [34] (Figure 11.2b). On the other side, the electrochemical functionalization, a continuous current or potential is provided to the CNT electrode engrossed in an appropriate reagent solution (Figure 11.2c). This led to producing reactive species through the reagent due to the transfer of electrons emerging from the CNT and reagent. In this regard, small

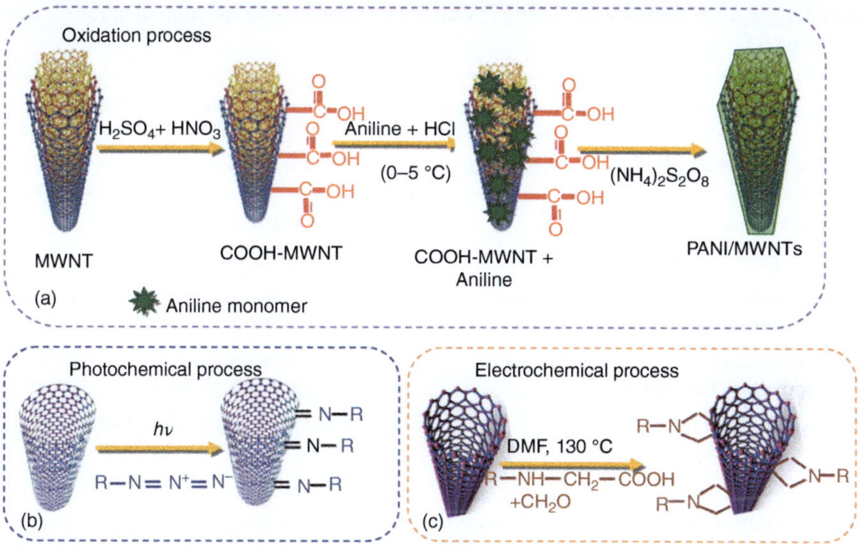

Figure 11.2 (a) Oxidation, (b) photochemical, and (c) electrochemical modification of CNTs under covalent functionalization approach.

diameter nanotubes were produced through the electrochemical reduction of aryl diazonium salts. It was reported that CNTs functionalized with 4-*tert*-butylbenzene species accomplished improved solubility in the organic medium [35]. Further, covalently functionalized CNTs-based biosensors have investigated the capacity to analyze and detect biological analytes. For example, a copper nanoparticle (CuNP) immobilized on functionalized multi-walled-carbon nanotubes (f-MWCNTs) was employed to develop an amperometric electrochemical sensor for the sensing of H_2O_2, which displayed good repeatability, high stability, and reproducibility [36].

11.2.2 Noncovalent Functionalization of CNTs Toward Sensing

In comparison to the covalent functionalization, noncovalent functionalization of the nanotubes does not escort to considerable modifications in the mechanical and electronic properties and chemical structure [14]. Since different organic molecules and functional groups could be connected to the CNTs' surface without hampering the π-system of the graphene backbone consequently protecting the physical and chemical features of nanotubes. Therefore, functionalization of CNTs through noncovalent reactions is relatively attractive and has imperative outcomes toward their electrochemical performances. Table 11.1 represents the comparison among noncovalent and covalent modification [34]. The comprehension of noncovalent functionalization is mostly attributed to the connection among the sidewalls of nanotubes and the hydrophilic section of the adsorbed molecules via van der Waals, π–π, CH–π attraction. The poor dispersibility of CNTs has been the main obstruction in the preparation and successful application of electrochemical sensors. Zhu and coworkers functionalized-CNTs using 1,10-phenanthroline-5,6-dione

Table 11.1 Typical comparison among covalent and noncovalent functionalization approaches.

Difference between covalent and noncovalent functionalization approaches	
Covalent functionalization	**Noncovalent functionalization**
Mode of hybridization is sp^3 type	Mode of hybridization is sp^2 type
Structure can be compromised	Retention of structure
Polymer/molecules or functional groups connected at the end or side walls of CNTs	Polymer/molecular or functional groups wrapping onto surface of CNTs
The procedure employs thioles, hydrogen, amides, oxygen, halogen, etc.	The procedure employs biomolecules, polymers, nanoparticles, surfactants, etc.
Employs covalent interaction	Employs van der Waals interaction

(a redox mediator) in order to enhance the electrocatalytic performances of nanotubes for the oxidation of nicotinamide adenine dinucleotide (NADH) [37]. They suggested that the investigation of NADH could be accomplished with a high sensitivity (8.77 mA mM^{-1}) and negligible interference at 0.0 V potential. By focusing on π–π interaction among CNTs and phenyl compounds, Wan and coworkers reported a sensitive material based on triphenylene-functionalized CNTs toward electrochemical sensing of 2,4,6-trinitrotoluene trace [38]. The obtained results revealed that triphenylene is a good receptor to 2,4,6-trinitrotoluene. The triphenylene-functionalized CNTs lead to an enhanced detecting unit with additional receptor positions to 2,4,6-trinitrotoluene and produced improved sensitivity and response.

11.2.3 Polymers Wrapping of CNTs Toward Sensing

Among organic material functionalized-CNTs, polymers have attained huge consideration because of their relatively proficient dispersity, which can interrupt the van der Waals force of attraction among the CNTs' side walls. Over the past decade, sincere efforts have been carried out to prepare polymer-functionalized-CNTs-based materials by synergistically combining the advantageous characteristics of individual materials [39]. Polymer wrapping shows the process of solubilizing and functionalizing CNTs, in which the synergistic interaction can offer a stable composition with the desired properties. Using the sonication approach CNTs could be proficiently dispersed in an aqueous medium in the existence of single-stranded deoxyribonucleic acid (DNA) due to the presence of multiple constructive π–π interactions [40]. Polysaccharides or saccharides are able to solubilize and functionalize-CNTs, whereas hydrophobic contact with polymers possessing characteristics of surfactant can provide solubility in aqueous medium. Basically, CNTs are insoluble in an aqueous starch solution but found miscible in the starch–iodine complex solution. Star et al. ascribed these findings as a consequence of the pre-organization of amylose in starch into a helical conformation through iodine. Iodine' substitution within the helix through CNTs using mechanism of "pea-shooting" produces amylose-wrapped nanotubes. The hydrophilic dispersions of starch-wrapped-CNTs might endure sparked disassembly through enzymatic hydrolysis by amyloglucosidase [41]. A wide variety of biopolymers [42, 43] have been utilized for the modification of CNTs [44]. Swager and coworkers described that conjugated polymers combined with selector side chains can offer selectiveness for the sensing of desired analytes and also able to determine closely similar structural isomers [45–47]. According to the literature survey, polymer-wrapped-CNTs (polymers such as poly(3-methylthiophene), poly(acrylic acid), polypyrrole (PPy), poly(neutral red) (PNR), and poly(methylene blue)) has become a favorable approach for the detection of DA [48–52]. In a report, onto the vinyl group f-MWCNTs' surface, an insulin-imprinted polymer was prepared through a cross-linker and phosphatidylcholine-comprising functional monomers [53]. Then, an electrochemical sensor has been engineered by modifying MWCNT-molecularly imprinted polymer over an electrode made of pencil graphite and utilized for sensing of insulin trace.

11.2.4 CNTs Decorated with Metal Nanoparticles Toward Sensing

Inorganic nanomaterial modified CNTs have also attained huge consideration. Typically, two types of inorganic nanomaterials i.e. noble metal nanoparticles (MNPs) for example, Pd, Au, Ag, Pt, , and metal oxide nanostructures including SnO_2, CuO, and ZnO have been used to functionalize CNTs. Moreover, various other nanomaterials have also been suggested as electrochemical sensing resources and used as electrode modifiers [54]. Although various types of electrodes based on un-modified pure CNTs have been studied. It is essential to communicate those residual particles and metal catalysts that remain from the fabrication procedure of CNT have been attributed for the detection of impure CNTs to definite compounds [55]. Therefore, the intended encapsulation of metal nanoparticles (NPs) shows a dynamic route for offering stronger or specific responses. MNPs have appeared as a novel set of nanomaterials that are mainly promising for materials engineering because of their sole magnetic, optical, catalytic, and electronic properties [56]. For fabricating the hybrid composition of nanotubes with metal oxide and MNPs, two key routes have been carried out. In the first route with the help of linkers, the pre-formed nanoparticles are connected to CNTs. In the second route, bare NPs are synthesized or directly dispersed over the nanotubes' surface. Herein; CNTs most probably act as templates in tailoring the size of MNPs or metal oxide as well behaving as a supporting matter. Due to the incorporation of nanotubes, MNPs were dispersed uniformly onto the CNTs' surface, which significantly enhanced their reactive surface area. Additionally, the exclusive characteristics of individual materials for such a developed hybrid might be combined or highly improved. Accordingly, the metal nanoparticles-based CNTs hybrid composite with electrochemical electrodes might be estimated to possess high sensitivity and enhanced catalytic activity. Unmodified nanotubes are insensible to hydrogen gas; however, evaporation of electrons beams over 5 Å of Pd onto chemical vapor deposition-grown nanotubes networks or single CNT offer sensitivity toward hydrogen gas at room temperature in air at a 4–400 ppm level. This enhanced sensitivity is due to the H_2 dissociation into atomic hydrogen onto Pd. In this progression, Pd work function is reduced and consequently stimulates electron transfer from palladium to nanotubes which led to a reduction in the carrier density and conductivity of p-type CNTs [57]. Sun et al. reported that CNT chemiresistive systems modified with palladium NPs can be constructed on the flexible substrate of poly(ethylene terephthalate). The fabricated system was found stable up to 1000 cycles of relaxing and bending without noticeable degradation of sensing activity [58].

11.3 Electrochemical Sensing Applications of CNTs

Biosensors and diagnostic devices based on CNTs and their composites have been utilized for very sensitive determination of analytes in environmental monitoring pharmaceutical industries, food quality, and biological examination. For example, CNTs-based electrochemical sensors have been engineered to detect metabolites, biomarkers of ions and proteins [59]. Sensors are miniaturized or small systems that have found widespread utilizations and are constructed for the regular monitoring of the biochemical or physicochemical characteristics of selective compounds so as to produce quantitative/or qualitative information. Typically, three-electrode system is involved in the electrochemical sensors; the working electrodes (WEs) (transfers/commute a bio-recognition response into an electrochemical response, the counter electrode [CE] and the reference electrode [RE]). Electrochemical biosensors are admired because of their relatively quick response times, small size, simple handling, and low-cost. CNT owing to its excellent electrochemical properties and small size have been established to be either superior or equal to most other electrodes to determine protein biomarkers, metabolites, and ions [34]. The enzyme immobilization on the WE is accountable for identification thus contributing to voltage or creating a current response [60]. According to the previous reports [61], the electrochemical sensors can be typically divided into bio-enzymatic (nanotubes-forest electrodes with immobilized enzymes, nanotubes-modified electrodes with immobilized enzymes, and nanotubes-paste enzyme electrodes) and bio-affinity biosensors (nanotubes-immunosensors and CNTs-DNA-based biosensors). Electrochemical sensors contain detectors such as conductimetric, amperometric, or potentiometric cells, which produce signals in response to the specific analytes present or through electrochemical reactions of the analytes. Since the last few decades, CNTs drew considerable attention and become the preference for the development of sensors because of their remarkable physicochemical properties and superior strength, as a consequence of their strangely large diameter-to-length ratios (as high as 1 : 132 000 000) and exclusive tubular nanostructure [62]. In addition, CNTs are coupled with excellent chemical stability, outstanding conductivity, good biocompatibility, and high sensitivity [63]. Moreover, the sidewalls and ends of CNTs can be straightforwardly functionalized through connecting nearly any preferred chemical moieties. In nanoscale sensors, CNTs could be brilliant transducers because of their remarkable sensitivity. According to the literature survey, CNTs could improve the electrochemical performance of vital biomolecules, and arbitrate rapid kinetics of electron transfer for a broad variety of electroactive materials [64]. There are two main approaches through which CNTs could be employed for electrochemical detection applications: (i) immobilization of the CNTs/or their derivatives onto the surface of an electrode, (ii) development of CNTs arrangement on a substrate and utilizing them as the sensor itself. The former one is fast, simple, controllable, and most frequent technique and commonly implies dispersing the preferred CNTs into an aqueous, organic, and other electrolyte medium followed by dispersing the nanomaterials through continuous mechanical agitation using an ultrasonic bath. An aliquot is then collected and deposited onto the desired WE, where the evaporation of solvent leaves the nanomaterials immobilized. When CNTs are immobilized onto the surface of electrode, since the defects along the tube axis and the end of the open tubes are the sources of electron transfer, the quantity of these positions significantly states the electrochemical activity of a sensor fabricated from CNTs. This shows that the location of the voltammetric signal related to the employed materials will move in the voltammogram from large to low potentials based on the load of edge plane-like defects or sites. An abundance of electrochemical sensor based on CNTs have been developed, and Table 11.2 represents a selection reported in present times where the CNTs response were found to be better than that of the fundamental electrode materials.

Table 11.2 CNT-based electrochemical sensors for the determination of different analytes.

Electrode	Analyte	LOD	References
MWCNT/poly(3,4-ethylenedioxythiophene)/GCE	Superoxide	1 μM	[65]
FePt/CNTs	Glutathione	1.0 nM	[66]
MWCNTs-COOH	Acephatemet	0.1054 pM	[67]
CNT aerogel electrode	DNA hybridization	1 pM	[68]
Polyaniline (PANI)/CNTs/Cu nanoparticles	Phthalate esters	0.03–0.08 nM	[69]
CNTs/tyrosinase	Phenol	35 nM	[70]
CNT-Co_3O_4	Glucose	0.16 μM	[71]
Polymers/CNT	Ethanol	100 μM	[72]
Laccase/CNT	Para-cresol	0.05 ppm	[73]
β-Cyclodextrin/MWCNT	Glucose in honey	0.42 μM	[74]
BiOI-Bi/CNTs	Micro ribonucleic acid (miRNA)	0.1 pM	[75]
Hemoglobin-collagen-CNTs	Hydrogen peroxide	0.91 μM	[76]
Au/MWCNT/glassy carbon electrode (GCE)	Cholesterol	0.5 μM	[77]

11.3.1 CNT-Based Sensors for Environment Protection

The pledge of low-powered, economical sensing system has projected the investigation of CNT-based electrochemical sensors for various applications, such as environmental health monitoring. Determination of toxic compounds that are seriously injurious to environmental and human fitness in both liquid and gas phase produces continuous objections toward sensitivity and selectivity of sensors. Pure CNTs have been exposed to combine through several gaseous analytes, and there are continued investigations are underway in the modification of nanotubes to inflate the performance and analyte range. Particularly the type of the contacts among the CNTs and the analytes is significant; therefore innovative functional synthetic selectors incorporated with nanotubes should be focused. Rahman fabricated cadmium sulfide (CdS) nanoparticles decorated CNT nanocomposites (CNT NCs) film in the basic phase using a simple solution method [78]. CdS-CNT NCs were employed to modify glassy carbon electrode (GCE) and the assembly was utilized to investigate lethal benzaldehyde (BZ). In electrochemical approach, CdS-CNT NCs were employed to construct the highly selective as well as sensitive benzaldehyde chemical sensor probe with an ultra-low limit of detection (LOD) (2.2 ± 0.1 pM) for the monitoring of healthcare and environmental area. Using the advantage of same technique, Rahman et al. synthesized CuS decorated CNT nanocomposite (CuS-CNT NC) again in an alkaline medium through the solution method [79]. This time harmful 1,2-diaminobenzene (DAB) opted as the desired analyte using the selectivity analysis that exhibited a fast signal and really low LOD (4.3 ± 0.05 pM) at the CuS-CNT/GCE sensor. They found this approach is budding as a proficient route in constructing an efficient electrochemical sensor to manage environmental pollution on a large scale. The bare drug waste (cytotoxic dosage) is regularly mixing on aquatic bodies through pharma industries and wastewater treatment plants. Therefore, prior sensing and proper dumping of these harmful drug remain with accurate and reliable are in immense need. In this regard, Alagumalai et al. engineered an electrochemical sensor based on Au doped oxidized-SWCNT and graphene oxide (Au@f-CNT/GO) and used for the sensitive detection of the Naproxen (NPZ) a therapeutic drug [80]. They reported that Au@*f*-CNT/GO/GCE represents higher electrocatalytic activity and low resistance toward sensing of NPZ. Considerable recovery outcomes of up to 99% in real sample and LOD of 14 nM (S/N = 3) strongly support that the Au@*f*-CNT/GO has great ability for the sensitive determination of NPZ in the different water and biological samples. Deshmukh et al. described the electrochemical investigations of Hg(II), Pb(II), and Cu(II), through ethylenediaminetetraacetic acid (EDTA), modified SWCNTs and polyaniline (PANI) based nanocomposite (PANI/SWNCTs) [81]. PANI/SWCNTs was electrochemically developed on stainless steel (SS) electrode through cyclic voltammetry technique followed by modifying with EDTA in the existence of an activating agent (1-ethyl-3(3-(dimethylamin propyl)carbodiimide). Analytical characteristic suggested high sensitivity and selectivity with LOD toward Cu(II), Hg(II), and Pb(II) was determined as 0.08, 0.68, and 1.65, respectively. Govindhan et al. [82] described a mediator-free electrochemical sensor constructed using

SWCNT-modified GCEs for the electrocatalytic oxidation of catechol, *p*-cresol, and *p*-nitrophenol. The engineered electrochemical sensor allow the sensitive determination of *p*-cresol, *p*-nitrophenol, and catechol, with the lowest detection limits of 3.7, 7.7, and 2.3 nM and enhanced sensitivities of 94.90, 110.38, and 135.08 $\mu A\,\mu M^{-1}$, respectively. They also found that the proposed electrochemical sensor exhibited good reproducibility and high stability along with quick response, offering new opportunities for the fabrication of sensing platforms toward the determination of multiple phenolic compounds together for environmental monitoring applications.

11.3.2 CNT-Based Sensors for Pharmaceutical Applications

Rapid and extremely sensitive detection of β-agonist, namely salbutamol can be accomplished through a micro-fabricated flow injection device using CNT/Au as a WE [83]. The acid-f-MWCNTs modified GCE was utilized for simultaneous detection of DA and acetaminophen (AP) without any interference from NADH, uric acid (UA), and ascorbic acid (AA) [84]. The proposed MWCNT-based electrode determined good reproducibility (2.3% [AP] and 1.3% [DA]), stability, high recovery, and repeatability (1.9%) in human serum (1.9% [AP] and 1.7% [DA]) and pharmaceutical preparation (2.7% [AP] and 1.7% [DA]). In biological fluids, for instance, urine and blood, UA exists together with DA and AA. Electrochemical investigation using unmodified GCE possessed limitations due to the overlapping of the oxidation voltages of UA, DA, and AA, and a prominent fouling consequence [85]. This can be overcome with the help of modified electrode for example utilization of poly(phenazine)/f-MWCNTs modified GCE for the determination of UA, with PNR [50] or poly(malachite green) (PMalG) [86]. These CNT-modified GCEs utilize differential pulse voltammetry (DPV) for the detection of epinephrine, in diverse pH (4.0 and 7.0). Using the fabricated electrode, the linear ranges detection was found to be 0.2 or 0.09 mM, with LOD of 0.3 mM. β-cyclodextrin incorporated CNT-modified electrodes (β-CD-CNT/Es) were developed for individual or simultaneous detection of adenine and guanine [87]. It was also found that β-CD-CNT/E allows efficient small-potential amperometric bio-detection of DNA with the LOD of 0.2 and 0.1 $\mu g\,ml^{-1}$ for adenine and guanine, respectively. For the selective determination of clindamycin, a unique sensitive molecularly imprinted biosensor was formed using the MWCNT layer with a thin molecularly imprinted sol–gel film [88]. The MWCNT imprinted sensor gratified the sensitivity, selectivity, simplicity, and speed necessities for the determination of clindamycin, and produce the probabilities of clinical trials in physiological fluids. MWCNT modified electrode based amperometric sensor fabricated using a mediator such as ferrocenedicarboxylic acid was employed for the concurrent detection of folic acid and 6-thioguanine [89]. The MWCNT modified electrode showed a proficient electrocatalytic potential for the oxidation of folic acid and 6-thioguanine at pH 9.0. In another report, an electrochemical sensor based on f-MWCNT modified GCE (f-MWCNT/GCE) was employed for the sensitive detection of flutamide (an anticancer drug) in artificial urine and pharmaceutical samples [90].

11.3.3 Monitoring of Biomolecular Compounds

To date, CNTs have been extensively reported with a very high affinity to immobilize different enzymes for definite biomolecules and the prepared composites with modified CNTs as electrochemical sensors could also be employed for biomolecular sensing [91]. Literature survey suggested that CNTs-based electrochemical sensors have effectively demonstrated the sensing of glucose, H_2O_2, microorganisms, DNA, biomarkers, and other biomolecules so on [91].

11.3.3.1 Glucose Sensor

Since the controlling of diabetes is of uttermost concern for society, individuals, and health-care personnel therefore monitoring of glucose levels is earned a great deal of interest, as it is taking a serious economic toll. The estimated count goes to around ~285 to 439 million individuals that will get affected through diabetes from 2010 to 2030, respectively [92]. The importance of glucose monitoring can be estimated from the fact that alone glucose sensors cover around 85% shares (~US$ 8.5 billion) of the biosensor market. A wide range of CNTs-based glucose biosensors has been constructed in recent times. For enzyme-free electrochemical detection of glucose, a sensing interface was constructed using the mixture of GO and MWCNTs decorated with gold–platinum (Au@Pt) core–shell nanoparticles [93]. In the modified GCE, the GO-dispersed MWCNTs efficiently avoid the aggregation of the MWCNTs and improve the specific surface area and Au@Pt NPs offer the synergistic electrocatalytic potential for the detection of glucose. The fabricated MWCNTs based sensor probe toward the detection of glucose demonstrated a low LOD (4.2×10^{-8} M) and two wide linear ranges from 1×10^{-4} M to 2.5×10^{-3} M and 5×10^{-8} to 1×10^{-4} M [93]. Wan et al. described an amperometric biosensor for glucose through PANI and chitosan (CS) coupled CNT-COOH (CS-CNTs) as the signal enhancers and glucose oxidase (GOx) for the detection of the glucose on Au electrode (Au-g-PANI-*c*-(CS-CNTs)-GOD) [94]. Further, GOD was covalently coupled with the pendant –OH of CS by 1,4-carbonyldiimidazole (Figure 11.3). They found that Au-g-PANI-*c*-(CS-CNTs)-GOD electrode offers an enhanced structural biocompatible microenvironment to improve the biocatalytic activity and amount of the enzyme immobilized and to enhance mediate the electron transfer. The experimental outcomes of the prepared biosensor demonstrated a linear response to glucose from 1 to 20 mM concentration range, good reproducibility, good sensitivity (21 $\mu A\,mM^{-1}\,cm^{-2}$), and preservation of >80% of the initial current response after 60 days (Figure 11.4). In another report, a highly densified carbon nanotube fiber (HD-CNTf) electrochemically modified with CuNPs-based microsensor was employed for the electrooxidation of glucose [95]. The microsensor was effectively employed for the quantification of glucose in diabetic urine of patients and normal human serum samples. Ni-based metal-organic framework (Ni-MOF) in the presence of functional CNTs solution (Ni-MOF/CNTs) was constructed via in situ self-assembly and employed as an efficient electrocatalytic oxidation candidate for the detection of glucose [96]. Ni-MOF/CNTs composite was employed as a sensitive nonenzymatic glucose sensor, which exhibits a low LOD (0.82 μM) over a broad

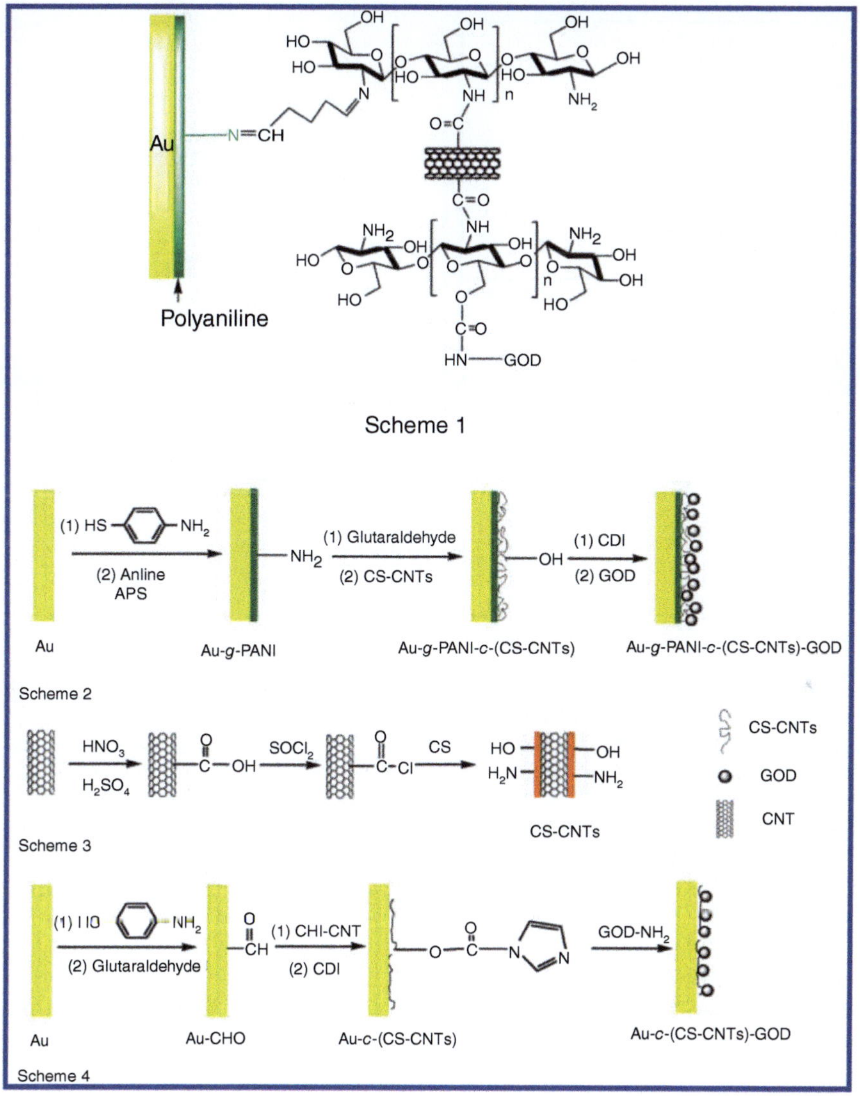

Figure 11.3 Possible illustration displaying the fabrication process of the (Schemes 1–3) Au-*g*-PANI-*c*-(CS-CNTs)-GOD, and (Scheme 4) Au-*c*-(CS-CNTs)-GOD biosensors. Source: Wan et al. [94]/with permission of American Chemical Society.

linear range (1 μM to 1.6 mM). An extremely sensitive biosensor for glucose based on nanobiocomposite film consisting of GOx, f-MWCNTs, and PPy was developed electrochemically [84]. The amperometric responses of the glucose biosensor attributed to a sensitivity of 95 nA mM^{-1}, a response time of about eight seconds, and a linear range up to 4 mM [97]. A flexible glucose biosensor is also fabricated using O_2 plasma-f-MWCNT films on polydimethylsiloxane (PDMS) substrate. The fabricated electrode demonstrates good electrocatalytic performance, a low LOD 0.01 mM, and a good sensitivity of 18.15 μA mm^{-2} mM^{-1} [98].

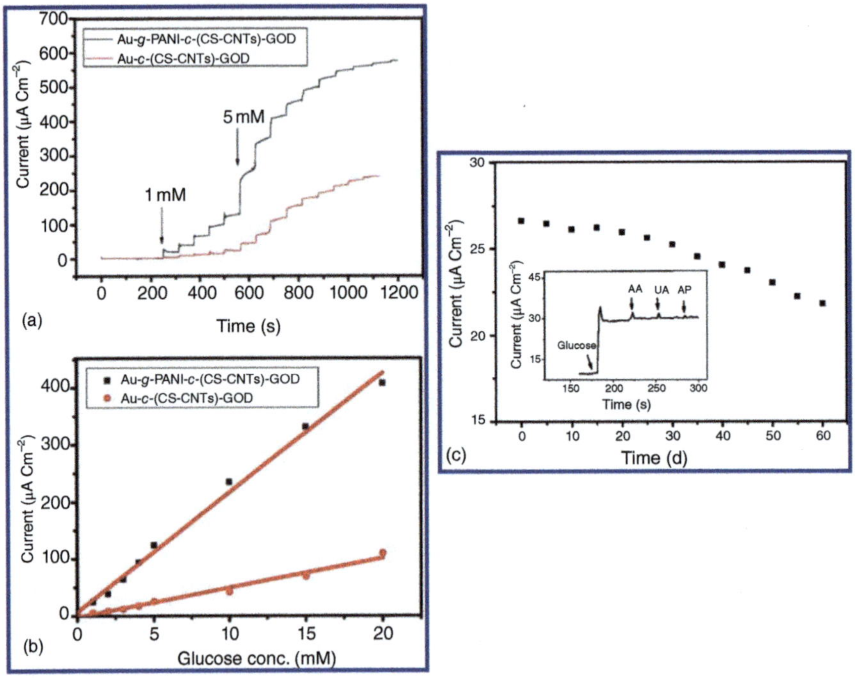

Figure 11.4 (a) Amperometric analyses of the fabricated biosensors, (b) the linear regression response (concentration–current curves) of the glucose, and (c) current response of the Au-*g*-PANI-*c*-(CS-CNTs)-GOD biosensor (at glucose concentration of 10 mM) as a function 60 days of store. The inset displays the interferants effect on the amperometric signals of the Au-*g*-PANI-*c*-(CS-CNTs)-GOD biosensor to 1.0 mM glucose, in the occurrence of acetaminophen (AP, 0.3 mM), UA (0.3 mM), and ascorbic acid (AA, 0.3 mM), in phosphate-buffered saline (PBS) solution of pH 6.0 under 0.5 V. Source: Wan et al. [94]/with permission of American Chemical Society.

11.3.3.2 DNA Sensor

Currently, the sensing of DNA is a part of great attention in food safety, pharmacogenetics, criminology, pathology, genetics, etc. The majority of DNA sensors are constructed by the single-stranded DNA immobilization over the surface of the electrode labeled by means of an electrochemical indicator to distinguish its corresponding objective progression. CNTs are potential candidates for the fabrication of electrochemical sensors for the detection of DNA. Noncovalent and covalent both linkage of CNTs with DNA have been studied where the latter offers the high selectivity, availability, and best stability throughout competing for hybridization [99]. Figure 11.5 illustrates an outline of the covalently linked progression. Through this linkage, it was attributed that DNA strands are attainable to the hybridization and firmly support hybridization among the moieties possessing complementary sequences in comparison to the noncomplementary sequences. Prakash et al. proposed for the first time, the fabrication of an ultrasensitive label-free biosensor device with a fully integrated standalone-CNT aerogel electrode for the determination of DNA hybridization [68]. The LOD for target DNA was found to be 1 pM and

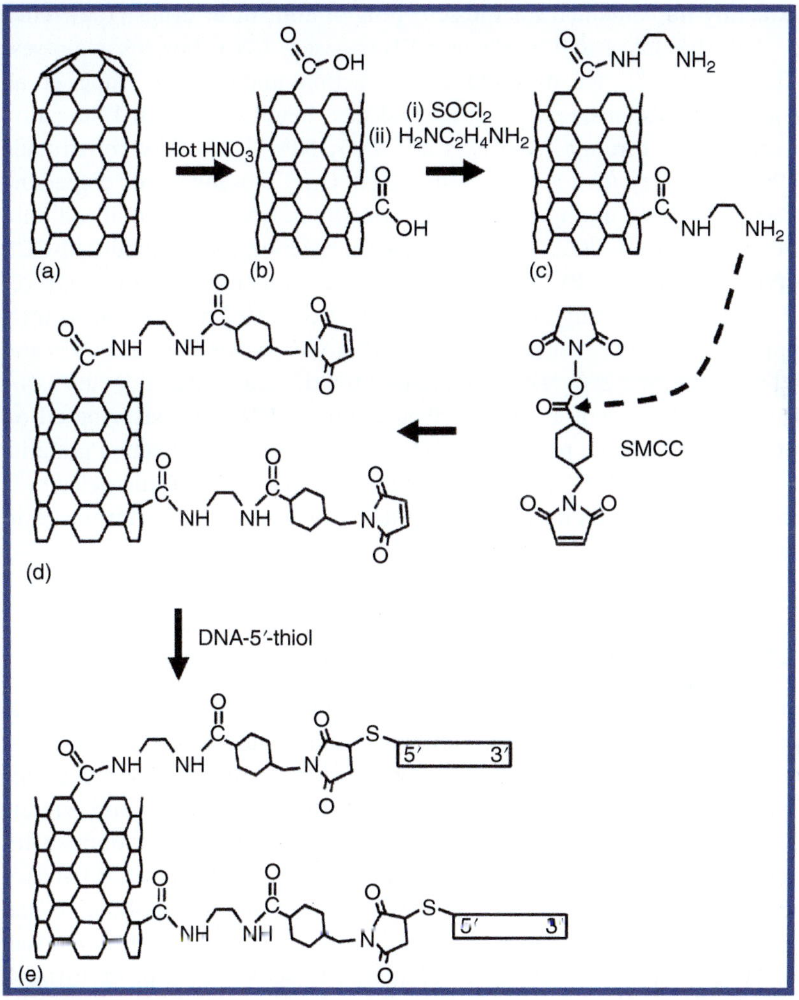

Figure 11.5 Overall illustration for construction of covalently linked DNA-COOH-SWCNT adducts. Source: Baker et al. [99]/with permission of American Chemical Society.

the total detection time required for a single measurement was 18 minutes. Ma et al. proposed CNT-transistors as biosensors for cancer biomarker determination, which has revealed greater selectivity and sensitivity [100]. The fabricated biosensors possessed concentration-dependent DNA detection with a theoretical LOD of 2 fM and a large linear detection range of magnitude up to 6 orders. Lu et al. studied the fabrication of flexible paper electrodes based on Ni-MOF composite/AuNPs/CNTs that were employed for specific human immunodeficiency virus (HIV) DNA detection via DNA hybridization through a redox indicator (methylene blue [MB]) [101]. Bolat proposed a biosensor for DNA based on electrodeposited cetyl trimethylammonium bromide-MWCNTs (CTAB-MWCNTs) nanocomposite on graphite electrodes for the straight sensing of double stranded DNA (dsDNA) and subsequently implemented for the screening of anticancer drugs [102]. They suggested that the fabricated surfactant-CNTs-based DNA biosensor possess the significant potential for different biointeraction applications. Yang et al. proposed a highly responsive and unique electrochemical DNA hybridization biosensor engineered through MWCNTs and ZrO_2 NPs [103]. They constructed the MWCNTs/nanoZrO_2/chitosan (CS)-modified GCE biosensor via dispersing ZrO_2 NPs and MWCNTs in oligonucleotides and chitosan were immobilized onto the surface of modified a GCE. The hybridization reaction on the modified-GCE was supervised through DPV response, which demonstrated a low LOD of 7.5×10^{-11} mol l^{-1} in the range of 1.49×10^{-10} to 9.32×10^{-8} mol l^{-1}.Following a similar approach, Jiao and coworkers proposed a DNA biosensor using ZnO NPs and MWCNTs by substituting ZrO_2NPs for enhanced hybridization detection and DNA immobilization [104]. Erdem et al. described a unique DNA biosensor based on the improvement of guanine response at MWCNTs-functionalized pencil graphite electrode (PGE) through DPV [105]. The modified PGE acted as a microelectrode probe associated with its high porosity and displayed enhanced performance in comparison to GCE.

11.3.4 CNTs-Based Sensors for Real Sample Analysis

Probably, toxic pollutants, sewage discharge, expired or unused products, and undisposed byproducts from the different industries, households, and health centers were directly disposing to ecosystems [2]. Long-term exposure of these pollutants into the environment causes a critical health risk to human beings and aquatic animals [3]. Therefore, prior detection of these desired chemical compounds with efficient reliability and accuracy is in great need. To meet this demand, CNTs-based electrochemical sensors could be successfully employed for the analyses of real samples in different fields for instance agriculture, food, fishing, and biomedical industries. Hudari et al. described a voltammetric sensor developed through MWCNTs and chitosan composites modified GCE for simultaneous determination of resorcinol (RSN) and *p*-phenylenediamine (PPD) in tap water samples [106]. The fabricated voltammetric sensor possessed an excellent catalytic performance toward PPD and RSN detection in tap water samples with a recovery rate of 97.08 to 102.19%. Rahman et al. described a MWCNT-based amperometric lactate biosensor for the sensing of the lactate concentration in commercial milk and the attained results were compared with a biochemical analyzer [107]. The obtained result of the fabricated biosensor was established to be in parallel with the outcomes from the biochemical analyzer (0.18 ± 0.006 and 0.174 ± 0.01 mM, respectively). Further, according to the literature survey CNTs-based electrochemical sensors perform better and are widely used for the application of commercial food sensing for the determination of undesired chemical residues, biomedical sensing applications, toxic dyes, and organic pollutants determination in real samples. Table 11.3 describes the utilization of CNTs-based electrochemical sensors toward real sample investigations.

Table 11.3 Reported CNTs-based electrochemical sensors toward sensing applications in real samples.

Sensor	Real sample	Analyte	LOD	References
ZnS/Au/f-MWCNT	Water samples	4-Nitrophenol	30 nM	[108]
SWCNT modified Au electrode	Drug	Rutin	0.10 nM	[109]
Modified-CNT electrode	Water samples	Hydroxylamine	1 μM	[110]
β-Cyclodextrin-MWCNTs-PANI-GCE	Injection	Dopamine	12 nM	[111]
MWCNT-film-modified electrode	Soft drinks	Quinoline yellow	0.5 μM	[24]
ZnO/MWCNTs-chitosan	Water samples	*p*-Nitrophenol	1.0 nM	[112]
MWCNT-SO_3H	Water	Cd^{2+}	0.08 μM	[113]
Hydroxylated-MWCNT	Water	Catechol and hydroquinone	0.074 and 0.076 μM	[114]
AuNP/MWCNT/GCE	Water	Bisphenol A	4 nM	[115]
$BMIMPF_6$-SWCNT/GCE	Apple/lake water	Methylparathion	1.00 nM	[116]
Pt nano-clusters/MWCNTs/GCE	Serum	Estrogen	0.18 μM	[117]
Acid chrome blue K/MWCNT	Water	Dihydroxybenzene	0.10 μM	[118]

11.4 Summary and Outlook

CNTs are emerging as potential dynamic candidates and are widely utilized in the construction of electrochemical sensors/biosensors. CNTs-modified electrodes have a lot of merits among the other carbon-based electrodes because of their great chemical stability, huge precise surface area, small size, brilliant mechanical strength, and excellent electrical and thermal conductivity. Although, there have a few limitations of CNT-modified electrodes for sensing application, the continuous emerging research progress in this area is contributing to conquer these drawbacks. Functionalized CNTs covering covalent and noncovalent modifications (inorganic, organic, organic–inorganic), polymer wrapping, and metal nanoparticles hybrids employed for electrochemical sensors applications have been briefly summarized in this book chapter. Through modifications, the electrochemical performances of nanotubes and their derivatives have been significantly improved. As transducers, immobilizers, carriers, and templates functionalized CNTs have emerged among the most promising sensing materials to functionalize electrodes toward electrochemical sensing. In this book chapter, functionalized-CNTs for the determination of several compounds have been described via summarizing lots of reported typical examples. In recent time, functionalized CNTs-based electrochemical biosensors for sensing several biomolecules is still a burning research field. Remarkably, it is apparent that functionalized CNTs-based electrochemical sensors have been an emerging trend for environmental monitoring and pharmaceutical applications. Notably, for the application of determining unfavorable analytes in real samples such as persistent organic pollutants, heavy metal ions, unused drugs, etc. functionalized CNTs have exhibited various promising advantages. Apparently, it is critical to evaluate efficient functionalization methods to heterogeneously modified CNTs and control their electronic characteristics without incorporating suitable compounds. Analytical detection at CNTs-modified electrodes reveals high sensitivities, low LOD, resistance to surface fouling, and reduction of overpotentials. These wonderful merits of CNTs produce them an exhilarating substitute for the engineering of unique electrochemical biosensors and sensors. However, despite having lots of progress in this field, CNTs still encounter several challenges that need to be investigated to execute the requirements of desired application of CNTs for sensors. The profitable development of defect-free and pure CNTs is expensive and complicated. Further, the production of CNTs is still not wholly controlled, for instance, tubes aggregations are not barred, nonspecific adsorptions of compounds to the nanotubes' walls are not interrupted and the uniformity in the tubes' lengths are not obtainable. Due to their harmful effects and toxicity in the lung, CNTs can be responsible for health risks because of agglomeration which leads to suffocation. CNTs toxicological effect is an obstruction for the utilization of CNT in bioelectronics and every subject incorporated through living biological structures. The insoluble nature of most of the solvents is another limitation of CNTs that significantly delayed their desired practical applications. Nevertheless, the employment of functionalized-CNTs improves the electrochemical sensitivity of electrodes for a variety of analytes detection and ultimately permit the construction of an efficient sensor, but the issues with stability and reproducibility are frequent in most works, and therefore, more precise research is needed on this aspect, in order to determine the complexities and their probable commercial implications. Various reports have confirmed that the results on a laboratory scale are encouraging. However, the real implications need more trials which have to be carried out efficiently for real-world samples. The development of actual technology based on functionalized-CNTs-based electrochemical sensors requires comprehensive trials in precise terms and the detailed matrix also desires to be optimized. It is recognized that the performance of functionalized-CNTs represents diverse electrochemical responses which could impact the determination of target analyte in the real matrix analyses. Therefore, understanding of these processes on the electrode surface needs to be investigated decisively as it facilitates feasible device development. Despite the above observations on current status, the extraordinary promising talent of modified-CNTs-based sensing applications is evident from the previous reports [119]. The literature survey evidently signifies the high sensitivity reachable for different analytes with a very low LOD [91]. It appears technology transfer activities to industry based on the transfer of this knowledge from universities are actively underway. The continuous research on the drawbacks of CNT discussed above and good industrial attention are expected to bring the potential of CNT-based sensors in near future. Therefore, it is so trustworthy to articulate that the genuine sensing applications of CNTs-based electrochemical sensors lounge ahead in the future.

Important Websites

https://www.nanocyl.com
https://www.ctimaterials.com/product-category/industrial-carbon-nanotubes-products
https://www.zdnet.com/article/5-surprising-uses-for-carbon-nanotubes
https://now.northropgrumman.com/carbon-nanotube-applications-in-daily-life
https://www.businesswire.com/news/home/20220105005766/en/Global-Carbon-Nanotubes-Market-Research-Report-2021-with-Forecasts-to-2031—ResearchAndMarkets.com

References

1 Yang, W., Guo, H., Xue, R. et al. (2021). 0.2CNT/$NiSe_x$ composite derived from CNT/MOF-74 as electrode material for electrochemical capacitor and electrochemical sensor. *Microchem. J.* 168: 106519. https://linkinghub.elsevier.com/retrieve/pii/S0026265X21006056.

2 Kant, T., Shrivas, K., Dewangan, K. et al. (2022). Design and development of conductive nanomaterials for electrochemical sensors: a modern approach. *Mater. Today Chem.* 24: 100769. https://linkinghub.elsevier.com/retrieve/pii/S2468519421003499.

3 Shi, H., Fu, L., Chen, F. et al. (2022). Preparation of highly sensitive electrochemical sensor for detection of nitrite in drinking water samples. *Environ. Res.* 209: 112747. https://linkinghub.elsevier.com/retrieve/pii/S0013935122000743.

4 Lu, H., He, B., and Gao, B. (2021). Emerging electrochemical sensors for life healthcare. *Eng. Regener.* 2: 175–181. https://linkinghub.elsevier.com/retrieve/pii/S266613812100027X.

5 Sheng, K., Jiang, H., Fang, Y. et al. (2022). Emerging electrochemical biosensing approaches for detection of allergen in food samples: a review. *Trends Food Sci. Technol.* 121: 93–104. https://linkinghub.elsevier.com/retrieve/pii/S0924224422000474.

6 Avelino, K.Y.P.S., Oliveira, L.S., Lucena-Silva, N. et al. (2021). Flexible sensor based on conducting polymer and gold nanoparticles for electrochemical screening of HPV families in cervical specimens. *Talanta* 226: 122118. https://linkinghub.elsevier.com/retrieve/pii/S0039914021000394.

7 Rivas, G., Rubianes, M., Rodriguez, M. et al. (2007). Carbon nanotubes for electrochemical biosensing. *Talanta* 74 (3): 291–307. https://linkinghub.elsevier.com/retrieve/pii/S0039914007006832.

8 De Volder, M.F.L., Tawfick, S.H., Baughman, R.H., and Hart, A.J. (2013). Carbon nanotubes: present and future commercial applications. *Science (80-)* 339 (6119): 535–539. https://doi.org/10.1126/science.1222453.

9 Chen, A. and Chatterjee, S. (2013). Nanomaterials based electrochemical sensors for biomedical applications. *Chem. Soc. Rev.* 42 (12): 5425. http://xlink.rsc.org/?DOI=c3cs35518g.

10 Zhu, L., Yang, R., Zhai, J., and Tian, C. (2007). Bienzymatic glucose biosensor based on co-immobilization of peroxidase and glucose oxidase on a carbon nanotubes electrode. *Biosens. Bioelectron.* 23 (4): 528–535. https://linkinghub.elsevier.com/retrieve/pii/S0956566307002801.

11 Tasis, D., Tagmatarchis, N., Bianco, A., and Prato, M. (2006). Chemistry of carbon nanotubes. *Chem. Rev.* 106 (3): 1105–1136. https://doi.org/10.1021/cr050569o.

12 Holzinger, M., Vostrowsky, O., Hirsch, A. et al. (2001). Sidewall functionalization of carbon nanotubes. *Angew. Chem. Int. Ed.* 40 (21): 4002. https://doi.org/10.1002/1521-3773(20011105)40:21<4002::AID-ANIE4002>3.0.CO;2-8.

13 Zhang, H., Guo, H., Deng, X. et al. (2010). Functionalization of multi-walled carbon nanotubes via surface unpaired electrons. *Nanotechnology* 21 (8): 085706. https://doi.org/10.1088/0957-4484/21/8/085706.

14 Tuncel, D. (2011). Non-covalent interactions between carbon nanotubes and conjugated polymers. *Nanoscale* 3 (9): 3545. http://xlink.rsc.org/?DOI=c1nr10338e.

15 Karousis, N., Tagmatarchis, N., and Tasis, D. (2010). Current progress on the chemical modification of carbon nanotubes. *Chem. Rev.* 110 (9): 5366–5397. https://doi.org/10.1021/cr100018g.

16 Santhosh, P., Manesh, K.M., Gopalan, A., and Lee, K.-P. (2007). Novel amperometric carbon monoxide sensor based on multi-wall carbon nanotubes grafted with polydiphenylamine—fabrication and performance. *Sens. Actuators, B* 125 (1): 92–99. https://linkinghub.elsevier.com/retrieve/pii/S0925400507000597.

17 Du, P., Liu, S., Wu, P., and Cai, C. (2007). Preparation and characterization of room temperature ionic liquid/single-walled carbon nanotube nanocomposites and their application to the direct electrochemistry of heme-containing proteins/enzymes. *Electrochim. Acta* 52 (23): 6534–6547. https://linkinghub.elsevier.com/retrieve/pii/S0013468607005816.

18 Sato, N. and Okuma, H. (2008). Development of single-wall carbon nanotubes modified screen-printed electrode using a ferrocene-modified cationic surfactant for amperometric glucose biosensor applications. *Sens. Actuators, B* 129 (1): 188–194. https://linkinghub.elsevier.com/retrieve/pii/S0925400507005898.

19 Balasubramanian, K. and Burghard, M. (2008). Electrochemically functionalized carbon nanotubes for device applications. *J. Mater. Chem.* 18 (26): 3071. http://xlink.rsc.org/?DOI=b718262g.

20 Zhang, M., Liu, K., Gong, K. et al. (2005). Continuous on-line monitoring of extracellular ascorbate depletion in the rat striatum induced by global ischemia with carbon nanotube-modified glassy carbon electrode integrated into a thin-layer radial flow cell. *Anal. Chem.* 77 (19): 6234–6242. https://doi.org/10.1021/ac051188d.

21 Senthil Kumar, A., Gayathri, P., Barathi, P., and Vijayaraghavan, R. (2012). Improved electric wiring of hemoglobin with impure-multiwalled carbon nanotube/nafion modified glassy carbon electrode and its highly selective hydrogen peroxide biosensing. *J. Phys. Chem. C* 116 (44): 23692–23703. https://doi.org/10.1021/jp3064933.

22 Winiarski, J.P., Rampanelli, R., Bassani, J.C. et al. (2020). Multi-walled carbon nanotubes/nickel hydroxide composite applied as electrochemical sensor for folic acid (vitamin B9) in food samples. *J. Food Compos. Anal.* 92: 103511. https://linkinghub.elsevier.com/retrieve/pii/S0889157520304452.

23 Yang, J., He, D., Zhang, N., and Hu, C. (2022). Disposable carbon nanotube-based antifouling electrochemical sensors for detection of morphine in unprocessed coffee and milk. *J. Electroanal. Chem.* 905: 115997. https://linkinghub.elsevier.com/retrieve/pii/S1572665721010249.

24 Zhao, J., Zhang, Y., Wu, K. et al. (2011). Electrochemical sensor for hazardous food colourant quinoline yellow based on carbon nanotube-modified electrode. *Food Chem.* 128 (2): 569–572. https://linkinghub.elsevier.com/retrieve/pii/S0308814611004481.

25 Britto, P.J., Santhanam, K.S.V., and Ajayan, P.M. (1996). Carbon nanotube electrode for oxidation of dopamine. *Bioelectrochem. Bioenerg.* 41 (1): 121–125. https://linkinghub.elsevier.com/retrieve/pii/0302459896050787.

26 Sarkar, B., Mandal, S., Tsang, Y.F. et al. (2018). Designer carbon nanotubes for contaminant removal in water and wastewater: a critical review. *Sci. Total Environ.* 612: 561–581.

27 Meyyappan, M. (2016). Carbon nanotube-based chemical sensors. *Small* 12 (16): 2118–2129.

28 Onyancha, R.B., Aigbe, U.O., Ukhurebor, K.E., and Muchiri, P.W. (2021). Facile synthesis and applications of carbon nanotubes in heavy-metal remediation and biomedical fields: a comprehensive review. *J. Mol. Struct.* 1238: 130462. https://linkinghub.elsevier.com/retrieve/pii/S0022286021005950.

29 Islam Rubel, R., Hasan Ali, M., Abu Jafor, M., and Mahmodul, A.M. (2019). Carbon nanotubes agglomeration in reinforced composites: a review. *AIMS Mater. Sci.* 6 (5): 756–780. http://www.aimspress.com/article/10.3934/matersci.2019.5.756.

30 Marchesan, S., Kostarelos, K., Bianco, A., and Prato, M. (2015). The winding road for carbon nanotubes in nanomedicine. *Mater. Today* 18 (1): 12–19. https://linkinghub.elsevier.com/retrieve/pii/S1369702114002594.

31 Abdulla, S., Mathew, T.L., and Pullithadathil, B. (2015). Highly sensitive, room temperature gas sensor based on polyaniline-multiwalled carbon nanotubes (PANI/MWCNTs) nanocomposite for trace-level ammonia detection. *Sens. Actuators, B* 221: 1523–1534. https://linkinghub.elsevier.com/retrieve/pii/S0925400515301805.

32 Dyke, C.A. and Tour, J.M. (2004). Covalent functionalization of single-walled carbon nanotubes for materials applications. *J. Phys. Chem. A* 108 (51): 11151–11159. https://doi.org/10.1021/jp046274g.

33 Morton, J., Havens, N., Mugweru, A., and Wanekaya, A.K. (2009). Detection of trace heavy metal ions using carbon nanotube-modified electrodes. *Electroanalysis* 21 (14): 1597–1603. https://doi.org/10.1002/elan.200904588.

34 Gupta, S., Murthy, C.N., and Prabha, C.R. (2018). Recent advances in carbon nanotube based electrochemical biosensors. *Int. J. Biol. Macromol.* 108: 687–703. https://linkinghub.elsevier.com/retrieve/pii/S0141813017324224.

35 Bahr, J.L., Yang, J., Kosynkin, D.V. et al. (2001). Functionalization of carbon nanotubes by electrochemical reduction of aryl diazonium salts: a bucky paper electrode. *J. Am. Chem. Soc.* 123 (27): 6536–6542. https://doi.org/10.1021/ja010462s.

36 Rezaei, B., Askarpour, N., Ghiaci, M. et al. (2015). Synthesis of functionalized MWCNTs decorated with copper nanoparticles and its application as a sensitive sensor for amperometric detection of H_2O_2. *Electroanalysis* 27 (6): 1457–1465. https://doi.org/10.1002/elan.201400612.

37 Mao, X., Wu, Y., Xu, L. et al. (2011). Electrochemical biosensors based on redox carbon nanotubes prepared by noncovalent functionalization with 1, 10-phenanthroline-5,6-dione. *Analyst* 136 (2): 293–298. http://xlink.rsc.org/?DOI=C0AN00402B.

38 Zhang, H.-X., Hu, J.-S., Yan, C.-J. et al. (2006). Functionalized carbon nanotubes as sensitive materials for electrochemical detection of ultra-trace 2,4,6-trinitrotoluene. *Phys. Chem. Chem. Phys.* 8 (30): 3567. http://xlink.rsc.org/?DOI=b604587c.

39 Ma, P.-C., Siddiqui, N.A., Marom, G., and Kim, J.-K. (2010). Dispersion and functionalization of carbon nanotubes for polymer-based nanocomposites: a review. *Composites, Part A* 41 (10): 1345–1367. https://linkinghub.elsevier.com/retrieve/pii/S1359835X10002009.

40 Zheng, M., Jagota, A., Semke, E.D. et al. (2003). DNA-assisted dispersion and separation of carbon nanotubes. *Nat. Mater.* 2 (5): 338–342. http://www.nature.com/articles/nmat877.

41 Star, A., Joshi, V., Han, T.-R. et al. (2004). Electronic detection of the enzymatic degradation of starch. *Org. Lett.* 6 (13): 2089–2092. https://doi.org/10.1021/ol0495826.

42 Cella, L.N., Chen, W., Myung, N.V., and Mulchandani, A. (2010). Single-walled carbon nanotube-based chemiresistive affinity biosensors for small molecules: ultrasensitive glucose detection. *J. Am. Chem. Soc.* 132 (14): 5024–5026. https://doi.org/10.1021/ja100503b.

43 Zhang, M. and Gorski, W. (2005). Electrochemical sensing platform based on the carbon nanotubes/redox mediators-biopolymer system. *J. Am. Chem. Soc.* 127 (7): 2058–2059. https://doi.org/10.1021/ja044764g.

44 Star, A., Liu, Y., Grant, K. et al. (2003). Noncovalent side-wall functionalization of single-walled carbon nanotubes. *Macromolecules* 36 (3): 553–560. https://doi.org/10.1021/ma021417n.

45 Fennell, J., Hamaguchi, H., Yoon, B., and Swager, T. (2017). Chemiresistor devices for chemical warfare agent detection based on polymer wrapped single-walled carbon nanotubes. *Sensors* 17 (5): 982. http://www.mdpi.com/1424-8220/17/5/982.

46 Ishihara, S., Azzarelli, J.M., Krikorian, M., and Swager, T.M. (2016). Ultratrace detection of toxic chemicals: triggered disassembly of supramolecular nanotube wrappers. *J. Am. Chem. Soc.* 138 (26): 8221–8227. https://doi.org/10.1021/jacs.6b03869.

47 Ishihara, S., O'Kelly, C.J., Tanaka, T. et al. (2017). Metallic versus semiconducting SWCNT chemiresistors: a case for separated SWCNTs wrapped by a metallosupramolecular polymer. *ACS Appl. Mater. Interfaces* 9 (43): 38062–38067. https://doi.org/10.1021/acsami.7b12992.

48 Hsueh, C. and Brajter-Toth, A. (1994). Electrochemical preparation and analytical applications of ultrathin overoxidized polypyrrole films. *Anal. Chem.* 66 (15): 2458–2464. https://doi.org/10.1021/ac00087a009.

49 Yogeswaran, U. and Chen, S.-M. (2008). Multi-walled carbon nanotubes with poly(methylene blue) composite film for the enhancement and separation of electroanalytical responses of catecholamine and ascorbic acid. *Sens. Actuators, B* 130 (2): 739–749. https://linkinghub.elsevier.com/retrieve/pii/S0925400507008672.

50 Yogeswaran, U. and Chen, S.-M. (2007). Separation and concentration effect of f-MWCNTs on electrocatalytic responses of ascorbic acid, dopamine and uric acid at f-MWCNTs incorporated with poly(neutral red) composite films. *Electrochim. Acta* 52 (19): 5985–5996. https://linkinghub.elsevier.com/retrieve/pii/S0013468607004203.

51 Wang, H.-S., Li, T.-H., Jia, W.-L., and Xu, H.-Y. (2006). Highly selective and sensitive determination of dopamine using a nafion/carbon nanotubes coated poly(3-methylthiophene) modified electrode. *Biosens. Bioelectron.* 22 (5): 664–669. https://linkinghub.elsevier.com/retrieve/pii/S0956566306000583.

52 Liu, A., Honma, I., and Zhou, H. (2007). Simultaneous voltammetric detection of dopamine and uric acid at their physiological level in the presence of ascorbic acid using poly(acrylic acid)-multiwalled carbon-nanotube composite-covered glassy-carbon electrode. *Biosens. Bioelectron.* 23 (1): 74–80. https://linkinghub.elsevier.com/retrieve/pii/S0956566307001820.

53 Prasad, B.B., Madhuri, R., Tiwari, M.P., and Sharma, P.S. (2010). Imprinting molecular recognition sites on multiwalled carbon nanotubes surface for electrochemical detection of insulin in real samples. *Electrochim. Acta* 55 (28): 9146–9156. https://linkinghub.elsevier.com/retrieve/pii/S0013468610011679.

54 Shamsipur, M., Najafi, M., and Hosseini, M.-R.M. (2010). Highly improved electrooxidation of glucose at a nickel(II) oxide/multi-walled carbon nanotube modified glassy carbon electrode. *Bioelectrochemistry* 77 (2): 120–124. https://linkinghub.elsevier.com/retrieve/pii/S1567539409001364.

55 Schnorr, J.M. and Swager, T.M. (2011). Emerging applications of carbon nanotubes. *Chem. Mater.* 23 (3): 646–657. https://doi.org/10.1021/cm102406h.

56 Georgakilas, V., Gournis, D., Tzitzios, V. et al. (2007). Decorating carbon nanotubes with metal or semiconductor nanoparticles. *J. Mater. Chem.* 17 (26): 2679. http://xlink.rsc.org/?DOI=b700857k.

57 Kong, J., Chapline, M.G., and Dai, H. (2001). Functionalized carbon nanotubes for molecular hydrogen sensors. *Adv. Mater.* 13 (18): 1384–1386. https://doi.org/10.1002/1521-4095(200109)13:18<1384::AID-ADMA1384>3.0.CO;2-8.

58 Sun, Y. and Wang, H.H. (2007). High-performance, flexible hydrogen sensors that use carbon nanotubes decorated with palladium nanoparticles. *Adv. Mater.* 19 (19): 2818–2823. https://doi.org/10.1002/adma.200602975.

59 Wang, Z. and Dai, Z. (2015). Carbon nanomaterial-based electrochemical biosensors: an overview. *Nanoscale* 7 (15): 6420–6431. http://xlink.rsc.org/?DOI=C5NR00585J.

60 Tîlmaciu, C.-M. and Morris, M.C. (2015). Carbon nanotube biosensors. *Front. Chem.* 3: http://journal.frontiersin.org/Article/10.3389/fchem.2015.00059/abstract.

61 Sireesha, M., Jagadeesh Babu, V., Kranthi Kiran, A.S., and Ramakrishna, S. (2018). A review on carbon nanotubes in biosensor devices and their applications in medicine. *Nanocomposites* 4 (2): 36–57. https://doi.org/10.1080/20550324.2018.1478765.

62 Wang, X., Li, Q., Xie, J. et al. (2009). Fabrication of ultralong and electrically uniform single-walled carbon nanotubes on clean substrates. *Nano Lett.* 9 (9): 3137–3141. https://doi.org/10.1021/nl901260b.

63 Hu, C. and Hu, S. (2009). Carbon nanotube-based electrochemical sensors: principles and applications in biomedical systems. *J. Sens.* 2009: 1–40. http://www.hindawi.com/journals/js/2009/187615.

64 Wang, J. (2005). Carbon-nanotube based electrochemical biosensors: a review. *Electroanalysis* 17 (1): 7–14. https://doi.org/10.1002/elan.200403113.

65 Braik, M., Barsan, M.M., Dridi, C. et al. (2016). Highly sensitive amperometric enzyme biosensor for detection of superoxide based on conducting polymer/CNT modified electrodes and superoxide dismutase. *Sens. Actuators, B* 236: 574–582. https://linkinghub.elsevier.com/retrieve/pii/S092540051630884X.

66 Karimi-Maleh, H., Tahernejad-Javazmi, F., Ensafi, A.A. et al. (2014). A high sensitive biosensor based on FePt/CNTs nanocomposite/*N*-(4-hydroxyphenyl)-3,5-dinitrobenzamide modified carbon paste electrode for simultaneous determination of glutathione and piroxicam. *Biosens. Bioelectron.* 60: 1–7. https://linkinghub.elsevier.com/retrieve/pii/S095656631400236X.

67 Li, X., Zhao, Y., Chen, J. et al. (2022). Ultra-sensitive acetylcholinesterase biosensor based on leaching $Al_{42}Cu_{40}Fe_{16}$ quasicrystal modified by MWCNTs-COOH for the determination of organophosphorus pesticide. *Microchem. J.* 176: 107214. https://linkinghub.elsevier.com/retrieve/pii/S0026265X2200042X.

68 Prakash, J., Dey, A., Uppal, S. et al. (2021). Label-free rapid electrochemical detection of DNA hybridization using ultrasensitive standalone CNT aerogel biosensor. *Biosens. Bioelectron.* 191: 113480. https://linkinghub.elsevier.com/retrieve/pii/S0956566321005170.

69 Annamalai, J. and Vasudevan, N. (2020). Detection of phthalate esters in PET bottled drinks and lake water using esterase/PANI/CNT/CuNP based electrochemical biosensor. *Anal. Chim. Acta* 1135: 175–186. https://linkinghub.elsevier.com/retrieve/pii/S0003267020309740.

70 Wee, Y., Park, S., Kwon, Y.H. et al. (2019). Tyrosinase-immobilized CNT based biosensor for highly-sensitive detection of phenolic compounds.

Biosens. Bioelectron. 132: 279–285. https://linkinghub.elsevier.com/retrieve/pii/S0956566319301897.

71 Çakıroğlu, B. and Özacar, M. (2018). A self-powered photoelectrochemical glucose biosensor based on supercapacitor Co_3O_4-CNT hybrid on TiO_2. *Biosens. Bioelectron.* 119: 34–41. https://linkinghub.elsevier.com/retrieve/pii/S0956566318305529.

72 Barsan, M.M., Pifferi, V., Falciola, L., and Brett, C.M.A. (2016). New CNT/poly(brilliant green) and CNT/poly(3,4-ethylenedioxythiophene) based electrochemical enzyme biosensors. *Anal. Chim. Acta* 927: 35–45. https://linkinghub.elsevier.com/retrieve/pii/S0003267016305220.

73 Zhao, K., Veksha, A., Ge, L., and Lisak, G. (2021). Near real-time analysis of Para-cresol in wastewater with a laccase-carbon nanotube-based biosensor. *Chemosphere* 269: 128699. https://linkinghub.elsevier.com/retrieve/pii/S0045653520328976.

74 Xia, J., Zou, B., Liu, F. et al. (2022). Sensitive glucose biosensor based on cyclodextrin modified carbon nanotubes for detecting glucose in honey. *J. Food Compos. Anal.* 105: 104221. https://linkinghub.elsevier.com/retrieve/pii/S088915752100421X.

75 Xiao, J., Deng, K., Liu, Z. et al. (2021). A photocathode based on BiOI-Bi/CNTs for microRNA detection coupling with target recycling strand displacement amplification. *Sens. Actuators, B* 348: 130691. https://linkinghub.elsevier.com/retrieve/pii/S0925400521012594.

76 Li, J., Mei, H., Zheng, W. et al. (2014). A novel hydrogen peroxide biosensor based on hemoglobin-collagen-CNTs composite nanofibers. *Colloids Surf., B* 118: 77–82. https://linkinghub.elsevier.com/retrieve/pii/S0927776514001672.

77 Haritha, V.S., Kumar, S.R.S., and Rakhi, R.B. (2022). Amperometric cholesterol biosensor based on cholesterol oxidase and Pt–Au/MWNTs modified glassy carbon electrode. *Mater. Today: Proc.* 50: 34–39. https://linkinghub.elsevier.com/retrieve/pii/S2214785321021477.

78 Rahman, M.M. (2020). In-situ preparation of cadmium sulphide nanostructure decorated CNT composite materials for the development of selective benzaldehyde chemical sensor probe to remove the water contaminant by electrochemical method for environmental remediation. *Mater. Chem. Phys.* 245: 122788. https://linkinghub.elsevier.com/retrieve/pii/S025405842030167X.

79 Rahman, M.M., Ahmed, J., and Asiri, A.M. (2022). Ultra-sensitive, selective, and rapid carcinogenic 1,2-diaminobenzene chemical determination using sol–gel coating low-dimensional facile CuS modified-CNT nanocomposites by electrochemical approach. *Microchem. J.* 175: 107230. https://linkinghub.elsevier.com/retrieve/pii/S0026265X22000583.

80 Alagumalai, K., Musuvadhi Babulal, S., Chen, S.-M. et al. (2021). Electrochemical evaluation of naproxen through Au@f-CNT/GO nanocomposite in environmental water and biological samples. *J. Ind. Eng. Chem.* 104: 32–42. https://linkinghub.elsevier.com/retrieve/pii/S1226086X21004573.

81 Deshmukh, M.A., Celiesiute, R., Ramanaviciene, A. et al. (2018). EDTA_PANI/SWCNTs nanocomposite modified electrode for electrochemical determination of copper (II), lead (II) and mercury (II) ions. *Electrochim. Acta* 259: 930–938. https://linkinghub.elsevier.com/retrieve/pii/S0013468617322533.

82 Govindhan, M., Lafleur, T., Adhikari, B.-R., and Chen, A. (2015). Electrochemical sensor based on carbon nanotubes for the simultaneous detection of phenolic pollutants. *Electroanalysis* 27 (4): 902–909. https://doi.org/10.1002/elan.201400608.

83 Karuwan, C., Wisitsoraat, A., Maturos, T. et al. (2009). Flow injection based microfluidic device with carbon nanotube electrode for rapid salbutamol detection. *Talanta* 79 (4): 995–1000. https://linkinghub.elsevier.com/retrieve/pii/S0039914009001507.

84 Alothman, Z.A., Bukhari, N., Wabaidur, S.M., and Haider, S. (2010). Simultaneous electrochemical determination of dopamine and acetaminophen using multiwall carbon nanotubes modified glassy carbon electrode. *Sens. Actuators, B* 146 (1): 314–320. https://linkinghub.elsevier.com/retrieve/pii/S0925400510001280.

85 Chen, J. and Cha, C. (1999). Detection of dopamine in the presence of a large excess of ascorbic acid by using the powder microelectrode technique. *J. Electroanal. Chem.* 463 (1): 93–99. https://linkinghub.elsevier.com/retrieve/pii/S0022072898004513.

86 Raoof, J.B., Ojani, R., and Baghayeri, M. (2013). Fabrication of layer-by-layer deposited films containing carbon nanotubes and poly(malachite green) as a sensor for simultaneous determination of ascorbic acid, epinephrine, and uric acid. *Turkish J. Chem.* 37 (1): 36–50.

87 Wang, Z., Xiao, S., and Chen, Y. (2006). β-Cyclodextrin incorporated carbon nanotubes-modified electrodes for simultaneous determination of adenine and guanine. *J. Electroanal. Chem.* 589 (2): 237–242. https://linkinghub.elsevier.com/retrieve/pii/S0022072806001057.

88 Zhang, Z., Hu, Y., Zhang, H., and Yao, S. (2010). Novel layer-by-layer assembly molecularly imprinted sol–gel sensor for selective recognition of clindamycin based on Au electrode decorated by multi-wall carbon nanotube. *J. Colloid Interface Sci.* 344 (1): 158–164. https://linkinghub.elsevier.com/retrieve/pii/S0021979709015598.

89 Ensafi, A.A. and Karimi-Maleh, H. (2010). Modified multiwall carbon nanotubes paste electrode as a sensor for simultaneous determination of 6-thioguanine and folic acid using ferrocenedicarboxylic acid as a mediator. *J. Electroanal. Chem.* 640 (1–2): 75–83. https://linkinghub.elsevier.com/retrieve/pii/S1572665710000135.

90 Farias, J.S., Zanin, H., Caldas, A.S. et al. (2017). Functionalized multiwalled carbon nanotube electrochemical sensor for determination of anticancer drug flutamide. *J. Electron. Mater.* 46 (10): 5619–5628. https://doi.org/10.1007/s11664-017-5630-6.

91 Pundir, M., Prasher, P., Vasić, K. et al. (2021). Enzyme modified CNTs for biosensing application: opportunities and challenges. *Colloid Interface Sci. Commun.* 44: 100506. https://linkinghub.elsevier.com/retrieve/pii/S2215038221001461.

92 Shaw, J.E., Sicree, R.A., and Zimmet, P.Z. (2010). Global estimates of the prevalence of diabetes for 2010 and 2030. *Diabetes Res. Clin. Pract.* 87 (1): 4–14. https://linkinghub.elsevier.com/retrieve/pii/S016882270900432X.

93 Wang, R., Liu, X., Zhao, Y. et al. (2022). Novel electrochemical non-enzymatic glucose sensor based on 3D Au@Pt core–shell nanoparticles decorated graphene oxide/multi-walled carbon nanotubes composite. *Microchem. J.* 174: 107061. https://linkinghub.elsevier.com/retrieve/pii/S0026265X21011474.

94 Wan, D., Yuan, S., Li, G.L. et al. (2010). Glucose biosensor from covalent immobilization of chitosan-coupled carbon nanotubes on polyaniline-modified gold electrode. *ACS Appl. Mater. Interfaces* 2 (11): 3083–3091. https://doi.org/10.1021/am100591t.

95 Gupta, P., Gupta, V.K., Huseinov, A. et al. (2021). Highly sensitive non-enzymatic glucose sensor based on carbon nanotube microelectrode set. *Sens. Actuators, B* 348: 130688. https://linkinghub.elsevier.com/retrieve/pii/S0925400521012569.

96 Zhang, X., Xu, Y., and Ye, B. (2018). An efficient electrochemical glucose sensor based on porous nickel-based metal organic framework/carbon nanotubes composite (Ni-MOF/CNTs). *J. Alloys Compd.* 767: 651–656. https://linkinghub.elsevier.com/retrieve/pii/S0925838818326768.

97 Tsai, Y.-C., Li, S.-C., and Liao, S.-W. (2006). Electrodeposition of polypyrrole–multiwalled carbon nanotube–glucose oxidase nanobiocomposite film for the detection of glucose. *Biosens. Bioelectron.* 22 (4): 495–500. https://linkinghub.elsevier.com/retrieve/pii/S0956566306002855.

98 Lee, J.-Y., Park, E.-J., Lee, C.-J. et al. (2009). Flexible electrochemical biosensors based on O_2 plasma functionalized MWCNT. *Thin Solid Films* 517 (14): 3883–3887. https://linkinghub.elsevier.com/retrieve/pii/S0040609009001631.

99 Baker, S.E., Cai, W., Lasseter, T.L. et al. (2002). Covalently bonded adducts of deoxyribonucleic acid (DNA) oligonucleotides with single-wall carbon nanotubes: synthesis and hybridization. *Nano Lett.* 2 (12): 1413–1417. https://doi.org/10.1021/nl025729f.

100 Ma, S., Zhang, Y., Ren, Q. et al. (2022). Tetrahedral DNA nanostructure based biosensor for high-performance detection of circulating tumor DNA using all-carbon nanotube transistor. *Biosens. Bioelectron.* 197: 113785. https://linkinghub.elsevier.com/retrieve/pii/S0956566321008228.

101 Lu, Q., Su, T., Shang, Z. et al. (2021). Flexible paper-based Ni-MOF composite/AuNPs/CNTs film electrode for HIV DNA detection. *Biosens. Bioelectron.* 184: 113229. https://linkinghub.elsevier.com/retrieve/pii/S0956566321002669.

102 Bolat, G. (2020). Investigation of poly(CTAB-MWCNTs) composite based electrochemical DNA biosensor and interaction study with anticancer drug Irinotecan. *Microchem. J.* 159: 105426. https://linkinghub.elsevier.com/retrieve/pii/S0026265X20317926.

103 Yang, Y., Wang, Z., Yang, M. et al. (2007). Electrical detection of deoxyribonucleic acid hybridization based on carbon-nanotubes/nano zirconium dioxide/chitosan-modified electrodes. *Anal. Chim. Acta* 584 (2): 268–274. https://linkinghub.elsevier.com/retrieve/pii/S0003267006022252.

104 Zhang, W., Yang, T., Huang, D.M., and Jiao, K. (2008). Electrochemical sensing of DNA immobilization and hybridization based on carbon nanotubes/nano zinc oxide/chitosan composite film. *Chin. Chem. Lett.* 19 (5): 589–591. https://linkinghub.elsevier.com/retrieve/pii/S1001841708000910.

105 Erdem, A., Papakonstantinou, P., and Murphy, H. (2006). Direct DNA hybridization at disposable graphite electrodes modified with carbon nanotubes. *Anal. Chem.* 78 (18): 6656–6659. https://doi.org/10.1021/ac060202z.

106 Hudari, F.F., de Almeida, L.C., da Silva, B.F., and Zanoni, M.V.B. (2014). Voltammetric sensor for simultaneous determination of *p*-phenylenediamine and resorcinol in permanent hair dyeing and tap water by composite carbon nanotubes/chitosan modified electrode. *Microchem. J.* 116: 261–268. https://linkinghub.elsevier.com/retrieve/pii/S0026265X14000836.

107 Rahman, M.M., Shiddiky, M.J.A., Rahman, M.A., and Shim, Y.-B. (2009). A lactate biosensor based on lactate dehydrogenase/nictotinamide adenine dinucleotide (oxidized form) immobilized on a conducting polymer/multiwall carbon nanotube composite film. *Anal. Biochem.* 384 (1): 159–165. https://linkinghub.elsevier.com/retrieve/pii/S0003269708006386.

108 Naik, S.S., Lee, S.J., Theerthagiri, J. et al. (2021). Rapid and highly selective electrochemical sensor based on ZnS/Au-decorated f-multi-walled carbon nanotube nanocomposites produced via pulsed laser technique for detection of toxic nitro compounds. *J. Hazard. Mater.* 418: 126269. https://linkinghub.elsevier.com/retrieve/pii/S0304389421012334.

109 Zeng, B., Wei, S., Xiao, F., and Zhao, F. (2006). Voltammetric behavior and determination of rutin at a single-walled carbon nanotubes modified gold electrode. *Sens. Actuators, B* 115 (1): 240–246. https://linkinghub.elsevier.com/retrieve/pii/S0925400505008014.

110 Foroughi, M.M., Beitollahi, H., Tajik, S. et al. (2014). Hydroxylamine electrochemical sensor based on a modified carbon nanotube paste electrode: application to determination of hydroxylamine in water samples. *Int. J. Electrochem. Sci.* 9 (6): 2955–2965.

111 Yin, T., Wei, W., and Zeng, J. (2006). Selective detection of dopamine in the presence of ascorbic acid by use of glassy-carbon electrodes modified with both polyaniline film and multi-walled carbon nanotubes with incorporated β-cyclodextrin. *Anal. Bioanal. Chem.* 386 (7–8): 2087–2094. https://doi.org/10.1007/s00216-006-0845-z.

112 Hu, Y., Zhang, Z., Zhang, H. et al. (2012). Sensitive and selective imprinted electrochemical sensor for *p*-nitrophenol based on ZnO nanoparticles/carbon nanotubes doped chitosan film. *Thin Solid Films* 520 (16): 5314–5321. https://linkinghub.elsevier.com/retrieve/pii/S0040609011020979.

113 Chen, K., Li, J., Zhang, L. et al. (2018). Facile synthesis of self-assembled carbon nanotubes/dye composite films for sensitive electrochemical determination of Cd (II) ions. *Nanotechnology* 29 (44): 445603. https://doi.org/10.1088/1361-6528/aadbf7.

114 Liu, Z., Liao, D., Yu, J., and Jiang, X. (2022). An electrochemical sensor based on oxygen-vacancy cobalt–aluminum layered double hydroxides and hydroxylated multiwalled carbon nanotubes for catechol and hydroquinone detection. *Microchem. J.* 175: 107216. https://linkinghub.elsevier.com/retrieve/pii/S0026265X22000443.

115 Ben Messaoud, N., Ghica, M.E., Dridi, C. et al. (2017). Electrochemical sensor based on multiwalled carbon nanotube and gold nanoparticle modified electrode for the sensitive detection of bisphenol a. *Sens. Actuators, B* 253: 513–522. https://linkinghub.elsevier.com/retrieve/pii/S0925400517311796.

116 Fan, S., Xiao, F., Liu, L. et al. (2008). Sensitive voltammetric response of methylparathion on single-walled carbon nanotube paste coated electrodes using ionic liquid as binder. *Sens. Actuators, B* 132 (1): 34–39. https://linkinghub.elsevier.com/retrieve/pii/S092540050800021X.

117 Lin, X. and Li, Y. (2006). A sensitive determination of estrogens with a Pt nano-clusters/multi-walled carbon nanotubes modified glassy carbon electrode. *Biosens. Bioelectron.* 22 (2): 253–259. https://linkinghub.elsevier.com/retrieve/pii/S095656630600011X.

118 Yang, P., Wei, W., and Yang, L. (2007). Simultaneous voltammetric determination of dihydroxybenzene isomers using a poly(acid chrome blue K)/carbon nanotube composite electrode. *Microchim. Acta* 157 (3–4): 229–235. https://doi.org/10.1007/s00604-006-0682-y.

119 Shukla, P., Saxena, P., Bhatia, V., and Jain, V.K. (2021). Electrostatic deposition and functionalization of CVD grown multiwalled carbon nanotubes for sensitive & selective detection of CO and NOx at room temperature. *Anal. Chim. Acta* 1177: 338766.

12

Chemically Modified Carbon Nanotubes for Lab on Chip Devices

Sudhakar Bansod[1], Tanvir Arfin[2], Reena Singh[3], Shaz Ahmad[3], and Krishna Neeti[3]

[1] *CSIR-Indian Institute of Chemical Technology, Fluoro-Agrochemicals Department, Uppal Road, Tarnaka, Hyderabad, 500007, India*
[2] *Hyderabad Zonal Centre, CSIR-National Environmental Engineering Research Institute (NEERI), Uppal Road, IICT Campus, Tarnaka, Hyderabad, 500007, India*
[3] *Department of Civil Engineering, National Institute of Technology Patna, Patna 800005, India*

Abbreviations

1D	one dimensional
AFM	atomic force microscopy
AMU	atomic mass unit
AOT	dioctyl sodium sulfosuccinate
C	carbon
CCVD	catalytic chemical vapor deposition
CNFET	carbon nanotube field-effect transistor
CNT	carbon nanotubes
CVD	chemical vapor deposition
DEP	dielectrophoresis
DGU	density gradient ultracentrifugation
DMF	dimethylformamide
DOS	disk operating system
DSSCs	dye-sensitized solar cells
DWCNTs	double-walled carbon nanotubes
EL	electroluminescence
FET	field-effect transistor
FTO	fluorine-doped tin oxide
HPA	high-power amplifier
LEDs	light-emitting diodes
LIBs	lithium-ion batteries
LOC	lab-on-a-chip
MNPs	nanomaterials
MWCNTs	multi-walled carbon nanotubes

Chemically Modified Carbon Nanotubes for Commercial Applications, First Edition.
Edited by Jeenat Aslam, Chaudhery Mustansar Hussain, and Ruby Aslam.

NMP	*N*-methyl-2-pyrrolidone
NRAM	nonvolatile random access memory
PCE	photoelectric capacity factor
PPy	polypyrrole
PV	photovoltaic
SPIO	saturation magnetization iron oxide
SWCNTs	single-walled carbon nanotubes
SWCNT-SAa	single-walled carbon nanotube saturable absorbers
VA-SWNT	vertically-aligned single-walled nanotube
VLS	vapor–liquid–solid

12.1 Introduction

The growth and acceptance of new revolutionary materials are advancing faster [1]. In sophisticated ways, essential raw materials are used in many subbranches of analytical chemistry to achieve precise quantitative and qualitative evaluations of hazardous and small amounts of critical compounds [2]. Materials science is regarded as the most crucial factor in increasing human life, as it leads to technological advancement and the creative improvement of substances using a variety of resources [3] Human beings' vital lifestyle and existence have given contemporary materials new specialized standards in the current age of expansion [4]. The substance is highly valued for its potential to impact people's living standards and global growth since life cannot exist without obstacles [5]. As a result, people's whole existence now relies only on the various materials and their fundamental properties [6]. It provided convenience, ease, and outstanding life-sustaining modules, and it met the acceptable criterion overall [7]. Most industrialized nations' economic development and long-term growth have benefited from such material [8]. The usage of raw materials and exciting compounds is investigated. Several methods, tools, and strategies are applied as a developmental component [9]. It is considered a vital vantage point for advancing current science, and it may later be applied to a variety of subjects [10].

The word carbon (C) is originated from the French word charbon and the Latin word "carbo," which means charcoal [11]. Carbon is placed in the fourth position in abundance by mass, and the element is represented by the symbol "C" with the atomic number 6. The twenty-first century is known as the "*carbon age*." Carbon may seem gloomy and black, yet it foreshadows happier days ahead, which might be termed "irony." From an early time, woods are used for cooking, consisting of carbon. Carbon dioxide causes the atmosphere to pollute, but it can clean our environment in its pure state, a piece of good news for future generations. The fourth most prevalent element on earth, C, appears in 15 isotopic systems, with ^{12}C, ^{13}C, and ^{14}C being the most stable. Radiocarbon dating uses the ^{14}C isotope, which has a half-life decay is 5730 years. ^{12}C makes up around 99% of all accessible C and is cast off to describe the atomic mass unit (AMU) [12]. A minimal energy gap exists between valence orbital, almost in the range of 4 eV. As a result, electrons present in the valance shell may effectively mix to produce sp, sp^2, and sp^3 hybrid bonds,

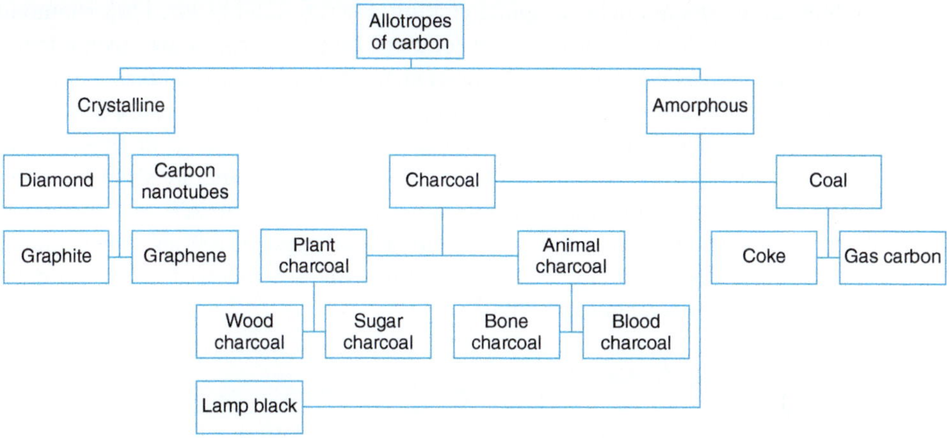

Figure 12.1 Crystallographic structure of the carbon allotropes.

resulting in the production of sigma and pie bonds. Amorphous carbon, diamond, fullerene, and graphite are the well-established allotropes of carbon. Figure 12.1 represents the crystallographic structure of the carbon allotropes.

12.2 Allotropes of Carbon

12.2.1 Diamond

A chemically inactive super insulator, with a melting point that is extremely high, is made up of carbon atoms having sp^3 hybridization. Diamond is well known for its hardness (10 on the Mohs hardness scale), which is achieved due to the intense nature of covalent bond present between the sp^3 hybridized atom of carbon and also has excellent thermal conductivity (900–2320 W $(mk)^{-1}$) [13] as well as optical dispersion having the range of 1–5.5 eV in the optical spectrum [14]. They are very valued in the ornament business because of their potential to produce dazzling, rainbow-like hues. Diamond is also used in drilling, cutting, and grinding industries because of its exceptionally high mechanical and chemical resistance and hardness. The application and properties of diamonds were renewed in many books [15]. The generation of a diamond from graphite at standard pressure and temperature is minimum favorable thermodynamically, owing to the positive value of ΔG^0 ($\Delta G^0 = +2.9\,\text{kJ mol}^{-1}$) [16].

Hannay documented the generation of an artificial diamond at elevated pressure and temperature with the help of a temperature source, light paraffin oil, and a catalyst (Mg, Li, K, and Na) [17]. In 1943, Hannay's result was supported by Bannister and Lonsdale [18]. Burton reported that the ambient pressure helps in the diamond synthesis from the carbon solution in lead–calcium alloy in the temperature range of 550–700 °C [19]. For the generation of diamond crystals at some high temperatures in the range of 900–1100 °C, Eversole suggests chemical vapor deposition (CVD). Atomic hydrogen plays an essential role in the CVD process by lowering the dimension of critical nuclei, graphite etching, surface stability, etc. [20]. Diamond synthesis using the CVD process was followed by discovering novel technologies using microwaves and plasma [21]. Luo et al. used spark plasma sintering to create a diamond from carbon fiber by converting carbon fiber to graphitic sheets and then forming onion-like carbon, which resulted in the development of single diamond crystals. This developed single-crystal combines to make suitable-sized diamond crystals. In the structure of a diamond, each carbon atom is linked with four other carbon atoms through a covalent bond, with a bond angle of 109°28′. The standard crystal is composed of a zinc blende structure with face-centered cubic and bond length of 154 pm between carbon [22].

12.2.2 Graphite

Graphite which has a covered structure, is also an allotrope of carbon, with each carbon atom linked to 2–3 other carbon atoms having sp^2 hybridization. The carbon atoms form a hexagonal arrangement in each layer, having a lattice parameter of 0.142 nm and a distance of 0.335 nm between each plane. The different layers are attached with the help of van der Waals force. Individual graphite layers are called graphene, and they are made of a carbon-forming honeycomb-like structure. There are many ways to synthesize graphene, such as CVD and mechanical exfoliation. [23]. Graphene oxide is created by oxidizing graphene with an appropriate oxidizing agent, forming hydroxyl and epoxy groups, carboxylic acid throughout the layer [24]. The addition of functionalities gives graphene oxide a hydrophilic property. As a result, they disperse readily in organic solvents and water. When these functional groups are removed from graphene oxide by chemical, thermal, or other means, the graphene oxide decreases [25]. Since different layers can easily slide on each other; graphite is mainly used for lubrication purposes.

12.2.3 Fullerenes

Kroto et al. discovered fullerenes, generally in the shape of hollow carbon spheres, tubes, or ellipsoids. They are composed of hexagonal and pentagonal carbon atom rings that are joined to one another, and the pentagonal ring separates them from graphene in terms of structural properties. Spherical fullerenes are named bulky balls, and the most common among them is C60 molecules composed of 20 hexagons and 12 pentagons [26]. Fullerenes have an organized cage structure with a carbon atom in even numbers. The following are some of the most common fullerenes synthesis methods

- Irradiation of carbon source (graphite) using a pulsed laser and helium as buffer gas [27].
- Generated directly with laser irradiation from polycyclic hydrocarbons [28].
- Resistive arc [29].
- Electric arc [30].

Fullerenes appeal to researchers because of their chemical stability, wide surface area, and excellent conductivity.

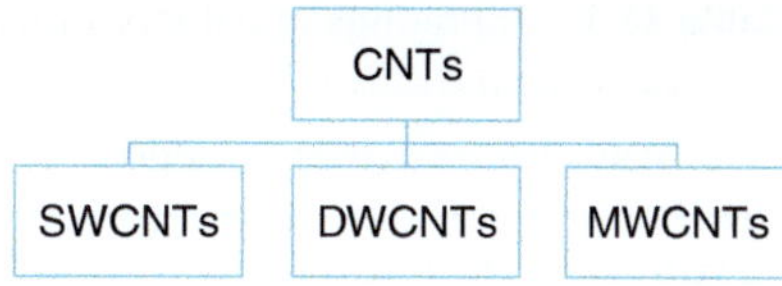

Figure 12.2 Classification of CNTs.

12.2.4 Carbon Nanotubes

Carbon nanotubes (CNTs) were also famous as bulky tubes in the early days. They are extended cylindrical fullerenes having a length in microns and sometimes in millimeters and having a diameter of the order nanometer. The carbon atom is linked together trigonally in a sheet having a curved structure that finally constitutes a hollow cylinder. Depending upon the number of graphite layers present, CNTs are mainly classified as single-walled carbon nanotubes (SWCNTs), double-walled carbon nanotubes (DWCNTs), and multi-walled carbon nanotubes (MWCNTs) as shown in Figure 12.2. The nature of bonding in CNTs is of sp^2 hybridized bonds, the same as in graphite. CNTs show a very high aspect ratio, more than 1000, and a large surface area of more than 1300 $m^2\ g^{-1}$ due to their unique structure of tubular form with minimal diameters. Moreover, the weight and density of CNTs are significantly less as they are made up of carbon atoms.

The development of published publications in biosensing applications for the most appealing carbon nanomaterials indicates the scientific community's interest in these materials, as shown in Figure 12.3.

CNTs have a worldwide commercial interest reflected in a manufacturing capacity that now surpasses several thousand tons per year. Bulk CNT powders are used in many commercial items, including rechargeable batteries, automotive components, sports goods, boat hulls, and water filters. Commercial activity with CNTs, on the other hand, has increased dramatically during the last decade. The total number of publications for the 14 years 1999–2003 (351 824 papers and 48 448 patents) was 351 824 papers and 48 448 patents, with a patent/paper ratio of 0.137. Between 1991

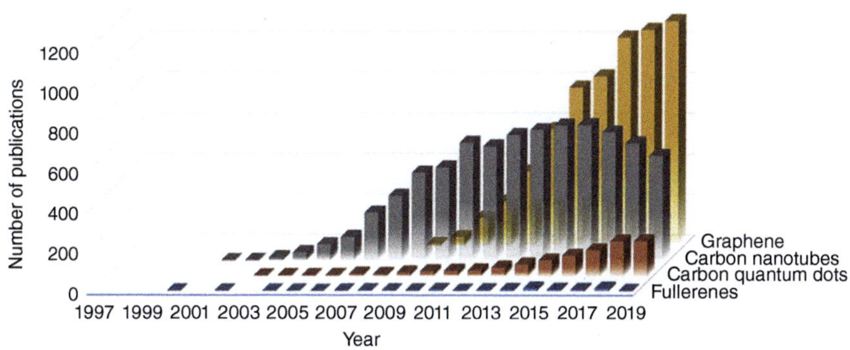

Figure 12.3 Graph depicting the number of scientific research publications dealing with carbon nanomaterials used in biosensors has been published each year since 1997. Source: Bianco et al. [31]/with permission of Elsevier.

Table 12.1 TCI ratings of patents with a high priority.

Patent number (real patent number)	TCI rating
2 (US Patent No. 6,401,526)	0.4479
22 (US Patent No. 7,374,793)	0.4271
39 (US Patent No. 7,514,116)	0.4375
77 (US Patent No. 7,744,843)	0.4896
80 (US Patent No. 7,763,230)	0.4375
96 (US Patent No. 7,851,064)	0.4479
97 (US Patent No. 7,854,991)	0.4688

and 2000, the average annual rise rate for publications was 10.0%, while the average annual increase rate for patents was 10.0%. From 2001 to 2010, the annual rise rate for publications was 8.26%, while the annual increase rate for patents was 9.14%. Both eras show an increase in cited journal publications and patents filed. On the other hand, patents are growing at a faster pace than published publications [32].

The technological centrality index (TCI) approach determines which patent in this patent network is comparatively relevant. Patent numbers 77, 97, 96, 2, 39, 80, and 22 are among the seven having TCI values more than 0.42. Table 12.1 demonstrates that these relatively significant patents have higher TCI ratings and are well ahead of other patents in the patent network. We can identify the relatively significant patents using the visual patent network and TCI ratings. These patents cover the fundamentals of CNT manufacture and use and emerging developments [33]. Table 12.2 gives the information of important companies, organizations, research group working on the application of chemically modified carbon nanotubes for lab on chip devices [34–37].

12.2.4.1 SWCNT: Various Synthesis Methods

SWCNTs were initially identified in the dirt created by an arc discharge employing graphite rod electrode in 1993. However, it was widely established in those early days that the arc-discharge process can only generate an inferior surrender of SWCNTs and many carbonaceous impurities. Following the finding, further research was conducted on several manufacturing techniques, including CVD and laser vaporization, which significantly increased SWCNT productivity. Progress in SWCNT manufacture has prompted many studies on the material's physical and chemical characteristics. In this regard, high-quality SWCNT manufacturing techniques are a critical essential technology for SWCNT research. SWCNT manufacturing technologies such as laser vaporization methods and arc-discharge produce high-crystalline graphene-like networks in SWCNTs because of the elevated growth temperature. As a result, SWCNTs using these approaches have shown much promise in applied research. Moreover, with the technology modification, the quality of SWCNTs generated by CVD has improved.

Table 12.2 List of various organizations and their applications in manufactures of nanotubes.

SN	Organizations	Applications
1	Adnano Technologies	Adnano Technologies is a supplier of various forms of graphene and multi-walled carbon nanotubes. They also provide analytical services such as FESEM, TEM, AFM, FTIR, XRD, XPS, contact angle, BET, zeta sizer, and master sizer.
2	Alfa Aesar	A global supplier of nanomaterials.
3	AVANSA Technology & Services	The company specializes in analytical characterization, consultancy, and synthesis of nanomaterials serving nanotechnology-based industries, universities, and institutes. They also manufacture carbon nanotubes, graphene, and various nanoparticles.
4	Beyond Pacific	Since 1996, the company has been perfecting the science of manufacturing large quantities of consistently high purity, ultra thin-walled carbon nanotubes (TWCNTs) and multi-walled carbon nanotubes (MWCNTs) under the brand name Nanophite®.
5	Bottom-Up Technology Corporation	The company manufactures graphene and carbon nanotubes.
6	Canatu	Canatu Oy's business is producing and selling a new class of versatile carbon-based components based on its novel nanobud nanomaterial.
7	Carbon Nanomaterial Technology	Manufacturer of carbon nanotubes and graphite nanofibers.
8	Carbon Solutions, Inc.	Research, development, and commercialization of single-walled carbon nanotubes, their chemistry, and their application to carbon-based nanotechnology.
9	CHASM Advanced Materials	The company acquired substantially all of SouthWest NanoTechnologies (SWeNT) assets and has reopened the state-of-the-art carbon nanotube (CNT) plant in Norman, Oklahoma. They manufacture high-quality single-wall, and few-wall CNT materials made using the patented CoMoCAT™ process under an exclusive license with the University of Oklahoma.
10	Cheap Tubes, Inc.	A supplier of various forms of carbon nanotubes.
11	Chengdu Organic Chemistry Co.	Producer of carbon nanotubes.
12	HeJi, Inc.	Producer of carbon nanotubes.
13	Marion Technologies	Specializes in the characterization, custom-made development, and the production of ceramic powders and nanostructured materials for industrial use.
14	Modern Synthesis Technology	The company manufactures graphene, fullerenes, and other carbon nanomaterials.
15	Nano-C	Producer of carbon nanotubes and fullerenes.

(Continued)

Table 12.2 (Continued)

SN	Organizations	Applications
16	NanoChemazone	A manufacturer and supplier of nanomaterials, graphene, carbon nanotubes, nanodiamonds, nanoparticles, nanoceramics, quantum dots, metal nanopowders, fullerenes, nanowires, nano- and micro-salts and derivatives.
17	Nanocomp Technologies	Production technology for carbon nanotubes and application-focused nanotube-based products.
18	Nanocyl S.A.	Nanocyl SA is a large-scale producer of multiwall carbon nanotubes, and associated carbon nanotubes concentrate or masterbatches.
19	Nanografi	A manufacturer of various nanomaterials such as carbon nanotubes and graphene.
20	NanographeneX	A manufacturer of graphene, carbon nanotubes, fullerenes, and various nanoparticles and nanoparticle dispersions.
21	NanoLab	Produces carbon nanotubes using the CVD growth process. The process has arrays of aligned carbon nanotubes on substrates.
22	NoPo Nanotechnologies	Manufacturer of carbon nanotubes.
23	Reade Advanced Materials	Reade advanced materials is a global distributor and value add toll chemical processor/packager of higher technology specialty chemical solids (metal, nonmetallic minerals, ceramic, composite, conductive, magnetic, nanomaterial, powder, fibers, abrasives, sheet, rod, additives, fillers, and reinforcements) with three customer care facilities in The Americas.
24	Reinste Nano Ventures	A manufacturer of nanomaterials.
25	SES Research	SES Research is the first manufacturer to provide commercial carbon nanomaterials. SES Research has provided high-quality carbon nanomaterials to the research community for over two decades. With orders as small as 100 mg and as large as 5 kg, SES Research is the largest supplier of carbon nanomaterials, often supplying other fullerene suppliers with their raw materials.
26	Stanford Materials	Worldwide supplier of rare earth, nonferrous, advanced ceramic materials, and carbon nanotubes.
27	TDA Research	R&D company develops new materials (polymers, carbons, and ceramics), catalytic and sorbent-based chemical processes, and military and aerospace components. Materials include fullerenes and carbon nanotubes. TDA is developing an exciting new class of nanomaterials, the metal–carbon nano encapsulated, for use in medical and electronics applications.
28	Texas Biochemicals	The company provides metal nanopowders, oxides and sulfides nanopowders, and nanotubes to industry and researchers worldwide.

Table 12.2 (Continued)

SN	Organizations	Applications
29	Thomas Swan	An established supplier of carbon nanotubes with a reputation for quality and reliability. Supplies multi-kg quantities of few-layer graphene nanoplatelets and non-carbon 2D materials such as boron nitride and molybdenum disulfide.
30	XinNano Materials	The company is a developer of carbon nanotubes and carbon nanotube inks.
31	Xintek, Inc.	Major products include carbon nanotube (CNT)-based field emission electron source, field emission grade CNT material, field emission X-ray source, and CNT AFM probes.
32	Zeon Corporation	In late 2015, Zeon Corporation completed and began operating the world's first mass-production plant for carbon nanotubes using the super-growth method developed by the National Institute of Advanced Industrial Science and Technology.

Bulky equipment setups and large reaction temperatures require colossal energy for laser vaporization and arc discharge methods [38]. Moreover, it is not easy to go from the laboratory to the commercial scale when using these methodologies for production at the mass level. In other words, the CVD process is renowned for its simplicity and suitability for large manufacturing. When carbon source molecules and catalyst nanoparticles are combined in a CVD process, SWCNTs may be synthesized from various raw materials, including the molecules and the catalysts that fuel their growth. Plasma hot and plasma filament methods were used to decompose carbon source molecules in the CVD development of SWCNTs for efficient C delivery on catalytic nanoparticles.

12.2.4.2 Growth Catalysts for SWCNT

In general, nanoparticles as a catalyst are required to form SWCNTs, irrespective of the manufacturing procedure, and nanoparticles belonging to the iron group, such as iron, nickel, and cobalt, are often utilized. The catalytic behavior of these iron-group metals for degrading hydrocarbons and generating graphite in the periphery of their majority crystals is well documented. Carbon atoms are generated on the catalyst nanoparticle's surface by the C-source molecules in the gas-phase decomposition, which dissolve in the catalyst metal. Consequently, depositing C nucleate on the nanoparticle's surface is used as a catalyst to produce SWCNTs. The oxidization or carbonization of nanoparticles owing to the tremendous demand for carbon sources is predicted to trigger the deactivation of the catalyst in the progress process. This growth process is known as the vapor–liquid–solid (VLS) model. Moreover, since the reaction temperature in the CVD development of SWCNTs is often smaller than that in the laser vaporization and arc-discharge process, it is debatable whether the catalyst nanoparticles are in the solid or liquid phases. Aside from the metals of the iron group, nanoparticles of noble metals such as gold, palladium, silver, platinum, copper, and also of aluminum were shown to act as catalysts for SWCNT (CVD) development. There have also been reports on SWCNT development employing non-metallic nanoparticles such as germanium, diamond, and alumina.

12.2.4.3 Approach of Introducing the Catalyst on SWCNTs (CVD) Growth

From the perspective of how the catalysts are injected into the reaction system for SWCNT development, we can categorize many CVD procedures broadly into two main categories: floating-supported-catalyst methods. Catalyst nanoparticles are stabilized on porous materials or inert substrates in the latter, whereas they reside as an aerosol inside a reactive atmosphere in the former. Since the supported catalyst technique benefits low-cost equipment and a broad process condition range, it makes many SWCNTs available. Moreover, the supported catalyst technique is known to have the following issues:

(i) It is essentially a batch process, and developing this process for continuous production is challenging.
(ii) In addition, the supported catalyst process often requires more porous materials or substrates than the quantity of SWCNTs produced, resulting in a considerable cost rise. Although intense research on technical improvement improves the supported catalyst approach, most of these difficulties are intrinsically challenging to fix.

On the other hand, the floating catalyst technique offers the advantage of assuring product homogeneity. HiPco represents floating catalyst CVD synthesis [39], a suitable continuous technique for commercializing SWCNTs. Cheng et al. presented the first investigation on SWCNT development using the floating catalyst approach in 1998 [40]. They generated great quality SWCNTs with the help of a floating catalyst CVD reactor, similar to what endo and Shikata described for VGCF synthesis, claiming reasonably large-scale and low-cost SWCNT manufacturing.

12.2.5 Double-Walled Carbon Nanotubes (DWCNTs)

DWCNTs are a novel carbon nanostructure class consisting of two nanotubes of carbon arranged concentrically. DWCNTs are the most straightforward system for exploring the impact of inter-wall interaction on the physical characteristics of CNTs because of their double-wall structure. DWCNTs offer more robust thermal stability and mechanical resistance than SWCNTs and have fascinating optoelectronic features.

12.2.5.1 Development of DWCNTs

Catalytic chemical vapor deposition (CCVD), arc discharge, and "peapod" processes commonly prepare DWCNTs. These procedures are almost similar to those used to make SWCNTs. Slight variations in catalytic composition, pressure, temperature, and carbon feedstock, on the other hand, may favor DWCNT development over SWCNT and MWCNT growth. It is worth noting that how CNTs are made (CCVD, arc discharge, or peapod) has a significant impact on their mechanical and thermal characteristics. This is most likely due to the variable quality of CNTs (length, defects, and diameter) synthesized by these distinct processes.

12.2.5.2 Purification of DWCNTs

Impurities such as amorphous carbon, SWCNTs, MWCNTs, and catalyst particles are present in DWCNTs when they are created. In a purification process, the elimination of metal catalysts with acids is usually followed by elevated-temperature oxidation to eliminate amorphous carbon and SWCNTs [41]. Since the rate of oxidation of DWCNTs is almost two times lower than that of SWCNTs having a similar diameter of 1.6 nm [42], hence a DWCNT containing purity of almost 95% may be produced from a fresh sample that contains only 10% DWCNT by the method of air oxidation at 500 °C. Even though heat oxidation may partly clean up DWCNTs, it is inefficient at eliminating MWCNTs, has a poor yield, and affects the optical and electrical properties of DWCNTs owing to the destructive power of oxidation reaction. M. Hersam and A. Green filter DWCNTs using the density gradient ultracentrifugation (DGU) method. The buoyant density of CNTs encased in surfactant changes with the wall number and diameter of the nanotube. This feature is utilized in DGU to sort liquid media with a density gradient using ultracentrifugation. MWCNTs, SWCNTs, and DWCNTs migrate separately during ultracentrifugation to sites where buoyant densities meet the gradients. A layer-by-layer collection of nanotube fractions allows sorting by structure and density (diameter selectivity and wall number). The CNTs are divided into four bands in a centrifuge tube, from beginning to end, which corresponds to SWCNTs, DWCNTs, MWCNTs, and a greater diameter CNT bundle. Repetition of the technique may provide ultrapure DWCNTs.

12.3 Carbon Nanotube Used in Solar Cells

Due to their unique physicochemical capabilities, nanomaterial has been investigated for application in brightness and photovoltaic (PV) systems. SWCNT optoelectronic properties but their use in organic solar cells such as absorber solar cells, polymers photovoltaics, and previous similar projects solar panels, whereby nanomaterials were used as photoanode materials. Electrical electrodes are made of CNTs, and illumination devices are made of CNTs. Geothermal, hydropower, and solar power are just a few healthy, renewable resources that can supply this electricity. Solar energy, on the other hand, is the most plentiful resource. Solar radiation reaches the globe at a rate of over 120 000 TW. Still, the next most significant asset, geothermal energy, can only produce 12 TW. When integrated across the whole land surface of the planet. The quantity of solar radiation impinging on the planet's surface is roughly 10 000 times the quantity of electricity used globally. On the other hand, solar energy currently provides less than 0.1% of global energy demand [42].

Because platinum is expensive, finding platinum-free counter-electrode materials has been a significant focus in the field of dye-sensitized solar cells (DSSCs). CNTs with single and multiwall appeared as front-runners, with duration and intensity to platinum [43]. They also provide the prospect of flexible DSSCs. When paired with flexible hours, electrode materials [44]. Because SWCNTs' semiconductor nanotubes restrict total conductivity, MWCNTs have received greater attention. MWCNTs being high frequency in methanol and drenched onto a fluorine-doped tin oxide (FTO) substrate in a 2006 research resulted in a platinum-free DSSC with a 7.59% complete competence below regular air mass (AM)1.5 [45]. By distributing the MWCNTs in a powder and applying them to FTO, the efficiency was increased to 7.67% AM1.5. In contrast, a similarly produced platinum counter electrode had 7.83% overall efficiency [46]. Recent studies using MWCNTs have sought to improve efficiency by enhancing the pseudorandom and alignment of MWCNTs in films [47] [48]. In this experiment, a system of MWCNTs was produced vertically using CVD and transferred to either a polymer or an optically transparent electrode (OTE). While the method had no net efficiency increase (7.04% overall vs. 7.34% for Pt), it did illustrate the use of functionalized nanotube films to improve shuttles and surface areas. The progress of sorted SWCNTs, wherein metallic SWCNTs provide benefits over MWCNTs in kinetic conductivity and greater surface area, is likely to impact this field substantially.

12.4 Carbon Nanotube Used in Optical Sensors

Since the discovery of SWCNT fluorescence, many research teams were striving to use the unique optical features of SWCNTs to develop biological and chemical sensors. Surface functionalization nanocrystals enable the capacity of SWCNT-based sensors to detect targets selectively with specific molecular elements. Integration or noncovalent bonding can be used to functionalize CNTs. In covalent chemistry, chemical bonds develop among nanotubes, whereas the – contact between adsorbate species accomplishes noncovalent, and the sidewall of the nanotube is noncovalent. Optical sensors require noncovalent chemistry because they retain SWCNTs' electrical properties, maintaining their luminous capabilities. Noncovalent surface modification of SWCNTs has been achieved using a range of substances, notably deoxyribonucleic acid (DNA) oligonucleotides [49] and peptides [50]. Variations in the adsorbed molecules or the surroundings affect the optical indications after the nanoparticles are functionalized noncovalently [51]. Table 12.3 shows the solid-state lasers with single-walled carbon nanotube saturable absorbers (SWCNT-SAs).

Table 12.3 Solid-state lasers with SWCNT-SAs.

Type of SWCNT	Gain medium (dopant/host)	Power (mW)	Wave length (nm)	Pulse duration (fs)	Repetition rate (MHz)	References
HiPco	Ti:Al_2O_3 (Ti:sapphire)	600	800	62	99.4	[52]
Arc-discharge	Nd:$GdVO_4$	3630	1063	8 400	122	[53]
Arc-discharge	Nd:Lu 0.15 Y 0.85 VO_4	902	1064	19 000	150	[54]
CO decomposition	Nd:Y_3Al_5O 12 (Nd:YAG ceramic)	130	1064	8 300	90	[55]
HiPco	Cr:Mg_2SiO_4 (Cr:forsterite)	202	1245	120	79	[56]

12.5 Carbon Nanotube Used in Light-Emitting Diodes

Developing high-efficiency light-emitting diodes (LEDs) has primarily focused on carbon-based optoelectronics. While substantial research has been done on CNT-based diodes [57], few studies have been on the features of CNT-based light-emitting strategies [58]. A field-effect transistor (FET)-based light emitters are the most effective CNT-based light emitters. A FET may be an interfacial sort, in which holes and electrons are inoculated into the CNT mediated and recombine to create electroluminescence (EL), or a unipolar type, in which only one kind of carrier is inserted into the CNT channel. A substantial bias, a strong electric field, or both may be created over local imperfections to cause influence resonances or ionizations, resulting in electrons. Therefore, EL differentiates the machines and create particle (n^-) and crevasse (p^-) areas. These devices need high bias and at least three separate electrodes, which adds to the complexity of the design and manufacture of combined microelectronics and photonic circuits.

12.6 Carbon Nanotube Used in Electronic Device Fabrication

Toluene, dimethylformamide (DMF), and *N*-methyl-2-pyrrolidone (NMP) were tested as solvents for SWCNT dispersion. The ideal solvent for dispersing SWCNTs for depositing and alignment was NMP. The dielectrophoresis (DEP) approach was successfully used to simultaneously deposit and align multi-walled nanotubes for four different types of gold electrodes. For four sets of electrodes, the electrical properties of DEP-aligned SWCNTs were measured and compared. The study is focused on investigating the wafer scale production of the integrated section device, hoping to present a pattern of thousands of SWCNT gadgets employing the DEP methods. The successful deposition and coordination of SWCNTs onto four pairs of anode and cathode at the same time suggest that the DEP process can be used to implementation of the national and alignment of SWCNTs onto thousands and hundreds of pairs of electrodes at the same time with just an ac voltage if thousands and hundreds of pairs of anode and cathode can be modeled in share information, which can be done quickly in microfabrication and nanofabrication. The DEP procedure is simple, affordable, and successful compared to other sophisticated processes for the depositing and alignment of SWCNTs. The objective of wafer-scale manufacture refers to the portion of integrated devices that might be achieved by improving device fabrication, electrode size downscaling, and the DEP process.

12.7 Carbon Nanotube Used in Lithium-Ion Batteries (LIBs)

CNTs, also known as graphene, have attracted much interest because of their nanoscale, high conductivity, vast electrochemically accessible contact area, and other properties. Many attempts were made to employ CNTs as a substitute for graphene in Li^+-ion cells, with human CNTs (SWCNTs) and cross CNTs (MWCNTs) being researched as electrode material in lithium-ion batteries (LIBs). SWCNTs have a specific capacitance ranging between 300 and 600 mAh g^{-1} [59], much more significant than graphite's (372 mAh g^{-1}). Mechanical and chemical treatments may boost the specific capacitance of SWCNTs up to 1000 mAh g^{-1} [60]. The utilization of hybrid synthetic structures, including CNT as a critical component, is a viable technique to improve the charge LIBs without lowering their irreversible capacity [61]. However, the existence of the binder reduces the anode's specific efficiency and capacity owing to its insulating properties. The capacity to preserve bifunctionality, wherein CNTs can work as both the active layer and the current collector, is the cutting-edge technology in this scenario, which entails employing gratis CNT sheets as an anode in LIBs. Light, elastic, binder-free, and thermally durable, these journal article materials are appropriate for elevated applications. Because of their high conductivity, suppleness, and low density, transfer, conductive CNT films bring up a new path for elevated flexible batteries [62]. Many different preparation processes are being used to make CNT films or sheets, including pressure filtering [62], layer-by-layer [63] rod coating [64], and CVD [65]. Table 12.4 shows the various Sn/graphene-based binary electrodes.

12.8 Carbon Nanotube Used in Chip Cooling

CVD was used to create nanotubes (CNT) array with matched likely to escalate and sponge CNTs, fluffy carbon fibers with microscopic CNTs on their surface. To explore their heat dissipation capabilities, heat sinks were created depending on

Table 12.4 Various Sn/graphene-based binary electrodes.

Materials	Cycle number	Current density	Capacity retention ($mAh\,g^{-1}$)	References
Graphene nanoribbons (GNRs) and SnO_2	50	$100\,mA\,g^{-1}$	825	[66]
Co_2SnO_4 hollow cubes@RGO nanocomposites	100	$100\,mA\,g^{-1}$	1000	[67]
SnS_2/RGO nanocomposites	200	0.1 C	1034	[68]
Graphene networks anchored with Sn@graphene	340	$200\,mAg^{-1}$	1076	[69]
SnO_2/graphene composites	1000	$2\,A\,g^{-1}$	1813	[70]

CNT arrays or sponge CNTs. Their microstructures were discovered to significantly influence thermal performance by altering the efficiency of air convection. Because of their aligned construction and vast contacting surface with the air, the long CNT arrays offer outstanding heat removal efficiency even under natural convection. At the same time, owing to their porous nature, sponge CNTs exhibit a more significant increase in heat dissipation capabilities under airflow. The findings may be used to produce low-cost, light, and high CNT-based passive heat sinks for chip chilling under various situations. The heat sink from CNT is handy for heat dissipation from a tiny heating element. However, operating conditions to attain their optimal performance may vary depending on the varying nanostructures of the sponges CNTs, the small CNT array, and the long CNT array. Because of its excellent heat dispersion efficiency even when under natural convection. However, when there is no circulation, the sponge CNTs slightly advantage in heat dissipation.

On the other hand, they are unneeded in environments with high airflow because a short CNT accumulator heat sink, which can be more cost-effective, performs even better. If there is any airflow, sponges CNTs also are a suitable alternative. Sponge CNTs may achieve the same heat removal effect as short CNT arrays, and sponges CNTs-based heat sinks are more likely to assemble fast owing to their flocculent nature. The linked network makes it difficult for the sponge CNTs to slip off, which is critical for the electromagnetic circuit's stability.

12.9 Carbon Nanotube Used in Photovoltaic Devices

Because of their remarkable electrical, mechanical, and thermal capabilities, human nanoparticles (SWCNTs) are projected to be promising nanomaterials. For solar cell devices with more excellent performance, SWCNTs' identity precision network structure of SWCNTs is produced through the vapor of water treatments of as created precipitously vertically-aligned single-walled nanotube (VA-SWNT). VA-SWNT was given a micro hexagonal lattice by merely exposing it to drying and vapor of water under normal circumstances. Triangular cell walls are vertically stacked microvascular SWCNTs with a densely bundled top section. Collapsed spaghetti-like SWCNTs keep in touch with the substrate inside each cell. The precision SWCNTs network film was placed on the upper of the layer with a 3 mm × 3 mm bare n-type semiconductor contact window center to create the SWCNT/n-Si heterojunction cell. SiO_2 surrounds the connection window as an insulating layer, while Pt serves as an electrode. According to our first testing, the ideal photoelectric capacity factor (PCE) with AM1.5 was 5.91%, with a throughput of 72%. This demonstrated a significant improvement compared to heterojunction solar cells utilizing spaghetti-like SWCNTs. Moreover, the performance improvement of anthocyanin solar cells was proven using micro hexagon SWCNTs.

12.10 Carbon Nanotube Used in Nonvolatile Random Access Memory

A rising electrostatic interaction reconfigurable wire arrangement memory module is a storage system's spontaneous nonvolatile random access memory (NRAM). A crossbar array composed of groups of CNTs on the surface and orthogonal CNTs hung on a periodical support array makes up the construction of the cellular device. The on/off switch is bistable. The two extremes on the thermodynamic efficiency vs. distance graph are connected to the principal at the cross points.

12.11 Carbon Nanotube Used in Potential Device

The continuous scalability of silicon-based will meet several hurdles that standard innovation may not overcome. While incorporating new substances into crystalline silicon may assist, connectivity difficulties must be addressed first. Original stuff configurations, such as nanomaterials generated via identity bottom-up processes rather than layering patterning techniques, may also provide improvements. The potential for microelectronics to leverage the enormous potential of carbon nanostructures is discussed in this study (CNTs). A quick overview of CNT fundamentals and growth strategies will be provided. Applications of CNT as connectors will be evaluated. A comparison of government silicon transistors and CNT–FETs reveals that CNT gadgets perform better. The designing of CNT devices, on the other hand, remains an unresolved challenge. In addition to combining silicon-like innovation, several of the most promising manufacturing processes are evaluated. Finally, a novel device is described that avoids the issues of CNT preemployment screening in independent transistors.

12.12 Carbon Nanotube Used in On-Chip Inductor

One of the essential considerations in increased schematic capture is that copper loss increases substantially as resistance rises at higher energies (due to skin and proximity effects). Because the capacitance of CNT couplers rises by a lesser amount or stays practically unaltered at high frequencies, CNT connectors may significantly improve the circuit's outstanding quality. An on resistor is examined in this section as an example of CNT interconnects' potential maximum performance.

12.13 Carbon Nanotube Used in Electronic Device

Single-walled nanoparticles (SWNPs), whether in the shape of nanosheets films of undirected graphs, associated groups, or anywhere in among, offer one novel sort of online substantial that may be incorporated into electronics in a traditional, ascendable manner. Such movies' electronic, kinematic, and optical features may approach the outstanding qualities of individual SWCNTs in some instances, producing them appealing for use in semiconductors, sensors, and many additional schemes. The manufacturing and gathering of SWCNTs into different thin-film structures are discussed in this chapter. It shows how they are used in integrated digital devices with management contractors reaching 100 transistors and analog RF devices with a frequency range up to the many gigahertz, such as consumer electronics. SWCNT transistors provide all active functionality. The findings are necessary steps toward building an SWCNT-based semiconductors innovation that might be useful in flexible electronics, rf analog devices, and other areas where existing systems could benefit.

Singular SWCNTs are a great way to learn about the thermodynamics of one-dimensional (1D) microscopic confined structures. SWCNT nanosheets provide more excellent connectivity that offers equally attractive research prospects and a materials approach that might lead to actual device implementations. SWCNT thin films have excellent electrical, structural, optical, pharmacological, and thermal capabilities, making them ideal for innovative portable systems. Some unique qualities make it feasible to provide functionality that can be difficult or impossible to accomplish with traditional materials. Transparency devices [71] and scalable, flexible, and sizable semiconductors are two examples [72]. Such SWCNT sheets may survive in the market materials for high circuitry or as more significant constituents of nonoverlapping connected systems are a type of ordered arrays [73]. The demos described below show these two separate applications in diverse areas. Despite recent advancements, significant difficulties in chemistry and materials research remain. Some of the most prominent approaches to eradicate them – SWCNTs are regulated systems to saturate the s-SWCNTs, and tactics to boost development and depositing. The viability of these advanced materials will be determined by how fast, and efficiently these obstacles can be addressed and advancements in rival materials classes such as inorganic nanostructured materials [74] and diamond flakes films [75]. Apart from these factors, we feel that the potential for real-world applications and the wide variety of linked fundamental issues point to a promising future for skinny SWCNT potential product development.

12.14 Carbon Nanotube Used in Quantum CNFETs

Nanotubes FETs may display quite unusual features when functioning in the atomic impedance limit, especially in the transition layer of semiconductor operation.

These innovative features are distinctive and not visible in ordinary semiconductors FETs with an extensive disk operating system (DOS), as we will see soon. To differentiate its unique qualities from the standard or contemporary FETs, we regard a carbon nanotube field-effect transistor (CNFET) that fulfills the fundamental impedance restriction in all functional areas (micro, linear, and reactive) as a quantitative FET. As a result of the intrinsic capacitive limit, $C_{ox} \gg C_q$, and thus also is VG. In a quantum CNFET, the gatekeeper controls the network's electrical characteristics. The oxide capacitive must be substantially more significant than the total capacitance, perhaps more important. To meet this criterion, it is worth getting an approximation of the oxide impedance. Keeping our emphasis on low energies (i.e. just the first subband is taken into account), the greatest C_q of highly conductive nanostructures is about two more than the fractional capacitive of metals nanotubes.

12.15 Carbon Nanotube Used in Schottky-Barrier Ballistic CNFETs

Nanostructured FETs with devices based (source/drain) metals interfaces are often observed in reality, not by choice or purpose, but solely because interfacial connections to conductive polymers nanotubes are significantly more straightforward achieve than the more desired ohmic sort. This is because engineering the material reference to be permeable to charge transport requires more time and work. Furthermore, conventional injection molding circumstances favor Schottky obstacles more frequently than not. Consequently, understand the device thermodynamics and IV properties of epitaxial wall CNFETs. The knowledge acquired from this work dramatically expands our understanding of epitaxial nanoscale semiconductors in general, encompassing CNR semiconductors and semiconductors nanoscale inverters. In reality, experiencing leakage current and the consequences for on/off conversion rate in (metal) dielectric breakdown meteoric CNFETs and similar 1D high-velocity FETs involves knowing of epitaxial device mechanics, as will have become apparent later in the story. Due to the intrinsically complicated device thermodynamics, the existence of Schottky barriers adds a significant amount of complication to modeling ionic conductivity in the carbon transistor. It is pretty simple to get disoriented or lose sight of the center device's mechanics. This is particularly true for new students learning about this topic. As a result, the talk will be more of a moderate and easy orientation to interfacial nanotube semiconductors.

12.16 Carbon Nanotube Used in Chemical Sensors and Biosensors

Nanostructures are perfect for chemistry and biosensors because of their antidote and high specific surface ratio. In an ideal world, every component is reasonably available in every environment. The key features that make CNTs and graphene so appealing to electrochemical sensors are the ability to disturb each particle and monitor the alterations electrically or physically are the key elements that make CNTs and graphene so attractive for electrochemical sensors. Furthermore, the dimensional stability and configurable electronic conductivity of CNTs motivate the research and production of a wide range of pesticide, subatomic, and genetic sensors that can benefit from the hundred years' knowledge about composites expressly embodied in organic chemistry. Although there are many ways to use CNTs as sensing devices, the underlying concept is as follows: super clean, great nanotubes are pasteurized by adding suitable covering or attaching a particular operational group to improve their specificity to the life forms printer activities of interest, or to modify them to the cultural context (e.g. for biosensors in solution phase). The latest CNT electrical modern or optical emission steps offer information about the target microorganism's interaction or existence. Many pharmacological and gas analyzers based on nanotubes have been developed [76]. CNTs are also utilized as electrodes for inductance sensors [77] or to ionize a gas by generating high voltage fields at their tips [78]. The character of the environmental gas is determined by the ensuing gas ionization or breakdown voltage. Furthermore, for electrical nose (e-nose) application, several co-sensors using CNTs were examined [79]. Enhanced selectivity and higher sensitivity, such as non-nanotube sensors, appear to be never-ending sensory research and development topics.

CNT-based sensing has been proven to provide parts per billion sensitivities for various gases [80]. Two-terminal CNT electronics or multiple transistor-like nanostructures are often used in nanotube-based sensors. The tiny nanoscale diameter of CNTs makes them excellent for interacting with macromolecules in biosensors. Nanotubes are usually processed (for example, enzyme-loaded or DNA wrapped) for various reasons, along with the necessity to be appropriate to the cellular environment or regulate specific proteins that send insight into the molecular events and functions [76]. Conversely, there have also been conflicting findings on nanomaterials' biomedical applications and toxicity *in vivo* [81], which has sparked much controversy in the scientific and general press. As nanotube detectors become more sophisticated, there is also a rising tendency to combine nanomaterials with silicon circuits to provide calibrated, enhanced, reliable sensor regulation, and performance [82]. Simulating sensor activity is also critical for the engineering and development of increased CNT-based electrochemical and biosensors [83]. In bio applications, CNTs are also being investigated for cancer therapy and graphene for drug carriers [84].

12.17 Carbon Nanotube Used in Field Emission of Electrons

Another of the oldest known attributes of CNTs scrutinized was the field release of photons from the recommendations of vertically stacked CNTs [85], resulting in a flurry of industrial and research exploration and discovery into a variety of potential implementations, including low-pressure electronics, frozen cathodes, transmission electron [86], spectroscopy displays [87], and X-ray inputs for clinical applications [88]. A property of a nanoelectrothermal device might be used in any system that uses an electron source. A dispersive energy X-ray of electrons occurs when a sufficiently enough electric field is applied to the nanotube's base to cause electrons to be emitted. Nanotubes have a slim, mechanically robust structure and good corrosion resistance and nontoxicity, which are desired features in a field emitter; also, their short radius allows for high game factors. The nanostructures are generally produced vertically on a substrate material by CVD and shaped by lithography. A relatively enough electric field (typically in a high vacuum) is applied between its electrodes (the positively top layer some reasonable distance from CNTs) and the cathode to produce continuous particle emission (the CNT conductive substrate). Today's wide variety of applications has progressed beyond the fundamental basis and is now primarily restricted only by one's imagination. Lightweight, increased composites with properties typically higher than metals and that, in certain situations, can be manufactured to be equivalent to durable spider silk [70], continuing to be apparent possibilities. The application catalog now includes novel applications such as sensors, ultracapacitors [89], superconductors for photovoltaic devices [90], properties of nano copper for increased interconnects in nanoscopic technology [91], CNT composite fibers for online textile products [92], CNT-coated dielectrics for investigating cells in the brain [93], and nanoparticle materials for support beams in tissue regeneration [94]. There is no scarcity of unique and inventive ideas. The moon elevator is the most well-known example of a brilliant application of graphene composites. Consumer items reinforced with CNTs, such as high-quality wooden sticks and bicycles, are always on the market.

12.18 Carbon Nanotube Used in Supercapacitor Devices with Enhanced Electrochemical Performance

A chemical copolymer approach was used to make polypyrrole (PPy) with a redox-active marigold dye as a protonated dopant. With the maximum capacity of 4.2 $F\,cm^{-2}$ at a scanned rate of 2 $mV\,s^{-1}$, PPy membranes with mass packing in the spectrum of 20–30 $mg\,cm^{-2}$ displayed almost box shape CVs, suggesting excellent capacitive behavior. A simple approach based on the polymers of PPy in the medium of chamomile and MWCNT, dispersion using photovoltaic (PV) dispersant, was used to make PPy coated MWCNT. Compared to simple PPy, the PPy-coated MWCNT electrodes demonstrated a significant enhancement in maintaining the highest standards at high scan intervals. At leakage current concentrations of 1–33 $mA\,cm^{-2}$, Ppy-coated MWCNT showed a reversible capacity of 1.3–1.6 $F\,cm^{-2}$ and excellent cycle stability.

12.19 Carbon Nanotube Used in Flip-Chip High Power Amplifiers

Heating and source irregularities for increased flip-chip amplification include a high power amplifier (HPA). The newly designed 15 m extended CNT shock's surface temperature is 1400 W $(mK)^{-1}$. The output power of a turn AlGaN/GaN HEMT HPA with a circuit width of 2.4 mm and CNT bumps, operated at 40 V, is 39 dBm at a range of 2.1 GHz without any heat-up deterioration. This is, to our recollection, the first description of practical use of CNTs that takes advantage of their thermal expansion [77].

12.20 Carbon Nanotube Used in Transistor Device Application

AC DEP with a biocompatible lubricant, dioctyl sodium sulfosuccinate (i.e. AOT), developed a wide range of distributed SWCNTs, was used to precisely put single-walled nanomaterials (SWNMs) in device topologies. The distribution of SWCNTs in AOT water solution was explored at various micellar similar to current to other cleansers better to serve the use of SWCNTs/AOT for technical operations. To analyze solubilized solitary SWCNTs, UV–Vis–NIR spectrometry and atomic force microscopy (AFM) microscopy were used SWCNTs in field-effect semiconductor geometrics were aligned using the scattered solution. We found that the AOT distinctly separate tubes are worthy candidates for optoelectronic devices because they are well-protected against disintegration during sonication. Standalone tubes are promising options for nanotube transistors.

Furthermore, AOT penetration enhancer lines are well against sonication disintegration. This is a consideration while selecting AOT for device alignment. We have shown that these long individual nanotubes can be precisely placed in device shapes. Conductivity transport for an SWCNTs wireless router has been demonstrated [78].

12.21 Carbon Nanotube Used in Supercapacitors and Batteries

Watery CNT films may be used as charge collecting and increased scaffolds for energy storage devices and lithium-ion banks' active components. Interconnected nanotubes generate conducting channels in dry, easily interpretable nanotube films. When combined with functional textiles, the resulting structure may effectively carry charges and potentially eliminate the need for typical metal charge collectors such as aluminum and copper foils in gadgets. In contrast, due to nanomaterials' isolation by materials in workaround properties of nano-active polymers [47], the promise of nanotubes in permeability improvement has not been completely realized, and charge collection is still necessary. Second, during galvanic operations, dry processable tube films may vary their porosity architectures to respond to the compaction of intelligent materials, avoiding the issue of active components detaching from metal foil-charged collectors. Finally, the large interfacial area of dry, easily interpretable nanotubes ensures excellent functioning material adherence to the nanotubes, eliminating the need for binders in nanowire composite electrodes.

Finally, dry-producible CNT films are challenging, foldable, translucent, and light (e.g. 1.5 mg cm^{-3} for torpedo films from the type of synthetic arrays) [95], making them potential candidates for portable, adaptive, and cheap power devices. The cycle batteries use different active implementations based on standard parameters [96, 97] coded as MnOx [98], NiO, and Co_3O_4 [99] for energy storage devices Fe_3O_4 [100], fibers [101], silicon [102], and SnO_2.

12.22 Future Scope

It is tough to predict the future of any key technology. On the one hand, there is a typical propensity to underestimate a technology's influence and rate of development. CNTs have several features that make them ideal materials for creating a lab-on-a-chip. CNTs were incorporated into various functional components to improve surface chemistry, biocompatibility, and electrical sensing. Furthermore, controlled synthesis of CNTs on diverse substrates utilizing CCVD or plasma-enhanced chemical vapor deposition (PECVD) allows these materials to be incorporated into existing systems.

The goal has been to use CNTs as a single microfluidic element with a single purpose. CNT components with many functionalities should be included in future work to improve lab-on-a-chip or build a complete system. To accomplish multiple/multiplexed tasks of physical filtering, electric field separation, and molecular identification, a microanalysis system may be built by incorporating a CNT membrane into capillary electrophoresis or DEP systems. This might lead to smaller lab-on-chip systems, have fewer functional parts, and use less power.

12.23 Conclusion

The present scenario turns into nanotechnology; hence, everybody will be trying to miniaturize the devices. The researchers try to make the devices as small as possible with improved performance day by day. Therefore, they want to use various updated technology by using chemically modified materials such as CNTs as the best option to overcome the problems. It is used in multiple applications such as energy devices means batteries, solar cells, medical appliances, optical sensors for chemical and biological molecules, LEDs, electronic-device fabrication, as an anode in LIBs, chip cooling, photovoltaic devices, random access memory, potential device, on-chip inductor, chemical sensors, and biosensors.

Acknowledgments

Author acknowledges the Knowledge Resource Centre, CSIR-NEERI, (CSIR-NEERI/KRC/2022/JAN/HZC/1) for their support. We are thankful to the KIM-Division (CSIR-IICT) for providing a library. The communication number for this chapter is IICT/Pubs./2021/385.

References

1 Arfin, T. and Rafiuddin (2009). Transport studies of nickel arsenate membrane. *J. Electroanal. Chem.* 636 (1): 113–122.

2 Arfin, T., Falch, A., and Kriek, R.J. (2012). Evaluation of charge density and the theory for calculating membrane potential for a nano-composite nylon-6,6 nickel phosphate membrane. *Phys. Chem. Chem. Phys.* 14 (48): 16760–16769.

3 Arfin, T., Bushra, R., and Mohammad, F. (2016). Electrochemical sensor for the sensitive detection of *o*-nitrophenol using graphene oxide-poly(ethyleneimine) dendrimer-modified glassy carbon electrode. *Graphene Technol.* 1 (1): 1–15.

4 Arfin, T. and Rangari, S.N. (2018). Graphene oxide-ZnO nanocomposite modified electrode for the detection of phenol. *Anal. Methods* 10 (3): 347–358.

5 Obulapuram, P.K., Arfin, T., Mohammad, F. et al. (2021). Surface-enhanced biocompatibility and adsorption capacity of a zirconium phosphate-coated polyaniline composite. *ACS Omega* 6 (49): 33614–33626.

6 Arfin, T., Bhaisare, D.A., and Waghmare, S.S. (2021). Development of PANI/Fe$(NO_3)_2$ nanomaterial for reactive orange 16 (RO16) dye removal. *Anal. Methods* 13 (44): 5309–5327.

7 Obulapuram, P.K.M., Arfin, T., Mohammad, F. et al. (2021). Adsorption, equilibrium isotherm, and thermodynamic studies towards the removal of reactive orange 16 dye using Cu(I)-polyaniline composite. *Polymers* 13 (20): 3490.

8 Arfin, T., Sonawane, K., and Tarannum, A. (2019). Review on detection of phenol in water. *Adv. Mater. Lett.* 10 (11): 753–785.

9 Arfin, T. and Tarannum, A. (2019). Rapid determination of lead ions using polyaniline-zirconium (IV) iodate-based ion selective electrode. *J. Environ. Chem. Eng.* 7 (1): 102811.

10 Mohammad, F., Arfin, T., and Al-Lohedan, H.A. (2019). Enhanced biosorption and electrochemical performance of sugarcane bagasse derived a polylactic acid-graphene oxide-CeO_2 composite. *Mater. Chem. Phys.* 229: 117–123.

11 Karthik, P.S., Himaja, A.L., and Singh, S.P. (2014). Carbon-allotropes: synthesis methods, applications and future perspectives. *Carbon Lett.* 15 (4): 219–237.

12 Andreas, H. (2010). The era of carbon allotropes. *Nat. Mater.* 9: 868–871.

13 Wei, L., Kuo, P.K., Thomas, R.L. et al. (1993). Thermal conductivity of isotopically modified single crystal diamond. *Phys. Rev. Lett.* 70 (24): 3764–3767.

14 Walker, J. (1979). Optical absorption and luminescence in diamond. *Rep. Prog. Phys.* 42 (10): 1605.

15 Coustel, N., Coq, B., Brotons, V. et al. (2002). Application of carbon nanotubes as supports in heterogeneous catalysis. *J. Am. Chem. Soc.* 116 (17): 7935–7936.

16 Ferro, S. (2002). Synthesis of diamond. *J. Mater. Chem.* 12 (10): 2843–2855.

17 Hannay, J.B. (1880). Artificial diamonds. *Nature* 22 (559): 241–241.

18 Bannister, F.A. and Lonsdale, K. (1943). Laboratory synthesis of diamond. *Nature* 151: 334–335.

19 Burton, C.V. (1905). Artificial diamonds. *Nature* 72: 397.

20 Anthony, T.R. (1990). Metastable synthesis of diamond. *Vacuum* 41 (4–6): 1356–1359.

21 Luo, B., Yuan, A., Yang, S. et al. (2021). Synthesis of diamond-like carbon as a dielectric platform for graphene field effect transistors. *ACS Appl. Nano Mater.* 4 (2): 1385–1393.
22 Luo, C., Qi, X., Pan, C., and Yang, W. (2015). Diamond synthesis from carbon nanofibers at low temperature and low pressure. *Sci. Rep.* 5: 13879.
23 Coroş, M., Pogăcean, F., Măgeruşan, L. et al. (2019). A brief overview on synthesis and applications of graphene and graphene-based nanomaterials. *Front. Mater. Sci.* 13 (1): 23–32.
24 Smith, A.T., LaChance, A.M., Zeng, S. et al. (2019). Synthesis, properties, and applications of graphene oxide/reduced graphene oxide and their nanocomposites. *Nano Mater. Sci.* 1 (1): 31–47.
25 Zhu, Y., Murali, S., Cai, W. et al. (2010). Graphene and graphene oxide: synthesis, properties, and applications. *Adv. Mater.* 22 (35): 3906–3924.
26 Kroto, H.W., Heath, J.R., O'Brien, S.C. et al. (1985). C60: buckminsterfullerene. *Nature* 318: 162–163.
27 Geckeler, K.E. and Samal, S. (1999). Syntheses and properties of macromolecular fullerenes, a review. *Polym. Int.* 48 (9): 743–757.
28 Shanbogh, P.P. and Sundaram, N.G. (2015). Fullerenes revisited. *Resonance* 20 (2): 123–135.
29 Ajie, H., Alvarez, M.M., Anz, S.J. et al. (2002). Characterization of the soluble all-carbon molecules C60 and C70. *J. Phys. Chem.* 94 (24): 8630–8633.
30 Montellano López, A., Mateo-Alonso, A., and Prato, M. (2011). Materials chemistry of fullerene C60 derivatives. *J. Mater. Chem.* 21 (5): 1305–1318.
31 Bianco, A., Chen, Y., Frackowiak, E. et al. (2022). Carbon science perspective in 2020: current research and future challenges. *Carbon* 161: 373–391.
32 Golnabi, H. (2012). Carbon nanotube research developments in terms of published papers and patents, synthesis and production. *Sci. Iran.* 19 (6): 2012–2022.
33 Chang, P.L., Wu, C.C., and Leu, H.J. (2011). Carbon nanotube fabrication: patent analysis. In: *Carbon Naotubes Applications on Electron Devices* (ed. J.M. Marulanda), 169–184. London: Intechopen.
34 Moser, I., Jobst, G., Aschauer, E. et al. (1995). Miniaturized thin film glutamate and glutamine biosensors. *Biosens. Bioelectron.* 10 (6–7): 527–532.
35 Navakul, K., Warakulwit, C., Yenchitsomanus, P.T. et al. (2017). A novel method for dengue virus detection and anti- body screening using a graphene-polymer based electrochemical biosensor. *Nanomed. Nanotechnol. Biol. Med.* 13 (2): 549–557.
36 Chan, C., Shi, J., Fan, Y., and Yang, M. (2017). A microfluidic flow-through chip integrated with reduced graphene oxide transistor for influenza virus gene detection. *Sens. Actuators, B* 251: 927–933.
37 Tseng, Y.T., Yang, C.S., and Tseng, F.G. (2009). A perfusion-based micro opto-fluidic system (PMOFS) for continuously in-situ immune sensing. *Lab Chip* 9: 2673–2682.
38 Endo, M., Muramatsu, H., Hayashi, T. et al. (2005). 'Buckypaper' from coaxial nanotubes. *Nature* 433 (7025): 476.

39 Karousis, N., Tagmatarchis, N., and Tasis, D. (2010). Current progress on the chemical modification of carbon nanotubes. *Chem. Rev.* 110 (9): 5366–5397.

40 Cheng, H.M., Li, F., Su, G. et al. (1998). Large-scale and low-cost synthesis of single-walled carbon nanotubes by the catalytic pyrolysis of hydrocarbons. *Appl. Phys. Lett.* 72: 3282.

41 Yoshida, H., Sugai, T., and Shinohara, H. (2008). Fabrication, purification, and characterization of double-wall carbon nanotubes via pulsed arc discharge. *J. Phys. Chem. C* 112 (50): 19908–19915.

42 Kamat, P.V. (2007). Meeting the clean energy demand: nanostructure architectures for solar energy conversion. *J. Phys. Chem. C* 111 (7): 2834–2860.

43 Lee, W.J., Ramasamy, E., Lee, D.Y., and Song, J.S. (2009). Efficient dye-sensitized cells with catalytic multiwall carbon nanotube counter electrodes. *ACS Appl. Mater. Interfaces* 1 (6): 1145–1149.

44 Pasquier, A.D., Unalan, H.E., Kanwal, A. et al. (2005). Conducting and transparent single-wall carbon nanotube electrodes for polymer-fullerene solar cells. *Appl. Phys. Lett.* 87 (20): 3.

45 Ramasamy, E., Lee, W.J., Lee, D.Y., and Song, J.S. (2008). Spray coated multi-wall carbon nanotube counter electrode for tri-iodide ($I^-{}_3$) reduction in dye-sensitized solar cells. *Electrochem. Commun.* 10 (7): 1087–1089.

46 Suzuki, K., Yamaguchi, M., Kumagai, M., and Yanagida, S. (2003). Application of carbon nanotubes to counter electrodes of dye-sensitized solar cells. *Chem. Lett.* 32 (1): 28–29.

47 Shen, L., Yuan, C., Luo, H. et al. (2011). *In situ* growth of $Li_4Tu_5O_{12}$ on multi-walled carbon nanotubes: novel coaxial nanocables for high rate lithium ion batteries. *J. Mater. Chem.* 21: 761–767.

48 Hu, L.B., Hecht, D.S., and Gruner, G. (2010). Carbon nanotube thin films: fabrication, properties, and applications. *Chem. Rev.* 110 (10): 5790–5844.

49 Cha, T.G., Baker, B.A., Sauffer, M.D. et al. (2011, 2011). Optical nanosensor architecture for cell-signaling molcecules using DNA aptamer-coated carbon nanotubes. *ACS Nano* 5 (5): 4236–4244.

50 Shim, M., Kam, N.W.S., Chen, R.J. et al. (2002). Functionalization of carbon nanotubes for biocompatibility and biomolecular recognition. *Nano Lett.* 2 (4): 285–288.

51 Strano, M.S., Zheng, M., Jagota, A. et al. (2004). Understanding the nature of the DNA-assisted separation of single-walled carbon nanotubes using fluorescence and Raman spectroscopy. *Nano Lett.* 4 (4): 543–550.

52 Khudyakov, D.V., Lobach, A.S., and Nadtochenko, V.A. (2010). Passive mode locking in a Ti:sapphire laser using a single-walled carbon nanotube saturable absorber at a wavelength of 810 nm. *Opt. Lett.* 35: 2675–2677.

53 Chen, H.R., Wang, Y.G., Tsai, C.Y. et al. (2011). High-power, passively mode-locked Nd:$GdVO_4$ laser using single-walled carbon nanotubes as saturable absorber. *Opt. Lett.* 36: 1284–1286.

54 Yang, K., Zhao, S., Zhang, G. et al. (2012). Diode-pumped passively mode-locked Nd:$Lu_{0.15}Y_{0.85}VO_4$ laser with a single walled carbon nanotube saturable absorber. *Opt. Commun.* 285: 158–161.

55 De Tan, W., Chen, F., Knize, R.J. et al. (2011). Passive mode locking of ceramic Nd:YAG using (7, 5) semiconducting single walled carbon nanotubes. *Opt. Mater.* 33: 679–683.

56 Cho, W.B., Yim, J.H., Choi, S.Y. et al. (2008). Mode-locked self-starting Cr:forsterite laser using a single-walled carbon nanotube saturable absorber. *Opt. Lett.* 33: 2449–2451.

57 Lee, J.U., Gipp, P.P., and Heller, C.M. (2004). Carbon nanotube p–n junction diodes. *Appl. Phys. Lett.* 85: 145–147.

58 Wang, S., Zheng, Q.S., Yang, L.J. et al. (2011). High-performance carbon nanotube light-emitting diodes with asymmetric contacts. *Nano Lett.* 11: 23–29.

59 Yan, J., Song, H., Yang, S. et al. (2008). Preparation and electrochemical properties of composites of carbon nanotubes loaded with Ag and TiO_2 nanoparticle for use as anode material in lithium-ion batteries. *Electrochim. Acta* 53 (22): 6351–6355.

60 Shimoda, H., Gao, B., Tang, X.P. et al. (2001). Lithium intercalation into opened single-wall carbon nanotubes: storage capacity and electronic properties. *Phys. Rev. Lett.* 88: 015502.

61 Reddy, A.L.M., Shaijumon, M.M., Gowda, S.R., and Ajayan, P.M. (2009). Coaxial MnO_2/carbon nanotube array electrodes for high-performance lithium batteries. *Nano Lett.* 9 (3): 1002–1006.

62 Chew, S.Y., Ng, S.H., Wang, J. et al. (2009). Flexible free-standing carbon nanotube films for model lithium-ion batteries. *Carbon* 47 (13): 2976–2983.

63 Lee, S.W., Gallant, B.M., Byon, H.R. et al. (2011). Nanostructured carbon-based electrodes: bridging the gap between thin-film lithium-ion batteries and electrochemical capacitors. *Energy Environ. Sci.* 4: 1972–1985.

64 Hu, L., Choi, J.W., Yang, Y. et al. (2009). Highly conductive paper for energy-storage devices. *Proc. Natl. Acad. Sci. U.S.A.* 106 (51): 21490–21494.

65 Zhong, X.H., Li, Y.L., Liu, Y.K. et al. (2010). Continuous multilayered carbon nanotube yarns. *Adv. Mater.* 22 (6): 692–696.

66 Lin, J., Peng, Z., Xiang, C. et al. (2013). Graphene nanoribbon and nanostructured SnO_2 composite anodes for lithium ion batteries. *ACS Nano* 7 (7): 6001–6006.

67 Zhang, J., Liang, J., Zhu, Y. et al. (2014). Synthesis of Co_2SnO_4 hollow cubes encapsulated in graphene as high capacity anode materials for lithium-ion batteries. *J. Mater. Chem. A* 2: 2728–2734.

68 Mei, L., Xu, C., Yang, T. et al. (2013). Superior electrochemical performance of ultrasmall SnS_2 nanocrystals decorated on flexible RGO in lithium-ion batteries. *J. Mater. Chem. A* 1: 8658–8664.

69 Qin, J., He, C., Zhao, N. et al. (2014). Graphene networks anchored with Sn@graphene as lithium ion battery anode. *ACS Nano* 8 (2): 1728–1738.

70 Chen, Y., Song, B., Chen, R.M. et al. (2014). A study of the superior electrochemical performance of 3 nm SnO_2 nanoparticles supported by graphene. *J. Mater. Chem. A* 2: 5688–5695.

71 Gruner, G. (2006). Carbon nanotube films for transparent and plastic electronics. *J. Mater. Chem.* 16: 3533–3539.

72 Reuss, R.H., Chalamala, B.R., Moussessian, A. et al. (2005). Macroelectronics: perspectives on technology and applications. *Proc. IEEE* 93 (7): 1239–1256.

73 Ahn, J.H., Kim, H.S., Lee, K.J. et al. (2006). Heterogeneous three-dimensional electronics by use of printed semiconductor nanomaterials. *Science* 314: 1754–1757.

74 Sun, Y.G. and Rogers, J.A. (2007). Inorganic semiconductors for flexible electronics. *Adv. Mater.* 19: 1897–1916.

75 Eda, G., Fanchini, G., and Chhowalla, M. (2008). Large-area ultrathin films of reduced graphene oxide as a transparent and flexible electronic material. *Nat. Nanotechnol.* 3: 270–274.

76 Besteman, K., Lee, J.O., Wiertz, F.G.M. et al. (2003). Enzyme-coated carbon nanotubes as single-molecule biosensors. *Nano Lett.* 3 (6): 727–730.

77 Snow, E.S., Perkins, F.K., Houser, E.J. et al. (2005). Chemical detection with a single-walled carbon nanotube capacitor. *Science* 307: 1942–1945.

78 Modi, A., Koratkar, N., Lass, E. et al. (2003). Miniaturized gas ionization sensors using carbon nanotubes. *Nature* 424: 171–174.

79 Chen, P.C., Ishikawa, F.N., Chang, H.K. et al. (2009). A nanoelectronic nose: a hybrid nanowire/carbon nanotube sensor array with integrated micromachined hotplates for sensitive gas discrimination. *Nanotechnology* 20 (12): 125503.

80 Li, J., Lu, Y., Ye, Q. et al. (2003). Carbon nanotube sensors for gas and organic vapor detection. *Nano Lett.* 3: 929–933.

81 Liu, Z., Davis, C., Cai, W. et al. (2008). Circulation and long-term fate of functionalized, biocompatible single-walled carbon nanotubes in mice probed by Raman spectroscopy. *Proc. Natl. Acad. Sci. U.S.A.* 105: 1410.

82 Akinwande, D., Yasuda, S., Paul, B. et al. (2008). Monolithic integration of CMOS VLSI and carbon nanotubes for hybrid nanotechnology applications. *IEEE Trans. Nanotechnol.* 7: 636–639.

83 Deng, J., Ghosh, K., and Wong, H.S.P. (2007). Modeling carbon nanotube sensors. *IEEE Sensors J.* 7: 1356–1357.

84 Bianco, A., Kostarelos, K., and Prato, M. (2005). Applications of carbon nanotubes in drug delivery. *Curr. Opin. Chem. Biol.* 9: 674–679.

85 de Heer, W.A., Chatelain, A., and Ugarte, D. (1995). A carbon nanotube field-emission electron source. *Science* 270: 1179–1180.

86 de Jonge, N., Lamy, Y., Schoots, K., and Oosterkamp, T.H. (2002). High brightness electron beam from a multi-walled carbon nanotube. *Nature* 420: 393–395.

87 Choi, W.B., Chung, D.S., Kang, J.H. et al. (1999). Fully sealed, high-brightness carbon-nanotube field-emission display. *Appl. Phys. Lett.* 75: 3129–3131.

88 Yue, G.Z., Qiu, Q., Gao, B. et al. (2002). Generation of continuous and pulsed diagnostic imaging x-ray radiation using a carbon-nanotube-based field-emission cathode. *Appl. Phys. Lett.* 81: 355–357.

89 Frackowiak, E. (2007). Carbon materials for supercapacitor application. *Phys. Chem. Chem. Phys.* 9: 1774–1785.

90 Kymakis, E. and Amaratunga, G.A.J. (2002). Single-wall carbon nanotube/conjugated polymer photovoltaic devices. *Appl. Phys. Lett.* 80: 112–114.

91 Sundaram, R.M., Sekiguchi, A., Sekiya, M. et al. (2018). Copper/carbon nanotube composites: research trends and outlook. *R. Soc. Open Sci.* 5 (11): 180814.

92 Dalton, A.B., Collins, S., Munoz, E. et al. (2003). Super-tough carbon-nanotube fibres. *Nature* 423: 703.

93 Keefer, E.W., Botterman, B.R., Romero, M.I. et al. (2008). Carbon nanotube coating improves neuronal recordings. *Nat. Nanotechnol.* 3: 434–439.

94 Rebecca, A.M., Brendan, F.L., Gunaranjan, V. et al. (2005). Collagen–carbon nanotube composite materials as scaffolds in tissue engineering. *J. Biomed. Mater. Res. Part A* 74A: 489–496.

95 Zhang, M., Fang, S., Zakhidov, A.A. et al. (2005). Strong, transparent, multifunctional, carbon nanotube sheets. *Science* 309 (5738): 1215–1219.

96 Lee, J.A., Shin, M.K., Kim, S.H. et al. (2011). Hybrid nanomembranes for high power and high energy density supercapacitors and their yarn application. *ACS Nano* 6 (1): 327–334.

97 Lin, H., Li, L., Ren, J. et al. (2013). Conducting polymer composite film incorporated with aligned carbon nanotubes for transparent, flexible and efficient supercapacitor. *Sci. Rep.* 3: 1353.

98 Zhou, R., Memg, C., Zhu, F. et al. (2010). High-performance supercapacitors using a nanoporous current collector made from super-aligned carbon nanotubes. *Nanotechnology* 21 (34): 345701.

99 Wu, Y., Wei, Y., Wang, J. et al. (2013). Conformal Fe_3O_4 sheath on aligned carbon nanotube scaffolds as high-performance anodes for lithium ion batteries. *Nano Lett.* 13 (2): 818–823.

100 Kang, K., Luo, S., Wu, Y. et al. (2013). Super-aligned carbon nanotube films as current collectors for lightweight and flexible lithium ion batteries. *Adv. Funct. Mater.* 23 (7): 846–853.

101 Fu, K., Yildiz, O., Bhanushali, H. et al. (2013). Aligned carbon nanotube-silicon sheets: a novel nano-architecture for flexible lithium ion battery electrodes. *Adv. Mater.* 25 (36): 5109–5114.

102 Zhang, H.X., Feng, C., Zhai, Y.C. et al. (2009). Cross-stacked carbon nanotuber sheets uniformly loaded with SnO_2 nanoparticles: a novel binder-free and high-capacity anode material for lithium-ion batteries. *Adv. Mater.* 21 (22): 2299–2304.

Part IV

Chemically Modified Carbon Nanotubes: Biomedical Applications

13

Chemically Modified Carbon Nanotubes in Cancer Therapy

Tejas Agnihotri[1,], Shiv K. Prajapati[2,*], Shyam S. Gomte[1], and Aakanchha Jain[1]*

[1] *National Institute of Pharmaceutical Education and Research (NIPER)-Ahmedabad, Department of Pharmaceutics, Palaj, Opp. Air Force Station, Gandhinagar, 382355, Gujarat, India*
[2] *Ram-Eesh Institute of Vocational and Technical Education, Department of Pharmacy, Greater Noida, 201310, Uttar Pradesh, India*

13.1 Introduction

Cancer has been considered one of the most terrible diseases affecting humans for almost a century. It poses not only a clinical and social burden on society but also a great economic burden. According to World Health Organization (WHO) reports in 2016, cancer has been ranked as the leading disease in the world affecting both men and women. In 2018, around 18 million people were diagnosed with one type or another of cancer, where lung, breast, and prostate cancers were three of the most common cancers among all types of cancers. Regarding mortality, cancer has been ranked second after ischemic heart disease and accounted for approximately 9 million deaths per year, worldwide, according to WHO updates in 2016. It is estimated that by 2060 cancer mortality will overtake mortality caused by ischemic heart diseases. Therefore, different therapeutic options for cancer treatment must be used in terms of a potent anticancer molecule, a therapeutic molecular target, or an effective drug carrier system. Current options revolve around chemotherapy, surgical removal of the tumor, and radiation therapy that seem to be inadequate to completely control the progression of metastasis and tumor growth [1].

Nanocarriers were one of the most remarkable options in the drug delivery system, especially for cancer treatment. Carbon nanotubes (CNTs) have sparked interest in nanotechnology from both a fundamental and advanced point of view [2, 3].CNTs primarily discovered by Iijima in 1991, have gained a lot of attention since then due to their excellent mechanical, electrical, and thermal properties, ability to functionalize, and increased stability [4]. Structurally, CNTs are made up of sp^2 hybridized carbon atoms arranged hexagonally with a diameter in the range of nanometers. Its inner surface is hollow and looks cylindrical in shape [3]. CNTs were found useful in the treatment of cancer due to their ability to carry drugs and other biological

* These authors contributed equally.

Chemically Modified Carbon Nanotubes for Commercial Applications, First Edition.
Edited by Jeenat Aslam, Chaudhery Mustansar Hussain, and Ruby Aslam.

molecules and also have the ability to interact with different ligands, proteins, antibodies, and enzymes. CNTs are essentially divided into two classes, single-walled carbon nanotubes (SWCNTs) wherein, a single sheet of graphene is rolled into a cylindrical shape and multi-walled carbon nanotubes (MWCNTs) wherein, multiple sheets of graphene are rolled into a cylindrical shape [4]. Figure 13.1 represents the recent developments of chemically modified CNTs in the cancer therapy

13.2 Carbon Nanotubes as Novel Nanocarriers

CNTs possess various attractive properties and distinctive physicochemical attributes, which potentiate them as an alternative carrier for the delivery of various kinds of drug substances. Iijima stated that the coiled graphene layers were joined by London's force of attraction [3, 10].

These types of interlinkage result in CNT packing together with the generation of substantial aggregates. CNTs have specific attributes such as greater tensile strength, lower weight, and exceptional transport conductivity with enhanced chemical and thermal stability. Since CNTs have a large surface area, they can easily be attached to larger molecular weight compounds such as enzymes, proteins, and various therapeutic agents [11–13].

13.2.1 CNTs as an Alternative to Conventional Carriers

CNTs are nanocarriers with the one-dimensional morphological arrangement and are studied as allotropes of carbon (second generation), after graphite and carbon. Chemically, these contain pure carbon molecules structured in a well-ordered way with hexagonal arrangement for tubes in a cylindrical shape [14, 15]. Moreover, they are also known as fullerene family compounds that differ from the rest compounds by their arrangement and amount of carbon molecules present [15]. There are several fascinating properties of CNTs such as their unique configuration, well-ordered architecture connected with high aspect ratio, exceptional chemical modification ability, and well-known surface functionalization attributes which make them suitable alternative nanocarriers for the applications of various therapeutic agents [16]. In addition, a larger inner volume and immobilization capacity help to make them biocompatible. Interestingly, CNTs are widely used in targeted drug delivery, controlled drug delivery, cancer therapy, transdermal drug delivery, and as carriers for various therapeutic agents such as DNA, RNA, siRNA, peptides, and proteins [14, 17, 18]. Figure 13.2 describes the overview of the distinct applications of functionalized CNTs with a variety of molecules including chemotherapeutic drugs, proteins, DNA, and siRNA. [19].

The exceptional conformational, biocompatible, and physicochemical attributes of CNTs with eased surface functionalization are useful in a wide range of applications. As compared to another conventional nanocarrier system, it provides a larger surface area for the loading of various therapeutic agents. Overall, with the applications mentioned above of CNTs, it can be considered as an alternative to other conventional nanocarrier systems [15].

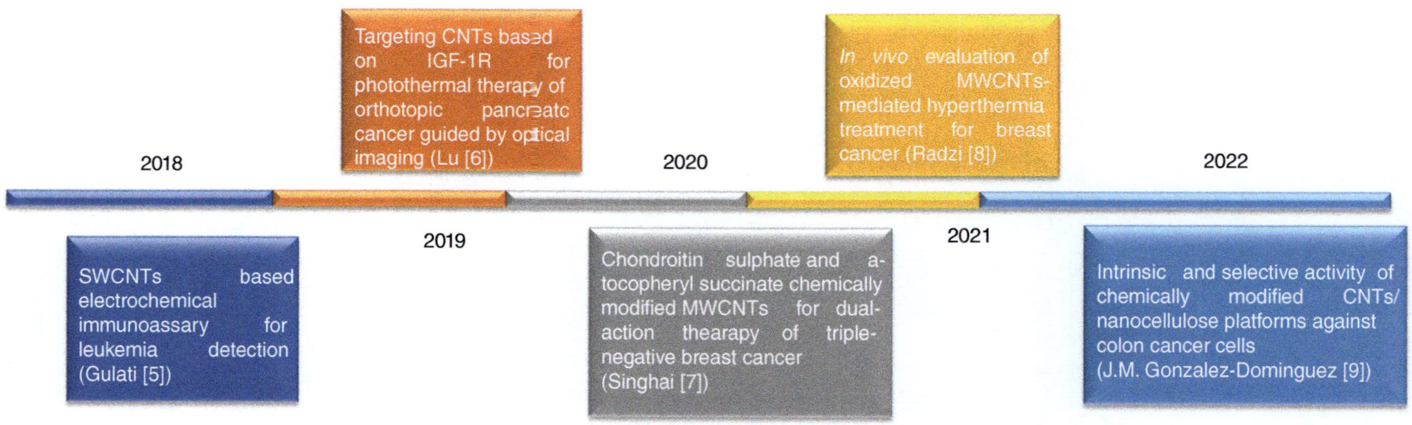

Figure 13.1 The recent development of chemically modified carbon nanotubes in cancer therapy. IGF-R = insulin-like growth factor-1.

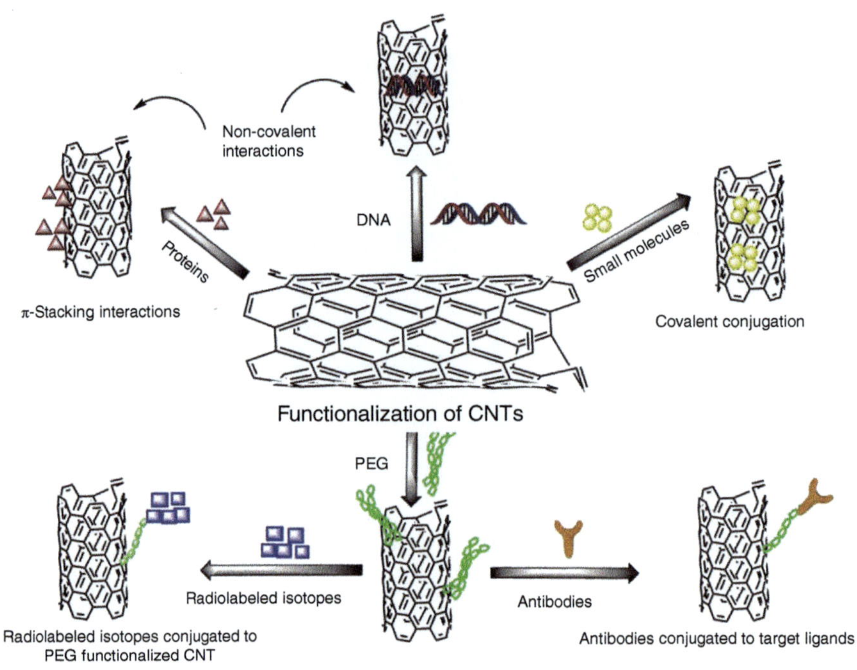

Figure 13.2 Overview of functionalization of carbon nanotubes using different molecules and their biomedical applications.

13.2.2 Biodistribution, Biocompatibility, and Cellular Uptake Mechanism of CNTs

The biodistribution of CNTs is usually not affected by the administration route. It has been revealed from the previous studies that these nanocarriers distribute quickly to the primary organs such as kidneys, bone, and stomach. In addition, the highest amounts of CNT were excreted from urine and a much smaller amount was excreted from feces. Interestingly, the functionalization approach of SWCNTs helps to increase their blood circulation time, decrease uptake by the reticuloendothelial system (RES), and total clearance from distinct organs. Functionalization helps to carry the therapeutic agents to the target sites. To accomplish the required therapeutic response, the cellular internalization of these nanocarriers needs to be addressed. Cellular internalization is primarily based on the surface charge, physical attributes, and chemical modifications of CNTs. Other parameters include the generation of supramolecular complexes and the degree of dispersion [20].

In context with biocompatibility, the original CNTs, residual transition metal elements such as cobalt, iron, and nickel may activate intracellular production of free radicals and finally oxidative stress (OS), resulting in cytotoxicity. For example, treating HEK293 (human embryonic kidney) cells with original SWCNTs promoted cell apoptosis and decreased cell proliferation [21]. Furthermore, conditioning of the original MWCNTs with calu3 human epithelial cells resulted in decreased cell viability. On the other hand, chemically modified CNTs exhibit greater biocompatibility when compared with pristine CNTs. It can be observed that the reaction of original CNTs with other chemicals aids in eliminating metal elements attached to the wall of CNTs through washing with organic solvents [22, 23]. Furthermore, L929 cells cultured with chemically modified MWCNT in greater amounts (25–100 $\mu g\,ml^{-1}$) had much-reduced multiplication compared to untreated cells. Similarly, incubation of human embryonic kidney (HEK) with acid-modified SWCNT resulted in a dose-based decrease in cell viability, as well as an increase in the production of pro-inflammatory cytokines interleukin-8 (IL-8, 19). Furthermore, the CNT cytotoxicity study on immune cells has been analyzed in many investigations. Images of transmission electron microscopy (TEM) revealed that carboxylated modified SWCNTs produced fewer intracellular aggregates and had a little influence on cell viability than pure SWCNTs after incubation with human monocyte-derived macrophages [24].

For the cellular uptake of CNTs, their needle-shaped structure allows them to penetrate cell membranes easily. As a result, the absorption capabilities of CNTs make them ideal for a variety of medicinal applications. Some of the mechanisms proposed for cell internalization are direct penetration into cells, passive uptake, and active uptake. In addition to these, there were five classes of endocytic processes by which CNTs can be taken up by cells. These are phagocytosis, pinocytosis, clathrin-mediated endocytosis, caveolin-mediated endocytosis, and clathrin/caveolae independent endocytosis. Phagocytosis is the process in which cells can take up particles as large as 1 μm and find them in macrophages, monocytes, and neutrophils. For clathrin/caveolin-mediated endocytosis, particles of around 100 nm diameter can be taken up, while particles greater than 300 nm can be taken up by micropinocytosis. Thus, depending on the structure, type, charge, diameter, and functionalization of CNTs, the uptake mechanism may vary [25]. Figure 13.3 represents a detailed mechanism of receptor-mediated cellular internalization of CNTs.

13.3 The Need for Chemical Modification and Its Importance

As mentioned earlier, CNTs possess varied fascinating properties for delivering various therapeutic agents. Besides this, the CNTs outer walls are chemically inert in nature and therefore, chemical modifications can be done for improving solubility and biocompatibility [26, 27]. The chemical modification of CNTs can be done by two primary approaches, covalent modification, and noncovalent modification. In a covalent modification, direct covalent linkages are involved, whereas, in noncovalent modifications, large polymer chains are twisted around the CNTs [28]. The application of CNTs has been restricted because of the low Van der Waals interaction and the interfacial interaction between the polymer system and CNTs. To overcome dispersion difficulties with CNTs, few techniques were developed to alter the surface attributes of CNTs. As stated earlier, these include covalent (chemical) and noncovalent (physical) modifications [29].

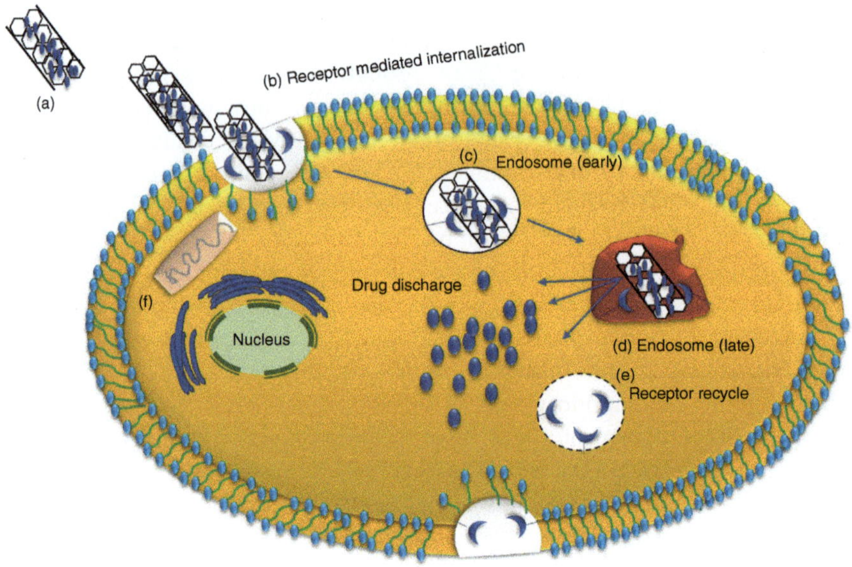

Figure 13.3 Schematic representation of cellular internalization of carbon nanotubes. (a) Drug-loaded carbon nanotubes for internalization of cells. (b) Receptor-mediated endocytosis. (c) Drug-loaded encaptured by endosome through receptor-mediated endocytosis. (d) The entry of carbon nanotubes into the late endosome and subsequent release of the drug via late endosomal pathway (e) receptor recycle (f) cellular organelles such as nuclear, endoplasmic reticulum, and mitochondria.

13.4 Chemically Modified Approaches of Carbon Nanotubes

CNTs can be surface modified to improve their solubility and biocompatibility with a biological environment, allowing them to be used as a drug carrier to their full potential. CNTs are attractive carriers for functionalization since they have a hexagonal compact sealed cylindrical shape with constrained $sp^2{}_z$ hybridized carbons [30]. An imbalance of orbitals is the cause of reactivity on the surface of CNTs. To make CNTs biocompatible, they are attached to polysaccharides, nucleic acids, and proteins [31]. The cap region of thin CNTs has the most reactive area, while the perpendicular axis of large diameter CNTs has the least reactive area [32, 33]. Chemical techniques that initiate new chemical moieties or result in new bond formation are used to alter or modify the surface of CNTs. The progression of improvement in encapsulation affinity, biocompatibility, and solubility are the key goals of functionalizing CNTs in cancer care. Covalent and noncovalent bonding can be used to functionalize CNTs [34, 35]. Figure 13.4 illustrates different strategies for the functionalization of CNTs for cancer treatment [36].

13.4.1 Covalent Chemical Modifications of CNTs

In CNTs, because of their curved and hemispherical shape similar to that of fullerene, the end of CNTs is potentially reactive as compared to the sidewalls

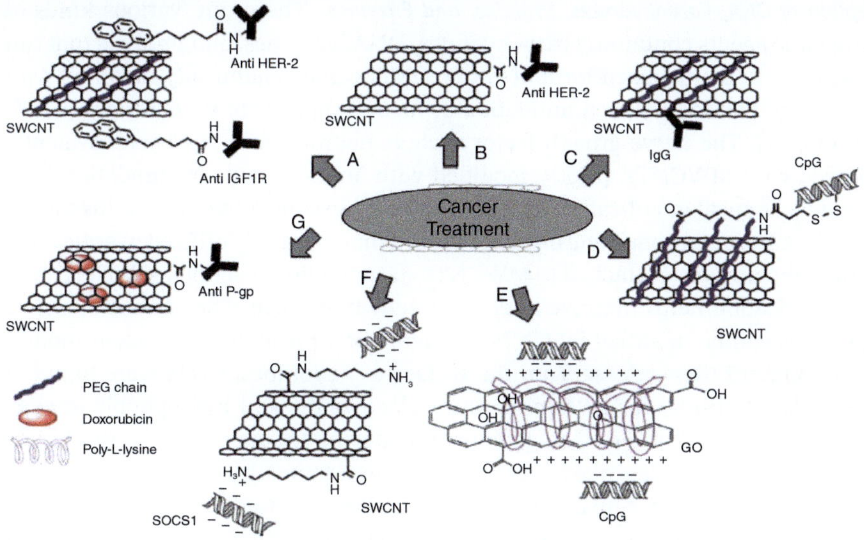

Figure 13.4 Different strategies using functionalized CNTs for the treatment of cancer. Source: Battigelli et al. [36]/The Royal Society of Chemistry/CC BY 3.0.

of CNTs. The defects such as stonewalls defects, sp^3-hybridized dislocations, and vacant spaces in the CNTs structure are all pentagon and heptagon pairings on these sidewalls. Chemical modification is mainly based on the formation of covalent bonds. CNT sidewalls possess an inert nature, and hence covalent chemical modification along CNTs fluorination happens over end caps. The chemical functionalization of CNTs entails the creation of covalent bonds between carbon and chemical moieties [34, 35]. Covalent chemical modifications on sidewalls are connected with alteration of hybridization via sp^2 to sp^3 with consequently, destruction of p-conjugation over the surface of graphenes. Covalent chemical modifications can be done by treatment with greatly reactive molecules. Additionally, cycloaddition reactions such as the addition of nitrene and carbene, the Diels–Alder reaction, bromination, chlorination, and hydrogenation were widely studied [35].

13.4.1.1 Oxidation of CNTs

For the biomedical applications of CNTs, their size must be reduced, as an increase in size greater than 10 μm leads to toxicological reactions [37]. Most techniques for cutting CNT are based on acid-catalyzed reactions such as H_2SO_4/HNO_3, HNO_3, $KMnO_4/H_2SO_4$, and others include OsO_4 [38–42]. In addition, reactions may be performed under sonication to enhance the reduction of the size of CNTs. In the defective location of CNTs, oxygen-consisting groups, primarily COOH, are established in these circumstances. The acid reaction time can regulate the length of MWCNTs and the degree of carboxylation of MWCNTs. Moreover, by amidation with amine and esterification with alcohol moieties, the –COOH consisting groups can be later derivatized. To ensure covalent bonding, carboxylic acids must be preactivated [38].

Coupling of DNA, Carbohydrates, Proteins, and Enzymes There are various kinds of therapeutic agents containing carbohydrates, DNA, enzymes, and proteins that can be attached to the oxidized form of CNTs. CNTs can be chemically modified with proteins by diamide-induced amidation at room temperature without loss of efficacy [38, 43]. The nerve growth factor such as neurotrophin has been covalently modified with MWCNTs, priorly modified with amino groups by amidation [44]. Covalent chemical modification has also been used in cancer therapy. In this aspect, tumor lysate protein consisting of different proteins commonly utilized in anticancer immunotherapy, have attached to MWCNTs via amidation reaction [45]. The chemical modification helps improve anticancer activity in mice. The various bioactive substances can be modified by CNTs through stimuli-responsive covalent modification to retard their release from the surface of nanotubes. CNTs were found to have a higher affinity for proteins through electrostatic and hydrophobic interactions and may result in protein agglomeration during the process. As a result, it is critical to thoroughly describe the conjugates to determine the nature of the protein link. A novel Staudinger–Bertozzi coupling has been devised as another approach for direct protein amidation reaction on oxidized CNTs, with no loss of protein function [38, 46]. By incorporating amino moieties and an azido tyrosine over CNTs containing a triaryl phosphine secondary, this can permit the development of a bond in azide groups which is available in protein.

Moreover, peptides can also be linked with CNTs through the amidation reaction. For example, gonadotropin-releasing hormone (GnRH), a peptide hormone upregulated in the plasma membrane of cancer cells, has been modified with covalent linkage over oxidized CNTs [38, 47]. The GnRH/MWCNTs were capable to infiltrate androgen-based prostate cancer cells, causing dose-dependent cytotoxicity. Furthermore, DNA can be chemically modified to CNTs via amide attachment in carbodiimide-stimulated coupling interactions [38]. The capacity of the DNA/CNT complex to hybridize with a complementary sequence of DNA opens up the possibility of biosensor applications. DNA can be covalently modified to SWCNTs, initially modified by amidation by amine function, which has been reacted with a linker having a ring of sulfhydryl-reactive maleimide. The maleimide group over the CNTs reacted exclusively with the 5′ end with a modified DNA thiol group [38, 48].

Enzymes can be attached to nanotubes through amidation of the enzymes' amine residues with the carboxyl moieties of the oxidized CNTs. CNT conjugates depend on enzymes that can be used as biosensors [38, 49]. Carbohydrates were covalently modified with SWCNTs for a wide range of applications as decontamination and detecting agents [38, 50, 51]. SWCNTs can be used as a framework to show many monosaccharide ligands for bacterial pathogen identification. In these aspects, through amidation, galactose-confined amino moieties were connected to oxidized SWCNTs. Mannose was linked to SWCNTs using a similar approach. In another interesting study, both mannose/SWCNT and galactose/SWCNT were made to attach to anthrax spores efficiently, inducing spore agglomeration and resulting in a significant decrease in colony-forming units [38].

Coupling of Small Drugs The various types of therapeutic agents can be covalently modified with CNTs through functionalization reactions. The anti-inflammatory, antimicrobial agent such as dapsone has been attached to MWCNTs through an amidation reaction. The reaction of rat peritoneal macrophages treated with dapsone/fluorescein isothiocyanate (FITC)/MWCNT was examined after the amine group over the dapsone-composed aromatic ring was modified via FITC reagent. Similarly, the antineoplastic agent can be attached via the covalent chemical modification technique. In this aspect, the dendrimer poly(amidoamine) (PAMAM) modified with folic acid and fluorescein, which target tumor cells upregulating great binding with folic acid receptors, has been attached to MWCNTs through an amidation strategy. The design was used to target cancer cells *in vitro* and was also used for imaging purposes. The multifunctional CNT-based complex can be made by modifying them using dendrimers [52].

Coupling of Ligands Similarly, CNTs can be chemically modified with ligands and fluorophores, and they were studied for *in vivo* and *in vitro* investigations [53]. For example, epidermal growth factor (EGF), has been linked to oxidized SWCNTs through amidation reaction to employ in squamous cell cancer. Likewise, the antineoplastic drug such as cisplatin, $Cl_2Pt(NH_2)_2$, was linked to SWCNTs through ligand exchange reaction via complexation with remaining –COOH function that did not react at the step of linking of EGF. Interestingly, they fabricated a material by linking amino polyethylene glycol (PEG)-quantum dot, in place of cisplatin, to the carboxyl group of SWCNTs via amidation linkage and this was further linked to EGF receptor ligand. When compared to the non-targeted conjugation, the EGF/cisplatin/SWCNT combination was able to efficiently and selectively target squamous cancer cells having EGF receptors being overexpressed on them [38, 54]. Figure 13.5 represents functionalized CNT with cisplatin and EGFR and their TEM images [55].

13.4.1.2 1.3-Dipolar Cycloaddition of Azomethine Ylides

The 1,3-dipolar cycloaddition of azomethine ylides is another tool for chemically modifying CNTs and has various applications specifically in nano-formulations. In particular, this type of reaction occurs between an aldehyde, α-amino acid, and results in the production of an iminium salt which further changes to azomethine ylide through thermally promoted decarboxylation. The ylides are formed *in situ* and incorporation on the surface of the nanotube consequently results in the production of pyrrolidine rings [38].

Coupling of Peptides CNTs chemically modified with 1,3-dipolar cycloaddition can be altered with peptides with wider applications such as synthetic peptide-dependent vaccine administration. In this regard, the peptide via the foot and mouth disease virus (FMDV) can be attached to the surface of the nanotube [38, 56]. To attach the peptide, NH_3^+-TEG/SWCNTs were incorporated with *N*-succinimidyl 3-maleimidopropionate. The acetylated FMDV peptide has cysteine at N-terminus and permits interact with maleimido-SWCNTs through chemo-selective ligation in the thiol function of cysteine and ring of maleimide. Later, a number of nuclear magnetic resonance (NMR) experiments were done to confirm the covalent bond link between CNTs and the peptide. The modified products were immunogenic and

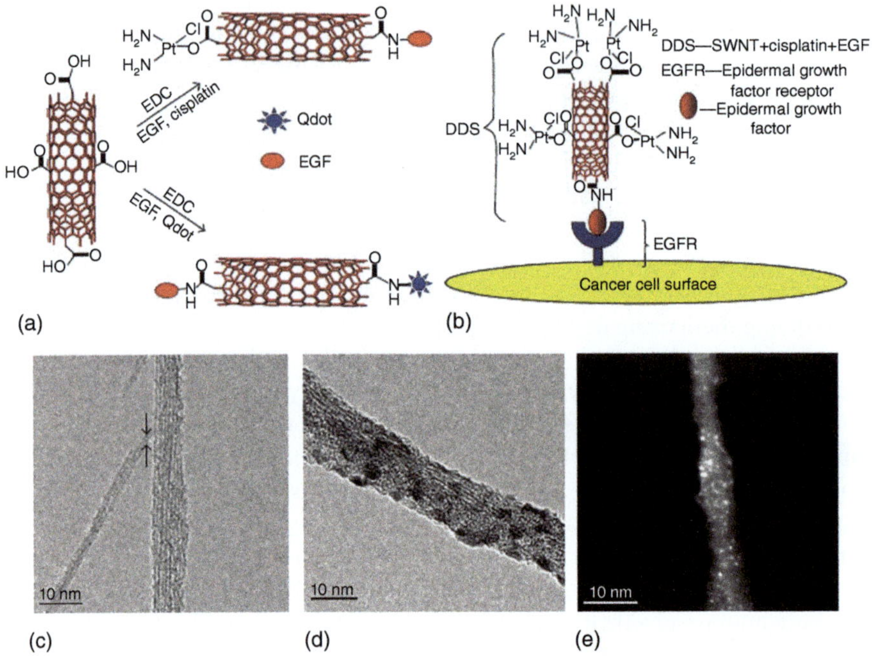

Figure 13.5 (a) Schematic representation of functionalization of SWCNTs with cisplatin. (b) SWCNT bioconjugated with cisplatin and EGF targeting EGFR. (c–e) TEM images with cisplatin as bright spot. Source: (c–e) Elhissi et al. [55]/with permission of John Wiley & Sons.

produced a higher antipeptide antibody response in mice. However, CNTs without peptides were non-immunogenic in nature.

Coupling of Antibodies Similarly, CNTs can also be linked to the antibodies such as monoclonal antibodies. The tumor-specific monoclonal antibody can be linked to NH_2-TEG/SWCNTs. The amino groups initially present in the SWCNTs were modified with 1,4,7,10-tetraazacyclododecane-1,4,7,10-tetraacetic acid (DOTA)-NCS and other bare amines were attached to the linker having a maleimide ring. Lintuzumab and Rituximab antibodies were interacted with 2-iminothiolane-HCl to bring in sulfhydryl functions for incorporations over the maleimide-SWCNTs followed by chelation of ^{111}In by DOTA. The antibody-modified radiolabeled SWCNTs were capable of particularly targeting a cancer xenograft *in vivo* and human tumor cell line *in vitro* [57].

Coupling of Drugs The CNTs can easily be linked to different classes of drugs. Double functionalization of MWCNTs with therapeutic agents and fluorescent probes was done by the amalgamation of amidation and 1,3-dipolar cycloaddition [42]. The carboxyl group of oxidized CNTs was initially derivatized by amidation with the diamino-TEG monoprotected by phthalimide moiety. The aminated MWCNTs were further modified by cycloaddition utilizing para-formaldehyde and α-amino acid with Boc mono-prevented diamino-TEG chain. Subsequently, with the suitable elimination of phthalimide, the amino groups were modified with FITC. The subsequent step was the breakage of the Boc group and the attachment of the ammonium group with amphotericin B, which is used in treatment of fungal infection. Modification of CNTs with amphotericin B resulted in a reduction in the cytotoxicity of the therapeutic agent toward mammalian cells and therefore helped to preserve the greater antifungal efficacy [58].

13.4.1.3 Halogenation

The SWCNTs possess the capability to react with peroxytrifluoroacetic acid under ultrasonication and can give oxygen-based chemical moieties and trifluoroacetic moieties covalently linked to the SWCNTs surface [59]. Fluorinated CNTs can be functionalized by α, ω-amino acids in the availability of base reactants such as pyridine. The solubility of functionalized amino acids in SWCNT can be regulated by the length of the aliphatic moieties of the substituent. In specific, 6-aminohexanoic acid derivative was used in enhancing the solubility in aqueous solution within pH range 4–11. However, 11-aminoundecanoic acid and glycine derivatives are not soluble at pH values of 1–14. In a similar study, fluorinated SWCNTs were chemically modified with urea, guanidine, thiourea, and aminosilane groups, which covalently linked to the sidewalls of CNTs as a result of the substitution of fluorine atoms. From the spectroscopic studies, it was revealed that guanidine and urea interact with fluorinated SWCNTs via their amino group producing C–N linkages with sidewalls of SWCNTs. However, thiourea was linked to the sidewalls of SWCNTs due to C—S bond interactions [60, 61].

13.4.1.4 Nucleophilic and Electrophilic Additions

Hirsch and coworkers used the nucleophilic addition of organomagnesium and organolithium compounds on the sidewalls of the SWCNTs forming negatively-charged intermediates, later due to air reoxidation, which resulted in unbundled, neutral and greatly modified SWCNTs by alkyl groups [61, 62]. This type of reaction exhibits high selectivity of metallic materials toward the semiconducting SWCNTs due to the presence of an electronic state closer to the Fermi level. Furthermore, the reactivity of SWCNTs toward the incorporation of organometallic substituents was established to be inversely proportional to the tube diameter and was based on the steric property that was added. Similarly, Syrgiannis and coworkers used the addition of sidewall amines to SWCNT, which consequently formed lithium amides *in situ* and resulted in amino-modified SWCNTs that increased solubility in organic solvents [61].

In the case of electrophilic reactions, alkyl halides addition to SWCNTs with the help of microwave irradiation results in the linking of hydroxyl and alkyl groups on the nanotube surfaces. This induction of alkyl moieties increases the reaction of SWCNTs and polymer-containing chains. In addition, the hydroxyl group can again be esterified by acyl halide to get an ester derivative of SWCNTs [61, 63]. Moreover, Tessonier et al. showed covalent linking of amino moieties toward the defective sites of the surface of MWCNTs, depending on the deprotonation–metalation mechanism, by using higher butyl lithium, with electrophilic substitution reaction.

13.4.1.5 Electrochemical Modifications

Electrochemistry is an exceptional technique for the chemical modifications of nanotubes. The suitable chemical modification of metallic SWCNTs with electrochemically formed aminophenyl radicals was attained and a later generation of CNT-based pH sensor was illustrated. In another interesting report, the electrochemical nitration of SWCNTs by nitro groups was obtained by utilizing a 6 M aqueous solution of potassium nitrite, that was oxidized by an anodic reaction [64]. Moreover, Sacher et al. investigated the morphological and chemical modifications of MWCNTs by using 2 keV Ar^+. The tubes after modifications were examined by high-resolution transmission electron microscopy (HRTEM), field emission scanning electron microscopy (FESEM), Raman spectroscopy, and X-ray photoelectron spectroscopy (XPS). These studies revealed the formation of free radical defects and amorphization of MWCNTs by Ar^+ dose *od* as small as 10^{13} ion cm^{-2}. Greater doses resulted in the destruction of material by fragmentation [61].

13.4.1.6 Radical Additions

Tour was the leading scientist to suggest that utilizing *in situ* produced aryl diazonium salts to covalently altered CNT sidewalls would result in chemically modified and suitable mixed hybrid mixtures in organic and aqueous solvents. Few theoretical tools and various experimental variants and refinements of already-used processes were offered in recent years, while investigations were conducted to understand the selectivity and mechanism of the reaction. Stephenson reported employing sodium nitrite in a combination of 96% sulfuric acid and ammonium persulfate to modify SWCNTs with aniline derivatives instead of the caustic and toxic oleum. Recently, the solvent-free approach was given by Chen et al. Furthermore, the Tour group illustrated that the chemical modification of SWCNTs in aqueous solutions was facilitated by employing organic triazene molecules as firm precursors to diazonium salts [65–67]. Covalently attached CNTs were spun into microfibers and used as hydrogen storage scaffolds in a recent study employing methylenedianiline as the diazonium salt precursor. The researchers discovered that these 3-D nanoengineered fibers physisorb two times as much hydrogen per unit surface area as standard macroporous carbon materials. Moreover, Darabi et al. showed the covalent linking of thioamide moieties to the sidewalls of CNTs in the initial two steps. Initially, the chemistry of diazonium salt was done by using 4-acetylaniline, for the formation and direct covalent linking of acetophenone groups on the SWCNTs. In a later procedure, by taking the steps of the Willgerodt–Kindler reaction, the acetophenone groups present in the SWCNTs interacted with morpholine and sulfur in dimethylformamide (DMF) to form the last thiomorpholides [61, 68].

13.4.1.7 Nitrene Cycloaddition

The nitrene cycloaddition, the C_2B_{10} carborane complex, was linked to the sidewalls of the SWCNTs by the nitrene cycloaddition reaction mechanism [61, 69]. In a basic reflux condition, the ring of aziridine was opened to formed hydrophilic SWCNTs in which the CNTs sidewalls were modified by ethoxide groups and the C_2B_{10}

carborane complex. The tissue distribution investigation suggested that boron atoms were aggregated in tumor tissues as compared to blood and different organs.

Moreover, in another interesting study of the nitrene cycloaddition reaction, many perfluoroaryl azides were used as precursors of nitrene for covalent modifications of vertically exhibited CNTs [70]. In addition, Bingel cyclopropanation was also used in CNTs in previous years. Therefore, very short SWCNTs were unbundled under strong reducing conditions and modified *in situ* with malonic acid bis-(3-*tert*-butoxycarbonylaminopropyl) ester. The formed nanomaterials were having 4–5 adducts per nm of the length of the nanotube. In a recent year, Bingel reaction was employed for the sidewall chemical modification of SWCNTs in microwave irradiation and was reported to decrease reaction time to 30 minutes, while enhancing the amount of covalently attached substituents on the SWCNTs sidewall in a predetermined way by altering the output power [61, 71].

13.4.2 Non-Covalent Modifications of CNTs

Noncovalent modification of CNTs is a valuable strategy for obtaining good dispersible and processable nanotubes while preserving the sp^2 hybridization. Using aromatic compounds, polymers, peptides, and proteins, CNTs can be functionalized by stacking and/or hydrophobic interactions. For biosensing, where electronic properties need to be preserved, a noncovalent approach is suitable for improving the dispersibility and debundling of CNTs. Following are some of the noncovalent reactions which were cited for different carrier molecules.

13.4.2.1 Electrostatic and π–π Interactions

Electrostatic interactions are one of the many interactions by which CNTs can be functionalized under the umbrella of noncovalent interactions. The charges can be introduced on the surface of CNTs to immobilize any charged molecules through these interactions [38]. For instance, positively charged chitosan nanoparticles were coated on oxidized MWCNT through electrostatic interactions. The formulated nanocomposites were found to be biocompatible and can be used for various biomedical applications due to their electrorheological properties together with their excellent mechanical and electrical characteristics [72].

Apart from electrostatic interactions, aromatic compounds can be conjugated to the CNT surface through π–π interactions. For conjugating proteins such as fibrinogen and hemoglobin to CNT, 1-pyrenebutyric acid *N*-hydroxy succinimide ester was utilized. The ammonium or carboxylic acid group in aromatic compounds has been exploited for conjugation [38]. Benda et al. investigated fluorene/carbazole copolymers and their interaction with CNT through π–π interaction. Through molecular dynamics simulation data, it was concluded that this noncovalent interaction was effective against these polymers and depended on factors such as the diameter of the CNTs, the number of polymers, and the initial geometry of CNTs. Thus, through this interaction, it is possible to tailor the polymer functionalization of CNTs that can be used for other biomedical applications [73]. π–π Interactions also play an important role in the functionalization of CNT in drug delivery applications. As π–π interactions maintain the structural and functional properties of drugs and are susceptible to pH changes, they play a vital role in stimuli-responsive drug delivery when drugs are loaded into CNTs. This interaction can also enhance the aqueous dispersibility of CNTs, and due to the large surface area of CNT, a large amount of drugs can be loaded into the system [74].

13.4.2.2 Adsorption of PEGylated Derivatives

For effective drug delivery through CNTs, CNTs are often conjugated to PEG. One of the biggest advantages of PEG coating is increased circulation time of the drug delivery system owing to its prevention of recognition of nanotubes from the RES. By adsorbing PEG on the surface of CNTs, protein interactions can be minimized. Through a molecular dynamics simulation study, it was found that regardless of the chain length of PEG molecules, SWCNTs can be coated effectively, and π–π interactions can be maintained regardless of CNT size. It was also revealed that longer PEG chains-coated CNT had low cytotoxicity in *in vivo* studies [75].

13.4.2.3 Adsorption of Enzymes, Proteins, DNA, Polymers, and Carbohydrates

DNA and other nucleic acids were known to interact with the carbon backbone of CNTs under the class of noncovalent functionalization. In the preparation of sensors, single-stranded DNA has often been used in combination with CNTs. By complexation of poly[3-(3-*N*,*N*-diethylaminopropoxy)thiophene] with CNT, the glucose oxidase enzyme has been reported to be incorporated into CNT. Similarly, a glutamate detector has been made by immobilizing glutamate oxidase on the surface of CNTs to assess the levels of enzymes *in vivo*. Oxidized MWCNT has been reported to interact with FITC-labeled human serum albumin to target human hepatocellular liver carcinoma cells (HepG2 cells) [38]. Khan et al. adsorbed the lipase enzyme on the surface of CNTs and assessed the activity and efficiency of the enzyme with the help of an assay solution. The impact of pH on the immobilization of enzyme was also checked along with FESEM and FTIR characterization. The enzyme concentration increased in direct proportion to its activity but had a negative impact on immobilization efficiency due to its saturation kinetics on the surface of MWCNT [76].

13.5 Chemically Modified Carbon Nanotubes as a Nanocarrier for Cancer Delivery System

13.5.1 Chemotherapeutic Agents (Small Drugs)

Several reports have talked about the use of modified CNTs as a delivery system for small drugs. Tavakolifard et al. demonstrated that SWCNTs when functionalized with folic acid and paclitaxel can be a viable option for the treatment of cancer. For this, scientists formulated a conjugation system of hemisuccinate derivative of paclitaxel and SWCNTs and further conjugated with folic acid. It was further subjected to FTIR analysis, TGA analysis, and scanning electron microscopy (SEM) analysis. An *in vitro* study was performed on MKN45 and MCF-7 cell lines and the cytotoxic effect was assessed with the MTT colorimetric assay. The results indicated that compared to paclitaxel alone, the paclitaxel-folic acid-SWCNT complex showed higher cytotoxicity through MTT assay and the successful conjugation of folic acid with SWCNT-paclitaxel through amidation between the carboxyl end of the folic acid and the amide of the SWCNTs was demonstrated by absorption peaks of 280 nm for the folic acid. No physical interaction was observed for the complex and was confirmed through FTIR, TGA, and atomic absorption spectroscopy; therefore, chemically functionalized CNTs with paclitaxel as an anticancer agent proved to be a good platform in anticancer therapy [77].

In other instance, the same anticancer agent, paclitaxel, was loaded in MWCNTs conjugated to riboflavin and thiamine and evaluated for its anticancer potential by *ex vivo*, *in vivo*, hemolytic potential, *in vitro* drug release, and stability study and was also characterized for FTIR, NMR, and SEM analysis. The results were promising enough to establish conjugated MWCNTs as an aid in anticancer treatment with entrapment efficiency of around 82–87% with greater cytotoxic potential for conjugated systems than unconjugated ones on MCF-7 cell lines. The conjugated systems showed sustained release *in vitro* (up to 80–90% drug release after 10 hours in phosphate buffer saline; pH 7.4). The biodistribution study was carried out in Sprague–Dawley rats after intravenous administration of the formulation and observed that plasma paclitaxel concentration was found to be higher in kidneys and liver, highlighting the mechanism of CNT decline through biodistribution and metabolism with no signs of toxicity. The formulation was found to be stable at 5 °C without observation of precipitation, viscosity, or color change; however, it was found to be the least stable at 35 °C. Further, when conjugated with riboflavin, it showed the least hemolytic potential compared to unconjugated MWCNTs. In conclusion, due to the overexpression of vitamins such as thiamine and riboflavin in cancer cells, the optimized formulation had great potential in cancer treatment when conjugated with riboflavin/thiamine in MWCNT [78]. Cao et al. reported polyethyleneimine modified MWCNTs and conjugated it with hyaluronic acid with the anticancer agent, doxorubicin incorporated into it. It was further characterized for drug loading, encapsulation, *in vitro* drug release, and cell viability. The doxorubicin conjugation system was found to have a loading of 72 % with a higher drug release at pH 5.8 than at pH 7.4 showing targeted delivery to the tumor where the pH of the tumor microenvironment is around 5.8, thus showing potential in cancer treatment [79]. Scientists investigated the functionalization of SWCNTs with chitosan derivatives namely, palmitoyl chitosan and carboxymethyl cellulose chitosan and it was used for the delivery of doxorubicin. Further, folate was conjugated with a view of targeting cancer cells. The drug release was carried out in a pH 5.5 and pH 7.4 phosphate buffer saline solution and the results indicated that it was greater, that is, 70% in pH 7.4 after 72 hours compared to 20% in pH 5.5, which mimics the cancer microenvironment. The drug loading was found to be around 75% for both carrier systems and was also found to be stable under physiological conditions. For the cell viability study, the MTT assay was performed and it was found that DOX-incorporated folic acid conjugated SWCNTs had a major impact on cell viability, which was reduced to 62% from 100% (unloaded folic acid conjugated

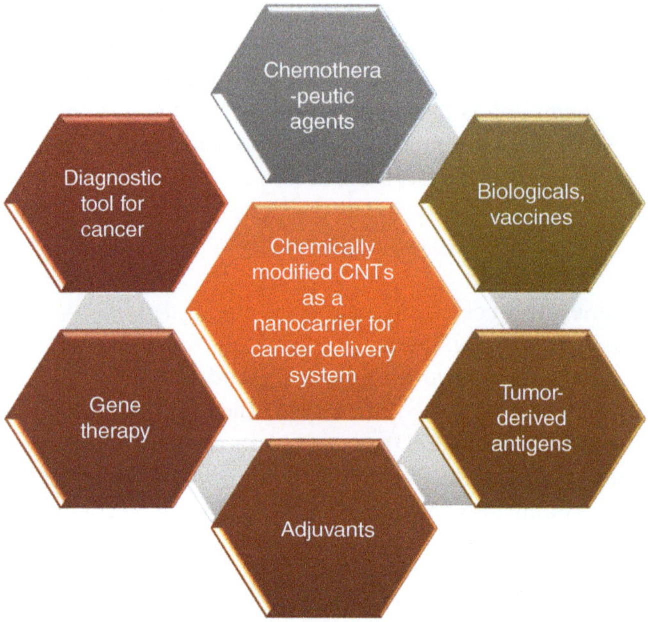

Figure 13.6 Chemically modified CNTs as a nanocarrier in various cancer delivery systems.

SWCNTs) [80]. Figure 13.6 represents chemically modified CNTs as a nanocarrier in various cancer delivery systems.

13.5.2 Biologicals, Vaccines, Tumor-Derived Antigens

Chemically modified CNTs were explored as a vector for the therapy of biologicals, vaccines, and tumor-derived antigens. In this regard, Sun et al. reported the application of carboxylated MWCNTs for the lysate protein derived from MCF-7 breast cancer called tumor P to the cells that present antigens (APC), usually dendritic cells (DCs). It was revealed from the flow cytometry study that, MWCNTs enhanced the uptake of covalently encapsulated tumor P via DCs in, *in vitro* conditions. Furthermore, DCs with MWCNT and tumor P were more promising than DCs with bare tumor P in enhancing lymphocyte-generated cytotoxicity toward MCF-7 cell lines [81, 82].

Similarly, Meng et al. investigated the efficiency of carboxylated MWCNTs to increase the anticancer immune effect produced against covalently immobilized H22 hepatic cancer cell-derived tumor lysate protein (H22P) [45]. H22 cells were implanted subcutaneously into BALB/c mice, and then the mice were injected subcutaneously with MWNT-H22P or free H22P beginning two days after the inoculation of the H22 cells. Additionally, inactivated H22 tumor cells were administered as vaccines to enhance therapy. The antitumor response was enhanced in mice that were given MWCNT-H22P. When lymphocytes were taken from MWCNTs-H22P administered mice, they exhibited greater cytotoxicity toward H22 cells than H22P-free lymphocytes given mice. Furthermore, mice with H22P were excluded, as a consequence of MWCNT-H22P therapy, successively suppressed H22 cell growth with repeated administration. This phenomenon occurred because of the initiation of antigen-specific memory T cells which had been seen in treated mice with different breast cancer cell line epithelial–mesenchymal transition (EMT) and led to potentiate cancer size.

13.5.3 Adjuvants

To enhance the full T cell activation ability, suitable stimulation of antigen-presenting cells is required [83]. However, natural activation of APCs was done with help of pattern recognition receptors (PRRs) such as C-type lectin receptors, toll-like receptors, and NOD-like receptors. These types of receptors identify ligands related to invading pathogens, called pathogen-associated molecular patterns (PAMPs). PAMPs were widely investigated in vaccine-based formulations as adjuvants to induce immunity promotion [84]. Specifically, synthetic TLR9 agonists in oligodeoxynucleotide form consisted of non-methylated deoxycytidine-deoxyguanosine dinucleotide motifs and were widely explored in cancer-based vaccine formulations [85, 86]. Transportation of the TLR9 agonist toward APCs has also been studied by utilizing CNTs. Zhao et al. studied SWCNTs transportation of CoG oligonucleotide (ODN) that may increase the cellular uptake and immune stimulatory attributes of vaccines and induce potent anticancer immune response. It was revealed from the studies that the uptake of fluorescently indicated CpG by bone marrow-derived monocytes *in vitro* was enhanced by noncovalent modification of SWCNTs. Additionally, these modifications also enhanced the generation of tumor necrosis factor (TNF-α), interleukin-12 (IL-12) by monocytes *in vitro*. Furthermore, SWCNT-CpG-modified intratumor injection effectively retarded the growth of GL261 glioma cells implanted intracranially in mice. However, when the injection was given alone as SWCNTs or as CpG, it did not result in retardation of the growth of glioma cells intracranially. Therefore, it can be concluded that the incorporation of SWCNTs with CpG is required for the induction of the anticancer immune response mechanism [82].

Similarly, Fan et al. investigated that intracranial injection of SWCNTs-CpG can suppress metastatic tumor growth in C57BL/6 mice. It can be illustrated by demonstrating that the administration of SWCNTs-CpG was more promising than bare CpG in retarding the growth of intracranial and subcutaneous inoculated B16F10 melanoma cells. Furthermore, investigating the therapeutic effectiveness toward a greater complexing glioma model, Quyang and his colleagues studied that SWCNTs-CpG with the combination of temozolomide, a chemotherapeutic agent, intratumorally administered to mice bearing intracranial K-luc glioma cell resulted in distinctive survival sustained in a manner which was distinct from the injection of bared temozolomide and CpG [82].

13.5.4 Gene Therapy

Gene therapy is the kind of molecular-based therapy that consists of the selection of DNA sequences and later generates potential ways to transport this genetic material into the cells. Gene therapy can be done using viral vectors, in which genes can be transported by viruses into cells. These viral vectors may be oncogenic, immunogenic, promote inflammations, and critical to modifications by ligand targeting mechanism. Another approach is by using nonviral vectors, in which fabricated liposomes and polymers would release bare DNA. However, these vectors are unable to permeate the nuclear membrane. To conquer these difficulties, CNTs are widely investigated as a nonviral delivery system, because of their capacity to penetrate the nucleus membrane and their suitable pharmacokinetic attributes [13].

The joining of DNA molecules to the surface of CNTs is done through different structural and functional groups that are present for interaction with CNTs. The chemically modified CNTs with DNA are used to enhance the solubility of CNTs in aqueous and organic media. Modified SWCNT-DNA has improved cell transport capacity and increased biostability compared to bare DNA, therefore to have wider applications in gene therapy [87].

13.5.4.1 Oligonucleotides (ODNs)

ODNs are the biologically functional compounds that prevent the expression of the gene in translational or transcriptional steps. Very few studies were reported on the application of CNTs for the transportation of ODNs [88]. Kiran et al. successfully studied the transportation of ODN to Hela cell lines by employing carbodiimide chemically modified SWCNTs [89].

Furthermore, Pan et al. designed encapsulated MWCNT with *C-myc* ODNs and delivered them into breast cancer MCF-7 cell lines, which promoted distinct apoptosis. Compared to siRNA, ODNs reduced specificity to target genes [89–91].

13.5.4.2 DNA/RNA Aptamers

Aptamers are huge, randomly arranged ODNs, compared to those that are usually attached to target molecules. Aptamers are easy to generate, modify, and are shorter in size when compared with proteins that are having complex nature. In addition, they are very promising for protein therapeutic delivery. Jeroen et al. illustrated an enhanced cellular uptake of MWCNTs loaded with MUC-1 aptamer on lung and breast cancer cell lines [92, 93].

Cancer antigen 125 (CA125), a transmembrane glycoprotein, is known to be the biomarker for ovarian cancer cells. The development of anti-CA125 ssDNA aptamer and its use in sensitive CA125 detection combined with CA125-specific antibody immobilized in a three-dimensional network of CNTs has been reported by Gedi et al. The chosen ssDNA aptamer has outstanding stability and is easy to modify compared to antibodies. Due to its remarkable stability and convenient modification of antibodies, the chosen ssDNA aptamer was found to have a higher binding affinity to recombinant CA125 enabling specific imaging in ovarian cancer cells [94].

The multidisciplinary study of nanobiotechnology consisting of biocompatibility, toxicology, internalization, and biomedical applications has emerged and has been studied quite extensively in the recent past. Barbosa et al. developed oxidized MWCNT noncovalently linked to DNA aptamer targeting tumor cells. It was found that even after functionalization, the nucleotide sequence of the aptamer was intact and the nanocomplex was stable through π-stacking along with good cell viability. It further elicited its use in constructing effective nanoplatforms [95].

Hemophilia, a bleeding disorder may result in arthritis and complications during surgery. It is important to know the level of factor IX (FIX), which is generally found less in hemophilia patients. Yao et al. developed SWCNT based interdigitated electrode sensor to quantitatively assess the level of FIX. Dual probing was evaluated with aptamer and antibody was found to have a detection limit of 1 pm and the detection limit for aptamer was found to be 100 fM. FIX detection was found to be improved by 10 times in the antibody–FIX–aptamer sandwich pattern, while the aptamer–FIX antibody sandwich pattern did not show improvement in detection. In summary, it was concluded that this assay proved to be useful for quantifying the level of FIX in patients suffering from hemophilia [96].

13.5.5 Diagnostic Tool for Cancer

CNTs not only deliver drugs, peptides, and nucleotides but also find a myriad of applications in the diagnosis of cancer. It can be used as a contrasting agent in ultrasonography where sound waves of 2–12 MHz are being utilized. Scientists have already explored the use of functionalized MWCNTs (with 1,3-dipolar cycloaddition of azomethine ylides) in cancer, such as in capturing ultrasound images and, in turn, targeting tumor markers. They showed that functionalized MWCNTs as a contrasting agent had a larger impact on the diagnosis compared to pristine MWCNTs and confirmed it by performing *in vitro* and *in vivo* studies. Photoacoustic imaging, another tool that is used in tumor assessment, has also been reported to have utilized CNT functionalization with PEG to cyclic Arg-Gly-Asp peptides. Another useful tool in the diagnosis of cancer is magnetic resonance imaging (MRI), which utilizes a magnetic field and aligns hydrogen atoms in water inside the body. The MRI contrast agents are divided into spin–lattice relaxation agents and spin–spin relaxation agents. Paramagnetic or superparamagnetic agents are conjugated to a CNT platform to form an active MRI CNT. For example, Gd-CNTs were administered intravenously to mice, and *in vivo* MRI revealed their absorption in the liver and spleen, confirming their ability to operate as a good CA despite their cellular internalization. Furthermore, such a complex can benefit from both intrinsicT2 contrast and oxidized intrinsicT2 contrast. To obtain a clear MRI image, CNTs, and T1 contrast derived from gadolinium ions were used. In the case of photothermal therapy, Hashida et al. formulated SWCNT composite using peptide with repetitive structure H-(Lys-Phe-Lys-Ala)-7-OH ((KFKA)7). (KFKA)7 was created to wrap SWCNTs while also being water-dispersible and stable. When hybrids were directly injected into implanted colon 26 tumors followed by NIR irradiation, it resulted in an increase in temperature up to 430 °C and a decrease in tumor size [97]. In this context, few clinical trials were under investigation. Table 13.1 enumerates clinical trials of CNTs in different phases. Table 13.2 describes patents regarding chemically modified CNTs.

Table 13.1 Clinical trials for chemically modified CNTs in the treatment of cancer.

Sr. no.	Name of the clinical study	Objective of the study	Study design	No. of participants	Outcome measures	Remarks	ID
1	Comparative investigation between CNTs based X-ray stationary breast tomosynthesis and 2D digital mammography in patient with having breast injury	To compare the radiologist confidence level in evaluating patients with known breast lesions using CNTs X-ray based stationary breast tomosynthesis and traditional mammography	Observational Model: Case-Only Time Perspective: Prospective. Study started on Jan. 2013	54	AUC and ROC investigation of the stationary breast tomosynthesis (SDBT) system for the examination of suspicious breast lesions in comparison to X-ray mammography. To evaluate Readers will be scored from −3 to +3	Completed on March 2018	NCT01773850
2	Sentinel lymph node sampling for patients with middle-high risk endometrial cancer confined to the uterus	The objective of study is to examine the SLN sampling on the prognosis of patients with middle-high risk endometrial cancer nearby uterus	Study started on February 2020. Randomized type, Intervention Model: Parallel Assignment Open label Treatment	780 (Estimated)	The percentage of patients who have first relapse within 2 years after surgery (SLN or LND) The percentage of patients who have first relapse within 5 years after surgery (SLN or LND) The occurrence rate of each adverse effects related to surgery (SLN or LND, the scores of each QOL survey	Estimated initial completion on 2025	NCT04276532

3	Buckypaper, device for surgery and solid tumor and hematopoietic cancer therapy and investigation of BP incorporation	To investigate the potential of BP in various tumors and its effects after incorporation	Interventional with phase I and II, randomized parallel treatment. Study started on October 2010	30	Capacity of the BP as bio-adhesive tape for prosthetic materials to avoid displacement of implanted prosthesis. toxicology of BP measured as number of BP fragments in bioptic samples of Bowman capsule of the nephron at 12 weeks and anticancer activity	Estimated initial completion on 2015	NCT02328352
4	Detection of gastric injury from exhaled saliva and breath by using CNTs based nanosensor	To investigate feasibility of gold nanoparticles with CNTs for the identification of gastric diseases	Observational study type with perspective	140 (Estimated)	Discriminate between malignant and benign gastric lesions with Na-nose and geographical comparison of VOCs between China and Latvia and Geographical comparison of VOCs between China and Latvia	Proof seeking from metabolomics and tran-scriptomics	NCT01420588
5	Investigation of CNTs based solid state head CT	The main objective of this study is to investigate stationary head CT (s-HCT) as a diagnostic tool in patients with known head trauma	Interventional, open label for diagnostic treatment	50 (Estimated)	Sensitivity of stationary head CT for the detection of a hemorrhage and specificity of stationary head CT for the detection of a fracture	Estimated initial completion on 2023	NCT04495634

Table 13.2 Patents related to the chemically modified CNTs.

Sr. No.	Patent number	Inventors	Chemical modification of CNTs	Reference
1	US 2010/0184669A 1	Harrison et al.	Chemical modification of protein or peptide to SWCNTs for cancer treatment	[98]
2	US 2012/0220921A 1	Chen et al.	Immunologically chemically modified CNTs for treatment of cancer	[99]
3	US 2009/0062785A 1	Harrison et al.	Protein-SWCNTs for detecting and destructing cancer tumors	[100]
4	US 2011/0098591 A 1	Haick et al.	SWCNTs modified with lipophilic small organic moieties in sensor apparatuses for investigating biomarkers in breath samples of lung cancer patients	[101]
5	US 2013/0102027 A 1	Mohajerzadeh et al.	Vertically based CNTs for detection of cancer cells	[102]
6	US 2017/0224840 A 1	Yao et al.	CNTs for selectively targeting and destroying cancer stem cells	[103]
7	US 2020/0217849 A 1	Wei et al.	CNTs sponges for capturing of circulating tumor cells	[104]

13.6 Limitations and Challenges of Chemically Modified CNTs

The chemically modified CNTs have greater applications in various drug delivery and therapeutics. Although, CNTs toxicity has become one of the limitations and challenging factors in CNT applications. This is due to its various features such as similar nature as asbestos fibers, inborn lipophilic nature, impurities, high surface area and greater aspect ratio, defects in surface, and chemical modifications. CNTs are responsible to produce different toxicities such as aggregation in tissues, enhanced oxidative stress, and damage to DNA. [89, 105, 106]. As its size is similar to microbes, CNTs promote phagocytosis and foreign body response. Consequently, it results in the induction of the myeloperoxidase pathway generating oxidative burst and, hence, it initiates OS (oxidative stress) by improving ROSs (reactive oxygen species) levels and induction of NF-κB signaling pathways. It has been revealed from the studies that shorter CNTs are less toxic as compared to larger CNTs. Moreover, fibrosis, acute inflammation, necrosis, and formation of granuloma were concerned with SWCNTs and eye irritation, hyperplasia of cells, and damage to Kupffer cells were associated with MWCNTs [63, 107]. Interestingly, different tools such as chemical modifications of CNTs with amphiphiles, decreasing their length, choosing a suitable CNT type, and optimizing the degree of chemical modification is currently available to conquer the limitations with CNTs [89]. The biocompatibility of CNTs is not certain because of their insolubility and toxicity. Some of the studies suggested that metals such as silver, iron, and titanium with CNTs at greater concentrations were responsible for enhancing cytotoxicity [28].

13.7 Conclusion and Future Perspective

Significant progress has been made in nanotechnology-based techniques for more than 20 years, especially in carbon-based materials, CNTs in particular. CNTs are a kind of nanomaterials with extensive applications in various fields and it has further spread their wings from engineering to biomedical applications. Chemically modified CNTs have attained a significant attraction from researchers due to their fascinating properties such as greater surface area, higher aspect ratio, and greater stability. The chemical modifications of CNTs by targeting agents or ligands is one of the interesting research areas of the recent past that helps to improve the therapeutic activity of the compounds. Moreover, chemical modification approaches to two-dimensional forms of carbon such as graphene-based materials are growing at a faster rate. These CNTs act like nanocarriers for targeting ligands, imaging agents, siRNA, chemotherapeutic agents, and many other types of drug substances. We have summarized various approaches toward chemical modification in CNTs along with CNTs cellular uptake with certain limitations and associated side effects of CNTs. The area of CNT applications in doping and biosensing is still in its nascent stage and needs further research. Toxicity issues of CNTs is another area that requires great efforts to counter and subside the toxicity to an acceptable level. Future investigations need to be driven along with toxicological and pharmacological properties of CNTs and the risk/benefit ratio must be explained thoroughly before their use for relatively unexplored clinical applications.

Well-Known Researchers/Organizations, Companies, Websites Involved in the Field of Carbon Nanotubes for Cancer Therapy

There are number of researchers/organizations continuously working on the developments of novel CNTs in cancer therapy. The Zhu research group is working on CNTs for many years. Zhu is a distinguished professor in the Department of Material Science and Engineering at NC State University. Hisao Haniu and his research group in Japan basically work on the responses of MWCNTs in tissue regeneration as a suitable scaffold material.

The different kinds of companies that used CNTs for their various applications include Arkema SA, Nanocyl SA, Carbon Solutions, Inc., Nanoshell LLC., Klean Commodities, SHOWA DENKO K.K., and Hyperion Catalysis International. The websites that are fully dedicated to CNTs development include http://www.photon.t.u-tokyo.ac.jp/~maruyama/nanotube.html, https://www.nanocyl.com, https://www.nanowerk.com/nanotechnology/introduction/introduction_to_nanotechnology_22.php, https://en.wikipedia.org/wiki/Carbon_nanotube, etc.

References

1 Mattiuzzi, C. and Lippi, G. (2019). Current cancer epidemiology. *J. Epidemiol. Global Health* 9 (4): 217–222.

2 Jhaveri, A., Deshpande, P., and Torchilin, V. (2014). Stimuli-sensitive nanopreparations for combination cancer therapy. *J. Controlled Release* 190: 352–370.

3 Rathinavel, S., Priyadharshini, K., and Panda, D. (2021). A review on carbon nanotube: an overview of synthesis, properties, functionalization, characterization, and the application. *Mater. Sci. Eng., B* 268: 115095. https://doi.org/10.1016/j.mseb.2021.115095.

4 Virani, N.A., McKernan, P., and Harrison, R.G. (2020). Photothermal therapy using carbon nanotubes for treating cancer. In: *Thermal Behaviour and Applications of Carbon-Based Nanomaterials* (ed. D. Papavassiliou, H. Duong and F. Gong), 325–344. Elsevier Inc.

5 Gulati, P., Kaur, P., Rajam, M.V. et al. (2018). Single-wall carbon nanotube based electrochemical immunoassay for leukemia detection. *Anal. Biochem.* 557: 111–119.

6 Lu, G.H., Shang, W.T., Deng, H. et al. (2019). Targeting carbon nanotubes based on IGF-1R for photothermal therapy of orthotopic pancreatic cancer guided by optical imaging. *Biomaterials* 195: 13–22.

7 Singhai, N.J., Maheshwari, R., Jain, N.K., and Ramteke, S. (2020). Chondroitin sulphate and α-tocopheryl succinate tethered multiwalled carbon nanotubes for dual-action therapy of triple-negative breast cancer. *J. Drug Deliv. Sci. Technol.* 60: 102080.

8 Radzi, M.R., Johari, N.A., Zawawi, W.F. et al. (2022). In vivo evaluation of oxidized multiwalled-carbon nanotubes-mediated hyperthermia treatment for breast cancer. *Biomater. Adv.* 134: 112586.

9 González-Domínguez, J.M., Grasa, L., Frontiñán-Rubio, J. et al. (2022). Intrinsic and selective activity of functionalized carbon nanotube/nanocellulose platforms against colon cancer cells. *Coll. Surf. B: Biointerf.* 212: 112363.

10 Ezzati Nazhad Dolatabadi, J., Omidi, Y., and Losic, D. (2011). Carbon nanotubes as an advanced drug and gene delivery nanosystem. *Curr. Nanosci.* 7 (3): 297–314.

11 Liu, Z., Tabakman, S., Welsher, K., and Dai, H. (2009). Carbon nanotubes in biology and medicine: in vitro and in vivo detection, imaging and drug delivery. *Nano Res.* 2 (2): 85–120.

12 Smart, S.K., Cassady, A.I., Lu, G.Q., and Martin, D.J. (2006). The biocompatibility of carbon nanotubes. *Carbon* 44 (6): 1034–1047.

13 Peretz, S. and Regev, O. (2012). Carbon nanotubes as nanocarriers in medicine. *Curr. Opin. Colloid Interface Sci.* 17 (6): 360–368. https://doi.org/10.1016/j.cocis.2012.09.001.

14 Bianco, A., Kostarelos, K., and Prato, M. (2005). Applications of carbon nanotubes in drug delivery. *Curr. Opin. Chem. Biol.* 9 (6): 674–679.

15 Beg S, Rahman M, Jain A, Saini S, Hasnain MS, Swain S, et al. Emergence in the functionalized carbon nanotubes as smart nanocarriers for drug delivery

applications. *Fullerenes, Graphenes and Nanotubes: A Pharmaceutical Approach*. Elsevier Inc.; 2018. 105–133 p. https://doi.org/10.1016/B978-0-12-813691-1.00004-X.

16 Beg, S., Rizwan, M., Sheikh, A.M. et al. (2011). Advancement in carbon nanotubes: basics, biomedical applications and toxicity. *J. Pharm. Pharmacol.* 63 (2): 141–163.

17 Bianco, A. (2004). Carbon nanotubes for the delivery of therapeutic molecules. *Expert Opin. Drug Delivery* 1 (1): 57–65.

18 Pantarotto, D., Singh, R., McCarthy, D. et al. (2004). Functionalized carbon nanotubes for plasmid DNA gene delivery. *Angew. Chem. Int. Ed.* 43 (39): 5242–5246.

19 Vardharajula, S., Ali, S.Z., Tiwari, P.M. et al. (2012). Functionalized carbon nanotubes: biomedical applications. *Int. J. Nanomed.* 7: 5361–5374.

20 Liu, Z., Cai, W., He, L. et al. (2007). In vivo biodistribution and highly efficient tumour targeting of carbon nanotubes in mice. *Nat. Nanotechnol.* 2 (1): 47–52.

21 He, Y., Hara, H., and Núñez, G. (2016). Mechanism and regulation of NLRP3 inflammasome activation. *Trends Biochem. Sci* 41 (12): 1012–1021.

22 Rotoli, B.M., Bussolati, O., Bianchi, M.G. et al. (2008). Non-functionalized multi-walled carbon nanotubes alter the paracellular permeability of human airway epithelial cells. *Toxicol. Lett.* 178 (2): 95–102.

23 Hou, P.X., Liu, C., and Cheng, H.M. (2008). Purification of carbon nanotubes. *Carbon* 46 (15): 2003–2025.

24 Porter, A.E., Gass, M., Bendall, J.S. et al. (2009). Uptake of noncytotoxic acid-treated single-walled carbon nanotubes into the cytoplasm of human macrophage cells. *ACS Nano* 3 (6): 1485–1492.

25 Zare, H., Ahmadi, S., Ghasemi, A. et al. (2021). Carbon nanotubes: smart drug/gene delivery carriers. *Int. J. Nanomed.* 16: 1681–1706.

26 Galano, A. (2010). Carbon nanotubes: promising agents against free radicals. *Nanoscale* 2 (3): 373–380.

27 Liu, R., Chen, Y., Ma, Q. et al. (2017). Noncovalent functionalization of carbon nanotube using poly(vinylcarbazole)-based compatibilizer for reinforcement and conductivity improvement in epoxy composite. *J. Appl. Polym. Sci.* 134 (26): 1–9.

28 Merum, S., Veluru, J.B., and Seeram, R. (2017). Functionalized carbon nanotubes in bio-world: Applications, limitations and future directions. *Mater. Sci. Eng., B* 223: 43–63. https://doi.org/10.1016/j.mseb.2017.06.002.

29 Jeon, I.-Y., Chang, D.W., Kumar, N.A., and Baek, J.-B. (2011). Functionalization of carbon nanotubes. In: *Carbon Nanotubes Polymer Nanocomposites*, vol. 17 (ed. S. Yellampalli), 91–110.

30 Williams, K.A., Veenhuizen, P., de la Torre, B.G. et al. (2002). Carbon nanotubes with DNA recognition. *Nature* 420 (6917): 761.

31 Vashist, S.K., Zheng, D., Al-Rubeaan, K. et al. (2011). Advances in carbon nanotube based electrochemical sensors for bioanalytical applications. *Biotechnol. Adv.* 29 (2): 169–188. https://doi.org/10.1016/j.biotechadv.2010.10.002.

32 Kuzmany, H., Pfeiffer, R., Salk, N., and Günther, B. (2004). The mystery of the 1140 cm^{-1} Raman line in nanocrystalline diamond films. *Carbon* 42 (5–6): 911–917.

33 Mahajan, S., Patharkar, A., Kuche, K. et al. (2018). Functionalized carbon nanotubes as emerging delivery system for the treatment of cancer. *Int. J. Pharm.* 548 (1): 540–558. doi: 10.1016/j.ijpharm.2018.07.027.

34 Cha, J., Jin, S., Shim, J.H. et al. (2016). Functionalization of carbon nanotubes for fabrication of CNT/epoxy nanocomposites. *Mater. Des.* 95: 1–8. https://doi.org/10.1016/j.matdes.2016.01.077.

35 Maheshwari N, Tekade M, Soni N, Ghode P, Sharma MC, Deb PK, et al. Functionalized carbon nanotubes for protein, peptide, and gene delivery. *Biomaterials and Bionanotechnology*. Elsevier Inc.; 2019. 613–637https://doi.org/10.1016/B978-0-12-814427-5.00016-0.

36 Battigelli, A., Ménard-Moyon, C., and Bianco, A. (2014). Carbon nanomaterials as new tools for immunotherapeutic applications. *J. Mater. Chem. B* 2 (37): 6144–6156.

37 Kostarelos, K. (2008). The long and short of carbon nanotube toxicity. *Nat. Biotechnol.* 26 (7): 774–776.

38 Battigelli, A., Ménard-Moyon, C., Da Ros, T. et al. (2013). Endowing carbon nanotubes with biological and biomedical properties by chemical modifications. *Adv. Drug Delivery Rev.* 65 (15): 1899–1920. https://doi.org/10.1016/j.addr.2013.07.006.

39 Kyotani, T., Nakazaki, S., Xu, W.H., and Tomita, A. (2001). Chemical modification of the inner walls of carbon nanotubes by HNO_3 oxidation. *Carbon* 39 (5): 782–785.

40 Liu, J., Rinzler, A.G., Dai, H. et al. (1998). Fullerene pipes. *Science* 280 (5367): 1253–1256.

41 Hwang, K.C. (1995). Efficient cleavage of carbon graphene layers by oxidants. *J. Chem. Soc. D* 04 (2): 173–174.

42 Wu, W., Wieckowski, S., Pastorin, G. et al. (2005). Targeted delivery of amphotericin B to cells by using functionalized carbon nanotubes. *Angew. Chem.* 117 (39): 6516–6520.

43 Huang, W., Taylor, S., Fu, K. et al. (2002). Attaching proteins to carbon nanotubes via diimide-activated amidation. *Nano Lett.* 2 (4): 311–314.

44 Matsumoto, K., Sato, C., Naka, Y. et al. (2007). Neurite outgrowths of neurons with neurotrophin-coated carbon nanotubes. *J. Biosci. Bioeng.* 103 (3): 216–220.

45 Meng, J., Duan, J., Kong, H. et al. (2008). Carbon nanotubes conjugated to tumor lysate protein enhance the efficacy of an antitumor immunotherapy. *Small* 4 (9): 1364–1370.

46 Yoshimura, S.H., Khan, S., Ohno, S. et al. (2012). Site-specific attachment of a protein to a carbon nanotube end without loss of protein function. *Bioconjugate Chem.* 23 (7): 1488–1493.

47 Yu, B.Z., Yang, J.S., and Li, W.X. (2007). In vitro capability of multi-walled carbon nanotubes modified with gonadotrophin releasing hormone on killing cancer cells. *Carbon* 45 (10): 1921–1927.

48 Baker, S.E., Cai, W., Lasseter, T.L. et al. (2002). Covalently bonded adducts of deoxyribonucleic acid (DNA) oligonucleotides with single-wall carbon nanotubes: synthesis and hybridization. *Nano Lett.* 2 (12): 1413–1417.

49 Patolsky, F., Weizmann, Y., and Willner, I. (2004). Long-range electrical contacting of redox enzymes by SWCNT connectors. *Angew. Chem. Int. Ed.* 43 (16): 2113–2117.

50 Gorityala, B.K., Ma, J., Wang, X. et al. (2010). Carbohydrate functionalized carbon nanotubes and their applications. *Chem. Soc. Rev.* 39 (8): 2925–2934.

51 Bandaru, N.M. and Voelcker, N.H. (2012). Glycoconjugate-functionalized carbon nanotubes in biomedicine. *J. Mater. Chem.* 22 (18): 8748–8758.

52 Vuković, G.D., Tomić, S.Z., Marinković, A.D. et al. (2010). The response of peritoneal macrophages to dapsone covalently attached on the surface of carbon nanotubes. *Carbon* 48 (11): 3066–3078.

53 Bhirde, A.A., Patel, V., Gavard, J. et al. (2009). Targeted killing of cancer cells in vivo and in vitro with EGF-directed carbon nanotube-based drug delivery. *ACS Nano* 3 (2): 307–316.

54 Bhirde, A.A., Patel, S., Sousa, A.A. et al. (2010). Distribution and clearance of PEG-single-walled carbon nanotube cancer drug delivery vehicles in mice. *Nanomedicine* 5 (10): 1535–1546.

55 Elhissi, A.M.A., Ahmed, W., Hassan, I.U. et al. (2012). Carbon nanotubes in cancer therapy and drug delivery. *J. Drug Delivery* 2012: 1–10.

56 Pantarotto, D., Partidos, C.D., Graff, R. et al. (2003). Synthesis, structural characterization, and immunological properties of carbon nanotubes functionalized with peptides. *J. Am. Chem. Soc.* 125 (20): 6160–6164.

57 McDevitt, M.R., Chattopadhyay, D., Kappel, B.J. et al. (2007). Tumor targeting with antibody-functionalized, radiolabeled carbon nanotubes. *J. Nucl. Med.* 48 (7): 1180–1189.

58 Benincasa, M., Pacor, S., Wu, W. et al. (2011). Antifungal activity of amphotericin B conjugated to carbon nanotubes. *ACS Nano* 5 (1): 199–208.

59 Liu, M., Yang, Y., Zhu, T., and Liu, Z. (2005). Chemical modification of single-walled carbon nanotubes with peroxytrifluoroacetic acid. *Carbon* 43 (7): 1470–1478.

60 Valentini, L., Macan, J., Armentano, I. et al. (2006). Modification of fluorinated single-walled carbon nanotubes with aminosilane molecules. *Carbon* 44 (11): 2196–2201.

61 Syrgiannis, Z., Hauke, F., Rohrl, J. et al. (2008). Covalent sidewall functionalization of SWNTs by nucleophilic addition of lithium amides. *Eur. J. Org. Chem.* 15: 2544–2550.

62 Wunderlich, D., Hauke, F., and Hirsch, A. (2008). Preferred functionalization of metallic and small-diameter single-walled carbon nanotubes by nucleophilic addition of organolithium and -magnesium compounds followed by reoxidation. *Chem. Eur. J.* 14 (5): 1607–1614.

63 Xu, Y., Wang, X., Tian, R. et al. (2008). Microwave-induced electrophilic addition of single-walled carbon nanotubes with alkylhalides. *Appl. Surf. Sci.* 254 (8): 2431–2435.

64 Maroto, A., Balasubramanian, K., Burghard, M., and Kern, K. (2007). Functionalized metallic carbon nanotube devices for pH sensing. *ChemPhysChem* 8 (2): 220–223.

65 Wang, H. and Xu, J. (2009). Theoretical evidence for a two-step mechanism in the functionalization single-walled carbon nanotube by aryl diazonium salts: comparing effect of different substituent group. *Chem. Phys. Lett.* 477 (1–3): 176–178. https://doi.org/10.1016/j.cplett.2009.06.076.

66 Stephenson, J.J., Hudson, J.L., Azad, S., and Tour, J.M. (2006). Individualized single walled carbon nanotubes from bulk material using 96% sulfuric acid as solvent. *Chem. Mater.* 18 (2): 374–377.

67 Chen, X., Wang, J., Zhong, W. et al. (2008). A scalable route to highly functionalized multi-walled carbon nanotubes on a large scale. *Macromol. Chem. Phys.* 209 (8): 846–853.

68 Darabi, H.R., Mohandessi, S., Aghapoor, K. et al. (2009). Thioamidation of single-walled carbon nanotubes: a new chemical functionalization protocol by the Willgerodt–Kindler reaction. *Aust. J. Chem.* 62 (5): 413–418.

69 Yinghuai, Z., Peng, A.T., Carpenter, K. et al. (2005). Substituted carborane-appended water-soluble single-wall carbon nanotubes: new approach to boron neutron capture therapy drug delivery. *J. Am. Chem. Soc.* 127 (27): 9875–9880.

70 Pastine, S.J., Okawa, D., Kessler, B. et al. (2008). A facile and patternable method for the surface modification of carbon nanotube forests using perfluoroarylazides. *J. Am. Chem. Soc.* 130 (13): 4238–4239.

71 Umeyama, T., Tezuka, N., Fujita, M. et al. (2007). Retention of intrinsic electronic properties of soluble single-walled carbon nanotubes after a significant degree of sidewall functionalization by the Bingel reaction. *J. Phys. Chem. C* 111 (27): 9734–9741.

72 Baek, S.H., Kim, B., and Do, S.K. (2008). Chitosan particle/multiwall carbon nanotube composites by electrostatic interactions. *Colloids Surf., A* 316 (1–3): 292–296.

73 Benda, R., Zucchi, G., Cancès, E., and Lebental, B. (2020). Insights into the π–π interaction driven non-covalent functionalization of carbon nanotubes of various diameters by conjugated fluorene and carbazole copolymers. *J. Chem. Phys.* 152 (6): 064708.

74 Zhuang, W.R., Wang, Y., Cui, P.F. et al. (2019). Applications of π–π stacking interactions in the design of drug-delivery systems. *J. Controlled Release* 294: 311–326.

75 Meran, M., Akkus, P.D., Kurkcuoglu, O. et al. (2018). Noncovalent pyrene-polyethylene glycol coatings of carbon nanotubes achieve in vitro biocompatibility. *Langmuir* 34 (40): 12071–12082.

76 Khan, A.K., Mubarak, N.M., Abdullah, E.C. et al. (2019). Immobilization of lipase enzyme carbon nanotubes via adsorption. *IOP Conf. Ser.: Mater. Sci. Eng.* 495 (1): 012055.

77 Tavakolifard, S., Biazar, E., Pourshamsian, K., and Moslemin, M.H. (2016). Synthesis and evaluation of single-wall carbon nanotube-paclitaxel-folic acid conjugate as an anti-cancer targeting agent. *Artif. Cells Nanomed. Biotechnol.* 44 (5): 1247–1253.

78 Singh, S., Mehra, N.K., and Jain, N.K. (2016). Development and characterization of the paclitaxel loaded riboflavin and thiamine conjugated carbon nanotubes for cancer treatment. *Pharm. Res.* 33 (7): 1769–1781.

79 Cao, X., Tao, L., Wen, S. et al. (2015). Hyaluronic acid-modified multiwalled carbon nanotubes for targeted delivery of doxorubicin into cancer cells. *Carbohydr. Res.* 405: 70–77.

80 Ali Mohammadi, Z., Aghamiri, S.F., Zarrabi, A., and Talaie, M.R. (2015). A comparative study on non-covalent functionalization of carbon nanotubes by chitosan and its derivatives for delivery of doxorubicin. *Chem. Phys. Lett.* 642: 22–28.

81 Sun, Z., Wang, W., Meng, J. et al. (2010). Multi-walled carbon nanotubes conjugated to tumor protein enhance the uptake of tumor antigens by human dendritic cells in vitro. *Cell Res.* 20 (10): 1170–1173.

82 Ouyang, M., White, E.E., Ren, H. et al. (2016). Metronomic doses of temozolomide enhance the efficacy of carbon nanotube CpG immunotherapy in an invasive glioma model. *PLoS One* 11 (2): e0148139.

83 Hassan, H.A.F.M., Smyth, L., Wang, J.T.W. et al. (2016). Dual stimulation of antigen presenting cells using carbon nanotube-based vaccine delivery system for cancer immunotherapy. *Biomaterials* 104: 310–322. https://doi.org/10.1016/j.biomaterials.2016.07.005.

84 Janeway, C.A. and Medzhitov, R. (2002). Innate immune recognition. *Annu. Rev. Immunol.* 20 (2): 197–216.

85 Gordon, S. (2002). Pattern recognition receptors: doubling up for the innate immune response. *Cell* 111 (7): 927–930.

86 Steinhagen, F., Kinjo, T., Bode, C., and Klinman, D.M. (2011). TLR-based immune adjuvants. *Vaccine* 29 (17): 3341–3355. https://doi.org/10.1016/j.vaccine.2010.08.002.

87 Abu-Salah, K.M., Ansari, A.A., and Alrokayan, S.A. (2010). DNA-based applications in nanobiotechnology. *J. Biomed. Biotechnol.* 2010: 715295.

88 Crinelli, R., Carloni, E., Menotta, M. et al. (2010). Oxidized ultrashort nanotubes as carbon scaffolds for the construction of cell-penetrating NF-κB decoy molecules. *ACS Nano* 4 (5): 2791–2803.

89 Kiran, A.R., Kumari, G.K., and Krishnamurthy, P.T. (2020). Carbon nanotubes in drug delivery: Focus on anticancer therapies. *J. Drug Deliv. Sci. Technol.* 59: 101892.

90 Pan, B., Cui, D., Xu, P. et al. (2009). Synthesis and characterization of polyamidoamine dendrimer-coated multi-walled carbon nanotubes and their application in gene delivery systems. *Nanotechnology* 20 (12): 125101.

91 Lima, W.F., Vickers, T.A., Nichols, J. et al. (2014). Defining the factors that contribute to on-target specificity of antisense oligonucleotides. *PLoS One* 9 (7): e101752.

92 Van Den Bossche, J., Al-Jamal, W.T., Tian, B. et al. (2010). Efficient receptor-independent intracellular translocation of aptamers mediated by conjugation to carbon nanotubes. *Chem. Commun.* 46 (39): 7379–7381.

93 Zhang, Y., Lai, B.S., and Juhas, M. (2019). Recent advances in aptamer discovery and applications. *Molecules* 24 (5): 941.

94 Gedi, V., Song, C.K., Kim, G.B. et al. (2018). Sensitive on-chip detection of cancer antigen 125 using a DNA aptamer/carbon nanotube network platform. *Sens. Actuators, B* 256: 89–97.

95 Barbosa, M.B., do Nascimento Martins, E.M., Teixeira, T.F. et al. (2019). A carefully designed nanoplatform based on multi walled carbon nanotube wrapped with aptamers. *Colloids Surf., B* 175: 175–183.

96 Yao, J., Li, S., Zhang, L. et al. (2020). Aptamer-antibody dual probes on single-walled carbon nanotube bridged dielectrode: comparative analysis on human blood clotting factor. *Int. J. Biol. Macromol.* 151: 1133–1138.

97 Sanginario, A., Miccoli, B., and Demarchi, D. (2017). Carbon nanotubes as an effective opportunity for cancer diagnosis and treatment. *Biosensors* 7 (1): 1–23.

98 Harrison, R.G., Resasco, D.E. (2009). Compositions and methods for cancer treatment using targeted carbon nanotubes. US Patent 0184669, filed 13 November 2009 and issued 27 August 2013.

99 Chen, W.R. (2011). Immunologically modified carbon nanotubes for cancer treatment. US Patent 0220921, filed 28 February 2011 and issued 04 March 2014.

100 Harrison, R.G., Resasco, D.E. (2008). Compositions and methods for cancer treatment using targeted single-walled carbon nanotubes. US Patent 0062785, filed 30 May 2008 and issued 5 March 2009.

101 Haick, H., Gang, P. (2009). Carbon nanotube structures in sensor apparatuses for analyzing biomarkers in breath samples. US Patent 0098591, filed 27 May 2009 and issued 05 February 2013.

102 Mohajerzadeh,S., Abdolahad, M., Sanaee, Z., and Abdollahi, M. (2012). Method for detecting cancer cells using vertically carbon nanotubes. US Patent 0102027, filed 19 April 2012 and issued 17 February 2015.

103 Yao, H., Zhang, Y., Sun, L., Liu, Y. (2015) Drug delivery system comprising a cancer stem cell-targeted carbon nanotube, preparation and use thereof. US Patent 0114840, filed 11 August 2015 and issued 10 August 2017.

104 Wei, B., Li, T., Lu, X.L. (2018). Capture of circulating tumor cells using carbon nanotube sponges. US Patent 0217849, filed 27 June 2018.

105 Sun, Z., Liu, Z., Meng, J. et al. (2011). Carbon nanotubes enhance cytotoxicity mediated by human lymphocytes in vitro. *PLoS One* 6 (6): e21073.

106 Bai, Y., Zhang, Y., Zhang, J. et al. (2010). Repeated administrations of carbon nanotubes in male mice cause reversible testis damage without affecting fertility. *Nat. Nanotechnol.* 5 (9): 683–689.

107 Tabet, L., Bussy, C., Amara, N. et al. (2009). Adverse effects of industrial multiwalled carbon nanotubes on human pulmonary cells. *J. Toxicol. Environ. Health Part A* 72 (2): 60–73.

14

Chemically Modified Carbon Nanotubes in Drug Delivery

Sougata Ghosh[1,2], Sirikanjana Thongmee[1], and Ebrahim Mostafavi[3,4]

[1] *Kasetsart University, Department of Physics, Faculty of Science, 50 Ngamwongwan Rd, Lat Yao, Chatuchak, Bangkok 10900, Thailand*
[2] *Kasetsart University, Department of Microbiology, Rajkot-Bhavnagar Highway, Rajkot, 360020, Gujarat, India*
[3] *Stanford Cardiovascular Institute, Stanford University School of Medicine, 240 Pasteur Dr., Stanford, CA 94305, USA*
[4] *Stanford University School of Medicine, Department of Medicine, 240 Pasteur Dr., Stanford, CA 94305, USA*

14.1 Introduction

Nanocomposites are attractive delivery systems for drugs as small dimensions of the material offer a large surface area for functionalization of large quantity of drugs. Further, nanomaterials can be co-functionalized with the targeting moieties for site-specific delivery, uptake, and accumulation of the drugs. This strategy, not only helps in the sustained release of the drugs but also longer retention of the drugs in the target tissues for enhanced effectivity [1]. The remarkable physicochemical properties of the carbon nanotubes (CNTs) include ultra-high mechanical strength, lightweight, high surface performance, and superior conduction have extended their applications in sensors, spacecraft, nanoelectronics, agriculture, cosmetics, and pharmaceutics [2, 3].

Among various nanomaterials, CNTs are considered as an ideal tool for transporting various therapeutic molecules such as drugs, bioactive peptides, nucleic acid, and other [4, 5]. Functionalization of drugs onto CNTs not only improve their pharmacological profiles but also helps in the reduction of both toxicity and immunogenicity [6]. Hence, CNTs-based systems are promising candidates nanomedicine.

Considering the appropriate physico-chemical and the optoelectronic properties expected in the CNTs according to their deserved applications, diverse mechanisms for synthesis [7]. The synthesized CNTs are then purified employing acid treatment, magnetic purification, and size exclusion chromatography for further functionalization of the suitable drugs. Highly hydrophobic surface of pristine CNTs is difficult to conjugate with drugs. Multifunctional CNTs are more preferred compared to single group functionalization for drug delivery. Various aryl diazonium salts are used for triple functionalization [8]. Several polymers can be functionalized

Chemically Modified Carbon Nanotubes for Commercial Applications, First Edition.
Edited by Jeenat Aslam, Chaudhery Mustansar Hussain, and Ruby Aslam.

together to achieve desired activity [9]. Surface of CNTs can also be functionalized by the surfactants [10]. In view of the background, this chapter gives an elaborate account of strategies involved in the modification of CNTs for drug delivery against infections, cancer, inflammation, and others.

14.2 Antibacterial

CNTs are reported to exhibit promising antibacterial activity which is majorly dependent on their size, surface area, functionalized antibacterial drug, and type of bacteria [11]. Among various nanomaterials, zinc oxide nanoparticles (ZnONPs) have gained considerable attention [12, 13]. ZnONPs are widely used in a diverse field of applications that include UV detectors, gas sensors, biosensors, solar cell, catalysis, and medicine [14–16]. MWCNTs–ZnO nanocomposite was 5–20 nm in outer diameter while the length was ~10 μm. The composite with optimum MWCNT content (10 wt%) exhibited promising photo-inactivation of *Escherichia coli* bacteria in presence of UV–visible light [17].

In another study, MWCNTs were conjugated with lysine under microwave irradiation using diazonium-assisted functionalization [18]. The conjugation was confirmed by Fourier transform infrared (FTIR) spectroscopy that displayed a prominent peak at 1554 cm^{-1} that was specific to the N–H bending vibration. Additional medium peaks at 3571 and 3198 cm^{-1} were due to the N–H stretching vibration, and the splitting that is specific to primary amine. It is important to note that lysine functionalization enhanced the antimicrobial efficacy of the MWCNTs as evident from the minimum inhibitory concentration (MIC) reduction against both Gram-negative and Gram-positive bacteria. Antibacterial activity of lysine functionalized MWCNTs increased up to 3.23, 2.27, and 2.55 times against Gram-negative bacteria *E. coli*, *Salmonella typhimurium*, and *Klebsiella pneumoniae*, respectively while it increased 2.1, 1.84, and 1.80 times against Gram-positive bacteria *Streptococcus agalactiae*, *Staphylococcus aureus*, and *Streptococcus dysgalactiae*, respectively.

Antibiotics like ciprofloxacin was also covalently functionalized on SWCNTs using polyethylene glycol (PEG) derivatives starting with triethylene glycol (TEG) that indicated almost 55% loading [19]. Each SWCNTs had an average diameter of 8 nm while the functionalized SWCNTs formed small dispersed bundles. Further, the functionalized SWCNTs exhibited 16-fold increase of antibacterial activity against *S. aureus* and *Pseudomonas aeruginosa* while eightfold activity enhancement was observed against *E. coli*. Also, impairment of the efflux pumps on treatment with nanoantibiotics was speculated to another potential mechanism of the bactericidal activity of the nanocomposite. Three novels modified carboxylated MWCNTs namely MWCNT-Dapsone, MWCNT-Dapsone-imine, and MWCNT-Dapsone-imine copper complex were synthesized for effective antibacterial activity [20]. Pristine MWCNTs displayed smooth surface while the functionalized MWCNTs were rough in nature that significantly inhibited both *E. coli* and *S. aureus*.

Biofilm-associated bacterial growth is a leading cause of nosocomial infections that are occasionally resistant to the available antibiotics [21]. Hence, various

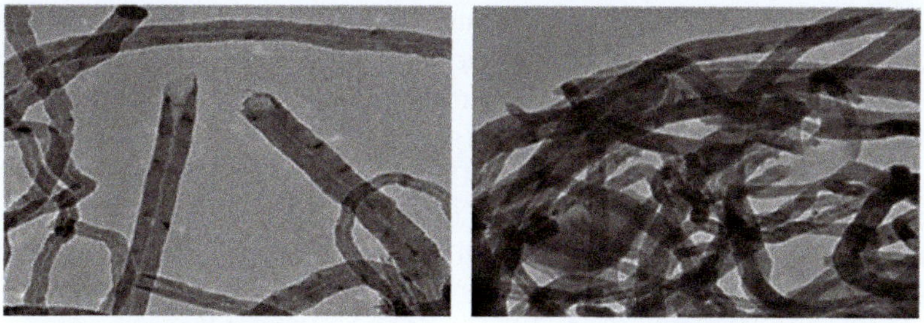

Figure 14.1 TEM images of derivative **4**/MWCNT composites. Source: Mohamed [29]/with permission of Springer Nature.

physical, chemical, and biological methods are employed to inhibit and eradicate bacterial biofilms [22]. However, the conventional methods are often not adequate for treating the biofilm-associated infections that rationalize the development of complementary and alternative strategies for the prevention of bacterial biofilms [23–25]. Bactericidal nanostructures are reported as the most potent biofilm inhibitory agents [26]. Functionalization of protoporphyrin IX (PPIX) on MWCNTs resulted in high phototoxicity against *S. aureus* [27]. This functionalization strategy not only enhanced the biocidal capacity of porphyrins but also provided mechanical strength to the MWCNTs. Irradiation in presence of the nanocomposite notably deactivated the *S. aureus* cells in solution. Even the nanocomposite films composed of the MWCNT-PPIX conjugates significantly resisted bacterial growth on the surface. Hence such surfaces can provide a facile strategy for the design of antimicrobial and antifouling coatings to prevent bacterial biofilms [28].

In another very interesting study, MWCNTs were incorporated in the epoxy chitosan Schiff's base derivative cross-linked by *p*-aminosalicylhydrazide [29]. Chemical reaction of chitosan with benzaldehyde, epichlorohydrine, and aminosalicylhydrazide resulted in formation of derivatives **1**–**4**. The aminosalicylhydrazide cross-linked chitosan, designated as derivative **4** was microporous in nature as evident from Figure 14.1. The pores may facilitate the water permeation and serve as the sites of interaction of the external stimuli with the hydrophilic functional groups of the cross-linked derivative **4**. Derivative **4**/MWCNT composites exhibited high antibacterial activity against *Enterococcus faecalis* and *Staphylococcus epidermidis*. It should be noted that the rational incorporation of functionalized moiety possessing electron-rich groups (imine, epoxide, OH, NH, CONHNH, and free NH_2) resulted in the enhancement of the antibacterial activity of chitosan. Additionally binding of the chitosan molecules with the DNA after entry can affect the Central Dogma (both transcription and translation) eventually killing the bacteria.

14.3 Antifungal

Conventional therapy to counter fungal infections includes antifungal agents, such as echinocandins, caspofungin, voriconazole, posaconazole, fluconazole, and others. Amphotericin B (AMB) was conjugated to CNTs for enhancement of its solubility, selectivity, therapeutic index, efficacy, and cellular uptake. Such strategy can also prevent aggregation thus resulting in more bioavailability and stability [30]. Both MWCNTs and SWCNTs were used in the study where the resulting composites functionalized carbon nanotube (f-CNT)-AMB 1 and f-CNT-AMB 2 showed 25% and 10% of AMB by weight, respectively. The f-CNT-AMB 1 exhibited a MIC value lower than 10 $\mu g\,ml^{-1}$ against the test fungal pathogens while the MIC value of f-CNT-AMB 2 was in a range between 5 and 20 $\mu g\,ml^{-1}$. Interestingly, the acapsular strains of fungi were more susceptible (two- to eight-folds) to the nanoconjugates compared to their capsulated counterparts. The drug-resistant fungal strains were inhibited significantly by the f-CNT-AMB 1.

In another study, the antifungal drug nystatin was conjugated to MWCNTs for enhanced antifungal effect [31]. Initially, the MWCNTs were oxidized by treating with concentrated sulfuric and nitric acid mixed in a ratio of 3 : 1. This resulted in the generation of carboxylic group, reduction in the length, and enhancement of the dispersibility in aqueous solutions. Further, ethylene-diamine was directly coupled with the carboxylic groups associated with the external walls of the oxidized MWCNTs, which eventually led to the introduction of the amino group via amide formation. Next diamide-activated amidation method was employed for conjugation of nystatin to the surface modified MWCNTs. UV–visible spectral peaks at 275 to 325 nm in the nanocomposites confirmed the successful conjugation of the nystatin. MIC value of the nanocomposite against *Candida albicans* (NCIM NO-3471), *Candida parapsilosis* (NCIM NO-3471), *Candida neoformans* (NCIM NO-3471) were 6, 4, and 2 $\mu g\,ml^{-1}$, respectively while the minimum fungicidal concentration (MFC) was 16, 14, and 11 $\mu g\,ml^{-1}$, respectively. Similar amide-functionalization of MWCNTs was also reported for successful functionalization of AMB [32]. FTIR spectra revealed the presence of the bands at 1573 and 1223 cm^{-1}, that correspond to N–H in-plane and C—N bond stretching, respectively, confirming the presence of the amide functional group.

Yet in another study, orthogonal functionalization was used to conjugate both fluorescein and AMB to SWCNTs and MWCNTs for simultaneous tracking and drug delivery as schematically represented in Figure 14.2 [33]. Interestingly, the length and the loading of the CNTs were the function of the duration of the acid treatment. It is important to note that with the increase in the duration of oxidation the reduction in the tube length and enhancement in the loading capacity was observed. The optimum duration for oxidation of the CNTs was eight hours that resulted in the most effective length and optimal loading. Also, stability to harsh acidic conditions to the nanocomposite was rendered by activating the modified CNTs as the acid chlorides followed by treatment with diaminotriethylene glycol (monoprotected as phthalimide). The UV–visible spectra in *N*,*N*-dimethylformamide (DMF) indicated absorption bands in the range 340–420 and 517 nm that were attributed to AMB and fluorescein isothiocyanate (FITC), respectively. The composite showed less toxicity with the human Jurkat lymphoma T cells were compared to free AMB. It was speculated that the novel material penetrated through the cell membrane as nanoneedles without causing cell death. The nanocomposites exhibited promising

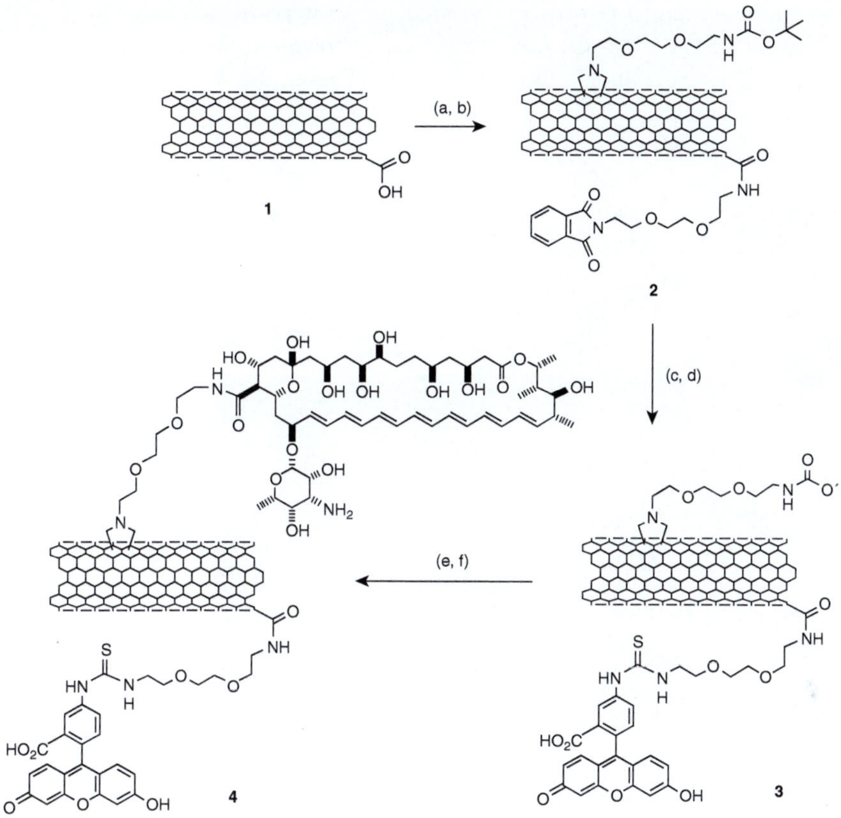

Figure 14.2 (a) Neat $(COCl)_2$;Pht-N$(CH_2CH_2O)_2$-CH_2CH_2-NH_2, dry THF, reflux; (b) Boc-NH-$(CH_2CH_2O)_2$-CH_2CH_2-$NHCH_2COOH/(CH_2O)_n$, DMF, 125 °C; (c) Hydrated NH_2-NH_2, EtOH, reflux; (d) FITC, DMF; (e) HCl 4_M in dioxane; (f) Fmoc-AmB, HOBt/EDC × HCl/DIPEA, DMF; 25% piperidine in DMF. BOC = tert butyloxycarbonyl; DIPEA = diisopropylethylamine; DMF = dimethyl formamide; EDC=*N*-Ethyl-*N′*-(3-dimethylaminopropyl)-carbodiimide; FMOC = fluorenylmethyloxycarbonyl; FITC = fluorescein isothiocyanate; HOBt = 1-hydroxybenzotriazole; Pht = phthalimide group. Source: Wu et al. [33]/with permission of John Wiley & Sons.

antifungal activity against *C. parapsilosis* ATCC 90118, *Cryptococcus neoformans* ATCC 90112 and clinical strains of *C. albicans*.

Covalent functionalization of lysine and arginine under microwave radiation was reported for reinforcement of the MWCNTs for better antifungal activity [34]. The functionalized nanomaterial showed variable degrees of antifungal activity against *Aspergillus niger, Aspergillus fumigatus, C. albicans, Penicillium chrysogenum, Saccharomyces cerevisiae, Fusarium culmorum, Microsporum canis, Trichophyton mentagrophytes, Trichophyton rubrum*, and *Penicillium lilacinum* as depicted in Figure 14.3. The highest zone of inhibition of MWCNTs-lysine and MWCNTs-arginine against *P. lilacinum* was equivalent to 15.7 ± 0.98 mm and 16.3 ± 1.3 mm, respectively, which was significantly higher compared to pristine MWCNTs (11.3 ± 2.10 mm). Hence, covalent linking of lysine and arginine to

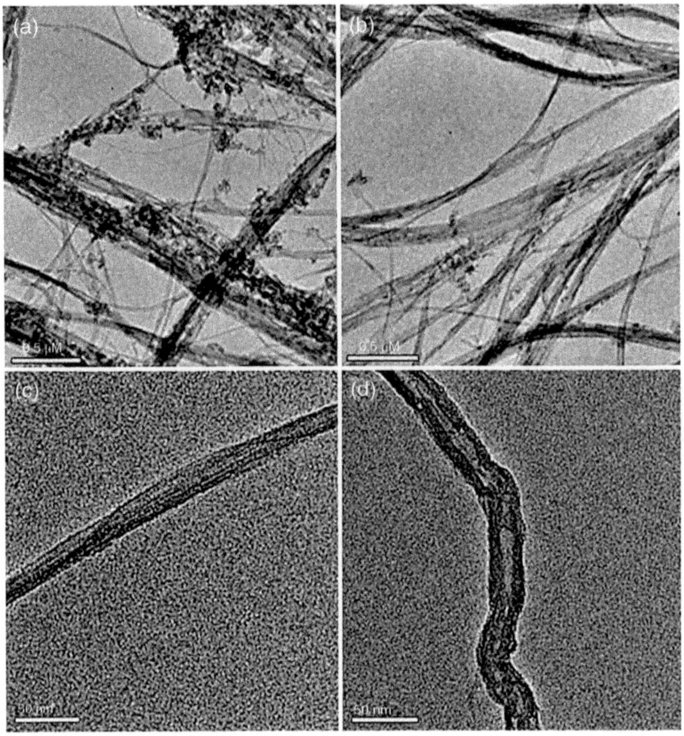

Figure 14.3 TEM images of (a) pristine SCWNTs (0.5 μm), (b) SWCNT-COOH (0.5 μm), (c) SWCNT-COOH (50 nm), and (d) SWCNT-PEI (50 nm). Source: Wang et al. [35]/with permission of Elsevier.

CNTs can be considered a significant strategy to enhance the activity of the antifungal drugs that will not only help in reduction of dosage but also prevent the development of multidrug resistance among the fungal pathogens.

14.4 Anticancer

Structural and functional modification of CNTs along with co-functionalization with drugs is reported as a powerful strategy to enhance the activity of the anticancer drugs against a broad spectrum of cancer types [36, 37]. Multifunctionalization approaches associated with rational conjugation of specific antibodies or receptor-specific ligands helps in targeted delivery of the drugs apart from improvisation of the bioavailability and cellular uptake [38–41]. Metal and metal oxide nanoparticles with promising anticancer activity are reported to provide an excellent surface for drug functionalization that can synergistically enhance the cumulative cytotoxicity against cancer cells [42, 43].

Molecular dynamics (MDs) behind conjugation of an anticancer drug paclitaxel (PTX) with pristine and PEG f-CNT were investigated employing in silico simulation [44]. The PTX molecules and CNTs could interact from a distance around 0.5–2.2 nm while the maximum interactions occurred when the distance was 0.92 and 2.08 nm. Interestingly, the van der Waals energy decreased when the PTX molecules reached to the proximity of the CNT surface. Modification of the CNTs with the PEG groups enhanced both the adsorption capacity and drug solubility in aqueous solution that might be attributed to the intermolecular hydrogen bonds between functional groups and solvent molecules and more polar interactions.

In another similar study, conjugation of both the inner and outer surfaces of the pristine and f-CNT with an anticancer drug dacarbazine (DAC) was explored using the MD simulations [45]. The interaction energies decreased from 0 kJ mol^{-1} at time of 0 ps to −158 kJ mol^{-1} at time of 22 000 ps in FCNT_DAC (out) and also decreased from 0 kJ mol^{-1} at time of 0 ps to −158 kJ mol^{-1} at time of 200 ps in CNT_DAC (out). Eventually, the energy of all of the systems reached to the minimum and equaled to −158 kJ mol^{-1}. Such variation was observed as the DAC molecule was located in the orientations with the most van der Waals energy (Lennard-Jones, LJ) that decreased remarkably at 2 ns time when the DAC molecule moved toward the surface of the CNTs' surface and positioned nearly along the wall surface of CNTs. Such an arrangement of the DAC molecule into and/or onto the CNTs and f-CNT is speculated to be controlled by π–π interactions.

Yet in another interesting study, a polyethylene glycol-doxorubicin (PEG-DXR) was functionalized on oxidized single-wall carbon nanohorns (oxSWNHs) [46]. This complex functionalization was accomplished by initially incubating a reaction mixture of both oxSWNHs and PEG-DXR in dimethyl sulfoxide (DMSO) or DMF where both the organic solvents featured relatively high electric dipole moments. Slowly the solution was made more aqueous by addition of water until final concentration of DMSO or DMF reached 10%. The recovery of the resulting PEG-DXR-oxSWNH composite was accomplished using a dextran-based size exchange chromatography (Sephadex G25) using a PD-10 gel filtration column equilibrated with water. Further purification was achieved by ultrafiltration. The mean hydrodynamic diameter of the composite was approximately 160 nm. The PEG-DXR-oxSWNHs were effective against NCI-H460 human nonsmall cell lung cancer cells even at very low concentration equivalent to 0.2 mg ml^{-1} DXR, which was attributed to induction of apoptosis. The probable mechanism behind the high anticancer activity was thought to be due to cellular uptake of the water-soluble PEG-DXR-oxSWNHs by endocytosis followed by hydrolysis of the PEG-DXR resulting in the release of the free DXR. Another possibility might be the release of the PEG-DXR into the extracellular milieu followed by apoptosis induction.

Not only chemotherapeutic drugs but radioactive analogues (i.e. metal halides) were delivered for anticancer activity by endohedral and exohedral functionalization of steam-purified SWCNTs that were heat sealed resulting in unique closed-ended CNTs with confined radioactive material inside [47]. The two selected payloads, $SmCl_3$ and $LuCl_3$ were loaded at 18.1 and 29.9 wt% into SWCNTs giving the composites represented as $SmCl_3$@SWCNTs and $LuCl_3$@SWCNTs, respectively. Further, nitrene cycloaddition was employed for external functionalization followed by the derivatization with a monoclonal antibody (Cetuximab) that is specific to the epidermal growth factor receptor (EGFR) that is overexpressed by various cancer cells. Target-specific anticancer activity was checked against U87 glioblastoma cells modified to overexpress EGFR (U87-EGFR+) and Chinese hamster ovary cells (CHO) that were devoid of EGFR. It is important to note that almost 80% of U87-EGFR+ were able to bind and internalize Cetuximab covalently or noncovalently bound to SWCNTs, unlike the CHO cells that showed less than 5% binding and internalization.

In another novel study, SWCNTs were initially functionalized with polyethylenimine (PEI) that were further surface modified with DSPE-PEG2000-maleimide for conjugating the tumor targeting NGR (Cys–Asn–Gly–Arg–Cys–) peptide via the maleimide group and sulfhydryl group of cysteine [35]. Also, hTERT siRNA was loaded onto this composite to obtain a novel complex represented as SWCNT-PEI/siRNA/NGR with notable tumor targeting siRNA delivering property. The PEI functionalization strategy using a combination of purification, oxidation, amination, and polymerization resulted in hydrophilic surface of the SWCNTs. Figure 14.3 reveals that amorphous carbons, nanoparticles, and metal catalysts appeared as impurities in the twisted pristine SWCNTs that were winded up forming a large bundle. Purification of the SWCNTs had a notable effect on the surface features. However, a rough surface with a layer of material attached to the sidewalls was evident in the TEM images of the SWCNT-PEI. The NGR tumor-targeted peptide led to complete cellular uptake within the PC-3 cells after four hours. The SWCNT-PEI/siRNA/NGR efficiently crossed the cell membrane inducing more prominent apoptosis and stronger suppression in proliferation of PC-3 cells. PC-3 tumor-bearing male BALB/c nude mice exhibited more tumor accumulation of the nanocomposite on treatment along with remarkable antitumor effect due to both RNAi and near-infrared (NIR) photothermal effect. Hence, the novel SWCNT-PEI/siRNA/NGR can have dual mode of cancer targeting and inhibition due to the synergistic combination of both gene and photothermal therapy.

14.5 Others

Various other diseases are also targeted by coupling appropriate drug with the surface-modified CNTs. A drug named theophylline, effective mostly against asthma and pulmonary disease was conjugated to the nanohybrid hydrogels developed using poly(methacrylic acid) (PMAA) networks and carboxyl-functionalized MWCNTs (MWCNT-COOH) [48]. An increase in the concentration of the MWCNT-COOH in the hydrogel resulted in the reduction in the micropore densities and increase in the large mesh sizes. High swelling in the hydrogels containing 10 wt% MWCNT-COOH was attributed to the collapse at pore walls because of large holes. Both pH-responsive property and the drug release were controlled by the hydrogen bonding and electrostatic repulsion resulting in the self-assembly of the MWCNT-COOH with the PMAA. The pH was directly proportional to the equilibrium swelling ratios and accumulative release of the functionalized theophylline.

SWNHs and oxSWNHs were used for the delivery of the anti-inflammatory glucocorticoid dexamethasone (DEX). Adsorption of the DEX onto the oxidized form, oxSWNHs was six times higher to the pristine SWNHs, which was mainly due to the higher affinity of the drug to the oxidized form [49]. The carrier was heat stable and treatment at temperature as high as 1200 °C under H_2, failed to diminish the high affinity for DEX in spite of removal of the oxygen-containing functional groups on oxSWNHs. Further, sustained DEX release was achieved at pH and 37 °C, which was mostly biphasic in nature. The release profile revealed that around 50% of the total bound DEX was released from the complex within a period of 14 days. The released DEX activated the glucocorticoid response associated transcription in mouse bone marrow stromal ST2 cells and also induced alkaline phosphatase in mouse osteoblastic MC3T3-E1 cells (Table 14.1).

14.6 Conclusions and Future Perspectives

CNTs based nanotherapeutics can revolutionize the pharmacokinetic and pharmacodynamic profile of the drugs which are active against various diseases. Effective functionalization strategies can bring about synergistic enhancement of the bioactivity as well as reduce the associated side effects [50]. Co-functionalization strategies of the CNTs based nanodrugs can help in simultaneous targeting, diagnosis, and bioimaging [36, 37]. Further functionalization with the superparamagnetic nanoparticles can help in external magnetic field mediated drug targeting and also for magnetic resonance imaging (MRI) [41]. Likewise, conjugation with appropriate nanomaterials with photothermal or photodynamic properties can effectively help in multimodal therapy against infections and cancer [25, 26]. CNTs-derived nanozymes may also help in the enhancement of the therapeutic index of the nanodrugs against cancer and other metabolic disorders. Numerous metal and metal oxide nanoparticles are biologically synthesized using microbes like bacteria, cyanobacteria, algae, and fungi [51]. Such nanoparticles are more biocompatible as their synthesis involves a green route free from any hazardous reducing or capping agents. Hence, it is of utmost importance to use these nanoparticles for surface modification of the CNTs for antimicrobial and antibiofilm activity for controlling the global burden of the nosocomial infections [28, 52]. Biogenic nanoparticles with promising anticancer activity are also often synthesized by medicinal plants [53, 54]. Such phytogenic nanoparticles should be used for enhancement of the anticancer activity of the multifunctionalized SWCNTs and MWCNTs. Several nanoparticles are reported to inhibit the α-amylase and α-glucosidase enzymes that play a key role in the carbohydrate metabolism and diabetes mellitus [16, 55–58]. Hence, such nanoparticles with promising enzyme inhibitory potential can be functionalized and explored against both type I and type II diabetes mellitus. Osteoporosis and traumatic bone tissue injury are key challenges that require novel nanotechnological solutions as the conventional therapeutic approaches are not always sufficient to ensure rapid effective healing and resist post-surgical multidrug-resistant infections [59]. Hence, the modified CNTs with attractive mechanical properties can be used

Table 14.1 Drug functionalized CNTs for biomedical applications.

Type of CNTs	Functionalized agent	Therapeutic application	References
MWCNTs	ZnONPs	Antibacterial	[17]
MWCNTs	Lysine	Antibacterial	[18]
SWCNTs	Ciprofloxacin	Antibacterial	[19]
MWCNTs	Carboxyl groups, dapsone, dapsone imine, dapsone imine copper complex	Antibacterial	[20]
MWCNTs	Benzaldehyde, epichlorohydrine and aminosalicylhydrazide	Antibacterial	[29]
MWCNTs and SWCNTs	Amphotericin B	Antifungal	[30]
MWCNTs	Nystatin	Antifungal	[31]
MWCNTs and SWCNTs	Amphotericin B	Antifungal	[32]
MWCNTs and SWCNTs	Amphotericin B, fluorescein isothiocyanate	Antifungal	[33]
MWCNTs	Lysine and arginine	Antifungal	[34]
CNTs	Paclitaxel	Anticancer	[44]
CNTs	Dacarbazine	Anticancer	[45]
Oxidized single-wall carbon nanohorns (oxSWNHs)	Polyethylene glycol-doxorubicin (PEG-DXR)	Anticancer	[46]
SWCNTs	$SmCl_3$, $LuCl_3$, monoclonal antibody (Cetuximab)	Anticancer	[47]
SWCNTs	Polyethylenimine (PEI), NGR (Cys-Asn-Gly-Arg-Cys-) peptide, hTERT siRNA	Anticancer	[35]
MWCNTs	Poly(methacrylic acid) (PMAA), theophylline	Asthma and pulmonary disease	[48]
SWNHs and oxSWNHs	Dexamethasone (DEX)	Anti-inflammatory, bone regeneration	[49]

as promising reinforcing materials in combination with the natural polymers such as collagen, gelatin, silk fibroin, hyaluronic acid, chitosan, sodium alginate, chondroitin sulfate, and carrageenan [60]. Developing osteogenic scaffolds using such composite hydrogels can definitely help in rapid osteogenesis and efficient osseointegration. In view of the background, chemically modified CNTs can be promising nanomedicine that can solve various unaddressed challenges in the medical practices. However, thorough investigation of their safety and the long-term effect should be investigated before introducing in regular therapeutic practice.

Acknowledgments

Dr. Sougata Ghosh acknowledges Kasetsart University, Bangkok, Thailand for Post-Doctoral Fellowship and funding under Reinventing University Program (Ref. No. 6501.0207/9219 dated 14th September, 2022).

References

1 Jha, R., Singh, A., Sharma, P.K., and Fuloria, N.K. (2020). Smart carbon nanotubes for drug delivery system: a comprehensive study. *J. Drug Delivery Sci. Technol.* 58: https://doi.org/10.1016/j.jddst.2020.101811.

2 Lehman, J.H., Terrones, M., Mansfield, E. et al. (2011). Evaluating the characteristics of multiwall carbon nanotubes. *Carbon* 49: 2581–2602. https://doi.org/10.1016/j.carbon.2011.03.028.

3 Mostafavi, E., Soltantabar, P., and Webster, T.J. (2019). Nanotechnology and picotechnology: a new arena for translational medicine. In: *Biomaterials in Translational Medicine*, 191–212. Academic Press https://doi.org/10.1016/B978-0-12-813477-1.00009-8.

4 Bianco, A., Kostarelos, K., and Prato, M. (2005). Applications of carbon nanotubes in drug delivery. *Curr. Opin. Chem. Biol.* 9: 674–679. https://doi.org/10.1016/j.cbpa.2005.10.005.

5 Zare, H., Ahmadi, S., Ghasemi, A. et al. (2021). Carbon nanotubes: smart drug/gene delivery carriers. *Int. J. Nanomed.* 16: https://doi.org/10.2147/IJN.S299448.

6 Dai, H. (2002). Carbon nanotubes: synthesis, integration, and properties. *Acc. Chem. Res.* 35: 1035–1044. https://doi.org/10.1021/ar0101640.

7 Kaur, J., Gill, G.S., and Jeet, K. (2018). Applications of carbon nanotubes in drug delivery: a comprehensive review. In: *Characterization and Biology of Nanomaterials for Drug Delivery: Nanoscience and Nanotechnology in Drug Delivery* (ed. S.S. Mohapatra, S. Ranjan, N. Dasgupta, et al.), 113–135. Elsevier https://doi.org/10.1016/B978-0-12-814031-4.00005-2.

8 Ravi Kiran, A.V.V.V., Kusuma Kumari, G., and Krishnamurthy, P.T. (2020). Carbon nanotubes in drug delivery: focus on anticancer therapies. *J. Drug Delivery Sci. Technol.* 59: https://doi.org/10.1016/j.jddst.2020.101892.

9 Star, S.A., Stoddart, J.F., Steuerman, D. et al. (2001). Preparation and properties of polymer-wrapped single-walled carbon nanotubes. *Angew. Chem.* 40: 1721–1725.

10 Vaisman, L., Wagner, H.D., and Marom, G. (2006). The role of surfactants in dispersion of carbon nanotubes. *Adv. Colloid Interface Sci.* 128–130: 37–46. https://doi.org/10.1016/j.cis.2006.11.007.

11 Mostafavi, E. and Zare, H. (2022). Carbon-based nanomaterials in gene therapy. *OpenNano* 7: 100062. https://doi.org/10.1016/j.onano.2022.100062.

12 Rohini, K., Kalyani, C., Shankar, G. et al. (2015). ZnO nanoparticles-red sandalwood conjugate: a promising anti-diabetic agent. *J. Nanosci. Nanotechnol.* 15: 4046–4051. https://doi.org/10.1166/jnn.2015.10323.

13 Cruz, D.M., Mostafavi, E., Vernet-Crua, A. et al. (2020). Green nanotechnology-based zinc oxide (ZnO) nanomaterials for biomedical applications: a review. *J. Phys. Mater.* 3: https://doi.org/10.1088/2515-7639/ab8186.

14 Asok, A., Ghosh, S., More, P.A. et al. (2015). Surface defect rich ZnO quantum dots as antioxidants inhibiting α-amylase and α-glucosidase: a potential anti-diabetic nanomedicine. *J. Mater. Chem. B* 3: 4597–4606. https://doi.org/10.1039/c5tb00407a.

15 Karmakar, S., Ghosh, S., and Kumbhakar, P. (2020). Enhanced sunlight-driven photocatalytic and antibacterial activities of flower-like ZnO@MoS_2 nanocomposite. *J. Nanopart. Res.* 22: https://doi.org/10.1007/s11051-019-4710-3.

16 Robkhob, P., Ghosh, S., Bellare, J. et al. (2020). Effect of silver doping on antidiabetic and antioxidant potential of ZnO nanorods. *J. Trace Elem. Med. Biol.* 58: https://doi.org/10.1016/j.jtemb.2019.126448.

17 Akhavan, O., Azimirad, R., and Safa, S. (2011). Functionalized carbon nanotubes in ZnO thin films for photoinactivation of bacteria. *Mater. Chem. Phys.* 130: 598–602. https://doi.org/10.1016/j.matchemphys.2011.07.030.

18 Amiri, A., Zardini, H.Z., Shanbedi, M. et al. (2012). Efficient method for functionalization of carbon nanotubes by lysine and improved antimicrobial activity and water-dispersion. *Mater. Lett.* 72: 153–156. https://doi.org/10.1016/j.matlet.2011.12.114.

19 Assali, M., Zaid, A.N., Abdallah, F. et al. (2017). Single-walled carbon nanotubes-ciprofloxacin nanoantibiotic: strategy to improve ciprofloxacin antibacterial activity. *Int. J. Nanomed.* 12: 6647–6659. https://doi.org/10.2147/IJN.S140625.

20 Azizian, J., Hekmati, M., and Dadras, O.G. (2014). Functionalization of carboxylated multiwall nanotubes with Dapsone derivatives and study of their antibacterial activities against *E. coli* and *S. aureus*. *Orient. J. Chem.* 30: 667–673. doi: 10.13005/ojc/300236.

21 Singh, B.P., Ghosh, S., and Chauhan, A. (2020). Control of bacterial biofilms for mitigating antimicrobial resistance. In: *Sustainable Agriculture Reviews 46. Sustainable Agriculture Reviews*, vol. 46 (ed. H. Panwar, C. Sharma and E. Lichtfouse), 147–176. Cham: Springer Online ISBN: 978-3-030-53024-2; Print ISBN: 978-3-030-53023-5. https://doi.org/10.1007/978-3-030-53024-2_7.

22 Singh, B.P., Ghosh, S., and Chauhan, A. (2021). Development, dynamics and control of antimicrobial-resistant bacterial biofilms: a review. *Environ. Chem. Lett.* 19: 1983–1993. https://doi.org/10.1007/s10311-020-01169-5.

23 Lahiri, D., Nag, M., Ghosh, S. et al. (2022). Electroactive biofilm and electron transfer in MES. In: *Scaling Up of Microbial Electrochemical Systems* (ed. D.A. Jadhav, S. Pandit, S. Gajalakshmi and M.P. Shah), 87–101. Elsevier https://doi.org/10.1016/B978-0-323-90765-1.00006-X.

24 Lahiri, D., Nag, M., Dey, A. et al. (2021). Exosome-associated host–pathogen interaction: a potential effect of biofilm formation. *J. Anal. Sci. Technol.* 12: https://doi.org/10.1186/s40543-021-00306-1.

25 Rashiya, N., Padmini, N., Ajilda, A.A.K. et al. (2021). Inhibition of biofilm formation and quorum sensing mediated virulence in *Pseudomonas aeruginosa* by marine sponge symbiont Brevibacteriumcasei strain Alu 1. *Microb. Pathogen.* 150: https://doi.org/10.1016/j.micpath.2020.104693.

26 Salunke, G.R., Ghosh, S., Santosh Kumar, R.J. et al. (2014). Rapid efficient synthesis and characterization of silver, gold, and bimetallic nanoparticles from the medicinal plant *Plumbago zeylanica* and their application in biofilm control. *Int. J. Nanomed.* 9: 2635–2653. https://doi.org/10.2147/IJN.S59834.

27 Banerjee, I., Mondal, D., Martin, J., and Kane, R.S. (2010). Photoactivated antimicrobial activity of carbon nanotube - porphyrin conjugates. *Langmuir* 26: 17369–17374. https://doi.org/10.1021/la103298e.

28 Ghosh, S., Turner, R.J., Bhagwat, T., and Webster, T.J. (2021). Novel and Future Treatment Strategies for Biofilm-Associated Infections. In: *Biofilm-Mediated Diseases: Causes and Controls* (ed. R.R. Ray, M. Nag and D. Lahiri), 239–276. Singapore: Springer https://doi.org/10.1007/978-981-16-0745-5_10.

29 Mohamed, N.A. and Abd El-Ghany, N.A. (2019). Synthesis, characterization and antimicrobial activity of novel aminosalicylhydrazide cross linked chitosan modified with multi-walled carbon nanotubes. *Cellulose* 26: 1141–1156. https://doi.org/10.1007/s10570-018-2096-5.

30 Bonincasa, M., Pacor, S., Wu, W. et al. (2011). Antifungal activity of amphotericin B conjugated to carbon nanotubes. *ACS Nano* 5: 199–208. https://doi.org/10.1021/nn1023522.

31 Uttekar, P.S., Kulkarni, A.M., Sable, P.N., and Chaudhari, P.D. (2016). Surface modification of carbon nano tubes with nystatin for drug delivery applications. *Indian J. Pharm. Educ. Res.* 50: 385–390. https://doi.org/10.5530/ijper.50.3.10.

32 Vossoughi, M., Gojgini, S., Kazemi, A. et al. (2009). Conjugation of amphotericin B to carbon nanotubes via amide-functionalization for drug delivery applications. *Eng. Lett.* 17: 4.

33 Wu, W., Wieckowski, S., Pastorin, G. et al. (2005). Targeted delivery of amphotericin B to cells by using functionalized carbon nanotubes. *Angew. Chem. Int. Ed.* 117: 6516–6520. https://doi.org/10.1002/ange.200501613.

34 Zare-Zardini, H., Amiri, A., Shanbedi, M. et al. (2013). Studying of antifungal activity of functionalized multiwalled carbon nanotubes by microwave-assisted technique, surface and Interface. *Analysis* 45: 751–755. https://doi.org/10.1002/sia.5152.

35 Wang, L., Shi, J., Zhang, H. et al. (2013). Synergistic anticancer effect of RNAi and photothermal therapy mediated by functionalized single-walled carbon nanotubes. *Biomaterials* 34: 262–274. https://doi.org/10.1016/j.biomaterials.2012.09.037.

36 Kitture, R. and Ghosh, S. (2018). Hybrid nanostructures for in vivo imaging. In: *Hybrid Nanostructures for Cancer Theranostics* (ed. R. Ashok and N.T. Bohara), 173–208. Elsevier https://doi.org/10.1016/B978-0-12-813906-6.00010-X.

37 Kitture, R. and Ghosh, S. (2021). Nanopharmacokinetics: key role in in vivo imaging. In: *Nano-Pharmacokinetics and Theranostics* (ed. N.D. Thorat and N. Kumar), 233–251. Elsevier https://doi.org/10.1016/b978-0-323-85050-6.00009-8.

38 Ghosh, S. (2019). Physically Responsive Nanostructures in Breast cancer Theranostics. In: *External Field and Radiation Stimulated Breast Cancer Nanotheranostics*. IOP Publishing 10.1088/2053-2563/ab2907ch2.

39 Ghosh, S., More, P., Derle, A. et al. (2015). Diosgenin functionalized iron oxide nanoparticles as novel nanomaterial against breast cancer. *J. Nanosci. Nanotechnol.* 15: 9464–9472. https://doi.org/10.1166/jnn.2015.11704.

40 Mallick, A., More, P., Ghosh, S. et al. (2015). Dual drug conjugated nanoparticle for simultaneous targeting of mitochondria and nucleus in cancer cells. *ACS Appl. Mater. Interfaces* 7: 7584–7598. https://doi.org/10.1021/am5090226.

41 Kitture, R., Ghosh, S., Kulkarni, P. et al. (2012). Fe_3O_4-citrate-curcumin: promising conjugates for superoxide scavenging, tumor suppression and cancer hyperthermia. *J. Appl. Phys.* 111: https://doi.org/10.1063/1.3696001.

42 Rokade, S.S., Joshi, K.A., Mahajan, K. et al. (2018). *Gloriosa superba* mediated synthesis of platinum and palladium nanoparticles for induction of apoptosis in breast Cancer. *Bioinorg. Chem. Appl.* 2018: https://doi.org/10.1155/2018/4924186.

43 Bloch, K., Parihar, V.S., Kellomäki, M., and Ghosh, S. (2022). Natural compounds from *Plumbago zeylanica* as complementary and alternative medicine. In: *Handbook of Oxidative Stress in Cancer: Therapeutic Aspects*, 1–28. Singapore: Springer https://doi.org/10.1007/978-981-16-1247-3_33-1.

44 Hashemzadeh, H. and Raissi, H. (2017). The functionalization of carbon nanotubes to enhance the efficacy of the anticancer drug paclitaxel: a molecular dynamics simulation study. *J. Mol. Model.* 23: https://doi.org/10.1007/s00894-017-3391-z.

45 Mirsalari, H., Maleki, A., Raissi, H., and Soltanabadi, A. (2021). Investigation of the pristine and functionalized carbon nanotubes as a delivery system for the anticancer drug dacarbazine: drug encapsulation. *J. Pharm. Sci.* 110: 2005–2016. https://doi.org/10.1016/j.xphs.2020.10.062.

46 Murakami, T., Fan, J., Yudasaka, M. et al. (2006). Solubilization of single-wall carbon nanohorns using a PEG-doxorubicin conjugate. *Mol. Pharmaceutics* 3: 407–414. https://doi.org/10.1021/mp060027a.

47 Spinato, C., De Garibay, A.P.R., Kierkowicz, M. et al. (2016). Design of antibody-functionalized carbon nanotubes filled with radioactivable metals towards a targeted anticancer therapy. *Nanoscale* 8: 12626–12638. https://doi.org/10.1039/c5nr07923c.

48 Zhang, C.H., Luo, Y.L., Chen, Y.S. et al. (2009). Preparation and theophylline delivery applications of novel PMAA/MWCNT-COOH nanohybrid hydrogels. *J. Biomater. Sci. Polym. Ed.* 20: 1119–1135. https://doi.org/10.1163/156856209X444466.

49 Murakami, T., Ajima, K., Miyawaki, J. et al. (2004). Drug-loaded carbon nanohorns: adsorption and release of dexamethasone in vitro. *Mol. Pharmaceutics* 1: 399–405. https://doi.org/10.1021/mp049928e.

50 Ghosh, S., Bloch, K., and Webster, T.J. (2021). Functionalized biogenic nanoparticles and their applications. In: *Nanobiotechnology* (ed. S. Ghosh and T.J. Webster), 303–322. Elsevier https://doi.org/10.1016/b978-0-12-822878-4.00019-5.

51 Ghosh, S., Bhagwat, T., Chopade, B.A., and Webster, T.J. (2021). Patents, technology transfer, and commercialization aspects of biogenic nanoparticles. In: *Nanobiotechnology*, 323–339. Elsevier https://doi.org/10.1016/b978-0-12-822878-4.00020-1.

52 Pormohammad, A., Monych, N.K., Ghosh, S. et al. (2021). Nanomaterials in wound healing and infection control. *Antibiotics* 10: https://doi.org/10.3390/antibiotics10050473.

53 Ghosh, S., Patil, S., Ahire, M. et al. (2012). *Gnidia glauca* flower extract mediated synthesis of gold nanoparticles and evaluation of its chemocatalytic potential. *J. Nanobiotechnol.* 10: 17.

54 Ghosh, S. and Patil, S. (2016). Gnidia glauca leaf and stem extract mediated synthesis of gold Nanocatalysts with free radical scavenging potential. *J. Nanomed. Nanotechnol.* 7: https://doi.org/10.4172/2157-7439.1000358.

55 Ghosh, S., More, P., Derle, A. et al. (2014). Diosgenin from dioscoreabulbifera: novel hit for treatment of type II diabetes mellitus with inhibitory activity against α-amylase and α-glucosidase. *PLoS One* 9: https://doi.org/10.1371/journal.pone.0106039.

56 Jamdade, D.A., Rajpali, D., Joshi, K.A. et al. (2019). *Gnidia glauca* - and *Plumbago zeylanica* -mediated synthesis of novel copper nanoparticles as promising antidiabetic agents. *Adv. Pharmacol. Sci.* 2019: https://doi.org/10.1155/2019/9080279.

57 Bhagwat, T.R., Joshi, K.A., Singh Parihar, V. et al. (2018). Biogenic copper nanoparticles from medicinal plants as novel antidiabetic nanomedicine. *World J. Pharm. Res.* 7: 183. doi: 10.20959/wjpr20184-10773.

58 Lawande, P.P., Sontakke, V.A., Kumbhar, N.M. et al. (2017). Polyhydroxylated azetidine iminosugars: synthesis, glycosidase inhibitory activity and molecular docking studies. *Bioorg. Med. Chem. Lett.* 27: 5291–5295. https://doi.org/10.1016/j.bmcl.2017.10.025.

59 Ghosh, S., Sanghavi, S., and Sancheti, P. (2018). Metallic biomaterial for bone support and replacement. In: *Fundamental Biomaterials: Metals* (ed. P. Balakrishnan and M.S. Sreekala), 139–165. Elsevier https://doi.org/10.1016/B978-0-08-102205-4.00006-4.

60 Ghosh, S. and Webster, T.J. (2021). Mesoporous silica based nanostructures for bone tissue regeneration. *Front. Mater.* 8: https://doi.org/10.3389/fmats.2021.692309.

15

Chemically Modified Carbon Nanotubes in Tissue Engineering

Richika Ganjoo[1], Shveta Sharma[1], and Ashish Kumar[2]

[1]*Lovely Professional University, Faculty of Technology and Sciences, Department of Chemistry, Jalandhar-Delhi, G.T. Road, Phagwara 144411, Punjab, India*
[2]*NCE, Department of Science and Technology, Government of Bihar, Nalanda, Patna 803108, India*

15.1 Introduction

Tissue engineering aims to repair and preserve normal function by replacing sick or damaged Tissue with biological alternatives [1]. Presently, tissue regeneration treatments include using discrete cells or cell scaffolds, tissue-induced biomolecules including proteins, medicines, oligonucleotides, and synthetic constructions [2]. The most often utilized tissue engineering method is the engineered constructs approach. Tissue engineering necessitates the development of biocompatible and bioactive biomaterials. CNTs have lately been studied for tissue engineering applications. In recent years, significant improvements in the understanding of cell and organ transplantation and the chemistry of carbon nanotubes (CNTs) have contributed to the further growth of tissue engineering using CNTs. CNTs may be utilized as fillers in this sector to increase the physical strength and conductivity of the tissue framework by distributing a tiny proportion of CNTs into a polymer or to boost the advantages of the extracellular matrix in its natural state [3]. The inclusion of CNTs into a polymer framework has several physicochemical and mechanical reinforcing properties, comprising improved strength, flexibility, biocompatibility, and gene regulation modulation for tissue healing. For specialized biological applications, CNTs may be changed to improve solubility and adjust the inclusion of functional groups. CNT's versatility and particular applications are governed by their unique characteristics, such as shape, size, and organization. CNTs are divided into two categories: single-walled carbon nanotubes (SWCNTs) and multi-walled carbon nanotubes (MWCNTs). As a result of the increased loading capability given by their larger surface area and their interactivity with biological sheaths, MWCNTs have appeared as potential options for delivery systems. Furthermore, their superior mechanical and electrical characteristics suggest that they might be useful as additives and osteogenic frameworks for bone formation and proliferation. The recent advances made in the field of tissue engineering employing modified CNTs have been illustrated in Figure 15.1.

Chemically Modified Carbon Nanotubes for Commercial Applications, First Edition.
Edited by Jeenat Aslam, Chaudhery Mustansar Hussain, and Ruby Aslam.

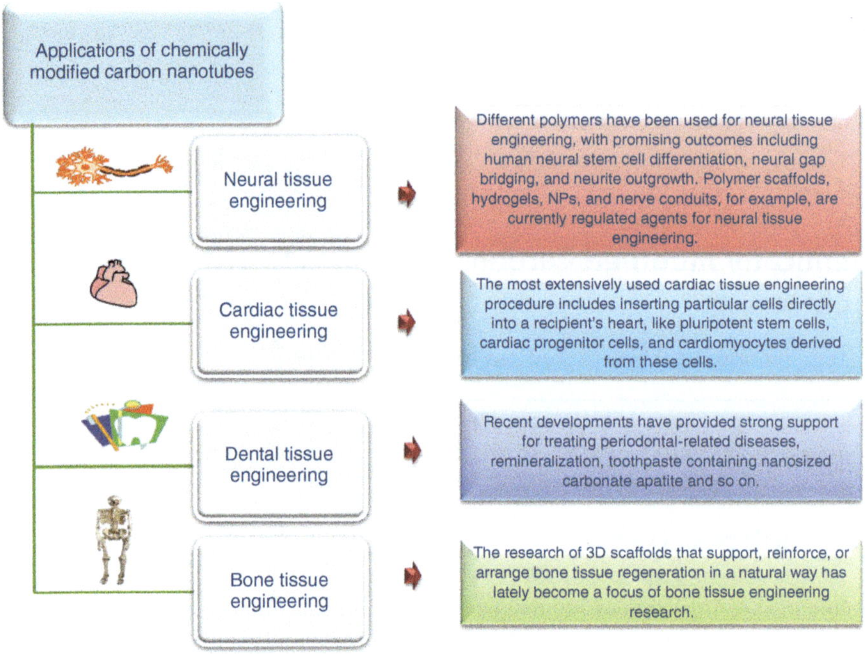

Figure 15.1 A schematic illustration of recent advancements made in the field of tissue engineering using chemically modified carbon nanotubes.

The creation of new materials to regulate, analyze, and assess tissue growth has been a key focus of tissue engineering research. The patent of Laurencin et al. describes a biocompatible composite material that can be formed into a scaffold with bone repair and regeneration capabilities. These scaffolds have enough mechanical strength to be beneficial for cortical bone repair and regeneration. The use of electrical stimulation as a therapy for injured bone tissues has been investigated. Electrically conducting SWCNTs, according to another patent, can improve cytocompatibility as well as specific cell functioning and proliferation [4]. Using electrically conductive SWCNT-based orthopedic/dental implants, Supronowicz et al. offer a method for boosting osteoblast growth. The SWCNT-based implant is in contact with osteoblasts in this manner. The implant is stimulated by an electric current, which stimulates the osteoblasts and promotes bone formation. Carroll et al. claim that electrically stimulated cells cultivated on SWCNTs proliferate faster than cells cultured on control surfaces, according to another SWCNT-based approach patented by them [5].

15.2 Applications of Modified Carbon Nanotubes in Tissue Engineering

15.2.1 Neural Tissue Engineering

The central nervous system (CNS) and the peripheral nervous system (PNS) are the two main components of the nervous system (PNS). CNS comprises the spinal cord and the brain, whereas sense neurons and motor neurons make up the PNS. Because the CNS and PNS cannot regenerate, they often exhibit long-term functional abnormalities due to illness or accidental damage. With the rapid aging of the population, the prevalence of neurological diseases is rising, posing a severe danger to human health [6]. Surgical sutures, allograft, and autologous transplantation are now the most common therapeutic therapies for neurological illnesses, all of which are used to heal the wounded nerve. Even though the sickness can be treated to some degree, there are still many drawbacks, including immunological rejection, frequent operations, and poor treatment outcomes. The advancement of nerve tissue engineering promises neurological disorder treatment [7]. CNTs have been used as substrates for neural tissue engineering in some studies, with findings showing that they can enable neurons adhesion, facilitate the production of longer and more elaborate neurites, and promote cell differentiation [2]. The potential toxicity of CNTs in biological systems and the difficulties of constructing controlled two-dimensional (2D)/three-dimensional (3D) structures have severely limited its use as a substrate for brain tissue engineering.

Layer-by-layer assembly was used by Shao et al. [8] to produce a membrane comprising CNTs and PDDA. Surprisingly, the findings revealed that positively charged PDDA and negatively charged PDDA were effective in biomanufacturing brain tissue engineering. Although polymeric frameworks provide friendly surroundings, they are electrically inert and have low tensile strength, limiting the use of tissue that has been propagated using electricity. The integration of CNTs with exceptional electrical properties and material performance, on the other hand, might provide an alternate channel for these nonconductive constituents in brain tissue engineering while also reducing CNT toxicity.

Vicentini et al. [9] generated a composite material comprised of poly(L-lactic acid) and employed electrospinning to construct a 2D layer for brain tissue engineering engaging three kinds of carbon nanostructures incorporating fragmented graphene, MWCNTs, and nanohorns as additives, with carbon loadings varying from 0.1 to 5 wt%. The composite material, including MWCNT, had a reduced electrical resistance but a more outstanding Young's modulus and tensile breakdown. These composite scaffolds are biocompatible and help proliferate the SH-SY5Y human neural precursor cell line.

To promote peripheral nerve regeneration, Arslantunali et al. [10] developed a MWCNT (pHEMA) composite. The addition of MWCNT increased both mechanical qualities and electrical conductivity, according to the results. The biocompatibility of the composite conduit was tested using SH-SY5Y neuroblastoma cells. There were no signs of cytotoxicity.

Serrano et al. [11] demonstrated that 3D substrates consisting of chondroitin sulfate (CS), a significant regulatory element of nervous tissue, and MWCNTs are preferential scaffolds for developing a sustainable network densely packed with neurons with a transient reduced glial level. Scaffolds were made from embryonic neural progenitor cells utilizing the ice segregation-induced self-assembly approach. Cell viability, morphology, vitality, neuron/glial differentiation, calcium signaling pathways, and mitochondrial activity were tracked over time on the scaffolds. The results were compared to 2D reference samples. The findings indicated the growth of feasible cultures augmented in neuron cells for approximately 20 days, even in the absence of poly-D-lysine covering, with the capability to create calcium transients and functional mitochondria. The critical factor behind these findings is a simultaneous neural-permissive reaction from both the structure and its constituents. They speculated that if these structures were incorporated during the acute period after injury, they could aid nerve regeneration because the most critical sequence of events, driving either neurological rejuvenation or fibroglial tissue repair, occurs during the first few days after graft implantation.

The yield strength, Young's modulus, and tensile power of allied MWCNT/chitosan composite frameworks were enhanced by 21.9%,12.7%, and 11.2%, respectively, as demonstrated by Gupta et al. [12] when compared to arbitrarily dispersed composite frameworks made of MWCNT and chitosan. MWCNT alignment resulted in extremely anisotropic electrical conductivity in comparison to the orientation of the scaffold.

Koppes et al. [13] used a combination of carboxylated SWCNTs and hydrogels to make composite scaffolds that were applied via direct current stimulation. When SWCNT loading was increased, the electrical conductivity and neurite outgrowth were dramatically enhanced compared to controls without SWCNT. In addition, substrates that contain SWCNT demonstrated up to a sevenfold rise in proliferation. When electrical stimulation was applied compared to not stimulated controls and devoid of nanofiller, on the other hand, the expansion of neurite needs further research into nano-filler qualities.

Numerous studies have recently focused on applying electrical stimulation to conductive biomaterials like CNT reinforced scaffolds to increase navigation and stimulation for neural cell regeneration.

According to Wang et al. [12], electrical stimulation promotes neurite outgrowth. They made PLGA composite frameworks using a blend of carboxylated MWCNTs and a poly-L-lysine layer to improve hydrophilicity. PC12 cells and DRG neurons were guided along the fiber path by scaffolds with aligned fibers. The results indicated that their neurite length increased when PC12 cells and DRG neurons were stimulated with a voltage of 40 mV. The electrical stimulation also enhances PC12 cell differentiation.

Because of their exceptional electrical conductivity, CNTs are very beneficial in the regeneration of neurons. Neurons may efficiently form on CNT beds, and changing the surface with 4-hydroxyonoenal, a substance linked to neuron growth, might increase the number of neurons and branching dimension across CNTs. Enhancing electron transmission through connections between neurons might also improve the functioning of neural circuits. Implanted electrodes are being used to trigger the neural system to treat various health issues, including Parkinson's syndrome, epilepsy, and severe pain. CNTs are an excellent alternative for electrodes and may be responsible for some notable drug delivery routes that enter cells through endocytosis. CNTs' uses include the transport of genes and proteins that can be used to create proteins by using the cells' biological functions; this might aid the body's healing mechanisms.

15.2.2 Cardiac Tissue Engineering

Cardiovascular diseases (CVDs) are the leading cause of mortality worldwide and in the United States. Many people with CVD will need transplantation. Nevertheless, transplantation has a significant threat of dismissal, and organ donors are in low supply. Vascular tissue regeneration has been recommended as a less complicated and more approachable substitute for repairing, regenerating, and improving damaged vessels. Tissue engineering creates natural alternatives by combining cells, scaffolds, and appropriate biochemical components. The scaffold is critical in providing a suitable environment for cell growth in tissue engineering. A perfect framework should allow for simple cellular attachment, proliferation, and differentiation without causing tissue toxicity.

Additionally, biocompatibility should be considered, allowing the host tissue to renew and replace it. To offer equivalent physical support *in vivo*, the scaffold's mechanical characteristics should be similar to the target tissues. CNT's exceptional mechanical and electrical capabilities have sparked interest in tissue engineering. Although no agreement on the biocompatibility of CNTs has been reached, multiple studies have shown that incorporating CNTs into other biomaterials may enhance the physical characteristics of the framework and enable cell adhesion, differentiation, and proliferation [14]. It is well established that the damage of cardiomyocytes, or particular cardiac muscle cells, is recognized to be involved in a range of heart problems [15]. Consequently, the most extensively used cardiac tissue engineering procedure is the best therapy, which includes inserting particular cells directly into a recipient's heart, like pluripotent stem cells, cardiac progenitor cells, and cardiomyocytes derived from these cells. Biomaterials for cardiac tissue engineering essentially be adequate in power to support myocardial tissue regeneration while being conductive to growth [16].

Although cardiac cell therapy has distinct benefits over conventional heart disease therapies, limitations like high risk of mortality, reduced cell seeding effectiveness, immunogenic reaction, and poor integration with host tissue have hampered the therapeutic success [17]. Cell reprogramming is a novel strategy that aims to solve some of the issues associated with cell treatment, but more study is required before being used in clinical settings. Biomaterials for cardiac tissue engineering have appropriate physical strength while also being conductive to myocardial tissue regeneration development. Polymeric materials may be biocompatible. However, they are often inert, limiting intercellular contact and signal transmission, and their mechanical qualities need to be improved. A wide variety of polymer/CNTs composites have been studied to enhance both electronic and structural conductivity properties of biopolymers.

To improve mechanical strength, Wickham et al. [18] used thiophene-conjugated carbon nanotubes (T-CNTs). They demonstrated that newly isolated CPCs from murine hearts did not adhere to or spread on polycaprolactone (PCL) films, regardless of whether T-CNT was present. Both PCL and PCL/T-CNT, on the other hand, promoted CPC adherence, growth, and development as electrospun meshes. T-CNT added to PCL culminated in a considerable improvement in structural rigidity without morphological alterations to the meshes. CPC proliferation was boosted into cardiomyocytes in T-CNT-containing meshes, but not differentiation.

Li et al. [19] fabricated a poly(*N*-isopropylacrylamide) (PNIPAAm) hydrogel comprising SWCNT. To form solid substrates, the hydrogel was lyophilized. The biocompatibility features were evaluated *in vitro* and *in vivo* testing. The addition of SWCNTs increases the biological performance of PNIPAAm substrates *in vitro* with BASCs. The SWCNTs reinforced with PNIPAAm greatly improved BASC engraftment in infract myocardium, according to *in vivo* experiments.

The intercalated discs (IDs) that link the cardiomyocytes are not sufficiently remodeled in current cardiac tissue engineering procedures. CNTs may be able to help overcome this constraint. A CNT/collagen cover was fabricated by Sun et al. [20] to help cardiac myocytes form IDs. 1-Integrin acts as a mechanotransducer in the myocardium, activating various intracellular signaling pathways such FAK, ILK, and Src. The composite substrate not only enhanced cardiomyocyte adhesion and maturation, but it likewise augmented ID formation as well as functioning by promoting ID-related protein expression. By stimulating the FAK/ERK/GATA4 pathway mediated by 1-integrin, composite substrates containing CNTs significantly speed up gap junction development.

The insulated walls of polymeric materials used to repair heart defects obstruct the transmission of electrical impulses between cardiomyocytes, resulting in arrhythmias when the biopolymeric materials are implanted.

To make electrospun cardiac conducive scaffolds, Mombini et al. [21] employed polyvinyl alcohol and chitosan with varied concentrations of CNTs (1, 3, and 5 wt%). According to the findings, the nanofiber with 1 wt% CNT exhibited the best physiochemical parameters (electrical conductivity 3.4×10^{-6} S cm^{-1}, elastic modulus 130 MPa). The application of electrical stimulation to a framework containing 1 wt% CNT increased the cardiac marker's gene expression.

Shin et al. [22] tackled this issue by developing patches of gelatin methacrylate (GelMA) and CNTs for cardiac tissue engineering. GelMA was coated on CNT, and photo-cross-linking created 50 m thick CNT-GelMA hybrid hydrogels. With neonatal rat cardiomyocytes, porous substrates were constructed and cultivated. CNT-GelMA had a threefold greater spontaneous beating rate and an 85% lower excitation threshold, significantly increasing cell adhesion, organization, and cell–cell interaction, according to the findings.

Pok et al. [23] synthesized a gelatin chitosan hydrogel containing SWCNTs that operate as electrical nanobridges for cardiomyocyte interaction. The composite hydrogels allowed for excitation conduction velocities comparable to those of normal cardiac tissue (22.9 cm s^{-1}), demonstrating that they can sustain cardiomyocyte activity.

CNTs incorporated in oriented poly(-glycerol sebacate):gelatin (PG) electrospun nanofibers were used by Kharaziha et al. [24] to develop robust and flexible hybrid substrates with improved electrical characteristics. MWCNTs were tested at various concentrations ranging from 0 to 1.5 wt%. Electrical and mechanical qualities improved with increasing MWCNT doses, as seen in the experiments. Cardiomyocytes were able to develop on the meshes that were created. Compared to meshes without MWCNTs, meshes with MWCNTs produced more impulsive and coordinated beating response, with a 3.5-fold reduction in the excitation threshold and a 2.8-fold increase in the highest capture rate.

CNTs have also been utilized to make 3D scaffolds when coupled with polymeric materials, in addition to 2D structures. Using an extrusion-based additive production method, Ho et al. [25] employed filament diameters and pore widths varying between 300 and 450 m and 425 and 675 m, respectively, to develop composite scaffolds of PCL and MWCNT. CNTs improved the mechanical properties of the PCL while also altering its morphology, increasing crystallinity. There were no harmful effects on H9C2 cells in biological experiments. The inclusion of MWCNTs slowed the deterioration process, according to the results. Izadifar et al. [26] developed a hybrid cardiovascular implant with nano reinforcement containing HCAECs that had a better electrical, structural, and physiological performance. For methacrylated collagen (MeCol) micropatterning, a harmless UV exposure period with no impact on cell feasibility was discovered. At various CNT mass ratios, the impact of carboxyl functionalized CNTs on MeCol and alginate matrix shape, structural characteristics, electrical behavior, and cellular reaction were studied. To make hybrid hydrogel structures with a CNT-reinforced alginate framework and cell-laden MeCol, a UV-incorporated 3D bioprinting approach was used. After 20 days of incubation in growth conditions at 37 °C, the tensile impedance, modulus, and swelling degree of hybrid constructs were determined for varied CNT mass ratios. The vitality, proliferation, and differentiation of HCAECs in the setting of bioprinted hybrid constructs were tested *in vitro* for 10 days. The functionalized CNTs created a strongly interlinked nanofibrous meshwork and enhanced the viscoelastic behavior and electrical conductivity of photo-cross-linked MeCol considerably. When compared to pure alginate, CNTs with an alginate film produced a nanofilamentous complex with fiber sizes ranging from 25 to 500 nm, which improved HCAEC adhesion and extension, in addition to mechanical and electrical properties. The hybrid 3D-printed with CNT infill constructions had dramatically improved toughness and electrical conductivity, especially in the medically appropriate frequency range (5 Hz). When compared to CNT-free hybrid constructions, the CNT-reinforced hybrid implants showed considerably more edoema after 20 days of culture. Over 10 days of *in vitro* incubation, HCAECs showed considerable cellular proliferation, migration, and differentiation (lumen-like development) for a certain CNT mass ratio. The findings of this research are crucial in the development of conductive, resilient, and possibly prevascularized hybrid cardiac patches.

Conductive scaffold materials have been shown to assist tissue engineering applications by controlling modulation, electroactive cell adhesion, and propagation. CNTs have presented considerable potential in tissue engineering owing to their outstanding structural characteristics, among other conductive materials. Due to the necessity for miscible solvents to dissolve CNTs and polymers, or surface functionalization of CNTs, current techniques for integrating CNTs in polymers limit the broad use of CNTs in tissue engineering. These approaches either restrict the types of polymers that may be used or have a negative impact on polymer/CNT composite characteristics.

Jiang et al. [14] proposed a new technique for making PCL/gelatin/CNT matrix yarns employed in CVD scaffolds. The effects of different PCL/gelatin ratios and CNT concentrations on the mechanical and biological characteristics of PCL/gelatin/CNT yarns were investigated and outlined to offer recommendations for managing yarn mechanical and biological attributes. To increase the orientation and lengthening of ECs on yarns, CNTs were employed. They used PCL/gelatin/CNT yarns to develop a scaffold based on textile for vascular tissue engineering to demonstrate the applicability of this finding. The scaffold has structure and mechanical properties similar to those of natural vessels. Cultured ECs grew in the scaffold lumen and displayed alignment morphology. Overall, the findings demonstrate that wet electrospinning-fabricated polymer/CNT composite yarns are excellent choices for vascular scaffold fabrication. The limitations of existing methods for integrating CNTs into polymers are solved by this new approach, enabling CNTs to be employed more broadly in tissue engineering. While this research focuses on CNTs, the method might be employed to analyze biomolecules in biopolymers or other functional materials.

15.2.3 Bone Tissue Engineering

Treatment of human bone abnormalities, such as those caused by tumor removal, trauma, and aberrant bone formation, has significant limits. Current treatments, including autographs, allographs, and metal prostheses, cannot repair bone, but they use an artificial substance to replace the removed bone. Tissue engineering's latest innovation is to generate a tissue substitute by growing bone cells on a nonnatural 3D framework or a living prosthesis [27]. Ideally, a bone tissue rejuvenation matrix should match the structural properties of the original bone tissue. It should have an extraordinary porosity and pore size and good interconnectivity for bone tissue ingrowth, making it biocompatible with surrounding tissue [28]. Biodegradable or nonbiodegradable synthetic scaffold material should be used. On top of the nonbiodegradable matrix, cells proliferate and deposit new living composite that develops working bone [29]. Still, no manufactured matrix was enough to be used with 100% quality. As a result, innovative biomaterials and scaffold production procedures are critical to the success of bone tissue engineering [27].

Tanaka et al. [30] developed a 3D block structure of CNTs and compared it to PET-reinforced collagen to see its effectiveness as a scaffold material for bone restoration. Comparative mechanical tests showed no significant variations in the compressive strengths of the produced structure and rat femoral bone. CNT scaffolds exhibited former cell adherence than PET-filled collagen scaffolds. The inclusion of recombinant human BMP-2 in the CNTs block increased ALP activity, suggesting excellent osteogenesis.

Lin et al. [30] isolated and grew MSCs on high-tensile-strength PLGA/c-MWCNT nanocomposites from rat bone marrow for the first time. Atomic force microscopy (AFM) and scanning electron microscopy (SEM) results revealed that the MWCNTs with carboxyl functionalization might give a more significant dispersed PLGA matrix, according to SEM and AFM results. Notably, the PLGA/c-MWCNTs have better tensile power than the PLGA approved by FDA. Results obtained from various studies like X-ray photoelectron spectroscopy (XPS), contact angle, and degradation in living systems showed that incorporating c-MWCNTs improved the PLGA composite's hydrophilicity. Furthermore, the rat MSC's adhesion, proliferation, and ALP secretion were aided by the nanocomposite films based on PLGA/c-MWCNT. MSCs attached to PLGA/c-MWCNT nanocomposite films produced more ALP, a bone marker. As a result, the performance of these nanocomposite films calls for further research into *in vivo* characterization and 3D scaffold production for bone regeneration.

Patel et al. [31] assessed the potential of u-CNTs for tissue engineering after they were effectively generated by chemical treatment. The produced u-CNTs showed distinct functional groups in their structures with increased diameters and zeta potentials. The X-ray diffraction spectroscopy (XRD) patterns show that the u-CNTs displayed partial characteristics of both pure and oxidized CNTs. The numerous functional groups present in the u-CNTs increase heat transport and decrease thermal stability. More viable cells were found in the u-CNT media than pure CNT media and the controls, indicating that they are more biocompatible.

Furthermore, the nanomaterial content in the culture medium had a positive impact on cellular activity. The u-CNT-treated groups formed more mineralized nodules than the control groups, suggesting their improved mineralization efficiency. Compared to the control, u-CNTs had a higher expression of osteogenic transcription factors, indicating a higher osteogenic potential. These findings indicate that the produced material might be employed in bone regeneration tissue engineering. This conclusion is a limitation of our research; more research is needed to substantiate our results for advanced applications.

Tanaka et al. [32] compared hydroxyapatite (HA) with CNTs *in vivo* and vitro. CNTs had improved protein absorption and release abilities, according to the findings. With the addition of recombinant human BMP-2, CNT porous assemblies showed more remarkable cell growth, improved osteoconduction, and more significant bone formation *in vivo* trials. The use of bioceramics like hydroxyapatite and bioglass in conjunction with CNTs for bone restoration has previously been described in some studies. The ability of the CNTp in the usage of bone regeneration therapy was tested in this research. Porous polymer CNTp has micropore diameters comparable to those found in multi-pore 3D HA matrices used for bone regeneration therapy. CNTp demonstrated excellent cytocompatibility, protein adsorption, and protein release control. When loaded with rhBMP-2, it showed great independent osteoconduction and outstanding new bone growth *in vivo* studies. Based on these findings, we believe the CNTp is a promising framework that can be used to fill the space in case of fractures and deformations, as well as to repair bones early and firmly in tissue engineering, and it is a matrix on which cell growth, retention, and release of growth factors can be controlled. The SEM and fluorescent microscopy images of the studies are displayed in Figure 15.2.

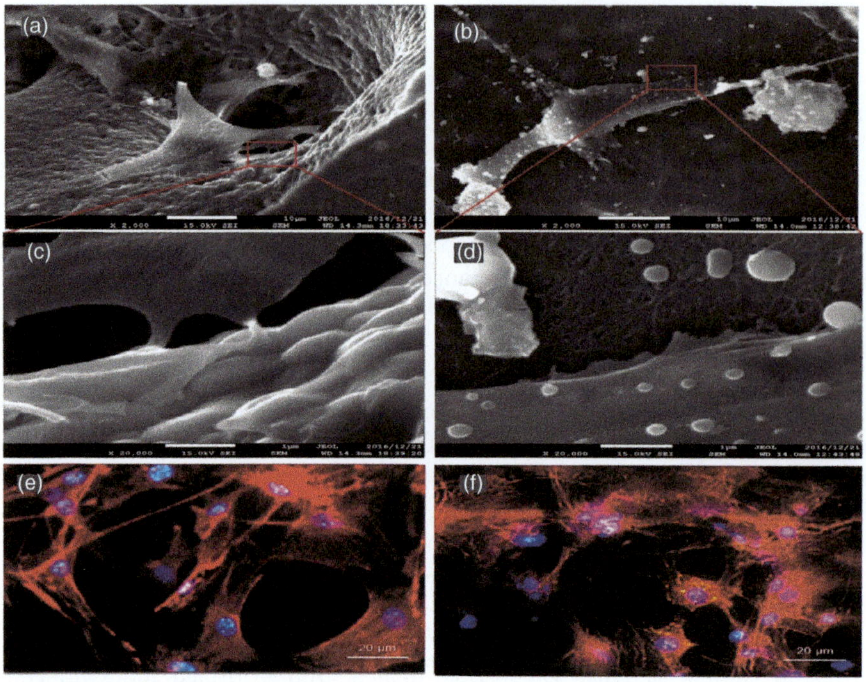

Figure 15.2 Spreading of MC3T3-E1 cells on IP-CHA and CNTp scaffolds. Cells on (a, c, e) IP-CHA or (b, d, f) CNTp. SEM images of (a, c) IP-CHA and (b, d) CNTp. Actin filaments (red) and nuclei (blue) were identified in cells of (e) IP-CHA and (f) CNTp. Source: Tanaka et al. [32]/with permission of MDPI.

CNTs were employed as reinforcement fillers by Liu et al. [33] to improve the stability against compression. They obtained was highest value with 3% of CNTs (37.32 MPa) and resistance from further rupture, and the highest value of 1.58 MPa $m^{1/2}$ with 3 wt% of CNTs. A powder-bed fusion additive manufacturing technology was used to create the scaffolds. According to cell tests, the composite displayed high cytocompatibility in human osteosarcoma MG-63 cells.

MWCNT/HA frameworks have varying CNT loads from were evaluated by Khalid et al. [34]. The living cell destruction caused by the composite depends upon the dose and living cell proportion in the sample. It reduced as the amount of CNT increased, according to results employing human osteoblast sarcoma cell lines.

PLAGA and SWCNT matrix microspheres were created by Gupta et al. [35], and properties of MC3T3-E1 cells like adhesion, cell growth, and gene expression were investigated. Compared to pure PLAGA scaffolds, new cells had good adhesion and fast development with clear morphology, which confirmed the addition of SWCNT increased cell growth and gene expression.

Mesgar et al. [36] created gelatin–chitosan scaffolds with varying concentrations of functionalized MWCNTs using freeze-dry methods. Adding MWCNTs increased modulus by 11 and 9.6 times, respectively; however, increasing MWCNT concentration resulted in mechanical property loss owing to agglomeration. The introduction of CNTs enhanced the production of a bone-like apatite layer in stimulated bodily fluid.

Employing a freeze-drying process, the performance of matrix of PLLA and CNTs was evaluated in the living system and laboratory by Duan et al. [37]. The framework understudy was implanted with bone marrow mesenchymal stem cells (BMSCs) in the living system and findings revealed that CNTs increased cell growth while encouraging bone-related differentiation.

Gonçalves et al. [38] also investigated 3D CNT framework with porosity in bone tissue engineering. They formed MWCNT-filled PCL-hydroxyapatite scaffolds by additive manufacturing system based on pressure-assisted extrusion. PCL pellets were dissolved in dichloromethane to make a slurry in a 50 wt% PCL composite, the MWCNT concentration ranged from 0 wt% (50% by weight hydroxyapatite) to 10 wt% (40% hydroxyapatite by weight). Fully interconnected squared-shaped porous matrices with pore dimensions 450–700 m were formed.

Wang et al. [39] used injection molding to introduce a bone cement made of polymethyl-methacrylate (PMMA), including varying doses of MWCNTs with inconsistent weight % of the order of 0.1, 0.25, 0.5, and 1 for high-load joint replacement. Findings in the investigation carried on rBMSCs adhered to each other, and improvement in the growth was noticed. The osteogenic differentiation was also elevated from both protein and gene expression levels. Adding 1 wt% MWCNTs into PMMA bone cement considerably improved bone regeneration (with a ratio 42% in 12 weeks after surgery), as was revealed by *in vivo* experiments.

Finally, CNTs might be an excellent component of the 3D framework for bone tissue engineering. Biopolymers and biominerals, bone-building hormones, and nanocarbon compounds could form a composite material that can benefit from these material's advantages [40]. An additional concern is that there have been insufficient pre-clinical and clinical investigations using nanomaterials for bone healing too far, despite the need for extensive toxicity research on these materials.

Numerous researchers have invested considerable effort in conducting extensive research and developing novel ways for using nanomaterials as scaffolds, coating films, or additives to enhance the performance of specific bone tissue substrates, as summarized in Table 15.1 [39, 41–45].

15.2.4 Dental Tissue Engineering

Bioengineering, cell transplantation, and drug delivery principles have all seen significant advancements thanks to nanotechnology research in the last 15 years, allowing for the development of numerous new biological alternatives for sick and weakened tissues to sustain and restore their functionalities [46]. Nanotechnology's many implementations have resulted in substantial advances in healthcare, particularly in the dental field. The use of revolutionary nanomaterials in oral health is one factor that has fueled new advancements in the dental field [47]. To stimulate and accomplish remineralization, biomimetic nanotechnology, for example, mimics the structure of tooth enamel and surrounding tissues. The goal of the remineralization procedure is to restore the hardness of the tooth by restoring amelogenin-based

Table 15.1 Scaffolds and implants made of carbon nanotubes for use in bone tissue engineering.

Substrate	Application of CNT	Uses
Hydroxyapatite (HA)	Coating material for implants	It enhanced apatite mineralization and stimulated new bone development.
Chondroitin sulfate (CS)	Bone tissue scaffold	It boosted water uptake capacity and porosity, promoting cell proliferation, protein content, alkaline phosphatase, and mineralization, all of which aid in the development of ectopic bone.
Gelatin–hydroxyapatite (HA)	Artificial bone grafts	It had a similar structure and composition to natural bone but had an enhanced tensile strength, elastic modulus, and elongation rate, which resulted in increased cell viability.
Collagen	Bone repair biomaterials	It improved the mechanical stability and effectiveness of the structure.
Polymethyl-methacrylate (PMMA)	Bone cements	It enhanced osteogenic differentiation and osseointegration by promoting cell adhesion.

peptides and hydroxyapatite [48]. The invention of toothpaste containing nanosized carbonate apatite, which effectively closes dentin tubules and helps to reduce dental sensitivity, is an example of these advancements in dentistry [49].

A bioactive substance must possess properties that include microbial infection control, tooth strengthening, and bio-promoting activities to be helpful in the oral cavity. Inflammation management, tooth remineralization, and/or tissue regeneration are examples of these [50]. These foreign materials may be employed in irreversible restorations in dentistry, such as pulp capping, dentinal tubule occlusion, scaffolding, and tooth remineralization [51]. The remineralization of early caries lesions is an example of bioactivity. The dentin and enamel underneath the restoration might benefit significantly from the encouragement of biomineralization, as long as adhesion is not disturbed [52]. Many factors determine the lifespan of dental restoration, including the type of tooth, size, location, therapeutic function, and makeup of the dental material. Because of current scientific advancements and a better knowledge of the caries process, minimum intervention treatments are used for preserving the bulk of the natural tissue, extending the restoration's life expectancy. Dentists strive to minimize the percentage of dental restorations that fail due to large or small fractures and to reduce the repetitive threat of caries, which would necessitate another repair and further tooth tissue loss.

Glass ionomer cements (GICs)) are regenerative and luting substances with appealing qualities. An essential (ion-leachable) glass, water, and acid that are polymeric and soluble in water are the three primary components of these products and were developed by Wilson and Kent in 1972 as novel transparent dental fillers.

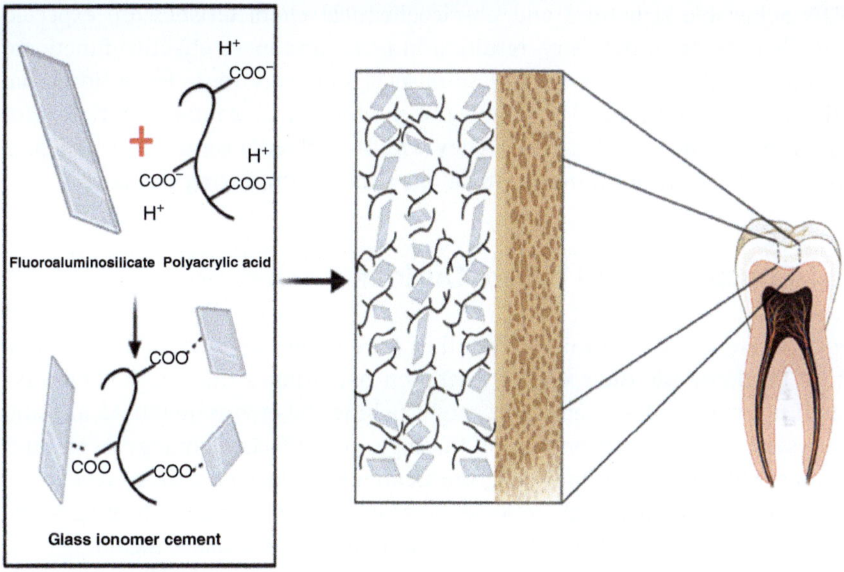

Figure 15.3 The fundamental components of a GIC and the interaction with a tooth. Source: Castro-Rojas et al. [49]/MDPI/CC BY 4.0.

Strength, stiffness, and fluoride release are all distinct characteristics of this kind of material compared to powdered fluoroaminosilicate. An acid–base interaction produces GICs. The acid damages and destroys the alumina-silicate glass, causing calcium and aluminum cations to be released. Cation chelation occurs between carboxylate groups, resulting in a cross-linked framework (Figure 15.3).

While GIC exhibits some desirable properties, they also have substantial limitations in reality, including poor stability in moisture, low tensile strength, low capacity to retain stress, and limited endurance. The utilization of GICs modified with resin has been suggested as a new way to partly resolve the composition problem, improve the material's physical qualities, and reduce fluoride ion release. Ongoing investigations for novel substances with superior and successful usage have resulted in nanotechnology advancement in the dental field.

MWCNTs at doses of 0–2% w/w were used as a reinforcing material, which resulted in dark-colored material suitable for posterior teeth only. The inclusion of 2% w/w MWCNTs improved the material's mechanical characteristics, increasing the hardness from 2.19 to 5.70 MPa. Additionally, this composite sustained more significant wear pressures than the other two formulations [53].

Restorative materials must be esthetically pleasing to all; because of that, GICs are often employed in a primary tooth. Therefore, color stability testing on new materials must be undertaken to ensure that they fit the criteria of the specific patient. CNTs have a higher color stability profile than other reinforcing materials such as silver nanoparticles when used in GIC [54]. On the basis of these color stability considerations, this material may be employed well in posterior replacements, particularly in primary teeth, where esthetically tooth treatments are less critical [55].

CNT's adjustable structural and physicochemical characteristics are expected to boost their usage in dentistry, resulting in novel and more effective functional applications. CNTs also have a lot of promise as a delivery route for antibacterial and tissue-regenerative medicines and the production of dental materials with unique features, such as bioactivity. They will undoubtedly serve as cutting-edge platforms for a variety of future investigations aimed at improving oral health.

15.3 Conclusion and Future Outlook

Since CNTs are made on a micro level, they have a high production and processing cost. Furthermore, their surface alteration necessitates the use of chemicals, which, if not managed appropriately, might pollute the environment. As a result, future research should concentrate on finding a cost-effective strategy to produce CNTs on a wide scale and improving the treatment technique used to regenerate CNTs and other nanomaterials to lower their cost. Despite these advantages, the number of CNTs used as filler or fortification in different materials, including polymers, is limited because of the aggregation of CNTs. This is a significant obstacle to their application and is often overcome by chemical functionalization, which enhances CNT dispersion and reduces clumping. This significantly increases the matrix's loading capacity, which positively influences the final composite's mechanical characteristics. When CNTs in a composite matrix are loaded optimally, stress is transferred to the matrix interface as efficiently as possible. The addition of CNTs into the polymer chain may boost the conductivity of framework related to nerve and, consequently, the sensibility of nerve cells.

Additionally, cardiac structures can also be improved in terms of conduction, structural rigidity, and biocompatibility by this addition. CNTs are employed to strengthen both the biological and mechanical aspects of bone tissue engineering. In the years ahead, nontoxic, safer, and more environmentally friendly CNTs and their composites (polymeric) may also be there, and further investigation is required on the different ways opted by CNTs that induce toxicity, along with potential factors responsible for cytotoxicity and genotoxicity. Long-term studies in living systems, comprising the breakdown of polymeric substrates that results in the emission of CNTs, must be addressed because most research focuses on the lab study of the early stages of cell development. Till date, the research examines the small CNT concentration in scaffolds as a result of fabrication constraints and toxicity while balancing the scaffold's electrical, mechanical, and biological capabilities. As a result, more significant CNT concentrations in the polymer matrix are anticipated. Although CNTs are becoming more appealing nanomaterials, there are still a number of major restrictions to overcome, and more research is needed to fully promote CNTs' translation into tissue engineering applications and ensure clinical success over an extended period. In the field of regenerative tissue engineering, developing nontoxic, secure, and environment friendly. CNTs and their polymeric composites remain a major problem.

Few Websites Related to CNTs Engineering

https://www.news-medical.net/life-sciences/Applications-of-Carbon-Nanotubes-(CNTs)-in-Regenerative-Medicine.aspx
https://nopo.in/web/artificial-bones-aided-with-single-walled-carbon-nanotubes
https://phys.org/news/2019-11-carbon-nanotubes-horizons-neuroscience-neural.html
https://www.nibib.nih.gov/science-education/science-topics/tissue-engineering-and-regenerative-medicine
https://en.wikipedia.org/wiki/Carbon_nanotube

References

1 Harrison, B.S. and Atala, A. (2007). Carbon nanotube applications for tissue engineering. *Biomaterials* 28 (2): 344–353.

2 Huang, B. (2020). Carbon nanotubes and their polymeric composites: the applications in tissue engineering. *Biomanuf. Rev.* 5 (1): 1–26. https://doi.org/10.1007/s40898-020-00009-x.

3 Zhang, Y., Bai, Y., and Yan, B. (2010). Functionalized carbon nanotubes for potential medicinal applications. *Drug Discovery Today* 15 (11–12): 428–435. doi: 10.1016/j.drudis.2010.04.005.

4 Supronowicz, P., Bizios, R., Ajayan, P. et al. (2003). Electricallyconducting nanocomposite materials for biomedical applications. US20030153965.

5 Carroll, D., Wagner, W., Levi, N. et al. (2009). Methods and compositions for printing biologically compatible nanotube composites of autologous tissue. US20090117087.

6 Xia, Y., Li, S., Nie, C. et al. (2019). A multivalent polyanion-dispersed carbon nanotube toward highly bioactive nanostructured fibrous stem cell scaffolds. *Appl. Mater. Today* 16: 518–528. https://doi.org/10.1016/j.apmt.2019.07.006.

7 Zheng, X., Zhang, P., Fu, Z. et al. (2021). Applications of nanomaterials in tissue engineering. *RSC Adv.* 11 (31): 19041–19058.

8 Shao, H., Li, T., Zhu, R. et al. (2018). Carbon nanotube multilayered nanocomposites as multifunctional substrates for actuating neuronal differentiation and functions of neural stem cells. *Biomaterials* 175: 93–109. https://doi.org/10.1016/j.biomaterials.2018.05.028.

9 Vicentini, N., Gatti, T., Salerno, M. et al. (2018). Effect of different functionalized carbon nanostructures as fillers on the physical properties of biocompatible poly(L-lactic acid) composites. *Mater. Chem. Phys.* 214: 265–276. https://doi.org/10.1016/j.matchemphys.2018.04.042.

10 Arslantunali, D., Budak, G., and Hasirci, V. (2014). Multiwalled CNT-pHEMA composite conduit for peripheral nerve repair. *J. Biomed. Mater. Res. Part A* 102 (3): 828–841.

11 Serrano, M.C., Nardecchia, S., García-Rama, C. et al. (2014). Chondroitin sulphate-based 3D scaffolds containing MWCNTs for nervous tissue repair. *Biomaterials* 35 (5): 1543–1551. https://doi.org/10.1016/j.biomaterials.2013.11.017.

12 Wang, J., Tian, L., Chen, N. et al. (2018). The cellular response of nerve cells on poly-l-lysine coated PLGA-MWCNTs aligned nanofibers under electrical stimulation. *Mater. Sci. Eng. C* 91: 715–726.

13 Koppes, A.N., Keating, K.W., McGregor, A.L. et al. (2016). Robust neurite extension following exogenous electrical stimulation within single walled carbon nanotube-composite hydrogels. *Acta Biomater.* 15 (39): 34–43.

14 Jiang, C., Wang, K., Liu, Y. et al. (2021). Using wet electrospun PCL/gelatin/CNT yarns to fabricate textile-based scaffolds for vascular tissue engineering. *ACS Biomater. Sci. Eng.* 7 (6): 2627–2637.

15 Nam, Y.J., Song, K., and Olson, E.N. (2013). Heart repair by cardiac reprogramming. *Nat. Med.* 19 (4): 413–415. https://doi.org/10.1038/nm.3147.

16 Chien, K.R., Domian, I.J., and Parker, K.K. (2008). Cardiogenesis and the complex biology of regenerative cardiovascular medicine. *Science (80-).* 322 (5907): 1494–1497.

17 Simon, J., Flahaut, E., and Golzio, M. (2019). Overview of carbon nanotubes for biomedical applications. *Materials (Basel)* 12 (4): 1–21.

18 Wickham, A.M., Islam, M.M., Mondal, D. et al. (2014). Polycaprolactone-thiophene-conjugated carbon nanotube meshes as scaffolds for cardiac progenitor cells. *J. Biomed. Mater. Res. Part B* 102 (7): 1553–1561.

19 Li, X., Zhou, J., Liu, Z. et al. (2014). A PNIPAAm-based thermosensitive hydrogel containing SWCNTs for stem cell transplantation in myocardial repair. *Biomaterials* 35 (22): 5679–5688. https://doi.org/10.1016/j.biomaterials.2014.03.067.

20 Sun, H., Lü, S., Jiang, X.X. et al. (2015). Carbon nanotubes enhance intercalated disc assembly in cardiac myocytes via the β1-integrin-mediated signaling pathway. *Biomaterials.* 55: 84–95. https://doi.org/10.1016/j.biomaterials.2015.03.030.

21 Mombini, S., Mohammadnejad, J., Bakhshandeh, B. et al. (2019). Chitosan-PVA-CNT nanofibers as electrically conductive scaffolds for cardiovascular tissue engineering. *Int. J. Biol. Macromol.* 140: 278–287. https://doi.org/10.1016/j.ijbiomac.2019.08.046.

22 Shin, S.R., Jung, S.M., Zalabany, M. et al. (2013). Carbon-nanotube-embedded hydrogel sheets for engineering cardiac constructs and bioactuators. *ACS Nano* 7 (3): 2369–2380.

23 Pok, S., Vitale, F., Eichmann, S.L. et al. (2014). Biocompatible carbon nanotube-chitosan scaffold matching the electrical conductivity of the heart. *ACS Nano* 8 (10): 9822–9832.

24 Kharaziha, M., Shin, S.R., Nikkhah, M. et al. (2014). Tough and flexible CNT-polymeric hybrid scaffolds for engineering cardiac constructs. *Biomaterials* 35 (26): 7346–7354.

25 Ho, C.M.B., Mishra, A., Lin, P.T.P. et al. (2017). 3D printed polycaprolactone carbon nanotube composite scaffolds for cardiac tissue engineering. *Macromol. Biosci.* 17 (4): 1–9.

26 Izadifar, M., Chapman, D., Babyn, P. et al. (2018). UV-assisted 3D bioprinting of nanoreinforced hybrid cardiac patch for myocardial tissue engineering. *Tissue Eng. Part C* 24 (2): 74–88.

27 Bosi, S., Ballerini, L., and Prato, M. (2014). Carbon nanotubes in tissue engineering. *Top. Curr. Chem.* 348: 181–204.

28 Onyancha, R.B., Aigbe, U.O., Ukhurebor, K.E., and Muchiri, P.W. (2021). Facile synthesis and applications of carbon nanotubes in heavy-metal remediation and biomedical fields: a comprehensive review. *J. Mol. Struct.* 1238: 130462. https://doi.org/10.1016/j.molstruc.2021.130462.

29 Hwang, J.Y., Shin, U.S., Jang, W.C. et al. (2013). Biofunctionalized carbon nanotubes in neural regeneration: a mini-review. *Nanoscale* 5 (2): 487–497.

30 Lin, C., Wang, Y., Lai, Y. et al. (2011). Incorporation of carboxylation multiwalled carbon nanotubes into biodegradable poly (lactic-co-glycolic acid) for bone tissue engineering. *Colloids Surf. B: Biointerfaces* 83 (2): 367–375.

31 Patel, D.K., Dutta, S.D., Ganguly, K. et al. (2021). Enhanced osteogenic potential of unzipped carbon nanotubes for tissue engineering. *J. Biomed. Mater. Res. Part A* 109 (10): 1869–1880.

32 Tanaka, M., Sato, Y., Zhang, M. et al. (2017). In vitro and in vivo evaluation of a three-dimensional porous multi-walled carbon nanotube scaffold for bone regeneration. *Nanomaterials* 7 (2): 1–17.

33 Liu, J., Gao, C., Feng, P. et al. (2015). A bioactive glass nanocomposite scaffold toughed by multi-wall carbon nanotubes for tissue engineering. *J. Ceram. Soc. Jpn.* 123 (1438): 485–491.

34 Khalid, P., Hussain, M.A., Rekha, P.D., and Arun, A.B. (2015). Carbon nanotube-reinforced hydroxyapatite composite and their interaction with human osteoblast in vitro. *Hum. Exp. Toxicol.* 34 (5): 548–556.

35 Gupta, A., Main, B.J., Taylor, B.L. et al. (2014). In vitro evaluation of three-dimensional single walled carbon nanotube composites for bone tissue engineering. *J. Biomed. Mater. Res. Part A* 102 (11): 4118–4126.

36 Mesgar, A.S., Mohammadi, Z., and Khosrovan, S. (2018). Improvement of mechanical properties and in vitro bioactivity of freeze-dried gelatin/chitosan scaffolds by functionalized carbon nanotubes. *Int. J. Polym. Mater. Polym. Biomater.* 67 (5): 267–276.

37 Duan, S., Yang, X., Mei, F. et al. (2015). Enhanced osteogenic differentiation of mesenchymal stem cells on poly(L-lactide) nanofibrous scaffolds containing carbon nanomaterials. *J. Biomed. Mater. Res. Part A* 103 (4): 1424–1435.

38 Gonçalves, E.M., Oliveira, F.J., Silva, R.F. et al. (2016). Three-dimensional printed PCL-hydroxyapatite scaffolds filled with CNTs for bone cell growth stimulation. *J. Biomed. Mater. Res. Part B* 104 (6): 1210–1219.

39 Wang, C., Yu, B., Fan, Y. et al. (2019). Incorporation of multi-walled carbon nanotubes to PMMA bone cement improves cytocompatibility and osseointegration. *Mater. Sci. Eng., C* 103 (May): 1–13.

40 Chen, L., Hu, J., Shen, X., and Tong, H. (2013). Synthesis and characterization of chitosan-multiwalled carbon nanotubes/hydroxyapatite nanocomposites for bone tissue engineering. *J. Mater. Sci.: Mater. Med.* 24 (8): 1843–1851.

41 Pei, B., Wang, W., Dunne, N., and Li, X. (2019). Applications of carbon nanotubes in bone tissue regeneration and engineering: superiority, concerns, current advancements, and prospects. *Nanomaterials* 9: 1501.

42 Tan, W., Twomey, J., Guo, D. et al. (2010). Evaluation of nanostructural, mechanical, and biological properties of collagen nanotube composites. *IEEE Trans. Nanobiosci.* 9 (2): 111–120.

43 Yoon, I.K., Hwang, J.Y., Seo, J.W. et al. (2014). Carbon nanotube-gelatin-hydroxyapatite nanohybrids with multilayer core–shell structure for mimicking natural bone. *Carbon* 77: 379–389. https://doi.org/10.1016/j.carbon.2014.05.041.

44 Venkatesan, J., Ryu, B.M., Sudha, P.N., and Kim, S.K. (2012). Preparation and characterization of chitosan-carbon nanotube scaffolds for bone tissue engineering. *Int. J. Biol. Macromol.* 50 (2): 393–402. https://doi.org/10.1016/j.ijbiomac.2011.12.032.

45 Park, J.E., Jang, Y.S., Bae, T.S., and Lee, M.H. (2019). Biocompatibility characteristics of titanium coated with multiwalled carbon nanotubes-hydroxyapatite nanocomposites. *Materials (Basel)* 12 (2): 224.

46 Kanaparthy, R. and Kanaparthy, A. (2011). The changing face of dentistry: nanotechnology. *Int. J. Nanomed.* 6: 2799–2804.

47 Ge, J., Cui, F.Z., Wang, X.M., and Feng, H.L. (2005). Property variations in the prism and the organic sheath within enamel by nanoindentation. *Biomaterials* 26 (16): 3333–3339.

48 Lee, S.Y., Kwon, H.K., and Kim, B.I. (2008). Effect of dentinal tubule occlusion by dentifrice containing nano-carbonate apatite. *J. Oral Rehabil.* 35 (11): 847–853.

49 Castro-Rojas, M.A., Vega-Cantu, Y.I., Cordell, G.A., and Rodriguez-Garcia, A. (2021). Dental applications of carbon nanotubes. *Molecules* 26 (15): 1–14.

50 Fernando, D., Attik, N., Pradelle-Plasse, N. et al. (2017). Bioactive glass for dentin remineralization: a systematic review. *Mater. Sci. Eng., C* 76: 1369–1377. https://doi.org/10.1016/j.msec.2017.03.083.

51 Imazato, S., Kohno, T., Tsuboi, R. et al. (2020). Cutting-edge filler technologies to release bio-active components for restorative and preventive dentistry. *Dent. Mater. J.* 39 (1): 69–79.

52 Vallittu, P.K., Boccaccini, A.R., Hupa, L., and Watts, D.C. (2018). Bioactive dental materials—do they exist and what does bioactivity mean. *Dent. Mater.* 34 (5): 693–694. https://doi.org/10.1016/j.dental.2018.03.001.

53 Bussy, C., Ali-Boucetta, H., and Kostarelos, K. (2013). Safety considerations for graphene: lessons learnt from carbon nanotubes. *Acc. Chem. Res.* 46 (3): 692–701.

54 Pani, S.C., Aljammaz, M.T., Alrugi, A.M. et al. (2020). Color stability of glass ionomer cement after reinforced with two different nanoparticles. *Int. J. Dent.* 2020.

55 Sun, L., Yan, Z., Duan, Y. et al. (2018). Improvement of the mechanical, tribological and antibacterial properties of glass ionomer cements by fluorinated graphene. *Dent. Mater.* 34 (6): e115–e127. https://doi.org/10.1016/j.dental.2018.02.006.

16

Applications of Chemically Modified Carbon Nanotubes for Tissue Engineering

Yasaman Esmaeili[1], *Seyede Z. Mirahmadi-Zare*[2], *Ashkan Bigham*[3], *Maryam Boshtam*[4], *Saba Gharghish*[5], *and Raha Zomorrodi*[6]

[1] *Isfahan University of Medical Sciences, Biosensor Research Center, School of Advanced Technologies in Medicine, Hezarjerib Ave, Isfahan 8174673461, Iran*
[2] *Royan Institute for Biotechnology, ACECR, Reproductive Biomedicine Research Center, Department of Animal Biotechnology, Isfahan, 8159358686, Iran*
[3] *Institute of Polymers, Composites and Biomaterials–National Research Council (IPCB-CNR), Viale J.F. Kennedy 54–Mostra d'Oltremare pad. 20, Naples, 80125, Italy*
[4] *Isfahan University of Medical Sciences, Isfahan Cardiovascular Research Center, Cardiovascular Research Institute, Isfahan, 8158388994, Iran*
[5] *Tabriz University of Medical Science, Department of Medical Nanotechnology, Faculty of Advanced Medical Science, Tabriz, 5154853431, Iran*
[6] *Islamic Azad University, Department of Chemistry, Science and Research Branch, Simon Bulivar Blvd, Tehran, 1477893855, Iran*

16.1 Introduction

Tissue engineering involves restoring or replacing damaged tissue in order to regain its original biological function. To achieve this, the development of biomaterials that can replace tissue or assist in tissue growth is essential for developing rational and efficient tissue engineering. Notably, any tissue engineering material must possess biocompatibility. Furthermore, materials with strong mechanical properties are required for hard-tissue applications such as bone tissue [1]. Scaffolding made of biodegradable materials (mainly polymers) is used in tissue engineering to modify new tissue functionalities along with living cells and bioactive substances. Sufficient mechanical qualities, interconnected porosity, biocompatibility, and biodegradability are the most significant characteristics of scaffolding systems [2]. Various investigations have widely used carbon nanotubes (CNTs) to reinforce matrix materials in hard-tissue implants because of their remarkable mechanical properties. CNTs are inorganic nanocarrier materials with excellent anisotropic mechanical and electronic properties as well as easy functionalization with chemical functional groups and protonation, which make them unique nanomaterials for scaffold construction and tissue engineering [3]. CNTs can also be a useful tissue engineering material for better cell tracking, microenvironment sensing, transfection agent delivery, and scaffolding to integrate with the host's body.

Chemically Modified Carbon Nanotubes for Commercial Applications, First Edition.
Edited by Jeenat Aslam, Chaudhery Mustansar Hussain, and Ruby Aslam.

The use of CNTs as optical, magnetic resonance, and radiotracer contrast agents might improve tissue evaluation [4]. Furthermore, monitoring and modifying intra- and intercellular activities might aid in the development of well-designed tissues. CNTs can also be incorporated into scaffolds to provide structural guidance in addition to new traits including electrical conductivity, which might also be a useful resource in cellular boom control. Apart from these useful biomedical applications, CNTs have exhibited toxicity in preclinical and clinical studies [5]. According to the literature, the toxicity of CNTs is caused by a variety of reasons and conditions. The size, shape, and aspect ratio of these nanomaterials all affect their toxicity when taken up by cells, interfering with cellular functions in both the cytoplasm and the nucleus. As an example, a research group demonstrated the toxicological effects of several pure and commercial single-walled carbon nanotubes (SWCNTs) and multi-walled carbon nanotubes (MWCNTs) [6]. Three commercial CNT samples significantly elevated reactive oxygen species, which contributed to oxidative stress, and lowered mitochondrial membrane potential. Acid-purified SWCNTs with low metal concentration displayed similar harmful effects to the negative controls, with little or no cytotoxicity. CNTs are also prone to contamination during synthesis by certain compounds, such as metal ions, which contribute to the overall cytotoxicity of the final products [7]. When CNTs are taken up by tumor cells, they induce reactive oxygen species and cause cell death, which can damage lipids and DNA. Nonetheless, the possible cytotoxic effects of these nanomaterials can be circumvented through suitable chemical functionalization of their surfaces, including polymers, hydrogels, and nanoparticles [8].

In light of this, this chapter focuses on presenting the main functionalization approaches and biomedical applications of CNTs in various tissue engineering applications, as well as on identifying the most effective strategies to overcome the shortcomings that impede their preclinical use.

16.2 Biochemical Modifications for Tissue Engineering

Despite the wide range of possibilities for CNTs in biomedicine and tissue engineering, two major drawbacks in a commercial scale should be considered: their intrinsic toxicity, which derives from the metal catalyst residue in the manufacturing process, and their low water solubility. Their high surface energy causes them to be barely dispersed in solution and not soluble in most biocompatible solvents, and, consequently, their biological applications are hardly limited. In light of this, so far, different approaches including surface modification have been established for pure and homogeneous CNTs with lower toxicity, better dispersion, and higher solubility (Figure 16.1a). Figure 16.1b represents a recent advancement of CNTs in tissue engineering.

16.2.1 Polymer-based Modification

Polymer modification can increase not only the solubility and dispersibility of CNTs but also improve their interfacial interaction with polymeric matrix in

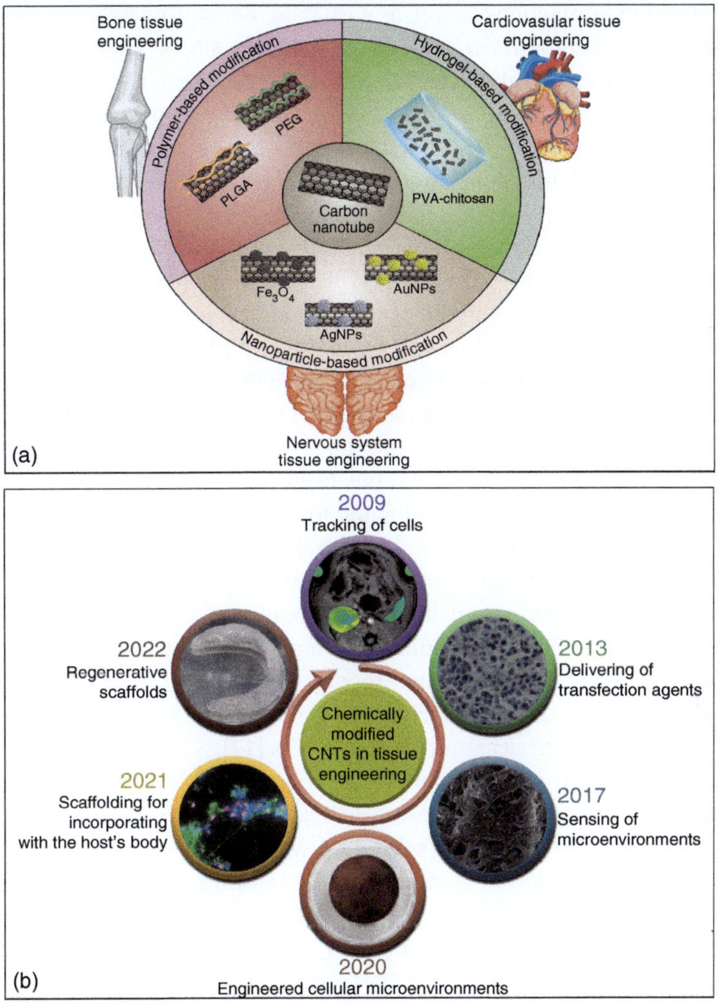

Figure 16.1 (a) Schematic representation of carbon nanotubes (CNTs) and their potential surface modification for use in tissue engineering, (b) timeline of recent advancements of CNTs in tissue engineering.

composites [9]. Although functionalized CNTs are more soluble in solvents, the sluggish evaporation of the solvent and the after-lie process take a long time, resulting in precipitated CNTs to inhomogeneous dispersion in the polymer matrix [10]. *In situ* polymerization is a method of explanation in which monomers or pre-polymers are opposed to functionalized CNTs or monomer-grafted CNTs, and then polymerization is completed by regulating temperature and time [11]. This approach uses covalent or noncovalent linkages to conceive a very strong force in the middle of two points: the polymer principle and CNTs.

Noncovalent attachment and covalent attachment are the two basic approaches for polymer modification of CNTs. Physical adsorption and/or wrapping of polymers to the CNT surface is a noncovalent technique [12]. CNT walls can interact with polymers via π–π stacking interactions as well as with those containing heteroatoms with free electron pairs. This approach does not degrade the CNTs' aromatic system, preserving their structure and characteristics. On the other hand, the "grafting" (chemical anchoring) of polymeric chains to CNTs is a covalent technique that can be established using "grafting to" or "grafting from" strategies [13]. It is based on the creation of a modified polymer capable of reacting with the edges or sidewalls of functionalized CNTs (e.g. acid-treated, amine-functionalized). However, in the presence of acids, the functionalization events generally cut the CNTs, causing significant surface degradation or even unrolling [14]. The presence of defects in their tube form can have a negative impact on their physicochemical properties. Another drawback of this strategy is that due to weak reactivity and high steric barrier of the polymeric segments, the amount of polymer grafted on the CNTs is limited. The polymer is produced from the surface of the nanotubes by *in situ* polymerization of monomers initiated by molecules placed on the tubes' edges and sidewalls in the "grafting from" technique [15]. This approach has been successful and controlled due to the strong reactivity of monomers, allowing for the synthesis of nanocompounds with high grafting levels. However, it requires careful monitoring of the reagent concentrations as well as the polymerization requirements [16]. Otherwise, CNT functionalization can be achieved using plasma treatment, a time-saving and nontoxic method of altering CNTs by immediately forming an increasing number of functional groups. Plasma treatment can easily modify the degree of hydrophilicity of the CNT surface, allowing for dispersion within the polymer matrix. Furthermore, this method does not alter the physical properties of CNTs [17].

So far, a wide range of synthetic and natural biopolymers have been used for functionalization of CNTs for scaffold designing. Biodegradable polymers including polylactic-*co*-glycolic acid (PLAGA), poly ethylene glycol (PEG), polylactic acid (PLA), and polyurethane (PU), as well as biodegradable ceramics such as bioactive glass, have all been reported to be used as scaffolds for tissue engineering [18]. A study teamwork created a poly(anilineboronic acid) (PABA) nanocomposite with DNA-functionalized carbon nanotubes (DNA CNTs) and nitrogen-doped graphene (NEG) using *in situ* chemical polymerization [19]. In the 3-aminophenylboronic acid monomer solution, DNA CNT NEG was shown to have good pre-dispersion and to be capable of forming a highly ordered and poly-conjugated structure with outstanding electrical conductivity (14 300 S m^{-1} at 3.0 wt% filler concentration). Another group used *in situ* polymerization to compare thermoplastic PU containing linear and branched CNTs [20]. They demonstrated that the composite including branched CNTs produced stronger networks and had better mechanical properties than linear CNT composites.

PLAGA is one of the most prevalent synthetic polymer-based materials, despite its inability to tolerate heavier components such as bone substitutes [21]. CNTs and PLAGA can be combined to improve their poor mechanical properties. Researchers discovered that a three-dimensional PLAGA membrane with only 3% liquid-dispersible MWCNTs had much improved compressive properties and a higher modulus than pure PLAGA scaffolds [22]. Absorption, multiplication, and mineralization of cells were also excellent. PLA is an environmentally friendly aliphatic polyester made from a readily available and renewable energy source [23]. PLA has a simple chemical structure, is biodegradable, and has high histocompatibility. It is a natural material used in the regeneration of bone tissue and bone engineering. PLA's intrinsic brittleness and low thermal stability, on the other hand, make it unsuitable for holding large items or encouraging cell growth. The incorporation of CNTs into a PLA matrix material can increase mechanical and surface properties, such as functionalization, and thereby cell survival. For bone tissue engineering, PLAGA and SWCNT composite microspheres were also developed (Figure 16.2a) [24].

Overall, the surface coating of CNTs with polymers can change surface characteristics and have direct effects on cell–material interactions. However, limitations on the scaffolds' general properties (i.e. mechanical capabilities) reduced overall effectiveness due to the absence of covering tiers coexist [26]. The polymer/CNT scaffolds were also made via a melting technique. This process is commonly used to make a mixture of thermoplastic polymers and CNTs [27]. Large shear forces are generated when air pressure or mechanical strength is applied to the molten-state polymer matrix, which aids in the de-agglomeration of CNT bundles and their homogeneous dispersion in the polymer matrix. Melt mixing was used to make a PLA-based CNT–carboxylic acid composite that had higher tensile strength and flexibility at breakage, as well as higher tensile strength and thermal conductivity than pure PLA [23]. Melt blending was successfully employed by another group to prepare polycaprolactone (PCL)/MWCNTs and modified with hyaluronic acid (PCL/hydroxyapatite [HA]/MWCNT) composites (Figure 16.2b) [25]. Through twist-assisted extrusion additive production, the composites were made into 3D porous scaffolds for bone tissue applications. The resulting scaffolds have improved mechanical characteristics and osteogenesis. The use of a combination of screw motions can help to align CNTs in a PCL matrix in a significant way.

16.2.2 Hydrogel-based Modification

Based on the type of cross-linking, hydrogels are divided into two groups: physical hydrogels formed by hydrogen or electrostatic interactions and chemical hydrogels adopting stronger bonds like covalent. Notably, the latter are more complicated and require different polymerization techniques and cross-linking agents [28]. The most noticeable characteristics of hydrogels are cell/biocompatibility, desirable degradation behavior, tailorable porosity, and mechanical properties. Apart from these advantages, hydrogels have some shortcomings, including low mechanical properties, weak degradation rate, and inducing cytotoxicity in tissue engineering and regenerative medicine, which lead to development of hybrid hydrogels [29]. Hybrid hydrogels have shown promising results in different applications, including scaffolding, gene therapy, and targeted drug delivery [30].

As various polymers have been used to develop CNT-incorporated hydrogels, the most well-known hydrogels are summarized in this section. Chitosan is a well-known natural polymer, biodegradable, and biocompatible and has been used in biomedicine extensively. Chitosan alone in any form (hydrogel, film, etc.) is

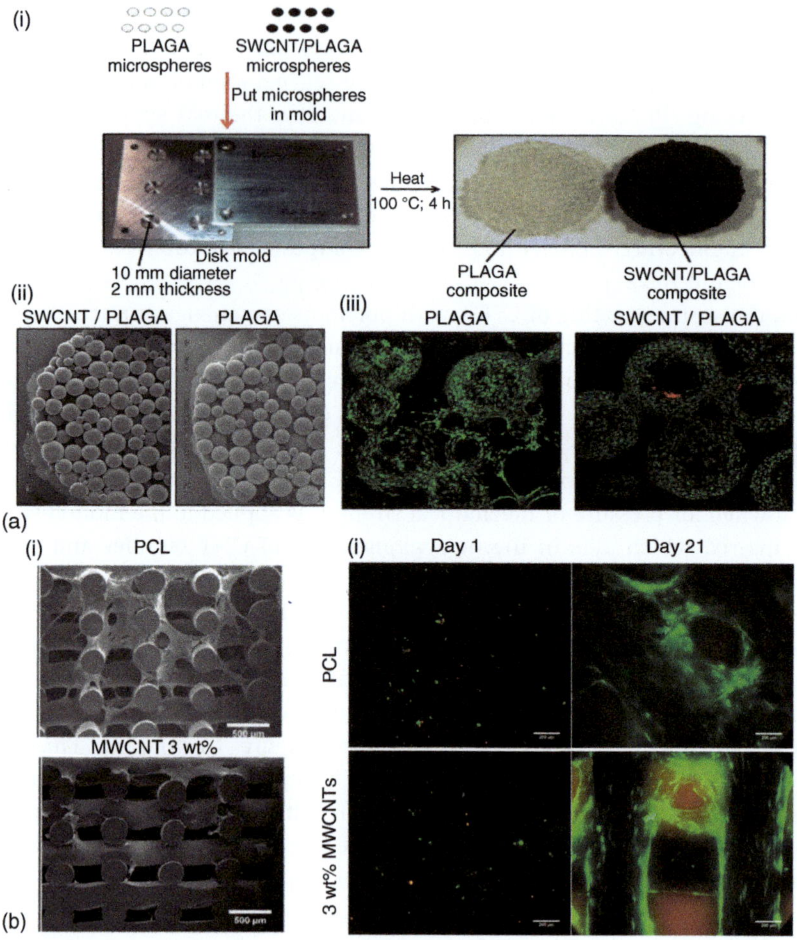

Figure 16.2 Recent studies on the polymer-based modification in combination with carbon nanotubes (CNTs): (a, i) thermal sintering was used to create 3D PLAGA and SWCNT/PLAGA composites, (a, ii) scanning electron microscope (SEM) images of PLAGA and SWCNT/PLAGA composites, (a, iii) growth and survival of MC3T3-E1 cells on PLAGA and SWCNT/PLAGA composites (at 310 3.1 zoom) at day 2. Source: Gupta et al. [24]/with permission of John Wiley & Sons (b, i) SEM images of cross-sectional views of PCL scaffold and PCL/MWCNT 3 wt%, (b, ii) live/dead assay at day 21 for scaffolds (green: live cell [flash], red: dead cell). Source: Huang et al. [25]/with permission of Elsevier.

brittle, requiring different fillers to strengthening it. Herein, we shed light on some interesting studies on CNT-reinforced chitosan hydrogels. A mechanically enhanced composite chitosan hydrogel with self-healing properties was developed. Upon the addition of CNTs to the hydrogel, the electrical and mechanical properties were significantly improved [31]. The combination of chitosan and CNTs has received considerable attention as hydrogel scaffold, due to chitosan antibacterial properties, biocompatibility, and biodegradability, besides all the improvements made by CNTs [32]. One of these advantages is piezoelectricity CNTs give to a scaffold; the combination of medium molecular weight of chitosan, MWCNTs, and β-glycerophosphate was adopted as a composite hydrogel scaffold for bone tissue engineering. The results implied that the piezoelectric nature of CNTs led to an improvement in cell compatibility [33]. Another pros is the photoresponsivity nature of CNTs, which can be applied in cancer therapy and antibacterial applications. A light-responsive doxorubicin-loaded hybrid hydrogel composed of PCL-PEG-PCL and chitosan–CNTs was synthesized for NIR-induced drug delivery. In the response to NIR irradiation, CNTs convert light into heat and disrupt the hydrogel's structure, resulting in the release of doxorubicin on the action site [34]. Somewhere else, this photoresponsivity was taken against the bacteria growth. The self-healing hydrogel wound dressing was developed based on *N*-carboxyethyl chitosan, Pluronic F-127, and CNTs. The existence of CNTs in the hydrogel's structure gave a great photothermal antibacterial activity and conductivity *in vitro* and *in vivo* (Figure 16.3) [35]. The chitosan–CNT hydrogels are found to possess great stretchability and folding capability, all of which are assisting for body motions purposes, e.g. skin-attached sensors, electronic devices [36].

Another natural polymer that has been used in combination with CNTs is guar gum. Guar gum has primarily been used as transdermal patches and drug carriers to the colon [37]. Based on carboxymethyl guar gum, a hybrid hydrogel was fabricated for transdermal drug delivery. As CNTs are hydrophobic in nature, a study took advantage of hydrophilic guar gum to improve the wettability of CNTs by making a covalent between both agents. The combination of guar gum and CNTs as a substrate to grow superparamagnetic iron oxide nanoparticles on its surface and the final structure showed a potential platform in biomedical applications [38].

Collagen, one of the most important biopolymers with a wide variety of applications in tissue engineering and regenerative medicine, is a structural protein being found in the extracellular matrix of different tissues, and its physical properties varies from one tissue to another [39]. As collagen can be formed in the fibrillary form, some interesting studies have added CNTs to this morphology to come up with improved mechanical and electrical properties. In one study, live muscle cells were added to a hydrogel composed of collagen and CNTs in which the cell viability was found around 85% up to seven days, showing the great potential of the hydrogel as either scaffold or biosensor [40].

As a synthetic polymer, polyvinyl alcohol (PVA) has desirable properties for biomedical applications – biocompatibility, degradability, and wettability. Different studies combined PVA with CNTs to achieve better mechanical properties for different types of applications [41]. The freeze/thawing method is one approach in which PVA and CNT hydrogels are combined with improvement in the mechanical properties of PVA and the swelling property of hydrogel [42]. However, this approach has some drawbacks, including weak dispersion of CNTs in the hydrogel's structure resulting in the aggregation of CNTs [36]. A hydrogel biosensor based on PVA and CNTs has recently been developed. Compared to freeze–thaw technique, this study adopted a one-step physical cross-linking approach. The CNTs were observed to exhibit good dispersion in the mixture of dimethyl sulfoxide/*N*,*N*-dimethylformamide after ultrasonication [43].

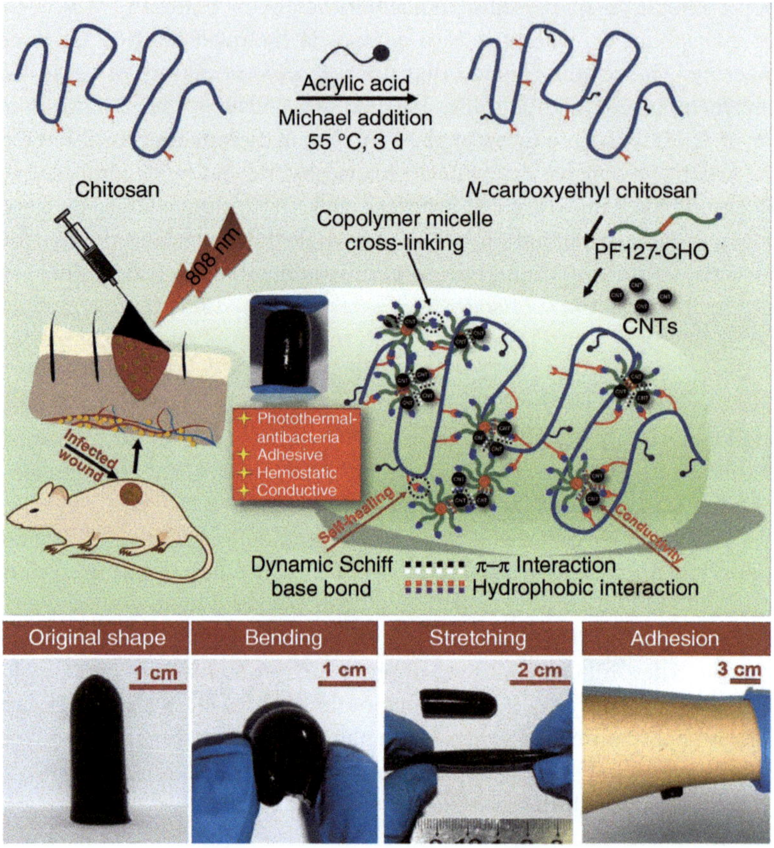

Figure 16.3 The multifunctional hydrogel for wound healing applications. An illustration of the hydrogel's preparation composed of *N*-carboxyethyl chitosan, Pluronic F-127, and CNTs with different interactions between the components and the hydrogel's original shape, stretchability, and adhesion capability. Source: He et al. [35]/with permission of Elsevier.

16.2.3 Nanoparticle-based Modification

Advances in nanoparticle manufacturing and the requirement for control over nanoscale systems have improved tissue engineering and regenerative medicine significantly. Nanoparticles with low toxicity, tailorable properties, different agent properties, targeted/stimuli–response delivery potential, and precise control over behavior (via external stimuli such as magnetic fields) have allowed them to be used to improve engineered tissues and overcome obstacles in tissue engineering technology. The biological events must be controlled spatially and temporally and monitored in real time for functional tissue and organ replacements. In light of this, nanoparticle-based modification has been used in tissue engineering scaffolds to administer numerous growth factors, provide contrast for imaging, and control the characteristics of scaffolds [44].

Based on the literature, gold nanoparticles (AuNPs), one of the most common metallic nanoparticles, are mostly used for CNT surface modification [45]. In the context of regenerative medicine, AuNPs could be used as a safety measure if the transplanted tissue is replacing a resected tissue or organ owing to tumor progression. As an example, a biocompatible composite of hydroxyapatite, CNTs, and AuNPs was developed to simulate the morphology of the extracellular matrix of bone [46]. In addition, AuNPs have been shown to induce angiogenesis and osteogenesis in stem cells. These particles affect the immune system's response to secrete osteogenic cytokines, stimulating macrophages to divide, which is beneficial for cell growth and proliferation [47].

Silver nanoparticles (AgNPs) are a colloid of nanometer-sized silver particles, which are one of the most extensively used metallic nanoparticles in the biomedical field, owing to their antibacterial capabilities [48]. Because bacterial infection is a major concern with synthetic tissues, AgNPs as a safe substance are a good option. The characteristics of AgNP have also been investigated in tissue engineering, mostly for wound dressing purposes. Infected chronic wounds and burn wounds are especially vulnerable [49]. Furthermore, the growing occurrence of multi-drug resistance to microbe-related infections is a serious public health issue that necessitates the inclusion of antimicrobial components in the wound healing scaffold design. Gelatin, PVA, PCL, cellulose acetate, and chitosan–alginate scaffolds have all been used to integrate these particles [50]. All of the scaffolds resulted have high antibacterial activity. In one study, a composite of Agion on CNTs inspired by mussels, dopamine (DA)-conjugated polymers (DA-polymers), and synthetic heparin-mimetic polymer (SHP) was prepared, converting silver ions into nanoparticles [51]. Due to the solid adhesive property of DA moieties, DA-SHP can readily anchor onto the CNT surface. This nanocomposite was able to inhibit the bacterial growth with good lethality.

Magnetic nanoparticles (MNPs) are iron oxide nanoparticles (typically Fe_3O_4 or Fe_2O_3), which have been investigated extensively in the biomedical field due to their low toxicity. Their applications include imaging, tracking stem cells *in vivo*, and monitoring transplanted/engineered tissues [52]. Several experiments have been carried out to deliver growth-factor-conjugated MNPs to cells, track their fate, or stimulate cells with MNPs in tissue engineering field. Taking advantage of the properties of MNPs and CNTs, magnetic carbon nanotubes (MCNTs) are synthesized in several procedures, including filling process [53], template-based synthesis [54], chemical vapor deposition [55], hydrothermal/solvothermal pyrolysis [56], sol–gel [57], and self-assembly [58]. A research group functionalized multi-walled carbon nanotubes (f-MWCNTs) with Fe_3O_4 MNPs and proposed a hybrid composite of MWCNTs/Fe_3O_4 for tissue engineering through the deposition–precipitation method [59]. *In vitro* investigation was done on hybrid composites by incubation with mesenchymal stem cells for one, three, and seven days, either in the presence or absence of a static magnetic field. Based on the initial results, modification of MWCNTs with iron oxide nanoparticles improved the biocompatibility of the oxidized MWCNTs. Moreover, a static magnetic field increased the cellular effect of Fe_3O_4/MWCNTs, indicating the good potential of the introduced platform for tissue engineering applications [59]. A solvent casting process was used to make PCL-based 3D composite scaffolds with functionalized CNTs and iron oxide in two different mass ratios (1 : 1 and 1 : 4) in another work (Figure 16.4a) [60]. The hydroxylated CNTs were combined with iron oxide to form hybrid nanoparticles, which were subsequently mixed with PCL matrix and fabricated into 3D composite scaffolds using the solvent casting/porogen leaching process. Cytotoxicity was assessed using the SAOS-2 human cell line [60].

In tissue engineering, graphene-based nanomaterials have also received considerable attention due to their exceptional mechanical strength and rigidity, better electrical conductivity, and surface usefulness. Indeed, as 2D materials with vast surface areas, graphene and its derivatives, such as graphene oxide (GO), may transport a wide range of biomolecules, including DNA, enzymes, proteins, and peptides, via covalent bonds or noncovalent interactions such as π–π stacking [64]. Thus, graphene and GO are great tissue engineering possibilities. Moreover, they can be applied for the modification of CNTs used in implants or scaffold materials, improving their biocompatibility, biodegradability, and mechanical properties [65]. In the Kim et al. study, the graphene did not change the CNT's mechanical properties even after 1×10^6 compressive cycles, while the coated CNTs quickly returned to their original shape after the compression release (Figure 16.4b) [61]. In addition, the coating has no effect on the integrity and porosity of CNTs while increasing the Young modulus. Therefore, it can be concluded that the coating of graphene on CNTs enhances their properties and thus makes them promising materials for bone tissue engineering scaffolds. Another study reported that compared to graphene–CNT uncoated scaffolds, graphene–CNT-coated hydroxyapatite substrates enhanced osteogenic development of a temperature-sensitive human fetal osteoblastic cell line (hFOB 1.19) (Figure 16.4c) [62]. Similarly, it was shown that the

Figure 16.4 Recent studies on the nanoparticle-based modification in combination with carbon nanotubes (CNTs): (a, i) SEM images of scaffolds with hybrid particles, (a, ii) the EDX spectrum of a selected point at the surface of scaffold, and (a, iii) confocal microscopy images (maximum intensity projection) of SAOS-2 cells with stained nuclei (center) and F-actin cytoskeleton (halo), cultivated for three days on different magnetic CNTs. Source: Świętek et al. [60]/with permission of Elsevier. (b) Reorganization of graphene-coated nanotubes under compression (isotropic orientation of struts within the aerogels at $\varepsilon = 0\%$ (a) and an increasing alignment at $\varepsilon = 40\%$), (b, ii) macroscopic visualization, showing that nanotube aerogels collapse and graphene-coated aerogels recover their original shape after compression by ≥90%. Source: Kim et al. [61]/with permission of Elsevier. (c) High-magnification SEM of carbon nanotube–hydroxyapatite (CNT-HA), and graphene nanosheet–hydroxyapatite (Gr-HA). Source: Oyefusi et al. [62]/with permission of Elsevier. (d, i) Enhancing mechanisms of carbon-based filler in the nanocomposite scaffolds, (d, ii) apatite formation on the scaffold surface after immersion in SBF, (d, iii) SEM images and fluorescence microscopy images (counterstained with calcein-AM and DAPI for live cell and cell nuclei in green and blue, respectively) of MG-63 cells on the scaffold containing polyetheretherketone–10 wt% nano-hydroxyapatite–0.2 wt% graphene nanosheet–0.8 wt% CNTs after seven days, (d, iv) fluorescence microscopy images of hBMSCs (human bone marrow mesenchymal stem cells) on S5 scaffold. Source: Feng et al. [63], Reprinted with permission from Dove Press Ltd.

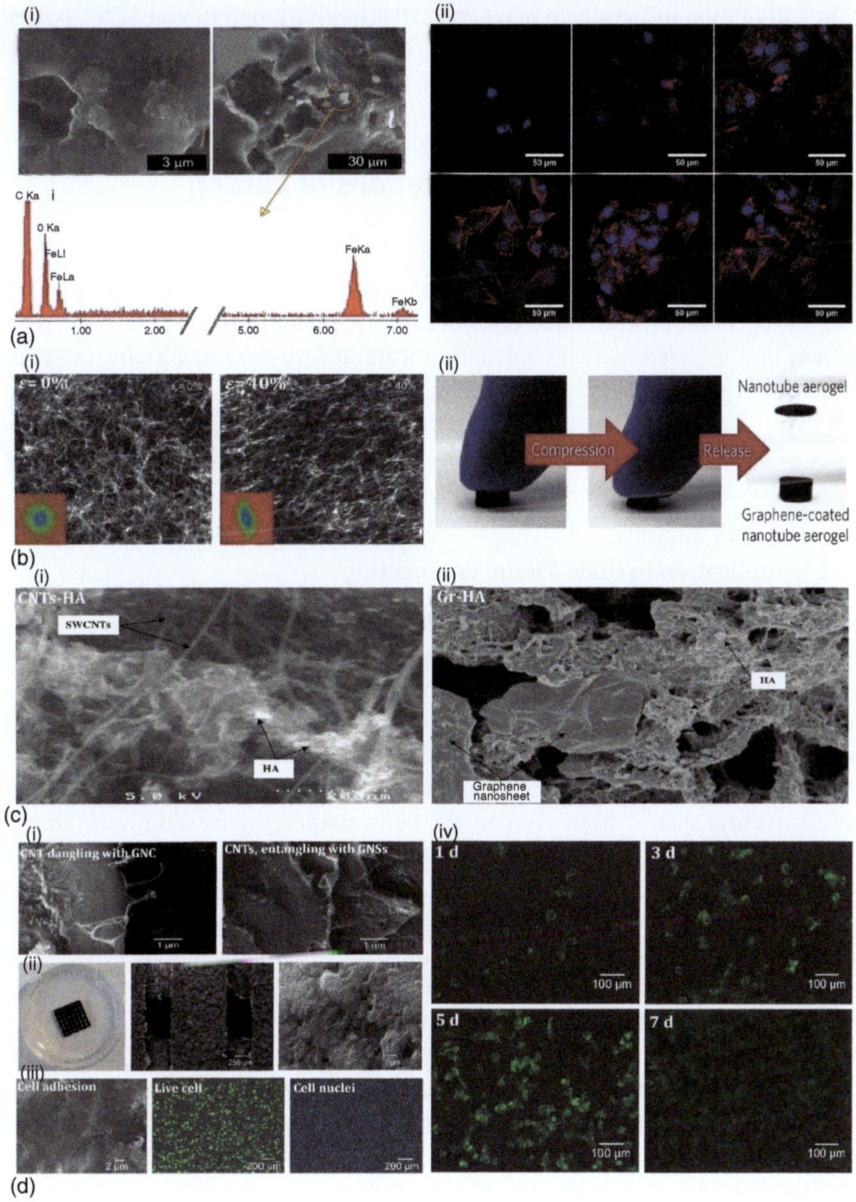
(i)
(ii)
3 µm
30 µm
C Ka
O Ka
FeLl
FeLa
FeKa
FeKb
1.00
2.00
5.00
6.00
7.00
50 µm
(a)
(i)
ε = 0%
ε = 40%
(ii)
Nanotube aerogel
Compression
Release
Graphene-coated nanotube aerogel
(b)
(i)
CNTs-HA
SWCNTs
HA
5.0 kV
(ii)
Gr-HA
HA
Graphene nanosheet
(c)
(i)
CNT dangling with GNC
CNTs, entangling with GNSs
1 µm
(ii)
250 µm
(iii)
Cell adhesion
Live cell
Cell nuclei
2 µm
200 µm
(iv)
1 d
3 d
5 d
7 d
100 µm
(d)

attachment of human mesenchymal stem cells (hMSCs) to graphene–CNT-coated hydroxyapatite/polyetheretherketone (HAP/PEEK) scaffolds was fast and had good vitality, proliferation, and differentiation toward bone (Figure 16.4d) [63].

16.3 Tissue Engineering Applications of Carbon Nanotubes

CNTs in composites can be created from biocompatible nanocrystals of collagen fibrils for the regeneration and engineering of various cells on a cellular structural basis. They may improve effective cellular communication with enzyme-binding proteins, influence cell physiology, and accelerate stem cell differentiation due to their beneficial high cell constants. The next part describes CNT tissue engineering applications in three of the most studied tissues: bone, neurological system, and cardiovascular system.

16.3.1 Application in Bone Tissue Engineering

As an important organ of the human body, bone has physiological roles besides providing a structural support. It is composed of hydroxyapatite as an inorganic component and mainly collagen type I as the organic phase. The bone's structure is mainly divided into two cancellous and cortical. The former's structure is porous in which various cells and the vasculature are accommodated. The latter has a dense structure, making it desirable for load-bearing applications [66]. Different types of materials have been studied for bone tissue regeneration, and among them hydroxyapatite alone or in combination with other materials has been the epicenter of research [67]. CNTs have been applied in bone tissue engineering not only as an implant and scaffold but also as drug delivery systems [68]. A recent study has made an interesting comparison between nanohydroxyapatite and CNTs to see which one exhibits better osteogenicity. As the first step, the adhesion of human adipose-derived mesenchymal stem cells on both samples were analyzed, revealing no significant difference in the quantity of attached cells, while the adhesion efficacy followed by proliferation was competitively better on the CNTs than nanohydroxyapatite. The stem cells were found to differentiate into osteogenic cells in the exposure of CNTs through activation of Notch signaling pathways. Through *in vivo* studies, it was observed that ectopic bone formation was induced by CNTs, whereas nanohydroxyapatite did not possibly due to the secretion of specific bone-inducing proteins from M2 macrophages triggering the induction of bone formation [69]. Except this recent study, most of the research made on the CNTs' potential in bone tissue regeneration were focused on their addition to various inorganic materials like hydroxyapatite, tricalcium phosphate for improving the total mechanical properties and biocompatibility [70]. CNTs were mixed with hydroxyapatite and subsequently coupled with collagen to stimulate the proliferation and migration of bone marrow stem cells, mimicking the natural structure and composition of bone. Increase in the expression of osteocalcin and bone salivary proteins was the sign indicating the promotion of stem cells [71]. In another study, the same combination was obtained with a three-step synthesis strategy. First, the chemical vapor deposition was adopted for *in situ* synthesis of CNTs in hydroxyapatite particles once the collagen molecules were added to the mixture to form collagen-encapsulated CNTs through amide bonds [72]. Finally, calcium phosphate nanoparticles were precipitated on the surface of collagen–CNTs through a chemical precipitation technique. The encapsulation of CNTs inside the collagen fibers avoided any potential cytotoxicity caused by CNTs due to the high similarity of tricalcium phosphate's composition to hydroxyapatite. Furthermore, it was combined with CNTs through a solution precipitation technique, indicating the formation of apatite on the samples as well as the improvement in calcification in the exposure of materials *in vitro*.

D bone scaffolds are well-known temporary matrices in which bone formation gets promoted. These scaffolds should have suitable biocompatibility, biodegradability, porosity, mechanical properties, etc. [73]. It was reported that the total porosity and pore size are the governing factors that influence cell attachment, migration, and proliferation. As porosity weakens the mechanical properties, it has always been an irreconcilable issue in bone scaffold fabrication. Therefore, the manufacturing technique is a key factor in producing highly porous scaffolds yet mechanically strong [74]. Freeze-drying as one of the scaffold's fabrication techniques has the potential to yield a highly porous structure resembling the cancellous bone morphology. In combination with the decellularized cell-derived extracellular matrix, silk fibroin scaffolds were developed through freeze-drying followed by reinforcement with CNTs. The scaffold's structure was hierarchical and showed *in vitro* bioactivity without hemolytic effects. Through assessment of expression of various osteogenic markers, the osteogenicity potential of CNT-reinforced scaffolds was proved [75]. 3D printing undergoes rapid development in technology, and 3D printing materials are being applied in various applications at an ever-increasing rate. This technique is of great importance in bone tissue regeneration as it holds great potential in developing highly porous scaffolds with great mechanical properties for load-bearing applications [39]. The electrical stimulation (ES) capability of CNTs was also studied to stimulate the bone formation rate at critical bone defects *in vivo*. For this purpose, PCL-MWCNTs 3D-printed scaffolds were developed. Both the MWCNT content and electrical stimulation culminated in the promotion of angiogenesis and mineralization. The *in vivo* studies performed on the critical sized bone defects are indicated through Figure 16.5 [76]. The study took up to 120 days while all the samples showed bone healing; thicker bones were formed in the exposure of the scaffold containing the highest amounts of MWCNTs exposed to electrical stimulation. It can be seen that vascularization has occurred in the vicinity of PCL scaffolds, while mineralization is low. On the other hand, the samples containing MWCNTs showed an increase in the mineralization through the tissues in which they had been implanted along with embedded osteocytes. As a result, the amount of MWCNTs had a direct relationship with the mineralization and more CNTs led to more precipitation.

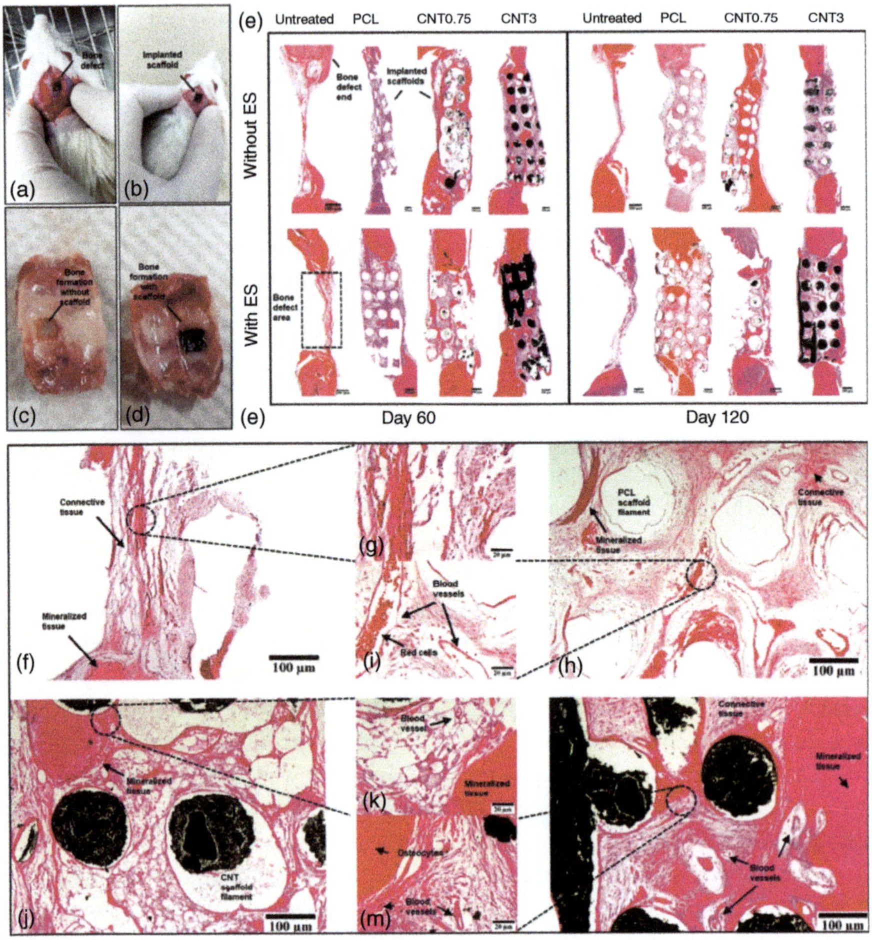

Figure 16.5 *In vivo* study of polycaprolactone-MWCNTs 3D-printed scaffold implanted in critical sized bone defects. (a) The defect made on the animal model. (b) Implantation of scaffolds at the defect's site. (c) The defect without being treated with any scaffold after 60 days. (d) The defect after being treated with the polycaprolactone-MWCNTs scaffolds after 60 days. (e) Cross sections made on the defects after 60 and 120 days of treatment with all samples. (f, g) The defect after 60 days only treated with electrical stimulation (ES) in two magnifications. (h, i) The bone formation in the vicinity of polycaprolactone scaffolds and ES after 60 days in two magnifications. (j, k) The bone formation in the vicinity of polycaprolactone–MWCNTs (0.75 wt%) scaffolds and ES after 60 days in two magnifications. (l, m) The bone formation in the vicinity of polycaprolactone–MWCNTs (3 wt%) scaffolds and ES after 60 days in two magnifications. Source: Huang et al. [76]/with permission of Springer Nature.

Another approach to take advantage of CNTs in bone tissue regeneration is their direct coating on the scaffold surface. Recently, a 3D-printed polymeric scaffold composed of poly(propylene fumarate) (PPF) coated with modified CNTs was fabricated. The CNTs were modified with water-soluble single-stranded deoxyribonucleic acid (ssDNA) to alter the surface charge to negative (Figure 16.6a). Notably, the 3D-printed scaffolds were surface-modified to arise amine groups ready to make an electrostatic bonding with the modified CNTs (Figure 16.6b). This method was successful enough to yield homogeneous distribution of modified CNTs on the scaffolds and an improvement in the attachment, migration, and proliferation of pre-osteoblast cells (Figure 16.6c). Moreover, as CNTs endowed the scaffolds with conductivity, electrical stimulation was applied to see whether the conductive layer could make any difference. It turned out that electrical stimulation enabled the modulation of cell behavior culminating in the expression of osteogenic markers (Figure 16.6d–n) [77].

16.3.2 Application in Nervous System Engineering

The human nervous system comprises two separate parts, the central nervous system and peripheral nervous system. The function of this system is to transmit signals, signs, and messages between nerves and other cells [78]. Nervous system tissue engineering is one of the most sensitive and challenging topics in biomedicine. Due to the nervous system's high sensitivity, particular complexity, and accuracy, any damage to it may cause different problems. Besides, the morphology and function of each neuron exhibit the complexity of the communication networks [79]. Investigations have shown that CNTs can stimulate regeneration, axonal growth, conduction, and transmission of electrical signals as a hollow, flexible nanomaterial with high electrical conductivity [80]. Both kinds of CNTs, single-walled and multi-walled, are great options for manufacturing implants. They have stable chemical properties that can easily coordinate with nerve tissues, increasing the branches and length of neurites and this procedure consequently facilitates the function of nerve tissues. They are used to develop new implantable neural devices and design scaffolds to modify the characteristics of neurons because of their high thermal expansion coefficient, high tensile strength, remarkable flexibility, bending capability, and great electrical and thermal conductivity nanotubes [81]. These nanoparticles are ineffective in the physiological environment of the body and have a high and suitable biocompatibility for long-term implantation. In addition, due to their nanoscale structures, they have a high surface-to-volume ratio that can reduce electrical impedance [82]. Notably, they can also conduct electricity well without altering or damaging their structures at densities above 10^7 A cm^{-2} [83].

Two alternative strategies for controlling neural cell activities using bio-functionalized CNTs have been reported in the literature. One way involves introducing stand-alone CNTs to neural cell culture medium, while another involves surface modification in preparation of subsequent attachment substrates such as scaffolds [84]. The first involves applying CNTs directly to neural tissue; the second involves using them as the surface or bulk modifiers of other materials. Properties of CNT scaffolds, including roughness, charge, polarity, and chemistry, can change the affinity of neurons attachment to CNT-modified surfaces [85]. The contacts between neurites and CNTs function as an exoskeleton, resulting in increased

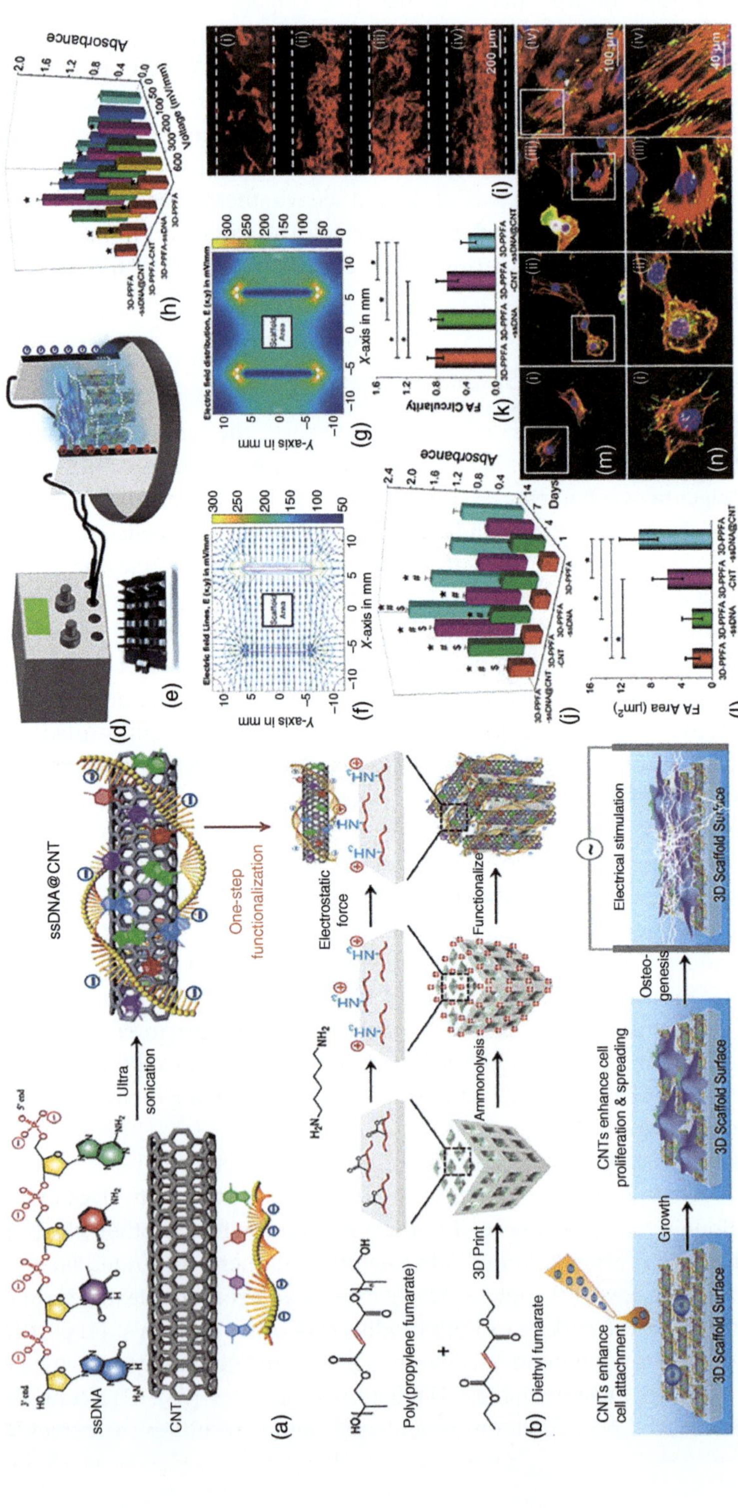
(a)
ssDNA
CNT
Ultra sonication
ssDNA@CNT
One-step functionalization
(b)
Poly(propylene fumarate)
Diethyl fumarate
3D Print
Ammonolysis
Electrostatic force
Functionalize
(c)
CNTs enhance cell attachment
Growth
CNTs enhance cell proliferation & spreading
Osteo-genesis
Electrical stimulation
3D Scaffold Surface
(d)
(e)
(f)
(g)
X-axis in mm
Y-axis in mm
(h)
Absorbance
Voltage (mV/mm)
(i)
200 µm
(j)
Days
(k)
FA Circularity
(l)
FA Area (µm²)
(m)
100 µm
(n)
40 µm

Figure 16.6 Modified-CNT-coated 3D-printed scaffolds for bone tissue regeneration. (a) The process of modification of CNTs with water-soluble single-stranded deoxyribonucleic acid. (b) Creation of amine surface groups on the scaffold and coating of modified CNTs on the scaffold through electrostatic bonding. (c) Schematic showing the positive effects of CNTs on the adhesion and proliferation of scaffolds plus the stimulating effects of electrical current on the osteogenicity. (d, e) Electrical stimulation process and the plate applied for doing the process and triggering the pre-osteoblast cells, which have been cultured on the scaffolds. (f) The simulation of electrical line. (g) The distribution of electrical field applied in the area of the scaffold. (h) The columns related to the cell number after being treated with different samples and various electrical voltages; the values were normalized to the values obtained for 3D-PPFA without any electrical stimulation. *: $p < 0.05$ denotes the significant difference between the same samples without being exposed to electricity. (i) The immunofluorescence images taken from the cultured cells on the different samples; red color represents F-actin. (j) The number of cells assessed through MTS assay; *: $p < 0.05$, #: $p < 0.05$, and &: $p < 0.05$ represent the significance difference between the samples and 3D-PPFA, 3D-PPFA-CNT, and 3D-PPFA-ssDNA groups at the same time interval, respectively. (k, l) Focal adhesion circularity and area related to the cells cultured on the scaffolds followed by being stimulated with electricity (*: $p < 0.05$). (m, n) Confocal microscopy adopted to take images in two magnifications relating to single cells on the scaffolds after being treated with electrical stimulation. (i)–(iii) Attributed to 3D-PPFA, 3D-PPFA-CNT, and 3D-PPFA-ssDNA scaffolds, respectively. Abbreviations: ssDNA, single-stranded deoxyribonucleic acid; CNT, carbon nanotube; PPF, poly(propylene) fumarate; PPFA, ammonolyzed poly(propylene) fumarate. Source: Liu et al. [77]/with permission of Elsevier.

membrane/material tight connection formation. CNTs with a larger surface area have a better charge injection capacity and a lower interfacial impedance, which help with electrochemical coupling via electron transfer between CNTs and neurites [86].

CNTs have proved to play an important role in the interactions between neurons and their surroundings. When employed as a scaffold, CNTs not only act as a reservoir for adsorbed proteins but also play a dynamic function in improving neuronal electrical performance. The idea of a direct electrical coupling is supported by the observation of discontinuous and tight contacts between MWCNT or SWCNT bundles and neuronal membranes (Figure 16.7a) [87]. Scientists have shown that a meshwork of MWCNTs outside neurons is in close contact with a tiny section of the neurotic membranes, indicating the electrical activities of nanomaterials in neuron excitability. Researchers also discovered an increase in paxillin expression in hippocampal neurons cultured on MWCNTs, a membrane protein involved in intracellular signaling pathways mediated by focal adhesions, demonstrating that electrophysiological signals from CNTs can be translated into specific neuronal signals [87]. MWCNT carpets produced on fused silica substrates were recently reported as optically transparent CNT-based substrates (tCNTs). tCNTs stimulate not only the growth and function of dissociated primary neurons' networks but also hippocampus neurons' spontaneous synaptic activity (Figure 16.7b) [88].

In another study, a poly(ethylene glycol) diacrylate (PEGDA) matrix containing aminated MWCNTs was developed [89]. The production method is represented in Figure 16.8a, which shows how CNTs can be easily included in developing 3D printing technologies to generate scaffolds that enable neural cell differentiation

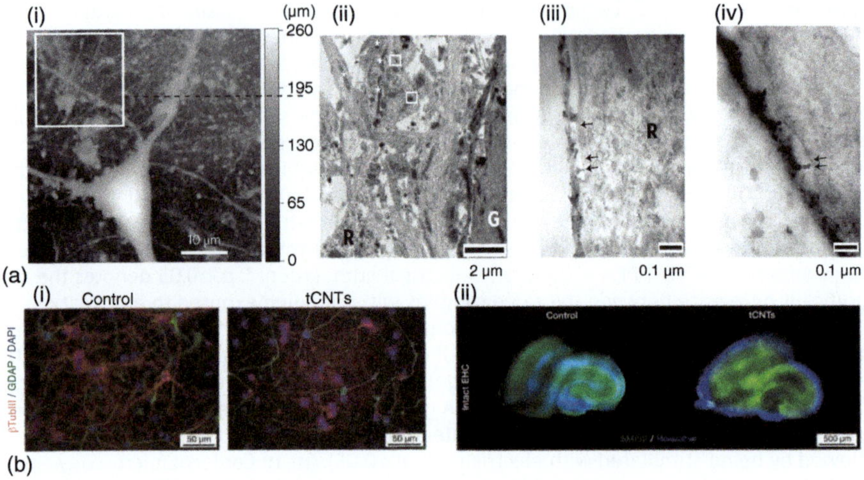

Figure 16.7 Modified-CNT scaffolds for neural tissue regeneration: (a, i) effect of conductivity and nanoscale features on neurons: AFM shows that RADA16 peptides form scaffolds similar to nanotubes and cells grow well on them without any effect on after-potential excitability, (a, ii) the ultrastructural interaction between multi-walled nanotubes and neurons. TEM planar sections of neurons grown on nanotubes show a healthy organization of neuronal networks, accompanied by the presence of synaptic contacts (rectangular boxes). The asterisk indicates mitochondria and R indicates clusters of ribosomes. Longitudinal sections of neurites containing microfilaments are clearly distinguishable. A portion of a presumed glial cell is indicated by "G," identifiable by the intense electron density of its cytoplasm, (a, iii, iv) TEM sagittal sections illustrate multi-walled nanotube–membrane contacts, indicated by arrows. Source: Cellot et al. [87]/with permission of Springer Nature. (b, i) tCNTs boost the spontaneous synaptic activity of hippocampal neurons: representative fluorescent micrographs depicting dissociated primary cells networks grown on glass control substrates (left) and on tCNTs substrates (right) stained against Tubulin III to point out neurons, GFAP to highlight astrocytes and DAPI to stain cell nuclei, (b, ii) tCNTs are suitable substrates for the development of healthy EHCs organotypic cultures. Source: Pampaloni et al. [88]/with permission of John Wiley & Sons.

and proliferation while having microelectroporous features. This proliferation did not occur until day 7, and the researchers assumed that this was due to the neural stem cells' adaptation to the substrate. Stronger MWCNT concentrations in the printed scaffolds resulted in a higher positive charge, which could aid in nerve cell formation and proliferation [89].

In clinical studies, peripheral nerve injuries are prevalent, however the nerve's ability to regenerate spontaneously varies depending on the severity of the injury, which is limited if the injury is too severe. Recently, a research group has overcome this problem by manufacturing a phosphate glass microfiber (PGF) scaffold with aminated CNTs (Figure 16.8b) [90]. The researchers employed the generated fibers as an interface material for neuron physical stimulation after surface connecting aminated CNTs to aligned PGF bundles. *In vivo* cell experiments were carried out after wrapping the CNT-PGF substrate around electrospun poly(L/D-lactic acid) (PLDLA) nanofibers. Finally, the implant is effective in restoring motor functions,

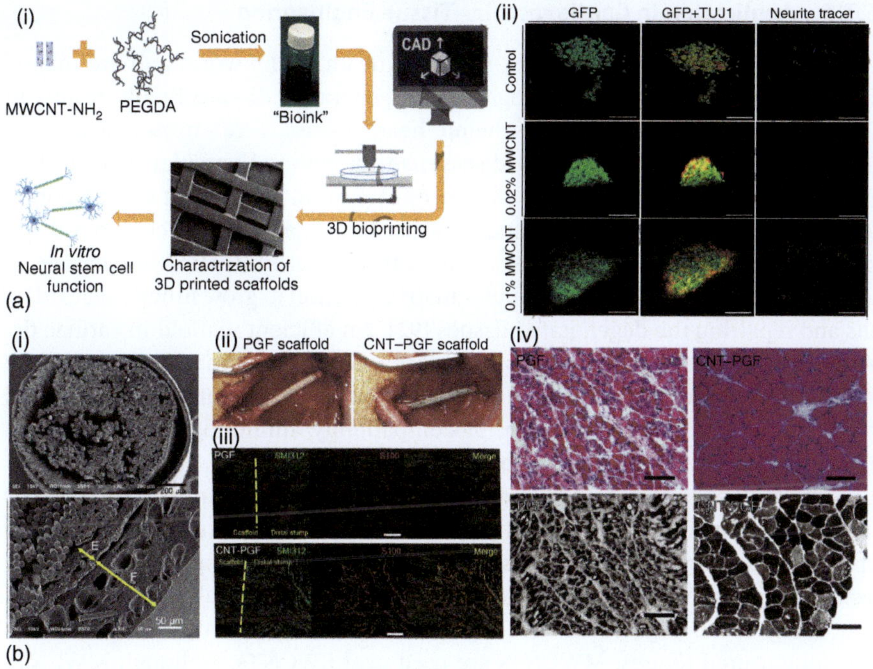

Figure 16.8 Modified-CNT-coated 3D-printed scaffolds for neural tissue regeneration: (a, i) representation of the producing method of 3D-printing nanoconductive MWCNT-PEGDA scaffolds, (ii) confocal micrographs of NSC growth and spreading on 3D printed scaffolds at day 8. Cells were stained with GFP (green) and TUJ1 (red). Source: Lee et al. [89]/with permission of IOP Publishing. (b, i) EM morphology analysis of CNT-interfaced PGFs (CNT-PGFs) and a scaffold for *in vivo* study, (ii) between the proximal and distal stumps of a totally transected rat sciatic nerve, a CNT-free PGF (PGF) or CNT–PG scaffold was implanted. Representative immunohistochemistry pictures of axons in the distal stump (11 mm from the proximal stump end) in the transverse section, (iii) PGF scaffold–implanted sciatic nerve (PGF) or CNT-PGF scaffold–implanted sciatic nerve (CNT-PGF) at 16 weeks post-implantation (yellow scale bar = 500 lm, white scale bar = 200 lm), (iv) representative images of H&E-stained gastrocnemius muscle following PGF scaffold (PGF) or CNT-PGF scaffold implantation (CNT-PGF) at 16 weeks post-implantation (black scale bar = 50 lm), and representative images of ATPase stained muscle (PGF and CNT-PGF) and examples of muscle fiber types (right, I = type I, Iia = type IIa, Iib = type IIb, black scale bar = 50 lm). Source: Ahn et al. [90]/with permission of Elsevier.

demonstrating that the CNT interface restored muscle function in the animal, as the scaffold-crossing axons reinnervated the gastrocnemius muscles.

Overall, while developing a functionally efficient brain prosthesis is still a long way off, a few promising nanomaterials appear to be quite useful for this goal, and CNTs are without a doubt one of them. CNTs are particularly attractive as ingredients in biomimetic scaffolds to direct axon regeneration and increase neural functions because of their good conductivity and effective supporting abilities, as well as their demonstrated biocompatibility with neuronal cells.

16.3.3 Application in Cardiovascular Tissue Engineering

Cardiovascular diseases are the first reason of death in the world and are mainly caused by atherosclerosis. This progressive degenerative disease finally results in other chronic disorders, such as ischemic heart disease, cerebrovascular disease, mesenteric ischemia, renovascular disease, and peripheral vascular disease [91]. In recent years, tissue engineering has been introduced as a potent contemporary approach for restoring damaged tissues. Cardiac tissue engineering tools commonly use cardiac progenitor cells (CPCs) along with natural/synthetic biomaterials as the scaffold, mimicking the extracellular matrix function to grow functional cardiac cells and repairing the degenerated tissues [92]. An efficient scaffold in cardiac tissue engineering should have appropriate mechanical and electrical properties [93]. Other necessary factors that should be considered in creating an ideal myocardial tissue are structure, biodegradation, biocompatibility, immunogenicity, elasticity, cell affinity, and scaffold inflammatory response [94].

An impressive resemblance between CNTs and proteins of cardiac extracellular dimension and electrical conductivity caused these nanocarriers to be beneficially applied as unique synthetic scaffolds in cardiac tissue engineering. Therefore, using electrical and mechanical stimulus, cellular alignment, metabolic activity, and contractility of the engineered tissues are developed [95]. To improve the efficacy of the engineered tissues, MWCNTs are used over SWCNTs. Although MWCNTs have more walls and a higher structural complexity than SWCNTs, numerous deficiencies in the MWCNTs' central layers allow them to function more efficiently [96]. By seeding H9c2 cardiac myoblasts into the scaffolds, the viability, activity, and morphology of the new tissue were checked. The most metabolic activity was related to the cells in the PU/collagen/CNT composite scaffold [97]. Moreover, a direct relationship between the collagen content of the composites and the diameters of the fibers was observed. Hence, the scaffolds with more collagen concentration revealed less elasticity and more biocompatibility, while the mechanical characteristics of the scaffolds were increased by the PU concentration. Hence, incorporation of CNTs in a high concentration of PU/collagen polymers offers an ideal scaffold for cardiac tissue engineering. Scientists incorporated SWCNTs into 3D-collagen hydrogel, which is capable of enhancing the stiffness and electrical characteristics of hydrogel matrices. Furthermore, after using the SWCNTs/collagen hydrogel for cardiac tissue engineering, they were able to improve cell adhesion and alignment with no toxicity, eventually obtaining the regenerated cardiac tissue with better function and contraction properties [98].

A recent work attained novel polymeric composites biomaterial via combining MWCNTs and chitosan to be applied for making scaffolds in cardiac tissue engineering. The synthesized MWCNTs/chitosan composite had better characteristics than pure chitosan. Afterward, endothelial cells and vascular myofibroblasts were separately grown on these scaffolds, illustrating the increased mechanical properties of the composites with 5% and 10% concentrations of MWCNTs erewhile 5% was the best. Moreover, both cell lines showed the most sufficient proliferation effects on MWCNT/chitosan scaffolds and MWCNT/chitosan (even CNTs) exhibited no toxicity [99].

The electrical properties of novel scaffolds made of various amounts of f-MWCNTs up to 0.1 wt% incorporated in a gelatin/CS mixture were reported in 2018. The conductivity and mechanical properties of the prepared scaffold increased to a sufficient level for use in cardiovascular tissue engineering [100]. In another work, researchers developed PVA-CS-CNTs using different concentrations of CNTs as scaffolds for cardiac tissue engineering [101]. The characteristics of scaffolds, such as conductivity, chemical stability, cross-linking properties, elastic modulus, and tensile strength, were further evaluated, and ultimately, the PVA-CS-CNT1 (with 1% CNT) was known to be the best nontoxic scaffold for cardiac tissue engineering [101].

Poly(*N*-isopropylacrylamide) (PNIPAAm) is another polymer used in cardiac tissue engineering. To overcome the limitations of PNIPAAm hydrogel, such as low electrical conductivity, mechanical properties, and modulus, it could be modified with SWCNTs. It was shown that the brown adipose–derived stem cells (BASCs) seeded on PNIPAAm/SWCNTs as a scaffold helped repair cardiac muscle. Regarding bioactivity, the cells on the modified hydrogel showed significantly more bioactivities than the ones on PNIPAAm. Furthermore, by incorporating the BASCs/PNIPAAm/SWCNTs into the infarcted myocardium of rats, repair benefits comprising cardiac cell adhesion and spreading were enhanced (Figure 16.9a) [102].

In a study, after fabricating scaffolds using fusing PCL to thiophene-conjugated carbon nanotubes (T-CNTs), scarred areas in murine hearts were regenerated [104]. The freshly isolated cardiac stem/progenitor cells (CPCs) from murine hearts were then embedded on the PCL/T-CNT backbone. As a result, cell adhesion, proliferation, and differentiation were observed. Compared to pure PCL mesh, mechanical properties and cell proliferation were significantly increased in PCL/T-CNT, while the morphological characteristics and differentiation of CPCs into cardiomyocytes did not vary [104]. Another study developed a cardiac tissue engineering scaffold by incorporating 3–6% W/W CNTs into aligned fibers of poly(ethylene glycol)–poly(D,L-lactide) copolymers (PELA) (Figure 16.9b) [103]. The increased electrical conductivity of the scaffolds enabled them to retain cardiomyocyte viability, improving their synchronous beating and motivating cell elongation. Among all the PELA/CNTs, the scaffold containing core-sheath fibers with 5% of CNTs in the fiber cores showed the best results. Considering the obtained results, the scaffold improved the growth of cardiac muscle through generating contractile proteins and modulating their beating rates nearly equal to the atrium [103].

An injectable reverse thermal gel (RTG)-poly-L-lysine is another polymer chemically conjugated to CNTs to fabricate a nanoengineered hybrid 3D-RTG-CNT for heart tissue repair without any toxicity [105]. The 3D niche associated with the

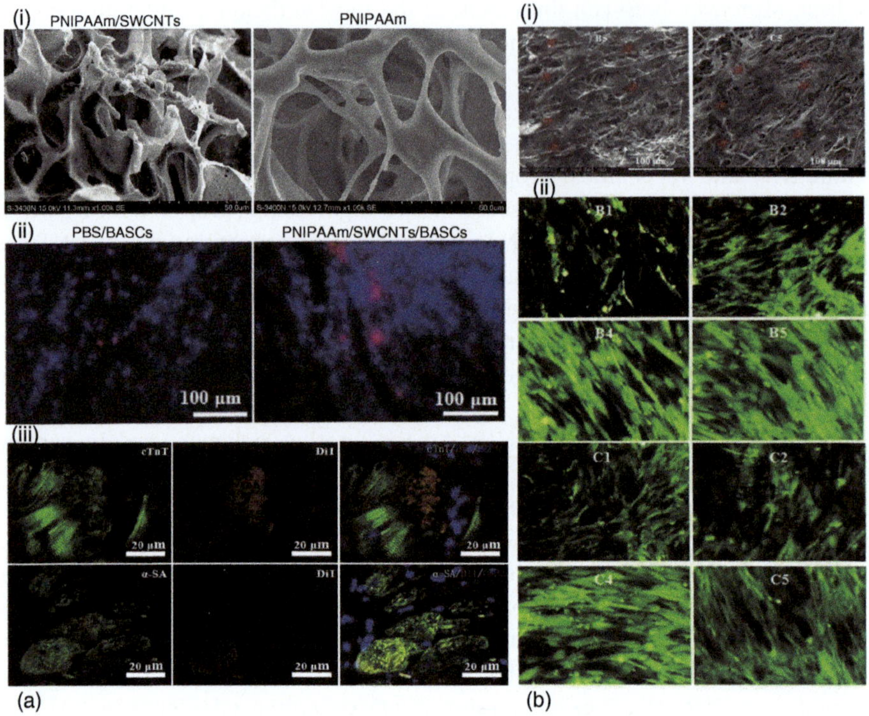

Figure 16.9 Modified-CNT scaffolds for cardiovascular tissue regeneration: (a, i) observation of PNIPAAm hydrogels with and without SWCNTs using a scanning electron microscope. The PNIPAAm/SWCNTs hydrogels had a rough structure, but the PNIPAAm hydrogel had a rather smooth structure, as observed by SEM, (a, ii) BASC survival in ischemic myocardium one week following surgery, the PNIPAAm/SWCNTs/BASCs group had significantly more DiI-positive cells than the PBS/BASCs group, (a, iii) BASCs differentiate into cardiomyocytes in a cardiac milieu. Immunohistochemistry against cardiac markers was done on the heart section four weeks after transplantation. DiI colocalization with cardiac proteins (cTnT and α-SA) was identified, showing that transplanted cells differentiated into cardiac lineages. Source: Li et al. [102]/with permission of Elsevier. (b, i) Typical SEM images of cardiomyocytes on blend and coaxial fibers containing 5% of CNTs. Red arrows show the cardiomyocytes and their orientation, (b, ii) typical immunofluorescent staining images of α-actinin synthesized by cardiomyocytes after incubation for 10 days on blend and coaxially electrospun fibrous mats containing different amounts of CNT. Source: Liu et al. [103]/with permission of Elsevier.

CNT can be a sufficient environment for stimulating cardiomyocyte proliferation while inhibiting fibroblast proliferation. A research group found that cardiomyocytes grew on the functionalized 3D gel-based matrix with high surveillance. In addition, the engineered cells were mature and bioactive with more alignment (Figure 16.10) [106].

Overall, CNTs incorporated into natural or synthetic polymers (PU, GelMA, gelatin methacryloyl, PELA, PMNT, PDMS, RTG)-poly-L-lysine, PNIPAAm, PVA, and CS) for modulating their electrical performance and mechanical functions are very close to the myocardial cells for use in cardiac tissue engineering.

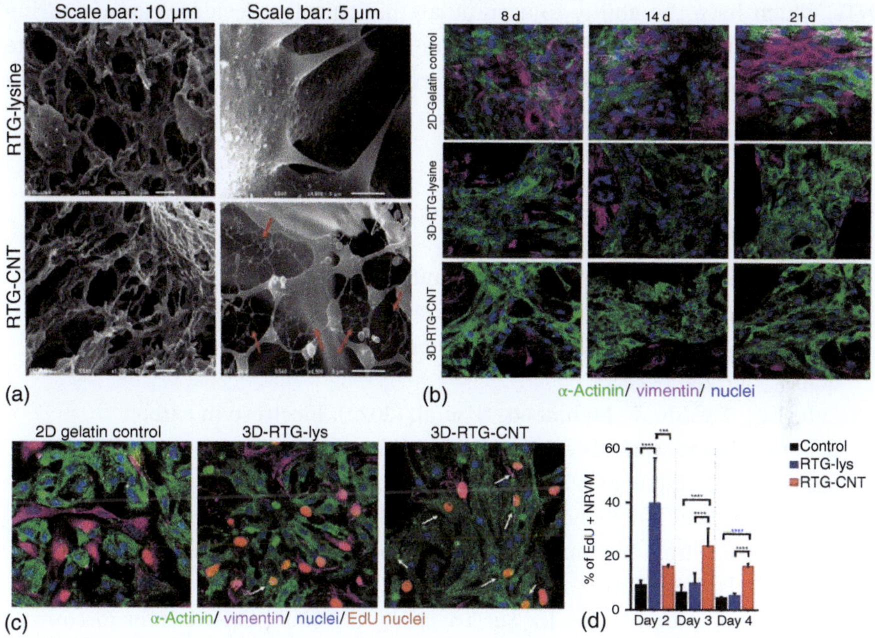

Figure 16.10 Modified-CNT-coated 3D-printed scaffolds for vascular tissue regeneration: (a) RTG systems cross-sectional micrographs at various magnifications. RTG-lysine (top) and RTG-CNT (bottom). In the RTG-CNT system, a fibrous CNT mesh network can be seen (arrows), (b) fluorescence staining of NRVMs, and fibroblasts cultured in different substrates. Sarcomeric α-actinin, vimentin, and DAPI (nuclei). (b, top) NRVMs cultured on 2D gelatin control. (b, middle) NRVMs cultured in 3D RTG-lysine. (b, bottom) NRVMs cultured in 3D RTG-CNT, (c) NRVM and fibroblast proliferation test. (From left to right, c) At day 4 of culture, NRVMs and fibroblasts were cultivated on 2D gelatin control, 3D RTG-lysine, and 3D RTG-CNT, (d) significant differences in dividing NRVMs on the RTG-lysine system were observed at day 2 when compared with the gelatin controls. Source: Peña et al. [106]/with permission of American Chemical Society.

16.4 Challenges and Future Perspectives

Unique combination of mechanical, optical, and electrical properties offered by CNTs makes them potential carbon nanomaterials in tissue engineering for improving biochemical, mechanical, and electrical properties. While CNTs are frequently used in small levels, their genotoxicity, cytotoxicity, carcinogenicity, and teratogenicity are highly dose and exposure dependent. Thus, although cell growth on CNTs preliminary confirmed their biocompatibility, advanced and rational approaches are needed to identify and evaluate toxicity, carcinogenic effects, and teratogenic effects of CNTs. Currently, there are no global standards for nano-specific risk assessments, which include applicable data criteria and research methods, and CNT risk evaluations are time-consuming and costly. Companies are devoted to assessing the safety of CNT-based products and enforcing the necessary safety precautions. Therefore, precautions should be taken before implementing

CNTs, which have the ability to accumulate biologically. Besides, manufacturing technologies will be employed in the future to produce CNTs involving composite scaffolds that closely mimic the natural environment for successful cell proliferation and differentiation in clinical applications.

References

1 Meredith, J.R., Jin, C., Narayan, R.J., and Aggarwal, R. (2013). Biomedical applications of carbon-nanotube composites. *Front. Biosci. (Elite)* 5: 610–621.
2 Saliev, T.J. (2019). The advances in biomedical applications of carbon nanotubes. *Carbon* 5 (2): 29.
3 Suh, T.C., Twiddy, J., Mahmood, N. et al. (2022). Electrospun carbon nanotube-based scaffolds exhibit high conductivity and cytocompatibility for tissue engineering applications. *ACS Omega*.
4 Pampaloni, N.P., Giugliano, M., Scaini, D. et al. (2019). Advances in nano neuroscience: from nanomaterials to nanotools. *Front. Neurosci.* 12: 953.
5 Negri, V., Pacheco-Torres, J., Calle, D., and López-Larrubia, P. (2020). Carbon nanotubes in biomedicine. In: *Surface-modified Nanobiomaterials for Electrochemical and Biomedicine Applications*, 177–217. Springer.
6 Pulskamp, K., Diabaté, S., and Krug, H.F. (2007). Carbon nanotubes show no sign of acute toxicity but induce intracellular reactive oxygen species in dependence on contaminants. *Toxicol. Lett.* 168 (1): 58–74.
7 Gaur, M., Misra, C., Yadav, A.B. et al. (2021). Biomedical applications of carbon nanomaterials: fullerenes, quantum dots, nanotubes, nanofibers, and graphene. *Materials* 14 (20): 5978.
8 Anzar, N., Hasan, R., Tyagi, M. et al. (2020). Carbon nanotube-a review on synthesis, properties and plethora of applications in the field of biomedical science. *Sens. Int.* 1: 100003.
9 Liu, P. (2005). Modifications of carbon nanotubes with polymers. *Eur. Polym. J.* 41 (11): 2693–2703.
10 Bugatti, V., Viscusi, G., Di Bartolomeo, A. et al. (2020). Ionic liquid as dispersing agent of LDH-carbon nanotubes into a biodegradable vinyl alcohol polymer. *Polymers* 12 (2): 495.
11 Chen, J., Liu, B., Gao, X., and Xu, D. (2018). A review of the interfacial characteristics of polymer nanocomposites containing carbon nanotubes. *RSC Adv.* 8 (49): 28048–28085.
12 Sireesha, M., Jagadeesh Babu, V., Kranthi Kiran, A.S., and Ramakrishna, S. (2018). A review on carbon nanotubes in biosensor devices and their applications in medicine. *Nanocomposites* 4 (2): 36–57.
13 González-Domínguez, J.M., Castell, P., Bespin-Gascon, S. et al. (2012). Covalent functionalization of MWCNTs with poly (p-phenylene sulphide) oligomers: a route to the efficient integration through a chemical approach. *J. Mater. Chem.* 22 (39): 21285–21297.

14 Abbasi, H., Antunes, M., and Velasco, J.I. (2019). Recent advances in carbon-based polymer nanocomposites for electromagnetic interference shielding. *Prog. Mater. Sci.* 103: 319–373.

15 Verma, P., Saini, P., Malik, R.S., and Choudhary, V. (2015). Excellent electromagnetic interference shielding and mechanical properties of high loading carbon-nanotubes/polymer composites designed using melt recirculation equipped twin-screw extruder. *Carbon* 89: 308–317.

16 Hayashida, K. and Matsuoka, Y. (2015). Electromagnetic interference shielding properties of polymer-grafted carbon nanotube composites with high electrical resistance. *Carbon* 85: 363–371.

17 Saka, C. (2018). Overview on the surface functionalization mechanism and determination of surface functional groups of plasma treated carbon nanotubes. *Crit. Rev. Anal. Chem.* 48 (1): 1–14.

18 Lekshmi, G., Sana, S.S., Nguyen, V.-H. et al. (2020). Recent progress in carbon nanotube polymer composites in tissue engineering and regeneration. *Int. J. Mol. Sci.* 21 (17): 6440.

19 Keteklahijani, Y.Z., Zeraati, A.S., Sharif, F. et al. (2020). In situ chemical polymerization of conducting polymer nanocomposites: effect of DNA-functionalized carbon nanotubes and nitrogen-doped graphene as catalytic molecular templates. *Chem. Eng. J.* 389: 124500.

20 Bonab, V.S., Maxian, O., and Manas-Zloczower, I. (2019). Carbon nanofiller networks-a comparative study of networks formed by branched versus linear carbon nanotubes in thermoplastic polyurethane. *Polymer* 175: 227–234.

21 Cheng, Q., Rutledge, K., and Jabbarzadeh, E. (2013). Carbon nanotube–poly (lactide-*co*-glycolide) composite scaffolds for bone tissue engineering applications. *Ann. Biomed. Eng.* 41 (5): 904–916.

22 Mikael, P.E., Amini, A.R., Basu, J. et al. (2014). Functionalized carbon nanotube reinforced scaffolds for bone regenerative engineering: fabrication, in vitro and in vivo evaluation. *Biomed. Mater.* 9 (3): 035001.

23 Zhou, J., Chen, J., Sun, H. et al. (2014). Engineering the heart: evaluation of conductive nanomaterials for improving implant integration and cardiac function. *Sci. Rep.* 4: 3733.

24 Gupta, A., Main, B.J., Taylor, B.L. et al. (2014). In vitro evaluation of three-dimensional single-walled carbon nanotube composites for bone tissue engineering. *J. Biomed. Mater. Res. Part A* 102 (11): 4118–4126.

25 Huang, B., Vyas, C., Roberts, I. et al. (2019). Fabrication and characterisation of 3D printed MWCNT composite porous scaffolds for bone regeneration. *Mater. Sci. Eng. C* 98: 266–278.

26 Ravanbakhsh, H., Bao, G., Latifi, N., and Mongeau, L.G. (2019). Carbon nanotube composite hydrogels for vocal fold tissue engineering: biocompatibility, rheology, and porosity. *Mater. Sci. Eng. C* 103: 109861.

27 Zhang, Q., Rastogi, S., Chen, D. et al. (2006). Low percolation threshold in single-walled carbon nanotube/high density polyethylene composites prepared by melt processing technique. *Carbon* 44 (4): 778–785.

28 Nelson, D.W. and Gilbert, R.J. (2021). Extracellular matrix-mimetic hydrogels for treating neural tissue injury: a focus on fibrin, hyaluronic acid, and elastin-like polypeptide hydrogels. *Adv. Healthc. Mater.* 10 (22): 2101329.

29 Kouser, R., Vashist, A., Zafaryab, M. et al. (2018). Biocompatible and mechanically robust nanocomposite hydrogels for potential applications in tissue engineering. *Mat. Sci. Eng. C* 84: 168–179.

30 Shin, S.R., Zihlmann, C., Akbari, M. et al. (2016). Reduced graphene oxide-gelMA hybrid hydrogels as scaffolds for cardiac tissue engineering. *Small* 12 (27): 3677–3689.

31 Karimi, A.R. and Khodadadi, A. (2016). Mechanically robust 3D nanostructure chitosan-based hydrogels with autonomic self-healing properties. *ACS Appl. Mater. Interf.* 8 (40): 27254–27263.

32 Esmaeili, Y., Khavani, M., Bigham, A. et al. (2022). Mesoporous silica@ chitosan@ gold nanoparticles as "on/off" optical biosensor and pH-sensitive theranostic platform against cancer. *Int. J. Biol. Macromol.* 202: 241–255.

33 Correa-Duarte, M.A., Wagner, N., Rojas-Chapana, J. et al. (2004). Fabrication and biocompatibility of carbon nanotube-based 3D networks as scaffolds for cell seeding and growth. *Nano Lett.* 4 (11): 2233–2236.

34 Dong, X., Wei, C., Liang, J. et al. (2017). Thermosensitive hydrogel loaded with chitosan-carbon nanotubes for near infrared light triggered drug delivery. *Coll. Surf. B Biointerf.* 154: 253–262.

35 He, J., Shi, M., Liang, Y., and Guo, B. (2020). Conductive adhesive self-healing nanocomposite hydrogel wound dressing for photothermal therapy of infected full-thickness skin wounds. *Chem. Eng. J.* 394: 124888.

36 Zeidner, M., Matthews, G., and Shemesh, D.O. (2016). Cognitive-social sources of wellbeing: Differentiating the roles of coping style, social support and emotional intelligence. *J. Happiness Stud.* 17 (6): 2481–2501.

37 Tekade, R.K., Maheshwari, R., and Tekade, M. (2017). Biopolymer-based nanocomposites for transdermal drug delivery. In: *Biopolymer-Based Composites*, 81–106. Woodhead Publishing.

38 Yan, L., Chang, P.R., Zheng, P., and Ma, X. (2012). Characterization of magnetic guar gum-grafted carbon nanotubes and the adsorption of the dyes. *Carbohyd. Polym.* 87 (3): 1919–1924.

39 Bigham, A., Foroughi, F., Rezvani Ghomi, E. et al. (2020). The journey of multifunctional bone scaffolds fabricated from traditional toward modern techniques. *Bio-Design Manufact.* 3 (4): 281–306.

40 MacDonald, R.A., Laurenzi, B.F., Viswanathan, G. et al. (2005). Collagen–carbon nanotube composite materials as scaffolds in tissue engineering. *J. Biomed. Mater. Res. Part A* 74 (3): 489–496.

41 Zhang, Y., Ren, E., Li, A. et al. (2021). A porous self-healing hydrogel with an island-bridge structure for strain and pressure sensors. *J. Mater. Chem. B* 9 (3): 719–730.

42 Tong, X., Zheng, J., Lu, Y. et al. (2007). Swelling and mechanical behaviors of carbon nanotube/poly (vinyl alcohol) hybrid hydrogels. *Mat. Lett.* 61 (8–9): 1704–1706.

43 Dong, X., Song, T., Ren, D. et al. (2022). Preparation of wearable strain sensor based on PVA/MWCNTs hydrogel composite. *Mater. Today Commun.* 31: 103278.

44 Fathi-Achachelouei, M., Knopf-Marques, H., Ribeiro da Silva, C.E. et al. (2019). Use of nanoparticles in tissue engineering and regenerative medicine. *Front. Bioeng. Biotechnol.* 7: 113.

45 Bodelon, G., Costas, C., Perez-Juste, J. et al. (2017). Gold nanoparticles for regulation of cell function and behavior. *Nano Today* 13: 40–60.

46 Ottman, N., Ruokolainen, L., Suomalainen, A. et al. (2019). Soil exposure modifies the gut microbiota and supports immune tolerance in a mouse model. *J. Allergy Clin. Immunol.* 143 (3): 1198–1206.

47 Boehler, R.M., Graham, J.G., and Shea, L.D. (2011). Tissue engineering tools for modulation of the immune response. *Biotechniques* 51 (4): 239–254.

48 Rajendran, S., Rathinam, V., Vallinayagam, S., and Sharma, V.K. (ed.) (2022). Nano biomaterials for tissue engineering applications—short review. In: *Proceedings of Fourth International Conference on Inventive Material Science Applications*, 131–140. Singapore: Springer.

49 Agnihotri, S., Mukherji, S., and Mukherji, S. (2013). Immobilized silver nanoparticles enhance contact killing and show highest efficacy: elucidation of the mechanism of bactericidal action of silver. *Nanoscale* 5 (16): 7328–7340.

50 Santos, F.G., Bonkovoski, L.C., Garcia, F.P. et al. (2017). Antibacterial performance of a PCL–PDMAEMA Blend nanofiber-based scaffold enhanced with immobilized silver nanoparticles. *ACS Appl. Mater. Interf.* 9 (11): 9304–9314.

51 Wang, G., Chen, X., Liu, S. et al. (2016). Mechanical chameleon through dynamic real-time plasmonic tuning. *ACS Nano* 10 (2): 1788–1794.

52 Pardo, A., Gómez-Florit, M., Barbosa, S. et al. (2021). Magnetic nanocomposite hydrogels for tissue engineering: design concepts and remote actuation strategies to control cell fate. *ACS Nano* 15 (1): 175–209.

53 Da Ros, T., Martín, N., and Nierengarten, J.-F. (2021). *Carbon Nanostructures for Biomedical Applications*. Royal Society of Chemistry.

54 Liu, Q., Chen, Z.-G., Liu, B. et al. (2008). Synthesis of different magnetic carbon nanostructures by the pyrolysis of ferrocene at different sublimation temperatures. *Carbon* 46 (14): 1892–1902.

55 Basheer, A.O., Hanafiah, M.M., Alsaadi, M.A. et al. (2021). Synthesis and characterization of low-cost carbon nanotubes by chemical vapor deposition for aluminum removal from aqueous solution. *Desalin. Water Treat.* 214: 440–451.

56 Ndlwana, L., Raleie, N., Dimpe, K.M. et al. (2021). Sustainable hydrothermal and solvothermal synthesis of advanced carbon materials in multidimensional applications: a review. *Materials* 14 (17): 5094.

57 Li, G., Zha, J., Niu, M. et al. (2018). Bifunctional monomer molecularly imprinted sol–gel polymers based on the surface of magnetic halloysite nanotubes as an effective extraction approach for norfloxacin. *Appl. Clay Sci.* 162: 409–417.

58 Salehabadi, A., Salavati-Niasari, M., and Ghiyasiyan-Arani, M. (2018). Self-assembly of hydrogen storage materials based multi-walled carbon nanotubes (MWCNTs) and $Dy_3Fe_5O_{12}$ (DFO) nanoparticles. *J. Alloys Compd.* 745: 789–797.

59 Cunha, C., Panseri, S., Iannazzo, D. et al. (2012). Hybrid composites made of multiwalled carbon nanotubes functionalized with Fe_3O_4 nanoparticles for tissue engineering applications. *Nanotechnology* 23 (46): 465102.

60 Świętek, M., Brož, A., Tarasiuk, J. et al. (2019). Carbon nanotube/iron oxide hybrid particles and their PCL-based 3D composites for potential bone regeneration. *Mater. Sci. Eng. C* 104: 109913.

61 Kim, K.H., Oh, Y., and Islam, M. (2012). Graphene coating makes carbon nanotube aerogels superelastic and resistant to fatigue. *Nat. Nanotechnol.* 7 (9): 562–566.

62 Oyefusi, A., Olanipekun, O., Neelgund, G.M. et al. (2014). Hydroxyapatite grafted carbon nanotubes and graphene nanosheets: promising bone implant materials. *Spectrochim. Acta Part A: Mol. Biomol. Spectrosc.* 132: 410–416.

63 Feng, P., Peng, S., Wu, P. et al. (2016). A nano-sandwich construct built with graphene nanosheets and carbon nanotubes enhances mechanical properties of hydroxyapatite–polyetheretherketone scaffolds. *Int. J. Nanomed.* 11: 3487.

64 Shang, L., Qi, Y., Lu, H. et al. (2019). Graphene and graphene oxide for tissue engineering and regeneration. In: *Theranostic Bionanomaterials*, 165–185. Elsevier.

65 Bahrami, S., Baheiraei, N., Mohseni, M. et al. (2019). Three-dimensional graphene foam as a conductive scaffold for cardiac tissue engineering. *J. Biomater. Appl.* 34 (1): 74–85.

66 Bigham, A., Aghajanian, A.H., Movahedi, M. et al. (2021). A 3D nanostructured calcium-aluminum-silicate scaffold with hierarchical meso-macroporosity for bone tissue regeneration: fabrication, sintering behavior, surface modification and in vitro studies. *J. Eur. Ceramic Soc.* 41 (1): 941–962.

67 Bigham, A., Salehi, A.O.M., Rafienia, M. et al. (2021). Zn-substituted Mg_2SiO_4 nanoparticles-incorporated PCL-silk fibroin composite scaffold: a multifunctional platform towards bone tissue regeneration. *Mat. Sci. Eng. C* 127: 112242.

68 Liu, H., Chen, J., Qiao, S., and Zhang, W. (2021). Carbon-based nanomaterials for bone and cartilage regeneration: a review. *ACS Biomat. Sci. Eng.* 7 (10): 4718–4735.

69 Du, Z., Feng, X., Cao, G. et al. (2021). The effect of carbon nanotubes on osteogenic functions of adipose-derived mesenchymal stem cells in vitro and bone formation in vivo compared with that of nano-hydroxyapatite and the possible mechanism. *Bioact. Mater.* 6 (2): 333–345.

70 Mirjalili, F., Mohammadi, H., Azimi, M. et al. (2017). Synthesis and characterization of β-TCP/CNT nanocomposite: Morphology, microstructure and in vitro bioactivity. *Ceram. Int.* 43 (10): 7573–7580.

71 Jing, Z., Wu, Y., Su, W. et al. (2017). Carbon nanotube reinforced collagen/hydroxyapatite scaffolds improve bone tissue formation in vitro and in vivo. *Ann. Biomed. Eng.* 45 (9): 2075–2087.

72 Li, H., Sun, X., Li, Y. et al. (2020). Carbon nanotube-collagen@ hydroxyapatite composites with improved mechanical and biological properties fabricated by a multi in situ synthesis process. *Biomed. Microdevices* 22 (4): 1–11.

73 Bigham, A., Aghajanian, A.H., Saudi, A., and Rafienia, M. (2020). Hierarchical porous Mg_2SiO_4-$CoFe_2O_4$ nanomagnetic scaffold for bone cancer therapy and regeneration: surface modification and in vitro studies. *Mater. Sci. Eng. C* 109: 110579.

74 Bigham, A., Rahimkhoei, V., Abasian, P. et al. (2021). Advances in tannic acid-incorporated biomaterials: infection treatment, regenerative medicine, cancer therapy, and biosensing. *Chem. Eng. J.* 134146.

75 Lemos, R., Maia, F.R., Ribeiro, V.P. et al. (2021). Carbon nanotube-reinforced cell-derived matrix-silk fibroin hierarchical scaffolds for bone tissue engineering applications. *J. Mater. Chem. B* 9 (46): 9561–9574.

76 Huang, B., Helaehil, J.V., Nalesso, P.R. et al. (2021). In vivo study of conductive 3D printed PCL/MWCNTs scaffolds with electrical stimulation for bone tissue engineering. *Bio-Design Manuf.* 4 (2): 190–202.

77 Liu, X., George, M.N., Park, S. et al. (2020). 3D-printed scaffolds with carbon nanotubes for bone tissue engineering: fast and homogeneous one-step functionalization. *Acta Biomater.* 111: 129–140.

78 Bardhan, N.M. (2017). 30 years of advances in functionalization of carbon nanomaterials for biomedical applications: a practical review. *J. Mater. Res.* 32 (1): 107–127.

79 Mezzasalma, S.A., Grassi, L., and Grassi, M. (2021). Physical and chemical properties of carbon nanotubes in view of mechanistic neuroscience investigations. Some outlook from condensed matter, materials science and physical chemistry. *Mater. Sci. Eng. C* 131: 112480.

80 Firkowska, I., Giannona, S., Rojas-Chapana, J.A. et al. (2008). Biocompatible nanomaterials and nanodevices promising for biomedical applications. In: *Nanomaterials for Application in Medicine and Biology*, 1–15. Dordrecht: Springer.

81 Facciolà, A., Visalli, G., La Maestra, S. et al. (2019). Carbon nanotubes and central nervous system: Environmental risks, toxicological aspects and future perspectives. *Environ. Toxicol. Pharmacol.* 65: 23–30.

82 Zare, H., Ahmadi, S., Ghasemi, A. et al. (2021). Carbon nanotubes: smart drug/gene delivery carriers. *Int. J. Nanomed.* 16: 1681.

83 Riley, P.R. and Narayan, R.J. (2021). Recent advances in carbon nanomaterials for biomedical applications: a review. *Curr. Opin. Biomed. Eng.* 17: 100262.

84 Redondo-Gómez, C., Leandro-Mora, R., Blanch-Bermúdez, D. et al. (2020). Recent advances in carbon nanotubes for nervous tissue regeneration. *Adv. Polym. Technol.* 2020.

85 Ma, P.-C., Siddiqui, N.A., Marom, G., and Kim, J.-K. (2010). Dispersion and functionalization of carbon nanotubes for polymer-based nanocomposites: a review. *Compos. Part A Appl. Sci. Manuf.* 41 (10): 1345–1367.

86 Gupta, P. and Lahiri, D. (2016). Aligned carbon nanotube containing scaffolds for neural tissue regeneration. *Neural Regen. Res.* 11 (7): 1062.

87 Cellot, G., Cilia, E., Cipollone, S. et al. (2009). Carbon nanotubes might improve neuronal performance by favouring electrical shortcuts. *Nat. Nanotechnol.* 4 (2): 126–133.

88 Pampaloni, N.P., Rago, I., Calaresu, I. et al. (2020). Transparent carbon nanotubes promote the outgrowth of enthorino-dentate projections in lesioned organ slice cultures. *Develop. Neurobiol.* 80 (9–10): 316–331.

89 Lee, S.-J., Zhu, W., Nowicki, M. et al. (2018). 3D printing nano conductive multi-walled carbon nanotube scaffolds for nerve regeneration. *J. Neural Eng.* 15 (1): 016018.

90 Ahn, H.-S., Hwang, J.-Y., Kim, M.S. et al. (2015). Carbon-nanotube-interfaced glass fiber scaffold for regeneration of transected sciatic nerve. *Acta Biomater.* 13: 324–334.

91 Flora, G.D. and Nayak, M.K. (2019). A brief review of cardiovascular diseases, associated risk factors and current treatment regimes. *Curr. Pharm. Des.* 25 (38): 4063–4084.

92 Nguyen, A.H., Marsh, P., Schmiess-Heine, L. et al. (2019). Cardiac tissue engineering: state-of-the-art methods and outlook. *J. Biol. Eng.* 13: 57.

93 Ahadian, S., Davenport Huyer, L., Estili, M. et al. (2017). Moldable elastomeric polyester-carbon nanotube scaffolds for cardiac tissue engineering. *Acta Biomater.* 52: 81–91.

94 Yu, H., Zhao, H., Huang, C., and Du, Y. (2017). Mechanically and electrically enhanced CNT-collagen hydrogels as potential scaffolds for engineered cardiac constructs. *ACS Biomater. Sci. Eng.* 3 (11): 3017–3021.

95 Liaw, N.Y. and Zimmermann, W.H. (2016). Mechanical stimulation in the engineering of heart muscle. *Adv. Drug Delivery Rev.* 96: 156–160.

96 Alagarsamy, K.N., Mathan, S., Yan, W. et al. (2021). Carbon nanomaterials for cardiovascular theranostics: promises and challenges. *Bioact. Mater.* 6 (8): 2261–2280.

97 S. A. (2016). Electroconductive and biocompatible carbon nanotube/polyurethane/collagen scaffolds for cardiac tissue engineering. *J. Tissue Sci. Eng.* 7 (2S): 156.

98 Sun, H., Zhou, J., Huang, Z. et al. (2017). Carbon nanotube-incorporated collagen hydrogels improve cell alignment and the performance of cardiac constructs. *Int. J. Nanomed.* 12: 3109–3120.

99 Kroustalli, A., Zisimopoulou, A.E., Koch, S. et al. (2013). Carbon nanotubes reinforced chitosan films: mechanical properties and cell response of a novel biomaterial for cardiovascular tissue engineering. *J. Mater. Sci. - Mater. Med.* 24 (12): 2889–2896.

100 Mesgar, A.S.-K.M.Z. and Khosrovan, S. (2018). Novel biopolymers/functionalized multi-walled carbon nanotube composite scaffolds for cardiac tissue engineering. *Transp. Phenom. Nano Micro Scales* 6 (1): 72–78.

101 Mombini, S., Mohammadnejad, J., Bakhshandeh, B. et al. (2019). Chitosan-PVA-CNT nanofibers as electrically conductive scaffolds for cardiovascular tissue engineering. *Int. J. Biol. Macromol.* 140: 278–287.

102 Li, X., Zhou, J., Liu, Z. et al. (2014). A PNIPAAm-based thermosensitive hydrogel containing SWCNTs for stem cell transplantation in myocardial repair. *Biomaterials* 35 (22): 5679–5688.

103 Liu, Y., Lu, J., Xu, G. et al. (2016). Tuning the conductivity and inner structure of electrospun fibers to promote cardiomyocyte elongation and synchronous beating. *Mater. Sci. Eng., C* 69: 865–874.

104 Wickham, A.M., Islam, M.M., Mondal, D. et al. (2014). Polycaprolactone-thiophene-conjugated carbon nanotube meshes as scaffolds for cardiac progenitor cells. *J. Biomed. Mater. Res. Part B* 102 (7): 1553–1561.

105 Pena, B., Martinelli, V., Jeong, M. et al. (2016). Biomimetic polymers for cardiac tissue engineering. *Biomacromolecules* 17 (5): 1593–1601.

106 Pena, B., Bosi, S., Aguado, B.A. et al. (2017). Injectable carbon nanotube-functionalized reverse thermal gel promotes cardiomyocytes survival and maturation. *ACS Appl. Mater. Interfaces* 9 (37): 31645–31656.

Part V

Chemically Modified Carbon Nanotubes: Construction Applications

17

Chemically Modified Carbon Nanotubes in Cement and Concrete Field

Ruby Aslam

Corrosion Research Laboratory, Department of Applied Chemistry, Aligarh Muslim University, Aligarh, 202002 U.P., India

17.1 Introduction

There is currently a lot of interest in the research and production of long-lasting building materials [1]. Because of the breakdown of construction materials, the number of repairs and replacements of structures has increased throughout time. It is estimated that the repair and maintenance of existing structures consume up to 50% of the entire construction industry resources in industrialized countries. Concrete is a common building material in the construction industry because of its low cost and consistent performance [2, 3]. Cement, aggregate, and water make up concrete. The comparatively low manufacturing costs, ease of production at the point of the application using locally available aggregates, high compressive strength, and ease of shaping the form due to its youthful thixotropic qualities are the most essential elements in its widespread use [4]. At the same time, typical concrete has weak mechanical performance, low toughness, easy cracking, and poor durability when compared to other construction materials such as metals and polymers, which affects the service life of the concrete and severely limits material growth. Furthermore, because concrete has limited tensile strength, it works best in conjunction with reinforcement in most circumstances. Because of its porous structure, it allows corrosive ions and gaseous contaminants to diffuse through its mass, corroding the reinforcement. Because concrete has chains of pores that allow access to the outside world, the quality of the concrete portion is harmed, putting the entire structure's stability in jeopardy. Even before an external force is applied, cement-based materials have defects and microcracks in the material and at the interfaces. Excess water, bleeding, plastic settlement, heat and shrinkage strains, stress concentrations imposed by external restrictions, and other factors cause these flaws and micro-cracks. Distributed microcracks propagate, consolidate, and align to become macrocracks when a load is applied. When loads are increased, even more, critical fracture growth occurs at the tips of the macro-cracks, resulting in unstable and catastrophic failure. Cement-based materials are prone to cracking

Chemically Modified Carbon Nanotubes for Commercial Applications, First Edition.
Edited by Jeenat Aslam, Chaudhery Mustansar Hussain, and Ruby Aslam.

under fatigue loads, and cracks provide easy access to harmful substances. Early saturation, freeze–thaw damage and scaling, discoloration, and other issues will result [5].

Researchers in a variety of fields have been focusing on improving the durability of concrete in recent years. To overcome the inadequacies of concrete, several solutions have been used, including the use of various functional fillers such as carbon nanotubes (CNTs), carbon fibers (CFs), steel fiber, graphite nanofibers, graphene, nickel, hybrid material [6], formulation changes, and polymer combinations. Pozzolanic fillers including rice husks, active silica, and metakaolin are commonly used. There is also a high demand for monitoring structural cracking and preventing future fracture propagation. These initiatives are vital for assuring critical structure repair, safety, and long-term durability. Nondestructive evaluations such as attaching or embedding foreign sensors (e.g. resistance strain gauges, optic sensors, piezoelectric ceramic, shape memory alloy, fiber-reinforced polymer bar, and among others) onto or inside structures have been employed in a variety of ways to fulfill requirements. Unfortunately, these sensors have some disadvantages, like poor robustness, inadequate sensitivity, increased cost, low survival rate, and/or incompatibility with certain structures. Ideally, structural materials should be able to sense their surroundings. These self-sensing cement-based materials could be useful for various applications, such as border security and highway traffic monitoring. Cement-based materials can also be used for structural health monitoring and condition evaluation for concrete structures [7].

Electrically conductive fillers are added to self-sensing (piezoresistive) cement-based materials to improve their ability to feel strain, stress, or cracks while keeping good mechanical qualities. Fibrous and particle fillers are two types of electrically conductive fillers. Short CFs, surface modification CFs, steel fibers, carbon-coated nylon fibers, and others are examples of acceptable fibrous fillers, whereas efficient particle fillers include carbon black, steel slag, nickel powder, and others [8]. The contact state between the fillers and the matrix changes as the piezoresistive cement-based materials are deformed or strained, affecting the electrical resistance of the cement-based materials. Electrical resistance measurements can thus be used to detect strain, stress, cracks, and damage.

CNTs are long, slender fullerenes with hexagonal carbon walls with a graphite-like structure and pentagonal rings in the end caps [9]. When compared to other typical fibers, CNTs have the highest overall qualities among carbon materials and offer several specific advantages as a reinforcing element for high strength and performance cementitious composites [10, 11]. CNT has Young's modulus of 1 TPa and tensile strength of 200 GPa, respectively. It can also be utilized as an electromagnetic shielding material because of its outstanding absorption properties. It can improve the durability and electrical conductivity of a product. CNTs have been successfully mixed into cement-making materials with little amounts of additional material. CNTs, when mixed with concrete, are projected to boost compressive strength beyond 200 MPa, allowing for the construction of mile-high skyscrapers. CNTs can prevent the occurrence of cracks, decrease porosity, and increase mechanical qualities, therefore extending cement durability, according to the findings. CNTs must be effectively disseminated within the matrix to allow good bonding between them and the surrounding hydrated cement matrix, which is the major factor determining the concrete microstructure of cement composites, to achieve the greatest results.

The progress of CNT research in the practical use of building materials was summarized in this chapter. The generally used methodologies for dispersion characterization of CNT are addressed first, followed by the functional applications of CNT in cement-based materials.

17.2 CNT Dispersion in Cement-Based Materials: Methodologies

The goal of the cement-CNT experiments is to find the optimal method to disperse CNTs. Various steps such as the use of various surfactants, the formation of covalent functionalized CNTs,,and the use of cement admixtures are studied. The use of various chemical techniques, such as the surface functionalization of CNTs, and the growing of CNTs onto cement particles, can also yield positive results. It was also discovered that the most effective way to disperse CNTs is by mixing them with a mortar paste.

17.3 Improvement of Concrete Properties by Addition of CNTs

In this chapter, we'll look at some of the most recent developments in carbon allotrope concrete/cement reinforcing. The characteristics and idiosyncrasies of CNT dispersion in cement mixtures are discussed.

17.4 Improvement in the Hydration Reaction

Carbon nanomaterials can speed up the pozzolanic reaction of cement, especially when used at an early age. Some carbon nanoparticles are absorbed by cement grains in fresh cement, while the majority are dispersed in pore solutions [12]. There is significant evidence that these nanomaterials, whether in single or agglomerate form, serve as nucleation sites for cement hydrates to develop and proliferate.

The researchers [13] used thermal analysis and X-ray diffraction spectroscopy (XRD) techniques to study the effects of 0.05 wt% GNPs on the hydration properties of cement products. The representative phases in Gross National Product (GNP)-cement composites were calcium hydroxide ($Ca(OH)_2$) and ettringite, according to XRD patterns. They found that the intensity of the typical peaks in the two phases of calcium hydroxide was higher in the GNP-cement composite than in the C3S phase. Moreover, Figure 17.1 exhibiting thermal gravimetric/differential thermogravimetry (TG/DTG) curves of GNP-cement composite indicated that the $Ca(OH)_2$ phase composition was higher. These findings support the idea that the cement made from this type of material has a higher hydration degree.

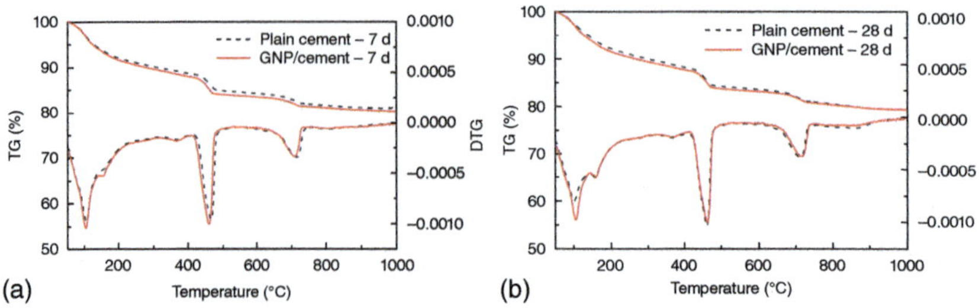

Figure 17.1 TG/DTG curves of plain cement and GNP-cement composite at the age of 28 days. Source: Wang et al. [13]/MDPI/CC BY 4.0.

17.5 Improvement in Mechanical Properties and Relevant Mechanisms

Using carbon nanoparticles as reinforcements to improve the mechanical properties of ordinary Portland cement (OPC) was one of the key goals of developing CN-complete blood counts (CBCs). At remarkably low amounts of carbon nanomaterials (usually 0.01–0.50 wt%), significant improvements in compressive, tensile, flexural, and fracture strengths, as well as the elastic modulus, have been discovered [14, 15].

The addition of CNTs to high-strength concrete significantly decreased the chloride diffusion coefficient by about 8.8–24.0% (Figure 17.2) [16]. When the percentage

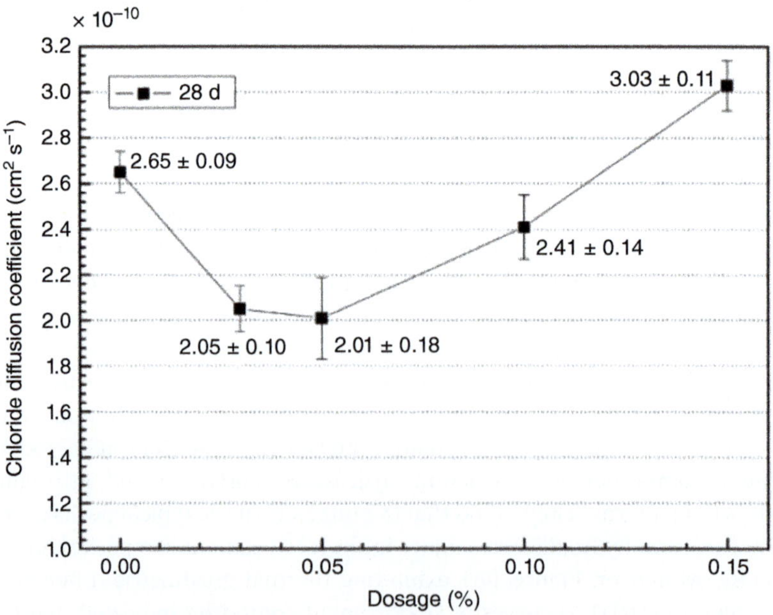

Figure 17.2 The chloride diffusion coefficient of UHSC with different content of MWCNTs. Source: Lu et al. [16]/MDPI/CC BY 4.0.

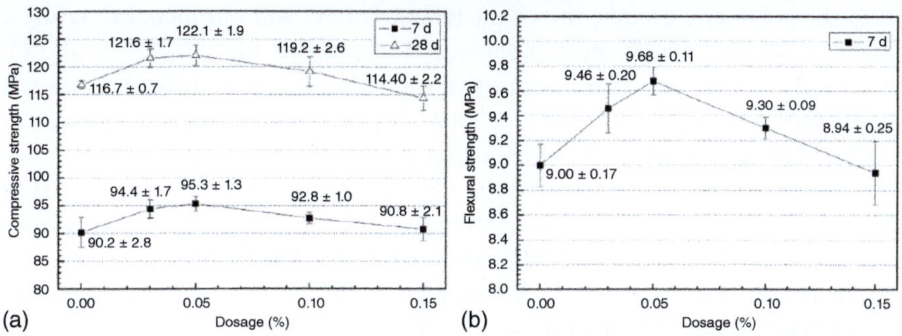

Figure 17.3 Mechanical properties of UHSC containing MWCNTs. (a) Compressive strength and (b) flexural strength. Source: Lu et al. [16]/MDPI/CC BY 4.0.

of these materials reached 0.05%, the cement's strength and compressive properties improved.

The intensities then decreased as the multi-walled carbon nanotube (MWCNT) percentage increased from 0.05% to 0.15% bwoc. When compared to the control ultra-high-performance-concrete (UHSC) sample, specimens containing 0.03% and 0.05% MWCNTs bwoc had a 4.66% and 5.65% increase in compressive strength at 7 days, 4.20% and 4.63% increase in compressive strength at 28 days, and 5.11% and 7.56% increase in flexural strength at 7 days (Figure 17.3). With 0.05% MWCNTs bwoc, the best results were observed. The morphology of the finished cement paste was examined using an field emission scanning electron microscope (FE-SEM) study to identify the process of improvement using MWCNTs.

Microstructural investigations show that MWCNTs operate as bridges across microcracks, ensuring that the tension load is transferred. Furthermore, many MWCNTs were extracted from the matrix without rupturing.

When up to 0.06% of CNTs were introduced to the samples, Qissab and Abbas discovered [17] that the splitting tensile strength improved by 3.1–37.5%. Both long and short CNTs were found in their samples. SEM images revealed CNT breakdown along the cracked surfaces of the samples, according to the researchers. During the fracture, no CNTs were extracted from the cement mixture. In comparison to normal mortar samples, Hunashyal et al. [18] discovered that mortar samples with 0.5% CNTs had a 19% increase in indirect tensile strength. They found that when CNTs samples failed, there was less strain. For example, incorporating 1% steel fibers (with a tensile strength of 2000 MPa) into a $w/c = 0.55$ concrete mix increased the splitting tensile strength by 36.3%. The rise was greater in lower w/c blends, and the samples were more ductile due to fiber inclusion. Another study [19] found that adding 0.5% 50 mm copper-coated crimped round steel fiber with an aspect ratio of 53.85 to concrete with a compressive strength of 25 MPa resulted in a 61.1% improvement in the splitting tensile strength. Shah et al. [20] discovered that adding 1.5% superplasticizer to a 34 MPa mix boosted its splitting tensile strength by more than 120% while maintaining all other ingredients the same.

Daghash [21] used multilayer CNTs in his PCs and looked at their mechanical properties. According to Daghash, if the multilayer nanotubes do not chemically bind to the polymeric groups, the characteristics of PC will not increase. When a hybrid of 0.025 wt% CNTs and 0.025 wt% GO was added to OPC, Lu et al. [22] showed a 21% improvement in compressive strength and a 24% increase in flexural strength. OPC reinforced with only 0.05 wt% CNTs increased compressive strength by 6.4% and flexural strength by 10.1%, whereas OPC reinforced with only 0.05 wt% GO increased compressive strength by 11.0% and flexural strength by 16.2%. After 28 days of age [23], cohesiveness and friction angle of cement-based composites containing 0.018 and 0.036 wt% MWCNTs were measured. The inclusion of 0.036 wt% CNTs increased cohesiveness and friction angle by 24% and 19%, respectively, implying better shearing resistance.

When MWCNTs were introduced to a cement composite material after being treated with an H_2SO_4 and HNO_3 mixed solution, Li et al. [24] discovered a 25% increase in flexural strength. It was discovered that CNTs and cement hydrations (such as C–S–H and calcium hydroxide) have interfacial interactions, resulting in high bonding strength between the reinforcement and cement matrix. Konsta-Gdoutos et al. [25] observed a similar gain in the flexural strength, using a smaller CNTs concentration. Lower concentrations of long MWCNTs (0.025–0.048 wt%) offer effective reinforcement, whereas greater concentrations of short MWCNTs (near to 0.08 wt%) are necessary to produce the same degree of reinforcement. The nanocomposites appear to have a higher amount of high stiffness C–S–H and less nanoporosity than the plain cement matrix. MWCNTs appear to lower the number of tiny holes due to their short sizes (20–40 nm).

Wang et al. [26] investigated the flexural toughness of cement-based composites with highly scattered MWCNTs. A FE-SEM was used to examine the morphology and microstructure of cement-based composites. The fracture energy of Portland cement pastes reinforced with various MWCNT concentrations rose considerably, peaking at 312.16 N m^{-1} for T2. The flexural toughness index of composites increased by up to 57.5% with a 0.08 wt% addition of MWCNTs by weight of cement. The porosity and pore size distribution data (Figure 17.4) data also revealed that the cement paste containing the MWCNT fraction has a more uniform and lower porosity. The total porosity of sample T4 is approximately 20.92% lower than sample PO.

The structure of the matrix reinforced with MWCNTs allows them to form bridges across voids and cracks. They also have a tube that can be pushed out of the matrix without failure as can be seen by the morphological structure of the samples (Figure 17.5).

Musso et al. [27] and Chaipanich et al. [28] found that increasing the CNT concentration (by 5–10 times) can boost compressive strength by about 20%. Cwirzen et al. [29] found that, despite a nearly 50% increase in compressive strength, the cement paste's flexural strength improved by just 10%. Lelusz [30] discovered that as the CNT dose was increased, the compressive strength of cement composites dropped. Manzur et al. [31] noted that the dosage rates of the MWCNTs varied by the weight of the cement used in the study. The study focused on developing

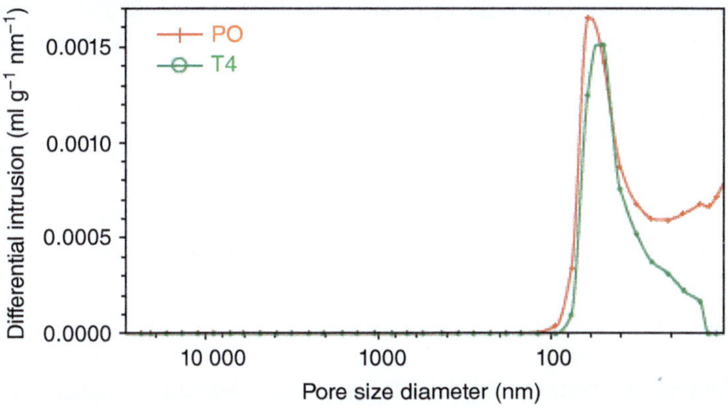

Figure 17.4 MIP analysis of pore size distribution. Source: Wang et al. [26]/with permission from Elsevier.

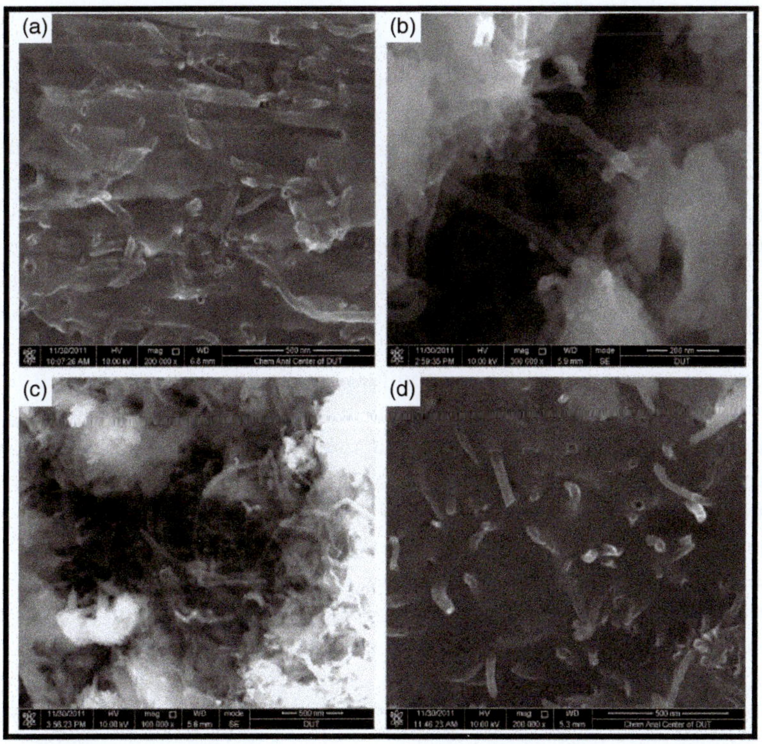

Figure 17.5 Typical FE-SEM images of samples after 28 days of curing: (a and b) 0.08 wt% MWCNTs of T2 and (c and d) 0.10 wt% MWCNTs of T3. Source: Wang et al. [26]/with permission of Elsevier.

surface-treated cementitious materials that are based on their flexibility and strength. Aside from being more durable, CNT reinforcement could also improve the performance of these materials.

Campillo et al. [32] tested a cementitious compound including CNTs under compression and discovered that at 6% and 30%, respectively, SWCNTs (single-walled carbon nanotubes) and MWCNTs enhanced the compression resistance in comparison to pure cement. In the study [33], the authors investigated the efficiency of CNTs incorporated for reinforcement of cementitious composites that were uniformly dispersed and randomly oriented.

The addition of 0.40 wt% CNTs (CN4) to cement mortar produced the best results when compared to the reference mixture, with a 40% increase in compressive strength, a 30% increase in tensile strength, 15% increase in flexural strength, and a denser microstructure [34].

17.6 Enhanced Durability

Because of their refined pore shape and reduced pore connectivity, CN-CBCs are more durable because they can prevent the movement of hostile ions or hazardous substances. Because the gain in durability reveals itself in strength qualities or coefficients, it extends the service life of concrete structures that are subjected to a variety of harsh situations. In a study, Li et al. [24] reported that adding 0.50 wt% MWCNTs to an OPC paste, they were able to reduce the mean radius of the pores and total porosity by 43.2% and 64%, respectively. The addition of MWCNTs also reduced pores with an equivalent radius of larger than 50 nm by 82% and had an added filling effect.

The macroscopic properties of CNTs modified reactive powder concrete (RPC) composites are significantly related to their pore structure characteristics, although the structural nature of RPC composites and the effect of CNTs on their pore structure are unclear [35]. Both the mercury intrusion porosimetry (MIP) and low-field nuclear magnetic resonance (LF-NMR) technologies were utilized to assess the pore structure features of RPC composites with surface-modified CNTs to fully understand and manage their properties. Many nano-scale pores (primarily interlayer and gel pores) with a radius range of 0.3–20 nm were discovered in RPC composites, according to experimental results. The pore content and size of RPC composites are reduced by all types of surface-modified CNTs used in this study, changing the pore structure of RPC. Varied CNTs, on the other hand, have different effects on pore (particularly nano-scale pore) structural properties. The hydroxyl functionalized and nickel-coated CNTs produce structural contraction of the C–S–H gel, whilst the carboxyl functionalized CNTs cause the alteration of some nanoscale pores within the gel. Surface-modified CNTs have mostly nano- and micrometer-scale effects on the pore structure of RPC composites. CNTs can induce pore water conversion inside C–S–H layers, as well as raise the matrix compactness of RPC composites and improve the network structure of hydration products.

Han et al. [36] demonstrated that adding 0.2% MWCNTs to reinforced cement-based composites can reduce the water sorptivity coefficient, water permeability, and gas permeability coefficient of cement mortar. These properties could improve the durability of cement-based structural materials. In a study, Manzur et al. [37] investigated the properties of MWCNTs in mortar. They found that the smallest size of the CNTs with an OD (outer diameter) of 20 nm or less had the highest compressive strength.

17.7 Improvements in Electrical and Thermal Conductivity

CNTs can greatly increase the electrical conductivity of cement-based materials. The high mobility of electrons and phonons on the surfaces of some carbon nanomaterials is linked to their remarkable conductivities, implying that both electronic charges and lattice vibrations can flow exceedingly quickly.

Jang et al. [38] discovered that increasing the volume proportion of moisture (by cement volume) by 39% boosted the electrical conductivity of CNT-cement by nearly 300%. In another report [39], the electrical conductivity of cementitious plain mortars was reported to be increased linearly with the increasing chloride content. On the other hand, the electrical conductivity of enhanced cementitious mortars increased significantly with the increasing weight content. The increasing chloride content of plain mortars resulted in a linear increase in the electrical conductivity of the mortar at a maximum rate of 0.12 $mS\,m^{-1}$ per weight content. On the other side, the electrical conductivity of CNTs enhanced mortar increased quadratically at maximum rates of 0.34 and 2.37 $mS\,m^{-1}$ per weight content. Qin et al. [40] discovered that as the temperature rose from 20 to 50 °C, the electrical resistance of cement containing 0.40% CNFs gradually decreased. However, as the temperature increased to 80 °C, the resistance increased again, and this variation was reversible during the cooling process. The presence of CNFs and CNTs in cement pastes caused the electrical resistance of the materials to gradually decrease as the temperature increased from 20 to 80 °C. The presence of 0.40 wt% CNFs and 1.5 wt% CNTs in the materials caused the temperature sensitivity of the composite materials to increase four times that of composites containing only 0.40 wt% CNFs.

MWCNTs were introduced to cement mortar in four different concentrations by the mass of cement: 0.00, 0.01, 0.015, and 0.02 wt% [41]. At 28 and 90 days after curing, resistivity tests in mortars with 0.01 and 0.015 wt% MWCNT loading showed up to a 10% reduction (Figure 17.6). Calculations of activation energy demonstrate that these claims are correct, implying that 0.01 wt% MWCNT looks to be the most effective loading technique for achieving specific conductivity improvement in cement mortar.

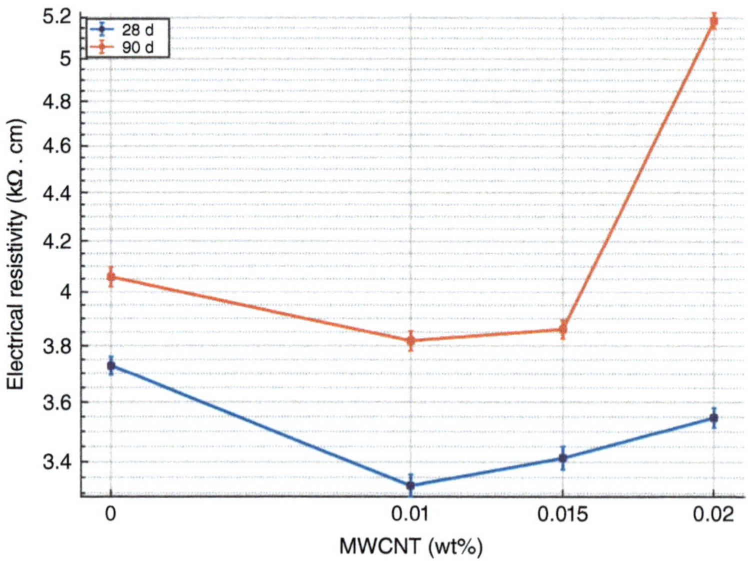

Figure 17.6 Average electrical surface resistivity of MWCNTs-mortar samples. Source: Cerro-Prada et al. [41]/MDPI/CC BY 4.0.

17.8 Improvements in Corrosion Resistance Properties of Cement/Concrete

Concrete's alkalinity (pH) is reduced when it is exposed to carbon dioxide (carbonation): the pH of the concrete, which previously ranged between 12.5 and 13.9, declines to values below 9.0 due to the hydrated lime $Ca(OH)_2$. As a result, as the embedded steel transitions from a passive to an active state, it corrodes rapidly [42]. Steel reinforcement corrosion products, due to their larger volume relative to steel, cause internal tensile strains in the concrete, resulting in the formation of internal cracks that eventually spread to cover concrete spalling while weakening the reinforcement/concrete connection.

CNTs increased the corrosion intensity of cement mortars containing up to 0.5% CNT when exposed to simulated seawater and hastened carbonation [43]. Konsta-Gdoutos et al. [44] found that adding 0.1 wt% CNTs to a steel alloy reduced corrosion rates and greatly boosted corrosion resistance by delaying the commencement of the corrosion response (initiation period). The scientists went on to investigate the impact of high CNT content (0.5 wt%) on mortar corrosion behavior, finding that reinforced mortars with higher CNT content had a higher corrosion rate than mortars with 0.1 wt% CNTs. As per the researchers, the observed delay in the commencement of corrosion was caused by a little reduction in permeability, which resulted in a reduction in aggressive agent infiltration. Consequently, it was determined that this enhanced corrosion propensity had to be investigated thoroughly.

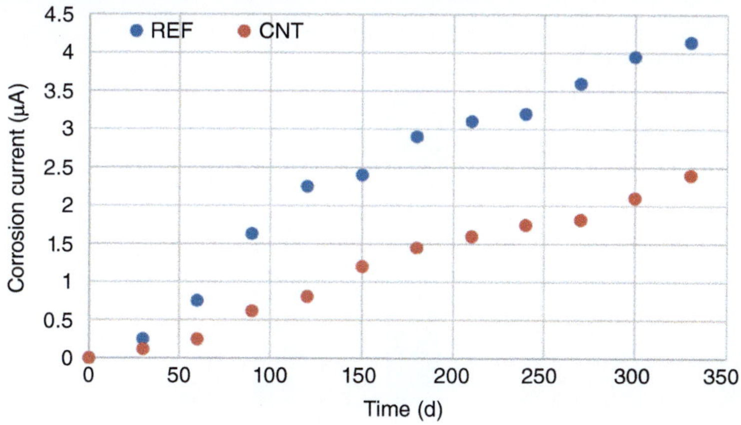

Figure 17.7 Corrosion current for the two mortar groups against time of exposure in NaCl. Source: Chousidis et al. [42]/Scientific Research Publishing Inc/CC BY 4.0.

The parameters of common strength mortars with and without MWCNTs were reported and compared [42]. The change in the density of test specimens and the porosity is responsible for the rise in flexural strength as a function of time spent in a corrosive environment. Both types of reinforced mortar specimens were corroded when exposed to Cl^- ions, according to corrosion current measurements (Figure 17.7).

Additionally, compared to the CNT combination, the reference (REF) group exhibited the greatest corrosion current values over the course of the 11-month measurement period. Figure 17.8 shows the average mass loss of three steel reinforcing bars following six and nine months of exposure to chloride salt, as measured by gravimetric and electrochemical methods. The REF specimens are more corroded than the CNT specimens, as can be shown from gravimetric/electrochemical measurements. The mass loss of the REF specimens was 50% more than the mass loss of the CNT specimens after a 6-month period in a corrosive environment. At the same time, the mass loss of the reinforcement in the REF group was 28% higher than in the CNT group after nine months of NaCl exposure.

The effect of introducing various concentrations of MWCNTs with diameters of 8–22 nm by the mass of cement on corrosion rates and pullout forces in concrete cubes [45]. Compressive strength increased by 7–20% when 0.01–0.03% CNTs were added, and tensile strength increased by 10–20% when 0.01–0.03% CNTs were added. When compared to a matching concrete specimen, CNTs concrete exhibits a higher binding strength. Furthermore, adding CNTs improves the bond strength of steel reinforcement bars with diameters of 12 and 16 mm by 36% and 21%, respectively. Steel reinforcement with a diameter of 16 mm is less corrosion resistant than steel reinforcement with a diameter of 12 mm (Figure 17.9).

The properties of the MWCNT cement mortar, such as its conductivity, strength, chloride migration coefficient, and anti-corrosive behavior were studied [46]. The results indicated that the steel corrosion intensity increased as the CNT

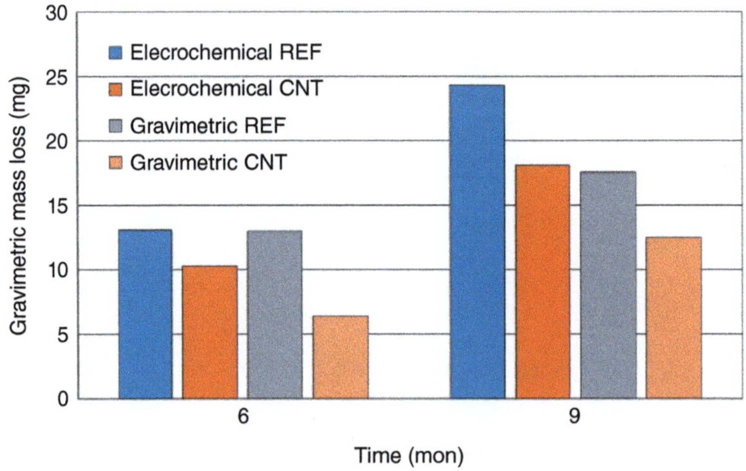

Figure 17.8 Gravimetric and electrochemical mass loss (mg) vs. time, of the steel reinforcing bars embedded in mortars and partially immersed in NaCl for nine months. Source: Chousidis et al. [42]/Scientific Research Publishing Inc/CC BY 4.0.

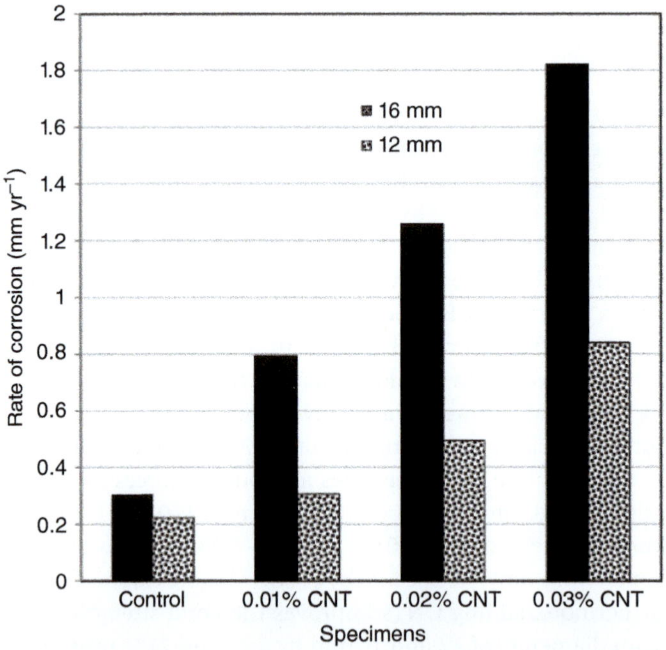

Figure 17.9 Effect of adding CNTs to corrosion rates of steel in concrete. Source: Hassan et al. [45]/Springer Nature/CC BY 4.0.

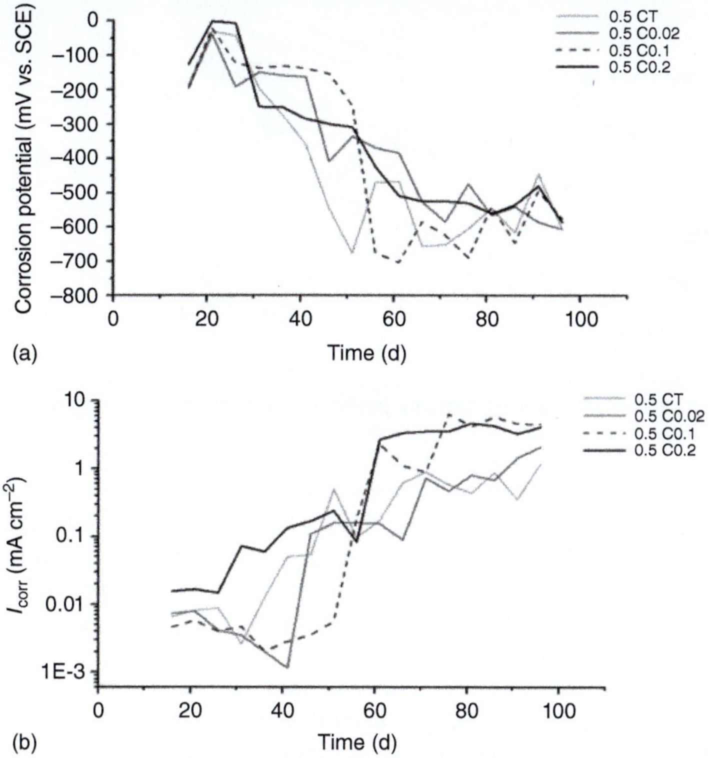

Figure 17.10 Changes in (a) corrosion potential (E_{corr}) and (b) corrosion rate (I_{corr}) of steel rebars buried in MWCNTs cement mortar under chloride erosion environment. Source: Li et al. [46]/De Gruyter/CC BY 4.0.

concentration rose. The effects of accelerated and natural chloride conditions on the mortar's corrosion behavior were also studied. The depassivation and chloride penetration of steel surfaces resulted in a significant reduction in the corrosion potential of the material as shown in Figure 17.10a. However, once the passive film of steel was removed, the corrosion potential did not change. The corrosion rate rose as the corrosion potential dropped, as seen in Figure 17.10b. The addition of a larger amount of MWCNTs to cement mortar resulted in a higher final corrosion rate of steel.

17.9 Potential Structural Applications of CNTs Reinforced Cement-Based Materials

The goal of developing intelligent and multifunctional CNT-based materials is to create new multipurpose and intelligent structures that can be used in various applications. Unfortunately, the piezoresistivity of these materials is not studied yet, which limits the development of these structures.

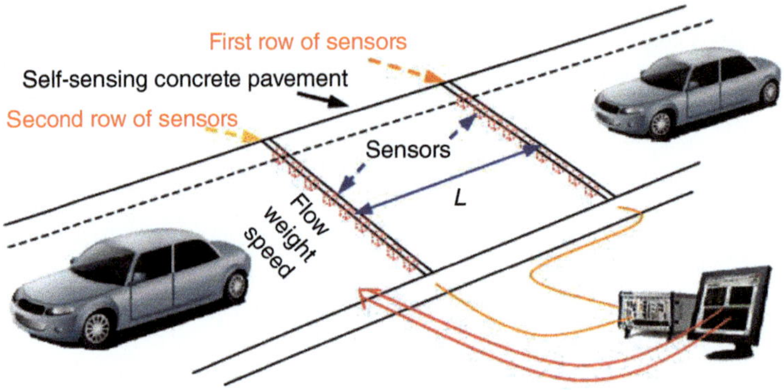

Figure 17.11 Illustration of self-sensing concrete pavement for traffic flow detection. Source: Han et al. [48]/with permission of IOP Publishing.

Veedu [47] described two types of multifunctional and smart CNTs reinforced cement-based materials that may be used in different ways. *In-situ* monitoring of structural health and integrity, as well as the weight, speed, and location of traffic passing over the bridge, is possible with these materials. Another potential use for nanocomposites is the construction of smart roadways that can measure traffic's location, weight, and speed. The researchers [48] built the self-sensing pavement using MWCNTs reinforced cement-based materials that have piezoelectric properties. The researchers tested the self-sensing pavement's ability to detect the presence of vehicles in traffic by performing vehicle loading experiments (Figure 17.11).

The researchers noted that the automotive loads could affect the electrical resistance of piezoresistive MWCNTs reinforced cement-based materials in the self-sensing pavement, which could allow them to monitor traffic flow as well as identify the distinct vehicular loading. Their findings suggest that these materials could be used for various traffic monitoring applications such as vehicle detection, weighing, and speed measurement.

Using piezoresistive SWCNTs reinforced concrete materials, Saafi et al. [49] created *in situ* wireless and embedded sensors that can monitor the damage caused by concrete constructions. These devices were placed inside the concrete beam to detect cracks and buildup. The constructed self-sensing concrete constructions may wirelessly detect crack propagation and damage buildup during loading by monitoring changes in the electrical resistance of the embedded sensors.

The findings of the study show that multifunctional and smart CNT-reinforced cement-based materials and structures can be used to monitor the structural integrity of concrete structures. To check the stability of concrete constructions, the use of cement-based materials and structures was recommended.

17.10 Challenges

CNTs exhibit various properties such as enhanced nucleation and pore structure improvement and nanoscale fracture management. It could pave the way for the development of new materials that can be used in the fabrication of multifunctional cement-based materials. However, there are still many obstacles that need to be overcome before these new materials can be used in concrete constructions.

(i) Because CNTs have a strong tendency to clump together, achieving a uniform CNT dispersion is extremely difficult. To obtain higher well-dispersed CNTs, there is also a scalability challenge. This dispersion can be achieved through a variety of means, including physical and chemical ones. However, because there is no single standard or specification for CNT dispersion, the application of CNT in construction materials is limited. For CNT to give a reference for developing a single dispersion standard, it is critical to summarize the drawbacks of various dispersion methods. The current research focuses mostly on macroscopic performance, with most of the content centered in the laboratory. In terms of practical applicability, related research is still in its infancy. There is a pressing need for more in-depth research, both theoretically and practically.

(ii) Numerical simulations and experiments should be used to develop a quantitative model that can explain the properties of nanocomposites and their potential applications in the future.

(iii) The piezoresistivity of CNT-reinforced cement-based materials under uniaxial compressive or tensile loading has been the focus of the previous study. Cement-based materials are subjected to complicated external stresses in real structures, such as bending, shear, and torsion. As a result, for structural applications, the overall piezoresistive properties of the nanocomposites should be established.

(iv) Current CNT costs are high enough that the manufacturing of substantial structures is cost-prohibitive even at very low rates of addition. Despite significant price reductions, CNTs are most likely to be utilized in specialized applications. CNTs reinforced cement-based materials can be used in important places of significant engineering constructions, depending on the application and design requirements. Aside from being used as structural components, sandwich, embedded, and coated structures can also reduce the consumption of CNTs in concrete constructions.

17.11 Conclusions and Future Scope

Because CNTs can operate as effective bridges to decrease and limit the propagation of microcracks across the matrix, adding them to cement-forming nanocomposites improves the mechanical properties of concrete significantly. Low density, higher compressive strength, decreased thermal conductivity, smaller average pore width, and more homogenous pore wall structure result from even adding 0.05% CNTs to cement. CNTs must be effectively disseminated within the matrix to allow good bonding between them and the surrounding hydrated cement matrix, which is the major factor determining the concrete microstructure of cement composites, to achieve the greatest results. The goal of each cement-CNT experiment is to find the optimal method for dispersing CNTs. Ultrasonication, the use of surfactants, the use of cement admixtures, and covalent functionalization can all help achieve this. The use of a combination of chemical approaches, such as surface functionalization with polymers, and other techniques, such as a new method of synthesizing cementitious nanocomposites by growing CNTs onto cement particles, also yields positive results. The importance of diameter, length, and the length-to-diameter ratio of CNTs, their concentration, functionalization, annealing, a combination of CNTs with other nanomaterials, and the water–cement ratio all have been discovered when it comes to dispersion of carbon allotropes in concrete.

References

1 Lea, F.M. (1970). *The Chemistry of Cement and Concrete*, 3e. London: Edward Arnold Ltd.

2 Taylor, H.F.W. (1998). *Cement Chemistry*, 2e. London: Thomas Telford Publication.

3 Balaguru, P. and Chong, K. (2006). Nanotechnology and concrete: research opportunities. In: *Proceedings of the ACI Session on Nanotechnology of Concrete: Recent Developments and Future Perspectives* (ed. K.S.S.P. Shah), 15–28. Farmington Hills, MI: American Concrete Institute.

4 Sheikh, T.M., Anwar, M.P., Muthoosamy, K. et al. (2021). The mechanics of carbon-based nanomaterials as cement reinforcement—a critical review. *Constr. Build. Mater.* 303: 124441.

5 Zheng, Y., Liu, J., Guo, P., and Gan, C. (2021). Fatigue characteristics of double damage reinforced prestressed hollow slab beams under freeze-thaw cycle erosion. *Appl. Sci.* 11: 7692.

6 Cui, K., Chang, J., Feo, L. et al. (2022). Developments and applications of carbon nanotube reinforced cement-based composites as functional building materials. *Front. Mater.* 9: 146.

7 Bekzhanova, Z., Memon, S.A., and Kim, J.R. (2021). Self-sensing cementitious composites: review and perspective. *Nanomaterials* 11 (9): 2355.

8 Zhang, H., Zhang, X., Fang, Z. et al. (2020). Recent advances in preparation, mechanisms, and applications of thermally conductive polymer composites: a review. *J. Compos. Sci.* 4 (4): 180.

9 Iijima, S. (2002). Carbon nanotubes: past, present, and future. *Physica B* 323 (1–4): 1–5.

10 Xing, W., Yang, W., Yang, W. et al. (2016). Functionalized carbon nanotubes with phosphorus-and nitrogen-containing agents: effective reinforcer for thermal, mechanical, and flame-retardant properties of polystyrene nanocomposites. *ACS Appl. Mater. Interfaces* 8 (39): 26266–26274.

11 Xu, S., Liu, J., and Li, Q. (2015). Mechanical properties and microstructure of multi-walled carbon nanotube-reinforced cement paste. *Constr. Build. Mater.* 76: 16–23.

12 Chen, S.J., Wang, W., Sagoe-Crentsil, K. et al. (2016). Distribution of carbon nanotubes in fresh ordinary Portland cement pastes: understanding from a two-phase perspective. *RSC Adv.* 6 (7): 5745–5753.

13 Wang, B., Jiang, R., and Wu, Z. (2016). Investigation of the mechanical properties and microstructure of graphene nanoplatelet-cement composite. *Nanomaterials* 6 (11): 200.

14 Shamsaei, E., de Souza, F.B., Yao, X. et al. (2018). Graphene-based nanosheets for stronger and more durable concrete: a review. *Constr. Build. Mater.* 183: 642–660.

15 Xu, Y., Zeng, J., Chen, W. et al. (2018). A holistic review of cement composites reinforced with graphene oxide. *Constr. Build. Mater.* 171: 291–302.

16 Lu, L., Ouyang, D., and Xu, W. (2016). Mechanical properties and durability of ultra high strength concrete incorporating multi-walled carbon nanotubes. *Materials* 9 (6): 419.

17 Qissab, M.A. and Abbas, S.T. (2018). Behaviour of reinforced concrete beams with multiwall carbon nanotubes under monotonic loading. *Eur. J. Environ. Civil Eng.* 22 (9): 1111–1130.

18 Hunashyal, A.M., Tippa, S.V., Quadri, S.S., and Banapurmath, N.R. (2011). Experimental investigation on effect of carbon nanotubes and carbon fibres on the behavior of plain cement mortar composite round bars under direct tension. *ISRN Nanotechnol.* https://doi.org/10.5402/2011/856849.

19 Vairagade, V.S. and Kene, K.S. (2013). Strength of normal concrete using metallic and synthetic fibers. *Procedia Eng.* 51: 132–140.

20 Shah, S.N.R., Aslam, M., Shah, S.A., and Oad, R. (2014). Behaviour of normal concrete using superplasticizer under different curing regimes. *Pak. J. Eng. Appl. Sci.* 15: 87–94.

21 Daghash, S.M., Tarefder, R., and Taha, M.M.R. (2015). A new class of carbon nanotube: polymer concrete with improved fatigue strength. In: *Nanotechnology in Construction*, 285–290. Cham: Springer.

22 Lu, Z., Hou, D., Meng, L. et al. (2015). Mechanism of cement paste reinforced by graphene oxide/carbon nanotubes composites with enhanced mechanical properties. *RSC Advances* 5 (122): 100598–100605.

23 Du, M.R., Chen, S.J., Duan, W.H. et al. (2018). Role of multiwalled carbon nanotubes as shear reinforcing nanopins in quasi-brittle matrices. *ACS Appl. Nano Mater.* 1 (4): 1731–1740.

24 Li, G.Y., Wang, P.M., and Zhao, X. (2005). Mechanical behavior and microstructure of cement composites incorporating surface-treated multi-walled carbon nanotubes. *Carbon* 43 (6): 1239–1245.

25 Konsta-Gdoutos, M.S., Metaxa, Z.S., and Shah, S.P. (2010). Multi-scale mechanical and fracture characteristics and early-age strain capacity of high performance carbon nanotube/cement nanocomposites. *Cem. Concr. Compos.* 32 (2): 110–115.

26 Wang, B., Han, Y., and Liu, S. (2013). Effect of highly dispersed carbon nanotubes on the flexural toughness of cement-based composites. *Constr. Build. Mater.* 46: 8–12.

27 Musso, S., Tulliani, J.M., Ferro, G., and Tagliaferro, A. (2009). Influence of carbon nanotubes structure on the mechanical behavior of cement composites. *Compos. Sci. Technol.* 69 (11, 12): 1985–1990.

28 Chaipanich, A., Nochaiya, T., Wongkeo, W., and Torkittikul, P. (2010). Compressive strength and microstructure of carbon nanotubes–fly ash cement composites. *Mater. Sci. Eng., A* 527 (4, 5): 1063–1067.

29 Cwirzen, A., Habermehl-Cwirzen, K., and Penttala, V. (2008). Surface decoration of carbon nanotubes and mechanical properties of cement/carbon nanotube composites. *Adv. Cem. Res.* 20 (2): 65–73.

30 Lelusz, M. (2015). Carbon nanotubes influence on the compressive strength of cement composites. *Czasopismo Techniczne* 1–11.

31 Manzur, T. and Yazdani, N. (2015). Optimum mix ratio for carbon nanotubes in cement mortar. *KSCE J. Civ. Eng.* 19 (5): 1405–1412.

32 Campillo, I., Guerrero, A., Dolado, J.S. et al. (2007). Improvement of initial mechanical strength by nanoalumina in belite cements. *Mater. Lett.* 61 (8, 9): 1889–1892.

33 Chan, L.Y. and Andrawes, B. (2010). Finite element analysis of carbon nanotube/cement composite with degraded bond strength. *Comput. Mater. Sci.* 47 (4): 994–1004.

34 Evangelista, A.C.J., de Morais, J.F., Tam, V. et al. (2019). Evaluation of carbon nanotube incorporation in cementitious composite materials. *Materials* 12 (9): 1504.

35 Wang, J., Dong, S., Dai Pang, S. et al. (2022). Pore structure characteristics of concrete composites with surface-modified carbon nanotubes. *Cem. Concr. Compos.* 128: 104453.

36 Han, B., Yang, Z., Shi, X., and Yu, X. (2013). Transport properties of carbon-nanotube/cement composites. *J. Mater. Eng. Perform.* 22 (1): 184–189.

37 Manzur, T., Yazdani, N., and Emon, M.A.B. (2014). Effect of carbon nanotube size on compressive strengths of nanotube reinforced cementitious composites. *J. Mater.* 2014 (1): 1–8.

38 Jang, S.H., Hochstein, D.P., Kawashima, S., and Yin, H. (2017). Experiments and micromechanical modeling of electrical conductivity of carbon nanotube/cement composites with moisture. *Cem. Concr. Compos.* 77: 49–59.

39 Kim, H.K. (2015). Chloride penetration monitoring in reinforced concrete structure using carbon nanotube/cement composite. *Constr. Build. Mater.* 96: 29–36.

40 Qin, J.J., Yao, W., and Zuo, J.Q. (2013). Temperature sensitive properties of hybrid carbon nanotube/carbon fiber cement-based materials. In: *Key Engineering Materials*, vol. 539 (ed. W. Yao), 89–93. Trans Tech Publications Ltd.

41 Cerro-Prada, E., Pacheco-Torres, R., and Varela, F. (2020). Effect of multi-walled carbon nanotubes on strength and electrical properties of cement mortar. *Materials* 14 (1): 79.

42 Chousidis, N., Zacharopoulou, A., Zeris, C., and Batis, G. (2022). Corrosion resistance and physical-mechanical properties of reinforced mortars with and without carbon nanotubes. *J. Mater. Sci. Chem. Eng.* 10 (1): 1–23.

43 del Carmen Camacho, M., Galao, O., Baeza, F.J. et al. (2014). Mechanical properties and durability of CNT cement composites. *Materials* 7 (3): 1640–1651.

44 Konsta-Gdoutos, M.S., Batis, G., Danoglidis, P.A. et al. (2017). Effect of CNT and CNF loading and count on the corrosion resistance, conductivity and mechanical properties of nanomodified OPC mortars. *Constr. Build. Mater.* 147: 48–57.

45 Hassan, A., Elkady, H., and Shaaban, I.G. (2019). Effect of adding carbon nanotubes on corrosion rates and steel-concrete bond. *Sci. Rep.* 9 (1): 1–12.

46 Li, W., Liu, Y., Jiang, Z. et al. (2020). Chloride-induced corrosion behavior of reinforced cement mortar with MWCNTs. *Sci. Eng. Compos. Mater.* 27 (1): 281–289.

47 Veedu, V.P. (2010). Multifunctional cementitious nanocomposite material and methods of making the same. US Patent 7666327 B1, filed 22 May 2008.

48 Han, B., Yu, X., and Kwon, E. (2009). A self-sensing carbon nanotube/cement composite for traffic monitoring. *Nanotechnology* 20 (44): 445501.

49 Saafi, M. (2009). Wireless and embedded carbon nanotube networks for damage detection in concrete structures. *Nanotechnology* 20 (39): 395502.

Part VI

Chemically Modified Carbon Nanotubes: Emerging Applications

18

Chemically Modified Carbon Nanotubes in 3D and 4D Printing

Krishna K. Yadav, Arushi Arora, Sushma Jangra, and Menaka Jha

Institute of Nano Science and Technology, Knowledge City, Sector 81, Mohali, Punjab, 140306, India

18.1 Introduction

Science and technology are on their wheels, moving forward nonstop and developing novel technologies to improve quality of life. 3D printing, also known as additive manufacturing, has sparked a lot of interest among these. It is an advanced technology that enables the production of limitless structures and shapes that are almost impossible with traditional technologies [1]. 3D printing is achieved by layer-by-layer deposition therefore known as additive manufacturing. The first step of 3D printing constitutes shredding or slicing of raw materials into layers for the required printable command sequence. In the next step, the file is transferred to the printer. This step deposits materials in X–Y planes according to the locations predefined in the source file. This process is recurred on Z-axis to build the predesigned 3D structure as per the system file [2]. 3D printing stands opposite the subtractive manufacturing process where the structures are carved based on material removal from solid blocks of materials by the processes of drilling, broaching, milling, and sawing [3]. Subtractive manufacturing produces 3D structures from raw materials such as rocks, metals, or wood, whereas additive manufacturing produces complex structural designs that include both hollow and solid pieces [4]. The latest advancement in 3D printing is the technique that can modify shapes after being removed from 3D printer [5]. Thus, a new era of 4D printing has started following the advancement. The main concept behind 4D printing is to design "stimuli-responsive 3D" materials [6]. To put it another way, 4D printing occurs when time is added to the 3D printing process. 4D printing allows for the creation of dynamically smart, stimuli-responsive, and self-healing materials, paving the way for large-scale smart device companies. 4D printing of smart materials and multi-material constructions now offers a number of advantages over 3D printing, which was previously available due to its rapid expansion. The 4D printed structures will be fully preprogrammed and offer time-dependent products [7]. Materials that are particularly flexible, extendable, and distorted can be used in the 4D printing

Chemically Modified Carbon Nanotubes for Commercial Applications, First Edition.
Edited by Jeenat Aslam, Chaudhery Mustansar Hussain, and Ruby Aslam.

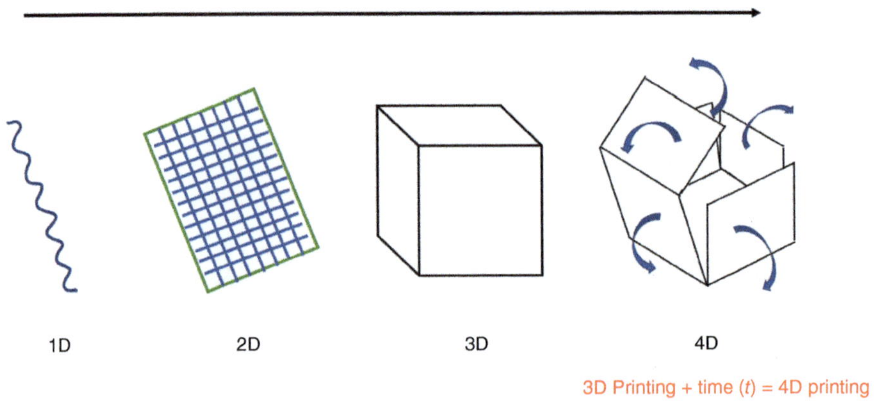

Figure 18.1 shows a diagram showing 1D, 2D, 3D, and 4D printing.

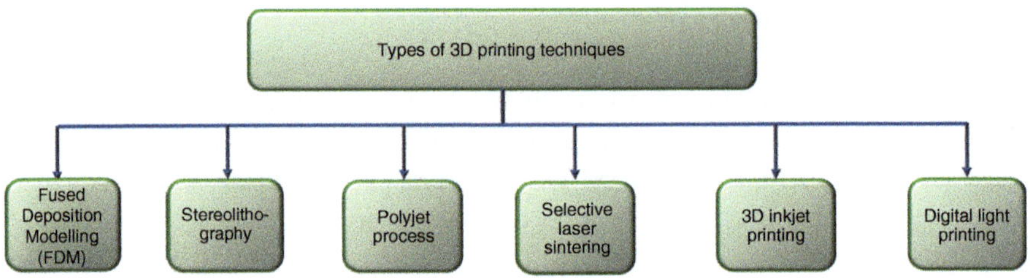

Figure 18.2 Types of 3D printing techniques.

process. Soft robotics, printed actuators in medical devices, electronics, smart textiles, and aerospace are just a few of the fields where 4D printing will make a significant difference in the future. Figure 18.1 depicts a schematic example of 1D, 2D, 3D, and 4D printing principles, with 1D representing a curve line, 2D representing a plane, and 3D representing a cubic structure. The 3D structure has changed over time in 4D, and the arrow points represent the direction of shape transition. The main fabrication processes of 3D printing are prevalent in the current scenario as given in Figure 18.2 and discussed briefly.

(a) *Fused deposition modeling (FDM)*: To make a 3D structure, thermoplastic polymers are liquefied and in hot molten and mushy conditions they are deposited by an extrusion head that follows a specific pathway predefined in software. FDM follows the principle of "additive manufacturing" by laying down material layer-by-layer. The nozzle from which the hot molten material comes out can move in both horizontal and vertical directions by the predefined controlled mechanism of computer-aided design (CAD) software mechanisms. A representative FDM image has been represented in Figure 18.3 [8].
(b) *Stereolithography*: A computer-controlled laser beam is used in this technique to create the patterns that are predefined by CAD software. Usually, ultraviolet (UV) light is aimed onto the surface of a thermoset resin. After imaging a

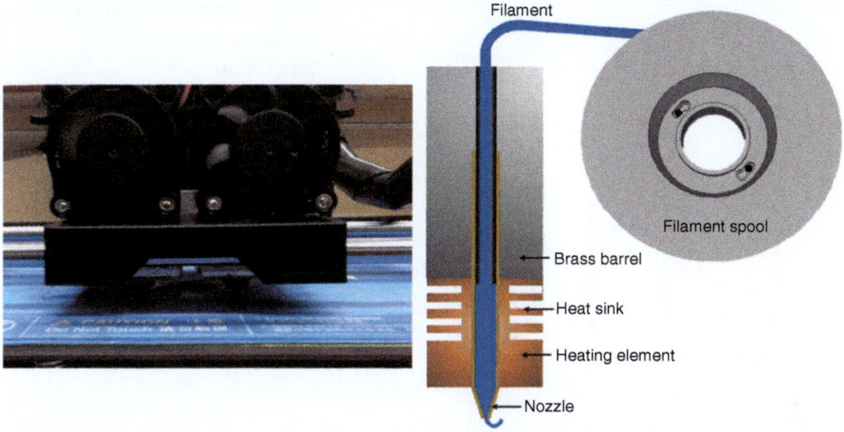

Figure 18.3 A schematic of the extruder and a photo of a twin extruder assembly from an FDM printer. Source: Acquah et al. [8]/IntechOpen/CC BY 4.0.

Figure 18.4 A typical view of stereolithography 3D printing technique of Formlabs. Source: Formlabs.

layer, the platform is shifted downwards, and recoating bar moves to apply the subsequent resin layer. After forming desired patterns, the lithographed resin is solvent treated to remove extra structures and cleaned thoroughly. Curing is also done to fully solidify the formed pattern [9]. Figure 18.4 represents the typical view of the Stereolithography 3D printing technique.

(c) *Polyjet process*: This is an excellent 3D printing technique that works the same as an inkjet printer. Droplets of photopolymer are extruded from the jet that solidifies when exposed to UV light. After curing the previous layer, the next

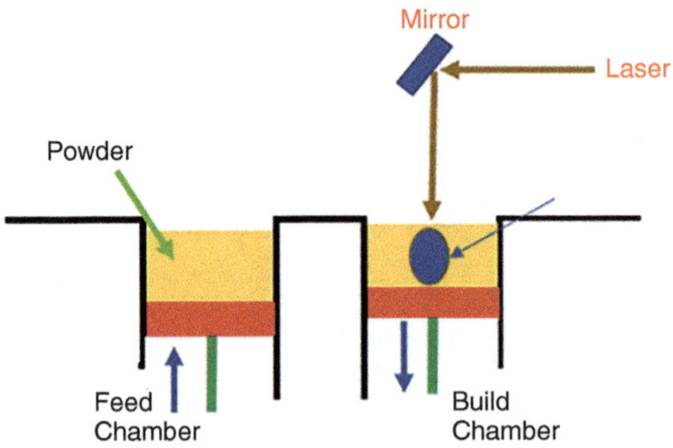

Figure 18.5 A schematic representation of SLA's operation.

layer is deposited followed by UV hardening and hence the layered deposition process continues.

(d) *Selective laser sintering*: In this type of additive manufacturing, a powered laser is used to sinter smaller particles of polymer powder into a solid 3D structure. The polymer powder is dispersed in a thin layer and heated to just below or equal to the constituent materials' melting temperature. The laser is then used to boost the temperature of certain portions of the material, thus fusing it into a solid layer. The process repeats for subsequent layers and cooling and post-processing is done and a representative image has been shown in Figure 18.5.

(e) *3D inkjet printing*: This additive manufacturing process operates at a low temperature and pressure and utilizes 3D printing of liquid materials and solid suspensions. Conductive nanoparticles or nanotubes such as carbon nanotubes (CNTs) can be deposited by this process. The printing suspension is extruded through a nozzle in a print head, which is scanned in a raster pattern over a surface forming multiple layers, and thereby forming a 3D structure.

(f) *Digital light printing*: Except for the photocuring process, this procedure is nearly identical to stereolithography. Here, light is exposed, to a vat of liquid polymer under safe conditions of light, and the image of the 3D model is projected onto the liquid polymer. The exposed part of the liquid polymer hardens followed by the downward movement of the build plate. Once again light is exposed to a new polymer plate and hence the process continues until a 3D model is formed (Figure 18.6).

3D–4D printed technologies are rapidly replacing traditional manufacturing processes. This innovative process has invaded its roots in almost all the fields of material science. Commercially, CNT filaments are procured from 3DXTech. Nanoacyl's PLASTICYL™ provides thermoplastic concentrates of CNTs to be exploited in applications requiring good electrical and mechanical properties. Commercially available 3D–4D printable CNT reinforced materials have also been made by companies such as Functionalize F-Electric, Filabot, and Cheap Tubes Inc. 3D printed CNTs were used to make high-performance concrete that can be used for

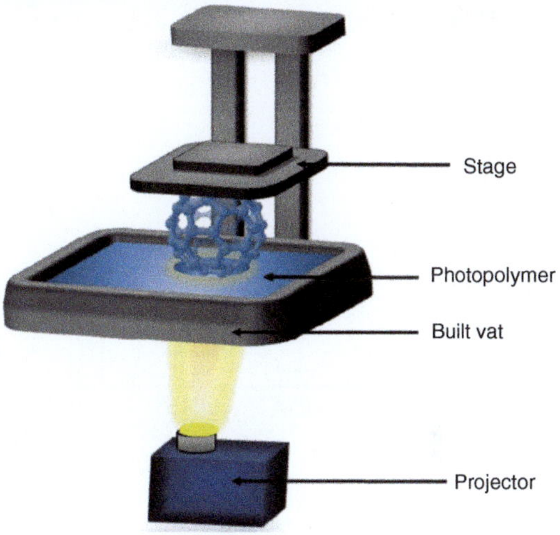

Figure 18.6 A digital light processing printer depicted as a diagram. Source: Adapted from https://www.3dsourced.com/3d-printing-technologies/digital-light-processing-dlp/.

construction purposes. They were reinforced along with other raw materials such as cement, ground blast furnace slag, steel fiber. The as-formed printed material showed excellent performance in terms of better cohesiveness and non-collapsible construction-based slag [10]. Since CNTs tend to absorb microwaves, they were incorporated into the polymer filaments followed by microwave irradiation. In comparison to their traditionally synthesized counterparts, the resulting 3D printed item has significantly increased strength and superior mechanical properties [11].

Carbon-based materials, notably CNTs, are used for 3D/4D printing techniques because of their superior electrical and thermal conductivity, and the resulting printed materials show the greatest results for their respective applications. A bar diagram of 3D/4D printing that has been worked on in the past few years has been shown in Figure 18.7, which is adopted from the earlier report of Hassan Alshahrani [12].

18.2 Method for the Carbon Nanotubes (CNTs) Modification

CNTs have unique features due to their morphology and structure. It has been considered that graphene sheets are rolled to form CNTs. The angle and curvatures of graphene sheets rolling to make CNTs are responsible for the exact structure of the nanotube. CNTs can be classified as "zigzag," "armchair," or "chiral" using a chiral vector. Single-walled carbon nanotubes (SWCNTs) were discovered to have a combination of metallic and semiconducting properties, whereas multi-walled carbon nanotubes (MWCNTs) are considered metallic conductors. Electrons are propelled through the tube without scattering thanks to 1D tube-like features, making them ideal for ballistic transport and allowing the nanotube to conduct without

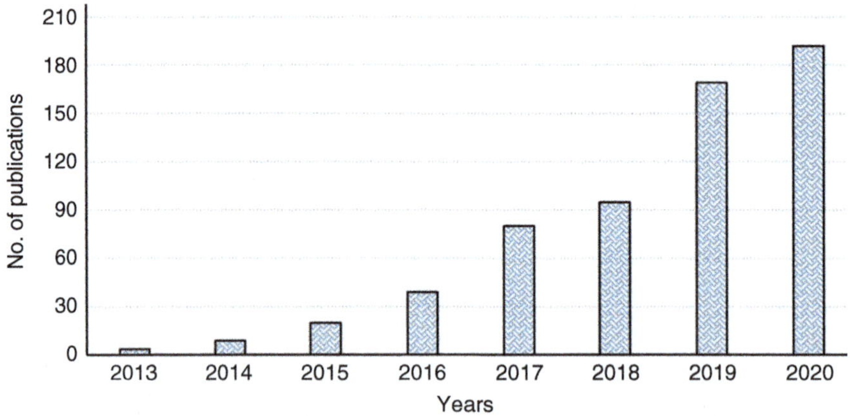

Figure 18.7 Number of publications related to 3D/4D printing from 2013 to 2020. Source: Snyder and Powers [11]/with permission of Elsevier.

squandering energy as heat. Apart from these intriguing features, many applications necessitate CNT modification to satisfy specific criteria, such as in biosensors, where biomolecules require electron mediators to facilitate electron transmission. In the same way, specific functional groups should be attached to CNTs for electrochemical metal ion sensors. In 1970, first time reported the modification over carbon substrate [13]. CNTs are frequently attached to a specific molecule or entity for chemical modification, which can be achieved easily via surface and bulk modification. Electrochemically induced procedures, polymer grafting, and metal nanoparticle deposition are examples of surface modification techniques and microwave-assisted modification, chemical reduction, and ball milling modification are examples of bulk modification techniques. Ajayan et al. described a novel polymer-CNT nanocomposite combination [14]. The researcher focused on changing the affinity of CNTs for different polymer matrices. Polymer-CNT has been displayed in a typical view of nanocomposite material that has been quickly advancing, Figure 18.8 [15].

The chemical modification of CNT can be done using the following way and briefly discussed.

18.2.1 Polymer Grafting

Polymers have chemical stability, a large number of active sites, good adhesion, and good electrical characteristics making them scientifically valuable. Because of the polymer properties combined with the CNT capabilities, polymer-grafted CNTs are a particularly important type of synthetic material. There are three approaches to make polymer grafted CNTs: in first step, it has been soaked in a polymer solution, the second is to use a chemical process that requires an oxidant, and the third is to use an electrochemical method. Various polymers were used to alter MWCNTs

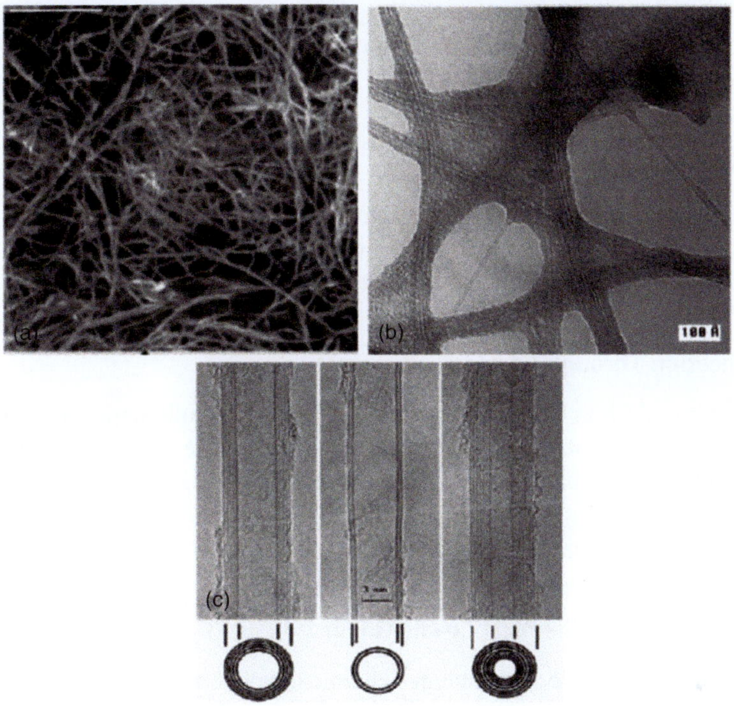

Figure 18.8 (a) CNT network-oriented at random in SEM micrograph. SWCNTs (b) and MWCNTs with distinct layers (c) are depicted in TEM pictures. Source: Agarwala et al. [15]/with permission of Elsevier.

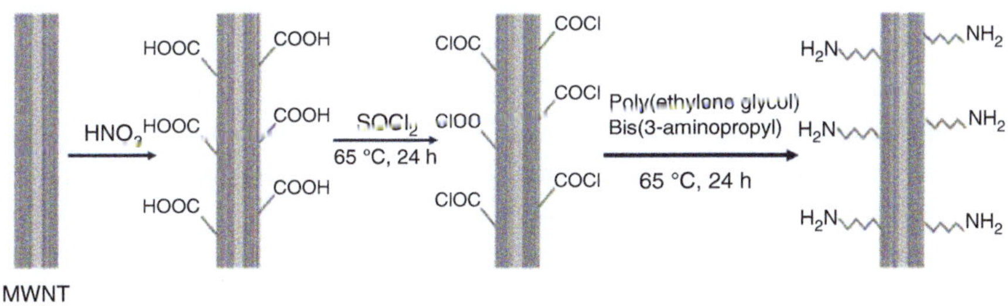

Figure 18.9 Modification of MWCNTs-polydiphenylamine with amine functionalization of MWCNTs. Source: Santhosh et al. [16]/with permission from Elsevier.

in this way here amine functionalization of MWCNTs-polydiphenylamine has been shown in Figure 18.9 [16].

18.2.2 Electrodeposition

Metal nanoparticles were placed over CNTs using electrodeposition in this procedure. The technique has a lot of possibility to control the nucleation and development of NPs. Noble metal salts of Au, Pt, and Pd were deposited on CNTs in the majority of cases. The interaction of CNTs with nanoparticles is caused by van der Waals interactions.

18.2.3 Electroless Deposition (ELD)

Without the need of any external help, such as a reducing chemical or applied voltage, nanoparticles adhere to the surface of nanotubes. A direct redox interaction between the tubes and the NPs speeds up CNT alteration.

18.2.4 Modification Through Covalent Attachment

Because the surface of CNTs has insufficient binding sites, simple deposition of nanoparticles on them is unevenly distributed, resulting in the lowest stability. To produce exceptionally stable and uniformly dispersed nanoparticles, a correct functionalization process must be used, which entails forming a covalent bond with the help of a linker.

18.2.5 Thermal-Assisted Chemical Covalent Functionalization

This three-step alteration procedure includes chemical oxidation, activation, and amidation reactions. The modified CNTs are highly stable and have a high surface coverage thanks to this multi-step method. These modified CNTs electrode have double layer formation over them, which makes them high hydrophilic.

18.3 Chemically Modified CNT for 3D Printing

Traditional 3D printing of polylactic acid (PLA) and acrylonitrile butadiene styrene (ABS) materials is reliant on preprogrammed printing parameters, including temperature and nozzle design. The fabrication of different kinds of polymers, gels, composites, and metallic materials that were done by 3D printing technology has bought considerable amendments to engineering and technology [17, 18]. As 3D printing has dominated its applications in the field of robotics, microelectrochemical systems, and tactile sensors whose major requirement is efficient conductivity. Therefore, conductive fillers that make up conductive nanocomposites such as CNTs or fibers, graphene, or metallic nanowires have been one of the evolving topics in 3D printing [19, 20]. CNTs are well-known for their electrical, thermal, and mechanical properties, making them a good contender for 3D printing polymers. Because of the presence of structural flaws that provide adequate nucleation sites, MWCNTs have a lot of interaction with polymers [21]. For CNT-based films a homogeneous dispersion of CNTs and the polymeric solution is required. The major problem encountered by using CNTs as additives in 3D printing is the problem of aggregation of CNTs that can have a deleterious effect on fused deposition modeling during 3D printing. This causes flux instability and possible blockage at the nozzle. As a result, scientists are frantically trying to figure out how to generate a CNT concentration high enough to pass past the percolation threshold. The power law is used to determine theoretically the concentration of CNTs for CNT-polymer composite to reach the limit of electrical percolation, by the following equation of power law:

$$\sigma = (\varphi - \varphi_c)^a$$

where σ is electrical conductivity, φ is the volume concentration, φ_c is the critical CNTs concentration at electrical percolation, and α is a critical exponent.

18.3.1 Commercially Available CNTs for Printing Technology

Filaments containing CNTs are commercially produced by the company 3DXTech. Their 3DXNano™ ESD ABS filaments that contain MWCNTs have a diameter of 1.75 and 2.85 mm. MG-94 premium ABS mixed with MWCNTs are used to tailor filaments and some other modifiers are used based on the application process. Nanocyl produces R&D grade CNTs and one of its leading experts in CNT reinforced materials. PLASTICYL are CNTs/thermoplastic concentrates, that contain 10–20% of CNTs, and are used in applications where the major focus is on electrical conductivity and strong mechanical characteristics. The CNT market is led by Nanocyl worldwide. PLASTICYL is one of their CNTs thermoplastic concentrates. This has good electrical conductivity and mechanical qualities and can be used in a variety of applications. The commercially available 3D printers available with different model has been shown in Figure 18.10.

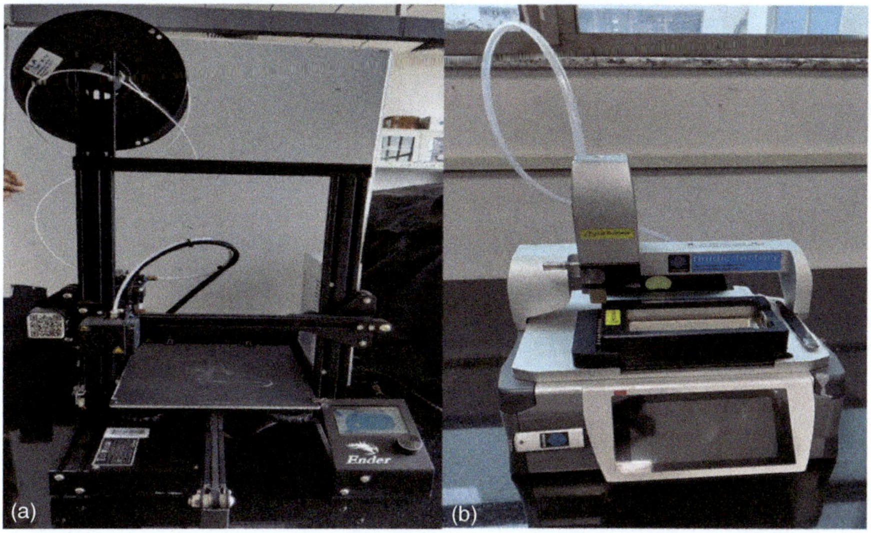

Figure 18.10 3D printers (a) Ender (Model 23525). (b) Dolomite (Model 4101638).

18.4 Application of Chemically Modified CNTs for 3D Printing

18.4.1 Mechanical Properties Enhancement

Various mechanical properties including tensile strength, elasticity, and stiffness are critical in biomedical devices. Tissue engineering, for instance, relies heavily on achieving desirable mechanical properties of tissue [22]. In orthopedics, the properties of bone should match with mechanical properties of specialized implants (prepared 3D-printed bones) [23]. If the mechanical properties of transplanted one are not similar then there will be a possibility of fracture due to stress shielding [24]. Implantation with mechanical qualities similar to ancient bones should be done to solve this problem. Sezer et al. demonstrated that incorporating multiwall CNTs into an acrylonitrile butadiene styrene polymer can result in a high tensile strength print [25]. Tensile strength was evaluated at various concentrations of MWCNTs, and it was discovered that dispersion of MWCNT leads to improved mechanical strength. Tensile strength tests reveal that specimens with a greater MWCNT ratio are stronger than those with a lower MWCNT ratio [25]. This material's applicability is broadened as a result of its increased mechanical strength. A schematic for the 3D printing of enhanced mechanical properties has been shown in Figure 18.11.

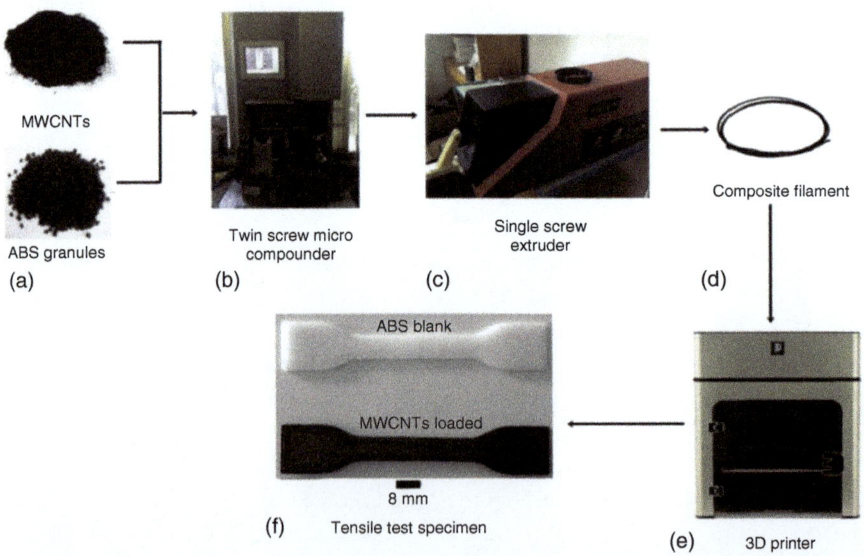

Figure 18.11 Preparation of nanocomposite filaments and 3D printing. (a) Digital images of MWCNTs and ABS granules. (b) Double-screw extruder. (c) 3D printer with a single-screw extruder. (d) Winding MWCNT/ABS nanocomposite filament in rolls. (e) The fused deposition modeling (FDM) 3D printing method. (f) A typical 3D-printed tensile test sample made with a MWCNTs–ABS composite filament. Source: Niinomi et al. [23]/with permission of Elsevier.

18.4.2 Energy Storage Devices

Energy storage devices are devices that capture the energy produced in any form and can be released in form of electrical energy [26–28]. A battery or accumulator is a device that stores energy. Energy can be stored in a variety of ways, including chemical, electrical, and kinetic storage. Some technologies can provide short-term energy storage, while others can provide long-term energy storage. The most common example of energy storage is rechargeable batteries that store chemical energy and during discharging it will provide electrical energy. Supercapacitors (SCs), in comparison to batteries, have a high charge and discharge rate as well as a long cycle life, making them an attractive alternative for energy storage. Films or networks of CNTs or graphene were utilized as electrode materials in previous stretchable supercapacitors. However, the transfer of the active material layer to a substrate is required for the manufacture of such a stretchy supercapacitor. As a result, there is a pressing need to develop strategies and integrate cutting-edge technology and production to manufacture high-performance supercapacitors. The use of printing techniques to make a flexible supercapacitor is an advantageous approach. Printed electronics have gained popularity in recent years for the manufacture of electronics such as sensors and energy storage devices. 3D printing technology has the benefit of being low-cost and capable of mass manufacturing with high efficiency. Researchers devised a method for printing high-quality thin films using graphene oxide; nonetheless, one of the significant challenges is the lack of stability in graphene films due to the uncontrolled pore structure. CNT additive between graphene sheets improved the rated capacity of printed graphene films by producing a porous structure, according to previous studies. The development of a high-performance energy storage device with a high energy density and a fast charging–discharging rate is critical for the development of flexible and wearable electronics [29]. Because of recent improvements in 3D printing technologies, it is now possible to fabricate 3D printed nanomaterial-based energy devices such as lithium-ion batteries, micro-supercapacitors, and pseudocapacitors [30]. Additive manufacturing's capacity to print multiple materials can reduce the number of processing steps required to create a battery. Unlike traditional technologies, which only operate with a single material at a time, 3D printing allows for the fusion of multifunctional nanomaterials. The performance of an energy storage device can be improved by using a printed electronics device with a hierarchical structure, and the existence of a hierarchy in the electrode can improve charge transfer and ion diffusion. Yu et al. printed the 3D micro supercapacitors (MSCs) using CNT [31]. A preprogrammed patterning method was used to regulate the printing. The base plate was heated to 80 °C during the printing followed by annealed at 120 °C for a night to eliminate any leftover solvent. To make the micro-capacitor, a polyvinyl alcohol (PVA)–H_3PO_4 electrolyte solution was cast onto the interdigital electrodes and air dried before being removed from the glass plate substrate. The specific capacitance of the 3D printed electrode (MSCs) is 2.44 F cm^{-2}, and it provides outstanding cycle stability and energy storage capacity [31]. Figure 18.12 shows a schematic representation of 3D printing of CNT-based MSCs.

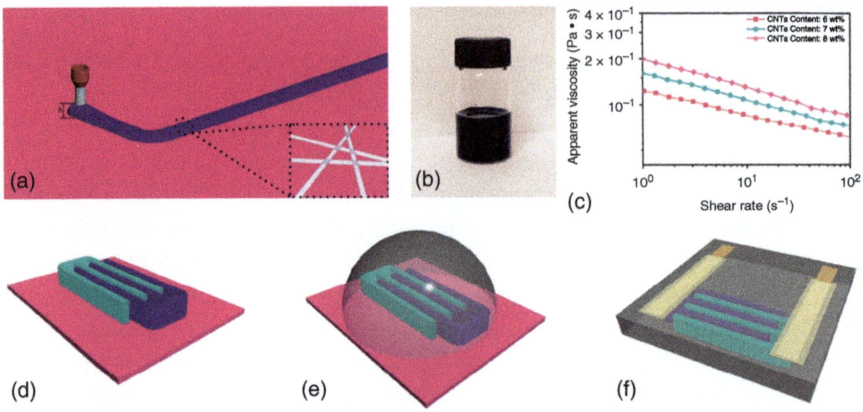

Figure 18.12 3D printing of carbon nanotube-based MSCs schematic illustration. (a) A 3D printing process diagram. (b) Optical pictures of ink. (c) Viscosity of apparent inks. (d) Interdigital electrodes as acquired. (e) The polymer gel electrolyte was solidified and cast. (f) MSCs in a package. Source: Sun et al. [29]/with permission of American Chemical Society.

18.5 Modified CNTs for 4D Printing Technique

Tibbits proposed the concept of 4D printing as a dynamic system that can change shape over time [32]. The fourth dimension is regarded as time due to response or change in function or shape over a while in response to stimulations. In addition to the 3D printing process, which involves layer-by-layer deposition of printing ingredients such as CNTs/polymers, 4D printing imparts new dimensions to the materials by being responsive to various external environmental stimuli [33]. So, 4D printing has been a rising topic of researchers' concern to be applicable in a wide variety of biomedical smart materials. Different types of chemical compounds are responsive to different types of stimuli some of which as summarized in Figure 18.13.

Although the technique of 4D printing is in the infancy stage and a lot of its science and applications are yet to be explored fully. 4D printing has exerted positive outcomes in the biomedical field yet a lot of challenges need to be overcomed related to the compatibility of 4D printed devices, as the microenvironment differs from person to person and hence the adaptability of formed devices. Tissue engineering, which repairs or replaces tissues in the human body, is a growing subject of study in biomedicine, thanks to the biocompatibility of CNT/polymer composites. The main advantage of this interaction is that a suitable environment with appropriate chemical composition is provided to the tissues that makes them suitable as an artificial podium for engineered tissues [34].

18.5.1 Shape Memory Functionalized for Liquid Sensors

Using 3D printing technology, conductive nanocomposites can be used to create a variety of sensors, including liquid sensors, capacitive pressure sensors, strain

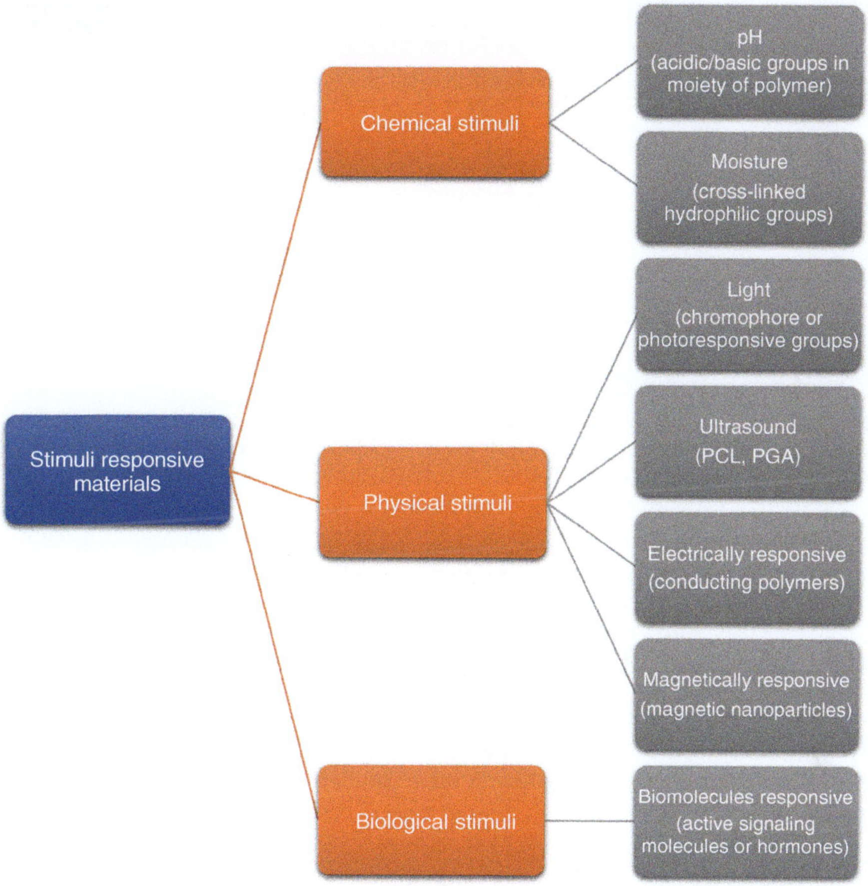

Figure 18.13 A representation of a list of stimuli and the properties and functionalities associated.

sensors, and temperature sensors. These sensors are based on the change in electrical conductivity when they are near or in contact with specific materials. When they moved from those specific materials the conductivity fluctuates and acts as sensors. When liquid enters the polymer network, however, swelling causes the distance between conductive fillers to grow. In addition to conventional techniques, Daniel et al. demonstrated 3D printing of PLA/CNT sensors using direct ink writing (DIW) and examined the effect of printed configurations on liquid sensitivity [35]. Here, the sizes of nozzles are in the micrometer range and will provide high resolution. poly(D,L-lactide-co-trimethylene carbonate) (PLMC) polymer ink was diluted with CNT in dichloromethane (DCM) solvent and used for direct ink writing fabrication of 3D printing. In addition, the crosslinking networks were triggered when the 3D-printed structure was photopolymerized in the UV chamber. A c-scaffold is a printed scaffold that has undergone UV crosslinking. The 3D-printed structures show the change in 25 V within 16 seconds [36]. These chemically modified CNTs

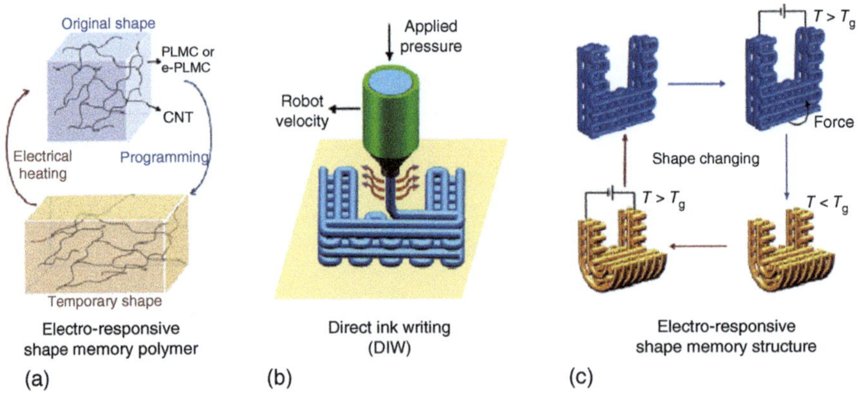

Figure 18.14 Shape memory nanocomposites printed in 3D and 4D (a) PLMC crosslinking reaction to produce c-PLMC; (b) DIW technology; (c) electro-responsive shape-changing behavior of 4D printed structure. Source: Wan et al. [36]/with permission of Elsevier.

were 3D printed for detecting solvent leakage in a variety of settings. A modified CNT-based 3D-printed sensor has been shown in Figure 18.14.

18.5.2 4D Printing of Stretchable Supercapacitors

As previously mentioned, energy storage in supercapacitors is gaining a lot of attention. Due to their intrinsic mechanical features and energy storage function, power-independent stretchable SCs have received substantial attention among alternative storage techniques. To create exceptionally flexible and high-performance SCs, Zhou et al. used 4D printing of composite electrodes for simple synthesis, a low-cost, and scalable technique [37]. The reduced graphene oxide (rGO)–CNT–poly(3,4-ethylenedioxythiophene)polystyrene sulfonate (PEDOT:PSS) solution that was utilized to make the 4D shape morphable flexible electrode [37]. The 4D printing has been done over polymer substrate and excellent adhesion between the substrate to the CNTs and PEDOT:PSS has been obtained. The CNT and rGO significantly enhance the conductivity and their impact will reflect in the supercapacitor performance. The rGO–CNT–PEDOT:PSS composite structure allows the 4D printed SCs to be charged at $5\,A\,g^{-1}$ with a specific capacitance of $20\,F\,g^{-1}$ and an 85% capacitance retention from 0.5 to $5\,A\,g^{-1}$ with substantial mechanical stretchability [37].

18.6 Boundaries for 3D/4D Printings and Prospects

18.6.1 Material Revolution

The properties of materials will always be at the core of 3D/4D printing technique. The combined properties of polymer/CNT hybrid encourage the research community and lead the printing technology over various other fabrication techniques. However, there will be always a possibility to search for new polymers for integration with CNTs, which have better materials properties such as conductivity and strength. At the same time, printing composite materials have various challenges during the process, including necking, inhomogeneity, and adhesion between the two components. Furthermore, composite material printing is more difficult than mono-material printing. Therefore, finding new composite materials having enhanced or similar properties free from above mention difficulties will be always in demand.

18.6.2 Interfacial Properties of Polymer-CNT and Homogeneous Distribution

Modifying the composition, structure, and distribution of polymer-CNT can alter its interfacial characteristics, which are crucial in determining the final attributes of printed structures. This topic has not been widely researched due to a lack of characterization approaches for the interface. The polymer interaction with CNTs is inhibited by the smooth surface of CNTs, and the surface of CNTs can be remodified by etching or functionalization. It will always be difficult to get the proper distribution of polymers across CNTs or vice versa since adhesion between polymers and CNTs is reliant on the structural structure of polymer chains and CNTs. In the final printed design, the homogenous distribution ensures appropriate stress management. Melt mixing, mechanical mixing, precipitation, and ultrasonication are only a few of the strategies for homogeneous distribution that were documented. Few of these approaches break the CNTs, which has an impact on the final 3D printing qualities. In analyzing the CNT dispersion in the final composite, quantitative analysis is also lacking.

18.6.3 Future Scope

3D–4D printing has advanced significantly in recent years, however, it is still not widely recognized by numerous businesses. To expand the horizons of its applications many limitations need to be prioritized. Since material properties form the heart of any commercial applications. Enhancement in properties of CNT/polymer is sure to take the progress in the field to the next level. The printable materials are a major applicability concern that needs to be addressed in this field. The restrictions of polymer materials with low T_g (glass transition temperature) and appropriate melting viscosity such as shape memory, thermoplastic polymer, and photopolymers do not extensively meet the set necessities. To tackle this limitation, diverse printable materials must be researched. Researchers are on their wheels to device new reinforcing agents along with novel polymers that possess far better properties to make printed materials versatile. Another reason for the limited usability of 3D–4D printed CNT/Polymer composites is the development of machinery that can fabricate enormous amounts of reinforced materials and useless cumbersome procedures and is also less time-consuming. Another key characteristic of 3D–4D printing is the development of spatial and lateral resolution of printed structures. Interfacial properties are another crucial factor that affects the residual properties of CNT/polymer composite in 3D–4D printing. Interfacial mechanical properties, such as frictional and residual stress, are worth investigating. Because CNTs have a flat and plain exterior surface, their interaction with polymer matrix is limited. Modifying the surface of CNTs to improve wettability has recently been a popular trend for improved interaction with the polymer in which they are reinforced. Furthermore, in 3D–4D printed devices, the homogenous distribution of CNTs in the polymer matrix offers overall stress management. Apart from many limitations and setbacks, there is still an immense scope to explore novel faster techniques that can provide high-grade 3D–4D printed materials. Distinct attention is required for the optimization of synthesis conditions and scalability to successfully achieve the final target of reinforced 3D–4D printed CNT/Polymer materials.

18.7 Conclusions

In the present chapter, a detailed discussion of 3D and 4D technology has been done. The various parts used to fabricate a 3D and 4D design has been also discussed. Most of the use of 3D/4D printing deals with the biomedical application therefore it has been discussed with appropriate examples. The application of 3D/4D printing in water purification, energy storage, and device manufacturing has been also discussed. The use of CNT in the modification of these properties has been also discussed in various applications. Further, the limitation of 3D/4D printing has been also discussed.

Important Website for 3D/4D Printings

The following website can be referred to get more about 3D/4D printing.

1. https://www.sculpteo.com/en
2. https://www.sigmaaldrich.com/IN/en/technical-documents/technical-article/materials-science-and-engineering/3d-printing/3d-and-4d-printing-technologies
3. https://all3dp.com/1/4d-printing
4. https://www.futurebridge.com/industry/perspectives-mobility/4d-printing-the-technology-of-the-future

References

1 Sanei, S.H.R. and Popescu, D. (2020). 3D-printed carbon fiber reinforced polymer composites: a systematic review. *J. Compos. Sci.* 4 (3): 98.

2 Ambrosi, A. and Pumera, M. (2016). 3D-printing technologies for electrochemical applications. *Chem. Soc. Rev.* 45 (10): 2740–2755.

3 Layani, M., Wang, X., and Magdassi, S. (2018). Novel materials for 3D printing by photopolymerization. *Adv. Mater.* 30 (41): 1706344.

4 Meng, T., Yi, C., Liu, L. et al. (2018). Enhanced thermoelectric properties of two-dimensional conjugated polymers. *Emergent Mater.* 1 (1, 2): 67–76.

5 Jung, K., Corrigan, N., Ciftci, M. et al. (2020). Designing with light: advanced 2D, 3D, and 4D materials. *Adv. Mater.* 32 (18): 1903850.

6 Chisholm, G., Kitson, P.J., Kirkaldy, N.D. et al. (2014). 3D printed flow plates for the electrolysis of water: an economic and adaptable approach to device manufacture. *Energy Environ. Sci.* 7 (9): 3026–3032.

7 Quanjin, M., Rejab, M.R.M., Idris, M.S. et al. (2020). Recent 3D and 4D intelligent printing technologies: a comparative review and future perspective. *Procedia Comput. Sci.* 167: 1210–1219.

8 Acquah, S.F.A., Leonhardt, B.E., Nowotarski, M.S. et al. (2016). Carbon nanotubes and graphene as additives in 3D printing. In: *Carbon Nanotubes – Current Progress of their Polymer Composites* (ed. M. Berber and I.H. Hafez), 506. InTech.

9 Waheed, S., Cabot, J.M., Macdonald, N.P. et al. (2016). 3D printed microfluidic devices: enablers and barriers. *Lab Chip* 16 (11): 1993–2013.

10 Wenxuan, H. (2018) High-strength and high-toughness active powder concrete doped with carbon nanotubes and preparation method thereof. CN104030634B, Li Qinghua, Xu Shiling, Liu Jintao, filed 12 June 2014 and issued 13 July 2016.

11 Snyder, L. and Powers, G. (2015) . Anti-Vista Antibodies And Fragments. cnt 2. WO2015/097536A2, filed 22 December 2014.

12 Alshahrani, H.A. (2021). Review of 4D printing materials and reinforced composites: behaviors, applications and challenges. *J. Sci.: Adv. Mater. Devices* 6 (2): 167–185.

13 Pandurangappa, M. and Kempegowda Raghu, G. (2011). Chemically modified carbon nanotubes: derivatization and their applications. In: *Carbon Nanotubes Applications on Electron Devices* (ed. J.M. Marulanda), 572. InTech.

14 Ajayan, P.M., Stephan, O., Colliex, C., and Trauth, D. (1994, 1979). Aligned carbon nanotube arrays formed by cutting a polymer resin–nanotube composite. *Science* 265 (5176): 1212–1214.

15 Agarwala, S., Goh, G.L., Goh, G.D. et al. (2019). 3D and 4D printing of polymer/CNTs-based conductive composites. In: *3D and 4D Printing of Polymer Nanocomposite Materials: Processes, Applications, and Challenges* (ed. K.K. Sadasivuni, K. Deshmukh and M.A.S.A. Al-Maadeed), 297–324. Elsevier.

16 Santhosh, P., Manesh, K.M., Gopalan, A., and Lee, K.-P. (2007). Novel amperometric carbon monoxide sensor based on multi-wall carbon nanotubes grafted with polydiphenylamine—fabrication and performance. *Sens. Actuators, B* 125 (1): 92–99.

17 Gross, B.C., Erkal, J.L., Lockwood, S.Y. et al. (2014). Evaluation of 3D printing and its potential impact on biotechnology and the chemical sciences. *Anal. Chem.* 86 (7): 3240–3253.

18 Muroi, H., Hidema, R., Gong, J., and Furukawa, H. (2013). Development of optical 3D gel printer for fabricating free-form soft & wet industrial materials

and evaluation of printed double-network gels. *J. Solid Mech. Mater. Eng.* 7 (2): 163–168.
19 Wei, X., Li, D., Jiang, W. et al. (2015). 3D printable graphene composite. *Sci. Rep.* 5: 11181.
20 Guo, S.Z., Qiu, K., Meng, F. et al. (2017). 3D printed stretchable tactile sensors. *Adv. Mater.* 29 (27): 1701218.
21 Mccarthy, B., Coleman, J.N., Curran, S.A. et al. (2000). Observation of site selective binding in a polymer nanotube composite. *J. Mater. Sci. Lett.* 19: 2239–2241.
22 Graham-Rowe, D. (2009). From dots to devices. *Nat. Photonics* 3 (6): 307–309.
23 Niinomi, M. and Nakai, M. (2011). Titanium-based biomaterials for preventing stress shielding between implant devices and bone. *Int. J. Biomater.* 2011: 836587.
24 Paul, J.P. (1999). Strength requirements for internal and external prostheses. *J. Biomech.* 32: 381–393.
25 Sezer, H.K. and Eren, O. (2019). FDM 3D printing of MWCNT re-inforced ABS nano-composite parts with enhanced mechanical and electrical properties. *J. Manuf. Processes* 37: 339–347.
26 Yadav, K.K., Wadhwa, R., Khan, N., and Jha, M. (2021). Efficient metal-free supercapacitor based on graphene oxide derived from waste rice. *Curr. Res. Green Sustainable Chem.* 4: 100075.
27 Nishanthi, S.T., Yadav, K.K., Baruah, A. et al. (2019). New sustainable and environmental friendly process of synthesis of highly porous Mo_2S_3 nanoflowers in cooking oil and their electrochemical properties. *Electrochim. Acta* 300: 177–185.
28 Yadav, K.K., Singh, H., Rana, S. et al. (2020). Utilization of waste coir fibre architecture to synthesize porous graphene oxide and their derivatives: an efficient energy storage material. *J. Cleaner Prod.* 276: 124240.
29 Sun, K., Wei, T.S., Ahn, B.Y. et al. (2013). 3D printing of interdigitated Li-ion microbattery architectures. *Adv. Mater.* 25 (33): 4539–4543.
30 Torres-Canas, F., Yuan, J., Ly, I. et al. (2019). Inkjet printing of latex-based high-energy microcapacitors. *Adv. Funct. Mater.* 29 (31): 1901884.
31 Yu, W., Zhou, H., Li, B.Q., and Ding, S. (2017). 3D printing of carbon nanotubes-based microsupercapacitors. *ACS Appl. Mater. Interfaces* 9 (5): 4597–4604.
32 Tibbits, S. (2014). 4D printing: multi-material shape change. *Archit. Des.* 84 (1): 116–121.
33 Momeni, F., Hassani, M., Liu, X., and Ni, J. (2017). A review of 4D printing. *Mater. Des.* 122: 42–79.
34 Ku, S.H., Lee, M., and Park, C.B. (2013). Carbon-based nanomaterials for tissue engineering. *Adv. Healthcare Mater.* 2 (2): 244–260.
35 Chizari, K., Daoud, M.A., Ravindran, A.R., and Therriault, D. (2016). 3D printing of highly conductive nanocomposites for the functional optimization of liquid sensors. *Small* 12 (44): 6076–6082.
36 Wan, X., Zhang, F., Liu, Y., and Leng, J. (2019). CNT-based electro-responsive shape memory functionalized 3D printed nanocomposites for liquid sensors. *Carbon NY* 155: 77–87.
37 Zhou, Y., Parker, C.B., Joshi, P. et al. (2021). 4D printing of stretchable supercapacitors via hybrid composite materials. *Adv. Mater. Technol.* 6 (1): 2001055.

Questions

18.1 Describe the various types of 3D printing techniques?

18.2 What are the main differences between 3D and 4D printing?

18.3 Discuss the major challenges that arise in 3D/4D printing?

18.4 How mechanical properties can be tuned by using CNTs reinforced 3D/4D printed materials?

18.5 What additional property do 4D printed materials acquire as compared to their counterparts (3D printed materials)?

18.6 How CNT additives in 3D/4D printing have influenced the energy storage abilities of various devices?

18.7 What can be the future prospects of CNTs reinforced 3D/4D materials in biomedical and smart devices application fields?

19

Chemically Modified Carbon Nanotubes for Tribology Applications

Aslam Hossain[1,2], R. Manu[3], and S.M.A. Shibli[1,4]

[1] *University of Kerala, Centre for Renewable Energy and Materials, Kariavattom Campus, Thiruvananthapuram, Kerala, 695581, India*
[2] *Southern Federal University, The Smart Materials Research Institute, Rostov-on-Don 344090, Russia*
[3] *Sri Vyasa NSS College, (Affiliated to University of Calicut), Research Department of Chemistry, Wadakkanchery, Thrissur, Kerala, 680623, India*
[4] *University of Kerala, Department of Chemistry, Kariavattom Campus, Thiruvananthapuram, Kerala, 695581, India*

19.1 Introduction

The demand for new versatile materials and sophisticated process techniques make it necessary to go for hybrid components and energy-efficient processes by optimizing materials properties for superior tribological performances. There was always a surge in tribology-related research reports that delved into property modification for optimum performance through improving the wear and lubrication aspect of the deposit. The vast amount of literature reports throws light into various process protocols and novel nanomaterials, which could be used as exotic materials in tribological applications. Notably, 3D, 2D, and refractory metal oxides have demonstrated extremely low frictional coefficients and wear rates, making them appealing materials for high-performance nano-lubricants and frictional applications.

Toward the increasing demand for energy-efficient and economic materials, contemporary research focused on the incorporation of various nanomaterials as lubricants and use of dimensionally hybrid materials/composites in different mechanical parts [1–11]. The purpose of this research is to provide a comprehensive review of the most recent developments in the tribological properties of various materials used as lubricants, additives, and fillers in a variety of applications. The essential physical, chemical, wear, and frictional properties of carbon nanomaterials as well as their production techniques, are first discussed. There are several methods such as thermal decomposition, chemical functionalization, additive modification, hydrothermal, and sol–gel, microwave for synthesizing and fabrication to improve the frictional/lubricating aspect of the materials and engineering components for their optimum performance.

As we know, the industry is thriving for improved techniques and modified design parts to curtail the energy loss happening due to frictional and wear mechanisms,

Chemically Modified Carbon Nanotubes for Commercial Applications, First Edition.
Edited by Jeenat Aslam, Chaudhery Mustansar Hussain, and Ruby Aslam.

every engineering application necessitates material property optimization for better tribological applications [12]. Over the last decade, the desire for new materials with enhanced surface engineering applications has driven the development of carbon-based nanomaterial-reinforced metal matrix composites, carbon nanotube (CNT)-incorporated lubricants, and various functionalized CNT as superior wear performance coating and other functional materials. Because of their superior mechanical, thermal, chemical, and electrical properties, carbon-based nanomaterials such as CNTs, graphene nanoplatelets (GNPs), and carbon fibers have piqued the interest of researchers as a potential solid lubricant reinforcement in the metal matrix [13–17].

CNT is used as an effective filler, grain refiner, and coating component that has got a pivotal role in the contemporary research work on lubricants and friction-free materials [18–20]. CNT aids in load transfer and improves the mechanical properties of printed parts, which effectively improves matrix material strength due to its large aspect ratio [21–26]. Besides, it has significantly higher mechanical strength and excellent heat conduction performance, which are considered essential characteristics for tribological application [27]. Certain hurdles that limit their application to use in tribological applications are lack of dispersion uniformity, compatibility between matrix and filler, and functionalization with certain functional groups that affect its properties [28–31]. Theoretically, nano-sized particles have been shown to result in significant improvement due to their extremely low proportion to the quantities added. However, the strength of many of these solidification processes and the powder metallurgical aspect of nanocomposites remain lower than those of conventional monolithic alloys due to nanoparticle agglomeration during solidification and material porosity. As a result, before such materials can be presented as an interesting alternative to conventional metals and alloys, these preparation challenges are to be clearly addressed. The present review elaborates on various possibilities and challenges using CNT incorporated reinforced composites, synthesis of nano-additive lubricants, etc. as superior wear materials and as supplementary materials in various tribological applications.

19.2 Tribological Phenomena – Principle, Mechanism, and Application

Tribology is an essential interdisciplinary area that delves into the design and development of basic components which can be static and dynamic machine parts that require performance and reliability. The word *tribology* was derived from the Greek word *tribos* by Peter Jost (1966) meaning rubbing. This is equivalent to friction and wear or lubrication science. Tribology is the intersection of science and engineering discipline that deals with the surfaces in relative motion. Tribology is not a single science, but rather a complex and multidisciplinary endeavor in which scientists from mechanical engineering, manufacturing, materials science, and computer science collaborate to conduct research. There are significant effects on tribological operational analysis in daily life. This type of surface-related operation is quite complex. To understand properly, convergence of several fields of knowledge is expected, viz., chemistry, applied physics, materials science, mathematics, thermodynamics, machine design, etc. Generally, the nature of the interaction between surfaces is controlled by the friction, wear, and lubrication of the materials. Forces such as mechanical, physical, and chemical nature are changed during these interactions. In this regard, the interfacial phenomenon such as tribology is an important subject for understanding the nature of interactions and solving related issues.

The interfacial nature can be significantly studied in the atomic scale by computational modeling, which is one of the most beneficial modern techniques. The evaluation of micro-electro-mechanical systems (MEMS) and nano electromechanical systems (NEMS) have enhanced the advancement of nanofabrication processes. These devices controlled the major sectors in modern science, such as information technology (IT), molecular biology, and semiconductor technology.

System analysis and system-related thinking are important aspects of tribological science. The fundamental constituents of tribology are tribo systems, which are physical systems of contacting surfaces. The tribosystems have subfields that include biotribology, nanotribology, space tribology, and tribotronics. It is a common knowledge that friction and wear are not substance characteristics. They are the result of a specific tribological system consisting of a bearing, shaft, and lubricant combination, and such are influenced by a variety of factors, including operating conditions. Many operational inputs, system structure and function, and loss outputs are associated with this tribological system. Technical and physical load parameters (such as load, sliding speed, and duration) are inputs, besides movement and temperature conditions that stress the system structure. The functional properties of the various mechanical systems and their run time output are of significance, considering the efficacy of tribological system.

Tribology plays an important role in modern machinery that is used on sliding and rolling surfaces. Brakes, clutches, wheels, and productive frictions include bolts and nuts. Pencil writing, polishing, and shaving are all examples of effective wear. Inefficient friction is caused by gears and bearings in internal combustion and aircraft engines. This necessitates expanding tribology research activities in order to minimize and eliminate losses caused by friction and wear technology. This will provide us with excellent efficiency, more negligible drawback, and significant saving.

19.3 Contemporary Research on Carbon Nanomaterials for Tribological Application

Friction and wear reduction are critical for any industry because they affect energy consumption, maintenance costs, and production. A high-quality lubricant is critical for machinery that operates at extreme temperatures and pressures in continuous operation. Under these aggressive working conditions, various types of additives of inorganic/organic nature were concomitantly employed to improve the tribological properties of lubricants. In recent years, there has been a greater advancement in high-performance materials and coatings for tribological applications, such as laminated-graded self-lubricating composites, composite-lubricating coatings, diamond-like carbon (DLC) films with surface texturing, and nano additives in lubricant oils [32–37]. For the development of these materials, carbon-based nanomaterials were widely researched for their suitability in various tribology applications [38–40]. The suitability of this type of material depends on how its inherent properties can be tailored by specific process sequence and modification by functionalization and other secondary treatment. The unique dimensional effects of nanomaterials reveal different tribological and mechanical properties when compared to conventional materials [14, 41–45]. A promising approach to lubricant material optimization is the combination of carbon nanomaterials and lubricating composites.

Traditional additives such as sulfides, chlorides, and phosphates were used to protect materials from severe wear and seizure, but they have limitations such as poor oil solubility, secondary product formation, and color. The darkening of the final lubricant, as well as increased levels of sulfur or phosphorus in the finished lubricant, pose an environmental risk. As a result, there is a need for the development of lubricant additives and compositions that can enhance overall lubricant output while also being more eco-sustainable and consistent with pollution control measures. Nanoparticles have emerged as potential lubricant additives in recent years due to their wide range of effects obtained in contexts of lubricant rheological, chemical, and anti-wear properties. Nanoparticles used as additives and filler materials in lubricants are based on the solid lubrication principle and were originally used as antiwear and anti-friction additives for machines operating in harsh environments. It has been demonstrated that adding nanoparticles to oil improves extreme-pressure properties and load carrying capacity while decreasing friction coefficients [16, 46–48]. CNTs, which were discovered in 1991, have sparked considerable interest as a result of their remarkable mechanical, physical, and other exceptional properties. Because of their diverse morphologies, high corrosion resistance, excellent mechanical properties, high thermal conductivity, and carbon nanomaterials are exciting candidates for a wide range of modern applications. The various carbon nanomaterials based on dimension are 0D fullerenes, 1D CNTs, 2D graphene, and 3D nanodiamonds that are useful in different modern applications. C_{60} is the most stable form of fullerenes as 0D nanomaterials. It has high electron mobility and easy chemical modification but the focus on CNTs and graphene diminished the attention on C_{60}. The discovery of fullerenes C_{60} and C_{70} initially explored their lubricating properties due to their specific structural orientation. The friction and wear behaviors of the fullerene coatings deposited on silicon Wafer were reported by Bhushan et al. [49, 50]. The carbon in graphene is sp^2-hybridized which has outstanding friction and wear behavior. The exceptional low surface energy, superior mechanical properties, and high thermal–electrical conductivity make them potential materials in both nano-scale and macro-scale tribological applications. The variation of graphene such as monolayer, few-layers, multi-layers, and reduced graphene oxide has unique characteristics to tune the friction and wear nature.

Nanodiamonds have distinct properties unlike fullerene, CNTs, and graphene while it has both sp^2 and sp^3 carbon atoms. The interest in nanodiamonds has

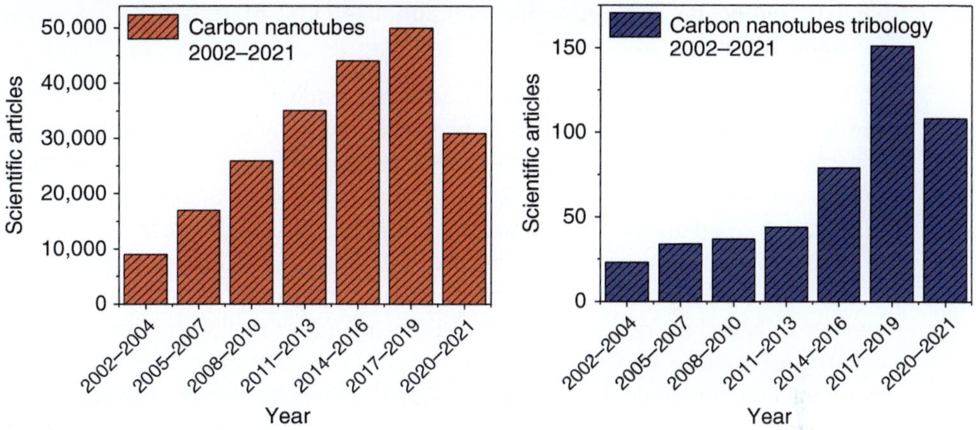

Figure 19.1 Statistical data on carbon nanomaterial-related tribology topics published between 2002 and 2021. On 7 December 2021, the data was compiled from the web of science (WOS) database.

been arising mainly due to their attractive high hardness and surface areas for biocompatible and wear-resistant coatings. There is a possibility to increase the use of nanodiamond to achieve comparable properties to that of CNTs. The use of carbon nanomaterials as a solid lubricant is a promising way to reduce energy consumption. Thus, extensive studies have been carried out with carbon nanomaterials in the last two decades. Figure 19.1a shows the published scientific articles since 2002 based on the research on CNTs which Figure 19.1b corresponds to research on CNTs on tribological applications. It is clear from the figures that the study with CNTs is growing day by day, but very little research has been done on tribological applications. This data was collected from the web of science database by both their titles containing the keywords "carbon nanotube" and "carbon nanotube tribological" on 7th December 2021.

19.4 Improvement of Tribological Behaviors of Bulk CNT Materials

Bulk materials such as polymer matrix composites and ceramic matrix composites are widely applied for various high strength and low fraction applications replacing conventional materials. Bulk materials can be used in several forms (metals, ceramics, and polymers) for tribological applications. The idea of mixing these basic matrices with suitable other materials, including nanomaterials to develop hybrid materials with improved shear strength and wear resistance, is a cumbersome job, as the homogeneous dispersion and the right process for the manufacture. Though the tribological properties have been widely studied the materials in the coating form but bulk materials reinforced by carbon nanomaterial (CNTs, graphene) also recently vastly studied.

19.4.1 Molecular Simulation for Functionalized CNT Materials for Nano Tribology

Molecular dynamics (MD) simulation studies are renowned theoretical method for simulating groups of atoms and molecules whose movements and interactions are governed by specific energy calculations. It is used to investigate the atomistic interactions that reflect the overall physical properties of materials. Statistical mechanics theory can be used to relate the microscopic properties obtained from MD simulation to the bulk material properties (i.e. mechanical, thermal, and electrical properties). In MD, atoms are represented as a point with an atomic mass. Each atom in the system has a distinct energy state as well as a moving trajectory governed by Newton's equation of motion. The attractive or repulsive forces between atoms are defined by interaction potentials. MD simulation data were widely employed to predict the change in the inherent properties of nano and compound materials during the formation process or in end-use application.

There are reports available on MD simulations under tensile strain to decide the elastic properties of polypropylene (PP) reinforced with functionalized CNTs. In the present case, the functionalization of CNT using a suitable functional group has its advantage in applying these materials as additives and as reinforcing materials in polymer matrix and high-wear application. The researchers have extensively studied CNTs and other nanomaterials using MD simulations with the development of the forcefield to explore the tribological properties. CNTs and polymer matrix incorporation has been successfully verified to be due to strong interfacial connections between nano-reinforcements and polymer matrix. MD simulations can support this experimental data for the investigation of inherent mechanisms of excellent mechanical properties of polymer composites from atomic viewpoints. Though limited reports are available on the investigation of tribological properties CNTs, several studies reported on polymer materials for applications in lubrication, dry friction, etc.

One of the recent works delves into the synthesis of epoxy nanocomposite systems with pristine and functionalized CNTs. The interfacial characteristics and mechanical responses of these CNT with other regular matrix structures were investigated using MD simulations. Functionalization, according to the findings, can enhance the interfacial shear stress and Young's modulus of CNT-epoxy blended nanocomposites. The functional groups improve interfacial adhesion with the epoxy matrix and serve as mechanical interlocking sites during deformation and high wear applications. The atomic-level understanding of the reinforcing mechanism provides physical new insights into material design for advanced nanocomposites.

One prime factor to be taken care of while using CNT in tribological applications is the load-bearing aspect when blended with various materials. Literature survey reports on analyzing various parameters such as Young's modulus and shear strength of the CNT incorporated composites using various MD simulation studies. The interfacial interaction between CNT and epoxy matrix, viz., weak van der Waals interaction, ionic or covalent bonding can be determined using this computational method. The inefficient load transfer, random orientation, and poor dispersion state of the CNTs materials could be the possible reason for the less effective reinforcement/negative reinforcement. The effectiveness of the load transfer is highly associated with the interfacial shear strength (ISS) and is also informative for reinforcing the effect of CNTs on polymer materials. These drawbacks are to be properly investigated with the aid of molecular simulation analysis. MD simulations are used to modeling of the interfacial phenomena in nano range that is recently studied in several material systems [51–54]. There have also been studies on the effect of CNT functionalization on the thermal and mechanical properties of the nanocomposites.

More research is needed because there have been few studies on the impact of functionalization on the mechanical properties of CNTs. The influence of various functional groups connected to the surface of single-walled carbon nanotubes (SWCNTs) such as –COOH, $-COOCH_3$, $-COSiCH_3$, and $-CH_2$ influences the elastic and wear properties of the composites studied using the MD approach. From an atomic standpoint, the mechanism of improved tribological properties of CNTs/polymer composites remains unknown. In the development of cavity pump technology, three-layer molecular frictional couple models between Fe atoms and CNTs/nitrile-butadiene rubber have been revealed as an anti-frictional material. During the frictional process, polymer materials were used to shear actions, and MD simulations revealed that CNTs/nitrile-butadiene rubber composites had a 60% higher shear modulus than pure nitrile-butadiene rubber matrix. Figure 19.2 depicts the simulated results of CNT/nitrile-butadiene rubber and pure nitrile-butadiene rubber for shear loading on the Fe layers in the intermediate and final regions of the

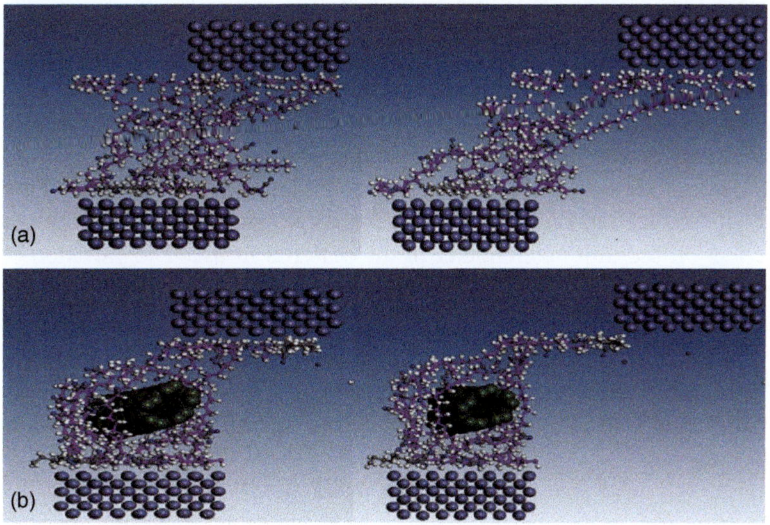

Figure 19.2 Snapshots of the composites in the middle and final states of the shear process for (a) the pure nitrile-butadiene rubber matrix and (b) the CNT/nitrile-butadiene rubber composites matrix. Source: Li et al. [55]/with permission from Elsevier.

shear process [55]. It is clear from the figures that nitrile-butadiene rubber in the composite bounded around the CNTs, leads to stronger bonds than pure ones. Hence, the clear idea of the bonding structure of functionalized CNT materials can be easily understood using the study of molecular simulation, which is very important to grow the field of tribology.

Although functionalization has been proposed as an effective method for improving tube solubilization, bundle exfoliation, reducing surface tension, and promoting interfacial bonding in CNTs, there is a lack of understanding of the developed materials in real-time analysis and their use in tribological applications.

19.4.2 Probing Complexities in Friction and Wear Properties of Functionalized CNT Materials

Polymers' mechanical, chemical, thermal, and tribological properties are influenced by fillers of various compositions, shapes, and sizes. poly-ether-ether ketone (PEEK) is a high-performance, temperature-stable polymer that is widely used as a replacement material for metal components in a variety of applications (automotive, aerospace, medical, and so on) and whose exciting mechanical and tribological properties can be further enhanced by the addition of a variety of fillers. Several studies have found that friction and wear properties have improved. Carbon-based fillers in the form of fibers, short fibers, nanotubes, nanofibers, and so on have been extensively studied, with researchers reporting improvements in properties such as wear rate reduction, high frictional coefficient, and mechanical property enhancement (modulus, yield stress, and so on), and changes in crystallinity of the developed material [56–60]. Similar results were reported for graphene, which is being used in a variety of applications, and research in these areas is expanding rapidly [61–64].

The cause for concern when we use CNT to improve various tribological characteristics is that solubilization, dispersion as an additive, blending with various composites, etc. are to be properly addressed. CNT atoms are sp^2 hybridized, which changes to sp^3 after functionalization. This alteration alters the structure of the CNTs, resulting in a change in their properties. During functionalization, there may happen tube rupture or modification in tubular morphology that can have a deleterious effect on the unique character of the material in use. The variation of Young's modulus with SWCNT volume fraction and the functional groups attached with SWCNT also modifies inherent tube characteristics. It is already established that functionalization reduces the elastic modulus. According to reports, MD simulations show a decrease in Young's modulus on vinyl functionalized SWCNTs, while atomic-scale finite element model (AFEM) simulations of hydrogen functionalized SWCNTs shows a similar decrease. Possible reasons for the reduction in modulus include (i) the functionalization, breaking the π-bond of the sp^2 hybridized carbon atoms and shifting the atom type from alkene to much more stable alkane, and (ii) the functionalization lowering energies connected with torsional and van der Waals components, which account for nearly 90% of the total potential energy. As the proportion of functionalized carbon atoms increases, the potential energy curves become nonmonotonic due to the increase in potential energy caused by van der Waals interaction. Only for the hydroxyl group, dipole–dipole interactions contribute significantly to the system's potential energy, resulting in a decreased modulus of elasticity with functionalization.

This decrease in modulus of elasticity could be attributed to nonuniformities in the nanotube structure. The elastic properties of the composites with CNTs functionalization were studied by means of the length of the nanotube, volume fractions, orientations, and polyethylene matrices. Other than this, the better adhesion and interaction between the matrix and reinforcement phases could be beneficial by which the mechanical properties of the fabricated nanocomposite can be enhanced.

19.4.3 Advances in Surface Engineering of CNTs for Tuning Friction and Wear Properties

In general, lubricants are used to control friction and wear on contact surfaces in industries (gears, scrolls, bearings, and seals). The fundamental principles of hydrodynamic lubrication theory have been studied for over a century and are still evolving as a result of commercial application. The viscosity and type of additive are the key parameters for engineering the friction and wear of the lubricant oil's surface. According to a recent report, friction accounts for nearly one-third of primary energy consumption, and wear accounts for approximately 80% of moving component failures each year. The key to bringing up the difficulties is to reduce friction and wear during the sliding process, which will not only reduce environmental impact but will also save money. Several studies have shown that nanoparticle additives have superior tribological properties to traditional solid lubricant additives. Experimental facts and theoretical studies proved that particle size, colloidal nature, protective film, and other materials' effects are the key factors that are to be improved to achieve low friction and wear by nanoparticle additives. The process of functionalization is to be carried out in such as way that it will not affect the inherent properties of CNT as that can lead to deleterious effects on CNT based additives. Covalent functionalization also disrupts CNT extended π networks, impairing their mechanical and electrical properties. The advancement in the surface treatments techniques including physical–chemical deposition and wet and dry coatings processing has kept pace with the advances in materials development. Novel techniques in surface engineering to be dealt with in detail and conventional process to be optimized to use CNTs for different applications.

19.5 Recent Advances in Hybrid CNT Materials in Tribological Application

Tribology research has led to the development of new fields such as superlubricity, bio-lubrication, and molecular lubrication. Green lubrication, low viscosity lubricants, and lubricants for harsh environments have significant commercial value and have recently received a lot of attention. The following is a summary of major lubrication developments.

19.5.1 Lubrication Properties of CNTs

The research on exploring the lubricating properties of CNTs has been advanced to a greater extent in recent years that are reflected in theoretical and experimental investigation. The tribology of the lubrication mechanism is evolving as a growing field that will attend a milestone in near future. The tribology technique is an important factor to reduce the friction coefficient by several orders of magnitude along with wear and noise friction. Currently, several research groups of tribologists are engaged in the study of lubrication properties of CNTs-based hybrid materials. Shinjo and Hirano introduced superlubricity by explaining the two contacting solid surfaces without resistance [65]. This is divided into two categories such as incommensurate crystalline contact-based superlubricity and disordered solid interface-based superlubricity [66]. The improved antiwear and anti-friction performance of nanofluids may be ascribed to the homogenous distribution of functionalized multi-walled carbon nanotubes (MWCNTs) and their intrinsic mechanical properties.

CNTs, graphene, graphite, and other 2D materials are ideal candidates for superlubricity due to their weak interlayer interaction [67]. Currently, different materials systems have reported their superlubricity on a large scale and their unique new physical mechanisms. Several activities were claimed in experimental-based nano-scale superlubricity, microscale frictionless sliding, and macroscale contacts in MWCNTs [66]. Recent reports of tribological superlubricity based on theoretical models and experimental explorations demonstrated the engineering of 2D materials down to atomic layer thickness, demonstrating several inherent benefits such as atomic smoothness, chemical inertness, and weak interlayer van der Waals interaction [68, 69]. It is clear that functionalization of CNT without affecting the tube morphology and π electron delocalization can deliver new materials that could pave the way for super lubricant fluids.

19.5.2 Tuning Additive Morphology and Wettability for Superlubricity in CNTs

The superlubricity of CNTs is due to their structural incommensurability between concentric nanotubes and this helps to displace comparatively better along their common axis [70]. The small interlayer friction was predicted using theoretical modeling recently, for double-walled CNTs materials [71]. The experiment with CNTs in a high vacuum showed interesting superlubricity at the atomic scale that proves the existence of defects, deformations, and contaminations in CNTs at ambient conditions, demonstrating the restrictions of superlubricity for engineering applications [71].

MWCNT is an ideal candidate for its superlubricity due to its coaxial cylindrical graphene layers that can rotate concerning one another. Few reports are available on the engineering of rotational nature for these materials because of their difficulties in performing nanomanipulation. These materials show higher practical values of intershell friction than theoretical ones because of defects. The modification of single CNTs also challenging due to its nanosize diameter. The recent report on producing centimeters-long double-/triple-walled CNTs opened a new way to explore superlubricity on the macroscale [72, 73]. According to Zhang et al., superlubricity in centimeter-long double-walled CNTs under ambient conditions exists [66].

19.5.3 Development of CNTs-Based Frictionless Surface Coatings

The tribology field cannot be fulfilled without coating technology. Several tribological coating technologies are there such as ceramic coating, metal coating, oxide coating, and composite coating. To control friction and wear, traditional coatings techniques were vastly used from small scale to large devices. Contemporary reports suggest the application of these types of techniques to MEMS and NEMS systems, which require low friction and high wear-resistant coating. To address these issues, tribologists focused on a carbon-based coating with low friction, high conductivity, and excellent mechanical and physical properties. Carbon nanomaterials outperformed bulk materials due to their well-organized structures at both the atomic and mesoscopic scales, which improved mechanical and electrical properties.

Initially, CNTs were prepared using an arc-discharge evaporation technique in the year 1991. This material is sp^2 hybridized and analogous to graphite structure. This can easily form the ideal cylindrical structure which can rotate easily during the nano-sliding. The special character of CNTs makes them the subject of research for friction and wear applications. Recently, Wang et al. fabricated CNT reinforced TiO_2 coatings while this coating remarkably enhanced the tribological properties as it influenced the microstructure of coating, structure strengthening, tribo-reorientation, tribo-protruding, tribo-film, and tribodefection. Further, we shall discuss the lubricating property of the CNTs as SWCNTs and MWCNTs along with their superior tribological performances.

19.5.3.1 Single-Walled CNTs (SWCNTs)

CNTs were widely explored for their application in different fields as functional materials, viz., as SWCNTs and MWCNTs. The reported SWCNTs coatings on the Si surface showed interesting wear resistance properties [74]. The effect of SWCNTs on the tribological properties of 6–7 m thick polyimide on a Si substrate at a contact pressure of 370 MPa was investigated in this study. SWCNTs containing composite coating revealed an excellent wear life due to fascinating mechanical properties that modified reinforced polyimide coating [74]. Polymer coatings have the advantage of their density and cost-effectiveness that makes them useful for lightweight tribological applications [75]. Because of the generation of oxygen-containing functionalized groups on the surface of the SWCNTs, this novel technique improved adhesion between the SWCNTs and the polymer matrix. There are reports on structural modification of SWCNTs forming more stable structures, effected through a transformation during sliding operation that encompasses low friction coefficient for the SWCNT solids. From these observations, we could predict that the process of formation and the specific treatment given to SWCNTs before being exposed to various functional applications largely influences their inherent properties.

19.5.3.2 Multi-Walled CNTs (MWCNTs)

Because of their low cost, MWCNTs in composites are expected to expand industrial applications. Besides, MWCNTs incorporated coatings have superior frictional coefficient and wear properties compared to SWCNTs-reinforced coatings. Several coating composites with MWCNTs have been already reported such as electroless MWCNTs/SiC, MWCNTs/Cu, MWCNTs/Al_2O_3, MWCNTs/epoxy, MWCNTs/Ni, MWCNTs/polyurethane, dual-layer MWCNTs/Ag, and MWCNTs/polyethylene oxide composite coatings. Because of their ability to inhibit crack propagation, MWCNTs were used in ceramic coating to improve toughness and wear resistance. Thus, many investigations have been done using MWCNTs-reinforced coating by various fabrication routes. Furthermore, CNTs were dispersed in a metal/polymer matrix to create novel coatings for tribological applications. MWCNTs are attached to the metal matrix by very weak van der Waals forces in this case. Because of the weak bonding between CNT and metal, they could simply slide between the contact surfaces, lowering the friction coefficient of the composite. MWCNT mixed with various types of mineral oil was found to be beneficial in reducing friction and wear due to rotating motion and better sliding mechanism in various processes. MWCNT incorporated metal matrix was employed as a composite for developing brake friction materials. These types of hybrid composite materials were analyzed for their efficacy through compression strength, density, and hardness to know their mechanical behavior.

19.5.4 Synergistic Effect CNT and Conventional Tribo-Filler Materials in Property Enhancement

There is a unique geometrical characteristic for each type of carbon nano-filler. As a result, several groups of tribologists studied the effect of different carbon nano-fillers as reinforcements on the mechanical properties of polymer materials. Potential candidates for improving the mechanical properties of neat polymers include CNTs and graphene-based nano-fillers. The hybrid fillers' specific shape has higher aspect ratios than other carbon nano-fillers, which is critical in obtaining optimized mechanical properties of the polymer nanocomposites. The studied synergistic effect of CNTs and graphene nanoparticles with a 1 : 9 filler ratio has higher tensile modulus, tensile strength, and elongation than other nanocomposites [76]. Later, CNTs has been grown on graphene nanoparticles and their tensile modulus, tensile strength, and fracture strain were higher than traditional CNT/GNP/epoxy trinary nanocomposites [77].

Similarly, graphene oxides and CNTs were used to enhance the tensile properties of polyamide, with the ratios of graphene oxides and CNTs of 3 : 1 and 1 : 9 exhibiting the best tensile strength and tensile modulus, respectively [78]. Recently, amine-functionalized multilayer graphene and CNTs were used as reinforcements to investigate the synergistic effect of these two nano-fillers on the mechanical properties of epoxy nanocomposites [79]. They claimed that a 1 : 3 graphene:CNT ratio had the greatest synergistic effect on the epoxy's tensile and flexural strengths, as well as its Young's modulus.

19.6 High-Temperature Tribology of CNTs

Several important aspects of the anti-frictional material should be deleterious once we consider the high-temperature interaction of two solid surfaces such as automotive, aerospace, power generation, and metalworking processes. This type of moving machine generates high temperature mainly due to frictional heating. These operated mechanical systems at high temperatures should have superior efficiency, performance, and reliability on friction and wear characteristics of contacting materials. Several reports have been published on the application of CNTs incorporated hybrid composites that can be utilized in high-temperature tribology.

19.6.1 Friction and Wear at High Temperature

The majority of studies on the effect of higher temperature operation on friction and wear have focused on hot forming and press hardening areas. This is due to the application of lightweight materials in vehicles to the reduction of fuel consumption. Recent reports emphasize the use of carbon nanomaterials as a friction material in commercial brakes as reinforcing fibers such as organic fibers, metal fibers, and ceramic fibers, which can have better wear resistance and friction stability at high temperatures. CNTs have shown property enrichment of nanocomposites as an alternative fiber in brake friction materials because of their high stiffness and high aspect ratios combined with low density. Despite the fact that much research has been done to enhance the properties of CNTs-based composites, controlling the dispersion and distribution of nano-scale fibers in a composite is a difficult problem. Numerous studies have been carried out to improve wear resistance by using different ratios of CNTs in composites or through modification of the CNTs surface [80]. However, the use of CNTs in friction materials at high temperatures is quite limited, whereas at low temperatures, CNTs as hybrid composites demonstrated potential candidates for tribological applications.

19.6.2 Control of Friction and Wear at High Temperatures

Conventional lubricating oils and greases are potential materials to control friction and wear. These materials degrade quickly and lose their lubricating properties above 300 °C. Speciality lubricants (mostly phase change materials), solid lubricant materials, composites, and surface coatings or claddings are the main approaches to control friction and wear at high temperatures. High-temperature tribology is currently focused on developing materials and surface engineering technologies to control friction and wear.

19.7 Biotribology of CNTs

The science of friction, lubrication, and wear in biological systems or natural phenomena is known as biotribology. It is a multifaceted and multidisciplinary system that affects every aspect of our daily lives, from prosthetic implants to personal care products. MWCNT is known as biocompatible and high-strength nanomaterials which attracted the attention of tissue engineering. It acts as a bridge between crack tips and prevents crack propagation. CNTs' unique structure allows them to transport organic compounds such as anti-inflammation drugs and bio-corrosion inhibitors. That also improves their cell production stimulation ability and lubricant features. MWCNT and Mg/PEO systems are considered for biomedical applications. The commonly used elements in orthopedic implants (screws and plates) and living tissues (bones and muscles) are in contact with physiological environment. As a result, dynamic conditions occur during surgical manipulations and post-healing periods. For this, an ideal biodegradable material for orthopedic implants should be wear resistant. MWCNTs incorporated into plasma electrolytic oxidation coatings on Mg alloys have been reported as potential biodegradable orthopedic implants [81]. Here, we have discussed several current reports on specific biotribology.

19.7.1 Joint Tribology

In the twentieth century, one of the greatest surgical successes was total joint arthroplasty when polytetrafluoroethylene (PTFE) was introduced for total hip replacements. Because of its low coefficient of friction, PTFE was later replaced by ultra-high molecular weight polyethylene, resulting in implant failure. Because of the wear particles generated by the polymer–metal interface, this type of joint replacement has a limited lifetime. The most common cause of implant failure is bone loss caused by the generated wear particles named "osteolysis-induced loosening." Later metal alloys or ceramics were replaced with ultra-high molecular weight polyethylene to improve from other problems such as pseudotumors, metal toxicity, and squeaking. Polyimide-CNTs nanocomposites were reported as lower friction material than pristine polyimide and claimed that polyimide-CNTs could replace ultra-high molecular weight polyethylene as a joint replacement material [82].

19.7.2 Inspiring Biomimetic Materials in Tribology Application

The term biomimetic comes from the Greek words bios, which means "life," and mimetikos, which means "imitative." Biomimetic materials are materials that have been inspired by nature. Examples of these materials include spider silk strength, bird flight mechanics, sharkskin water repellency, and so on. There are very few reports on the biomimetic nature of CNTs in tribology applications. Superhydrophobic surfaces with high and low contact angle hysteresis exhibit self-cleaning and low fluid drag. The lotus leaf is one of the natural specimens for superhydrophobic surfaces. Chae et al. demonstrated superhydrophobicity, self-cleaning, and low drag in a CNT composite with hierarchical structures [83]. Recently, ultra-high molecular weight polyethylene-based biomimetic materials were modified by orientation treatment and doping with fluorinated MWCNTs to improve their tribological behavior in order to simulate the wear of dog cartilage (Figure 19.3) [84].

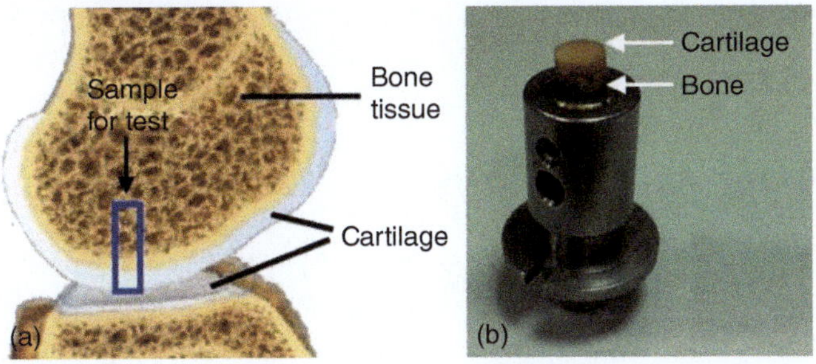

Figure 19.3 (a) Location of dog cartilage sample extraction (b) attachment for sample mounting for pin-on-disk tribological testing. Source: Maksimkin et al. [84]/with permission of Elsevier.

19.8 Modified CNTs for Commercial-Scale Tribology Applications

Several studies on the tribological application of CNT-based materials in small-scale research have already been published, but vast commercial applications remain limited. Some of the patents reported that claim the commercial tribology applications of CNT-based materials. The invention reported on CNTs in liquid fuels and lubricants to accelerate the burning by changing electrostatic interactions and viscosity [85]. Another invention reported on the reduction of friction in lubricating oils by the addition of lubricating oil containing CNTs [86]. In this case, SWCNTs have received more attention than the functionalized other carbon nanomaterials. CNT reinforced metal composite coating is also proposed for commercial tribological application [87]. According to this invention, CNTs (preferably SWCNTs) reinforced metal nanocomposites showed significantly higher thermal conductivity and electrical conductivity than pure metal phases. It is also observed that mechanical strength compared with pure metal phases improved, along with tailorable thermal expansion coefficient with the changing concentration of CNTs in the nanocomposite. It is expected that further small scale research will allow using CNTs-based materials for commercial tribological applications. Recent research describes the use of –COOH functionalized MWCNT materials as an additive in improving the tribological performance of two types of synthetic lubricants commonly used in lubricant applications: polyalphaolefin (PAO 100) and polypropylene glycol (PPG 2000). Studies are also evolving on the effect of the addition of surfactant functionalized MWCNT that could improve the film thickness of surfactant-MWCNT-oil that improved lubrication performance. One other work reports on the industrial application of MWCNT-ionic liquid hybrids as an additive combination for enhancing tribological performance of lithium grease. Various research groups delve into the use of these carbon nanomaterials in tribological applications and their commercial feasibility to make the process more efficient and economic.

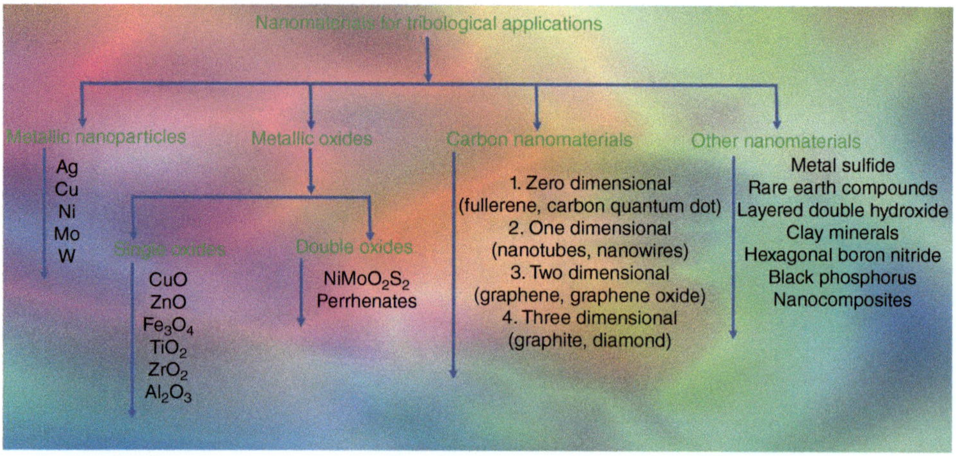

Figure 19.4 Summary of inorganic nanomaterials investigated for tribological applications.

19.9 Other Nanomaterials for Tribological Applications

With the evolution of nanotechnology, contemporary research focused on applying a range of nanomaterials for tribological applications. Therefore, varieties of nanomaterials based on, viz., inorganic, organic, refractory, and polymeric nanomaterials have been investigated for their use in tribological applications. Recently, several inorganic nano-additives have been reported for tribological applications such as metal nanoparticles [88, 89], metal oxides [90, 91], sulfides [92, 93], and layered silicates [94]. The interesting properties of nanomaterials attract investigation on it, especially as lubricant additives. Particles ranging in size from 2 to 100 nm are easily able to penetrate the friction area and replenish the microroughness. This intriguing feature prevents slipping on the contacting surface. Nanoparticles can smoothly precipitate or generate protective films on the friction surface due to their high surface energies [95]. They have also another outstanding feature of self-repairing effect on the worn surface. A summary of various inorganic nanomaterials is presented in Figure 19.4, which have been recently reported for tribological applications.

19.10 Summary and Outlook

CNTs have sparked considerable interest due to their unique properties, which include a large surface area, exceptional optical properties, superior mechanical strength, and high thermal and electrical conductivity. Because of their unique properties, they are suitable candidates for a wide range of applications, including electronics, biotechnology, and environmental applications. CNT and composite have also fascinated consideration due to their applications as lubricants and friction-free materials which are also used as effective filler, grain refiners, and coating components. Several experimental studies support that individual CNT showed much higher friction and wear properties than those of its coatings. Accordingly, new experimental strategies should be explored to tune and bridge between nanoscale and the macro-scale natures of CNTs. CNTs nano-lubricants, on the other hand, are likely to have improved heat transfer properties due to their nonspherical shape. CNT superlubricity is highly anticipated for engineering applications at the macroscale, but it is only understood at the nanoscale with perfect structures free of defects. The role of wear and adhesion of the nano-rolling mechanism of CNTs should be explored utilizing MD simulation. The nanofriction mechanism of CNTs at different atmospheres (N_2, H_2, Ar, etc.) can be further investigated experimentally along with the support of computational simulation to understand the potential route of friction control.

The development of CNT composites inspires both academia and industry due to their self-lubricating properties, high thermal, mechanical, and wear resistance. Understanding the nature, processing, and tailoring of these composites remains a challenge. However, most of the reports dealing with such chemically modified CNTs have focused on small-scale-based laboratory experiments. Further research is necessary for the application on large scale along the reuse technique of CNTs needs to be explored. The main limitation is the cost-effective operation of CNTs on large scale. Hence, some focused research on the CNTs by chemical modification is recommended, which can further reduce the cost of operation on a large scale and the toxicity of CNTs.

The critical drawback in this field, as found in many reports, is that the limitation of the CNTs during a specific application is solely attributed to the specific characteristics of the CNT used. But it is not applicable in many cases. The surface/coating should be developed in such a way that it holds the CNT to perform in the best way it can during long-term applications. Accordingly, a secondary tuning of the coating characteristics during its development is specifically required. It is hoped that the introduction of several CNT composites, their internal engineering, and stimulation for fundamental and impactful discovery can offer the possibility of next-generation materials for tribological applications.

Important Websites (Companies, Organizations, Research Groups)

1. www.nanowerk.com
2. www.electrichybridvehicletechnology.com
3. Kimaya Vyavhare and Pranesh B. Aswath, Materials Science and Engineering Department, University of Texas at Arlington, Arlington, TX, USA
4. W. Zhang, G.J. Ma, and C.W. Wu, Dalian University of Technology, State Key Laboratory of Structural Analysis for Industrial Equipment, Faculty of Vehicle Engineering and Mechanics, Dalian, 116024, PR China
5. Jie Chen, Wen Wang, and Sheng Zhang, Southwest Jiaotong University, School of Mechanical Engineering, Chengdu, 610031, China

References

1 Cornelio, J.A.C., Cuervo, P.A., Hoyos-Palacio, L.M. et al. (2016). *J. Mater. Res. Technol.* 5: 68–76.

2 Bhaumik, S., Prabhu, S., and Uppala, N. (2013). *J. Chem. Eng. Mater. Sci.* 4: 123–127.

3 Joly-Pottuz, L., Martin, J.M., Dassenoy, F. et al. (2005). *Carbon Nanotubes and Onions as Lubricant Additives*, 611–612. Paper No: WTC2005-63220, https://doi.org/10.1115/WTC2005-63220.

4 Khalil, W., Mohamed, A., Bayoumi, M., and Osman, T.A. (2016). *Fullerenes Nanotubes Carbon Nanostruct.* 24: 479–485.

5 Salah, N., Abdel-Wahab, M.S., Alshahrie, A. et al. (2017). *RSC Adv.* 7: 40295–40302.

6 Arkharova, I.V. and Zaporotskova, I.V. (2015). *Eurasian Chem. Technol. J.* 17: 295–300.

7 Kinoshita, H., Nishina, Y., Alias, A.A., and Fujii, M. (2014). *Carbon* 66: 720–723.

8 Song, H.-J. and Li, N. (2011). *Appl. Phys. A* 105: 827–832.

9 O. Elomaa (2015), Carbon nanomaterials and carbon coatings in lubrication, *Aalto University publication series Doctoral Dissertations,* ISBN:978-952-60-6231-0.

10 Kiu, S.S.K., Yusup, S., Chok, V.S. et al. (ed. A. Saptoro, W. SiawKhur, L.S. Wei, et al.) *Comparison on Tribological Properties of Vegetable Oil upon Addition of Carbon Based Nanoparticles*, vol. 206, 012043. IOP Publishing.

11 Jeng, Y.-R., Huang, Y.-H., Tsai, P.-C., and Hwang, G.-L. (2014). *J. Tribol.* 136: 041801.

12 Reinert, L., Green, I., Gimmler, S. et al. (2018). *Wear* 408: 72–85.

13 Pierson, H.O. (2012). *Handbook of Carbon, Graphite, Diamonds and Fullerenes: Processing, Properties and Applications*. William Andrew.

14 Moghadam, A.D., Omrani, E., Menezes, P.L., and Rohatgi, P.K. (2015). *Composites Part B* 77: 402–420.

15 Xu, Z., Zhang, Q., Shi, X. et al. (2015). *J. Mater. Eng. Perform.* 24: 1926–1936.

16 Chen, C.S., Chen, X.H., Xu, L.S. et al. (2005). *Carbon* 43: 1660–1666.

17 Peng, Y., Hu, Y., and Wang, H. (2007). *Tribol. Lett.* 25: 247–253.

18 Treacy, M.M.J., Ebbesen, T.W., and Gibson, J.M. (1996). *Nature* 381: 678–680.

19 Ebbesen, T.W., Lezec, H.J., Hiura, H. et al. (1996). *Nature* 382: 54–56.

20 Song, K., Zhang, Y., Meng, J. et al. (2013). *Materials* 6: 2543–2577.

21 Scharf, T.W., Neira, A., Hwang, J.Y. et al. (2009). *J. Appl. Phys.* 106: 013508.

22 Tjong, S.C. (2013). *Mater. Sci. Eng., R* 74: 281–350.

23 Reinert, L., Suárez, S., and Rosenkranz, A. (2016). *Lubricants* 4: 11.

24 Chen, X.H., Chen, C.S., Xiao, H.N. et al. (2006). *Tribol. Int.* 39: 22–28.

25 Arai, S., Fujimori, A., Murai, M., and Endo, M. (2008). *Mater. Lett.* 62: 3545–3548.

26 Lin, C.B., Chang, Z.-C., Tung, Y.H., and Ko, Y.-Y. (2011). *Wear* 270: 382–394.

27 Ahmad, I., Kennedy, A., and Zhu, Y.Q. (2010). *Wear* 269: 71–78.

28 Jang, J., Bae, J., and Yoon, S.-H. (2003). *J. Mater. Chem.* 13: 676–681.

29 Gong, X., Liu, J., Baskaran, S. et al. (2000). *Chem. Mater.* 12: 1049–1052.

30 Sen, R., Zhao, B., Perea, D. et al. (2004). *Nano Lett.* 4: 459–464.

31 Moniruzzaman, M. and Winey, K.I. (2006). *Macromolecules* 39: 5194–5205.

32 Mohamed, A., Osman, T.A., Khattab, A., and Zaki, M. (2015, Paper No: TRIB-13-1131, 011801 (5 pages)). *J. Tribol.* 137.

33 Kamel, B.M., Mohamed, A., El Sherbiny, M., and Abed, K.A. (2016). *Ind. Lubr. Tribol.* 68 (6): 723–728.

34 Mobasher, A., Khalil, A., Khashaba, M., and Osman, T. (2019). *Ind. Lubr. Tribol.* 72 (1): 9–14.

35 Mohamed, A., Ali, S., Osman, T.A., and Kamel, B.M. (2020). *J. Mater. Res. Technol.* 9: 2054–2062.

36 Mohamed, A., Tirth, V., and Kamel, B.M. (2020). *J. Mater. Res. Technol.* 9: 6178–6185.

37 Pei, X., Hu, L., Liu, W., and Hao, J. (2008). *Eur. Polym. J.* 44: 2458–2464.

38 Jia, X., Huang, J., Li, Y. et al. (2019). *Appl. Surf. Sci.* 494: 430–439.

39 Min, C., He, Z., Song, H. et al. (2019). *Appl. Sci.* 9: 170.

40 Wang, Z., Ren, R., Song, H., and Jia, X. (2018). *Appl. Surf. Sci.* 428: 630–639.

41 Coleman, J.N., Khan, U., Blau, W.J., and Gun'ko, Y.K. (2006). *Carbon* 44: 1624–1652.

42 Hong, H., Thomas, D., Waynick, A. et al. (2010). *J. Nanopart. Res.* 12: 529–535.

43 Chen, H., Wei, H., Chen, M. et al. (2013). *Appl. Surf. Sci.* 283: 525–531.

44 Yujun, G., Zhongliang, L., Guangmeng, Z., and Yanxia, L. (2014). *Int. J. Heat Mass Transfer* 74: 358–367.

45 Christensen, G., Yang, J., Lou, D. et al. (2020). *Synth. Met.* 268: 116496.

46 Joly-Pottuz, L., Dassenoy, F., Belin, M. et al. (2005). *Tribol. Lett.* 18: 477–485.

47 Neville, A., Morina, A., Haque, T., and Voong, M. (2007). *Tribol. Int.* 40: 1680–1695.

48 Huang, H.D., Tu, J.P., Gan, L.P., and Li, C.Z. (2006). *Wear* 261: 140–144.

49 Gupta, B.K., Bhushan, B., Capp, C., and Coe, J.V. (1994). *J. Mater. Res.* 9: 2823–2838.

50 Bhushan, B., Gupta, B.K., Van Cleef, G.W. et al. (1993). *Appl. Phys. Lett.* 62: 3253–3255.

51 Hwang, C.C., Wang, Y.C., Kuo, Q.Y., and Lu, J.M. (2010). *Physica E* 42: 775–778.

52 Zang, J.-L., Yuan, Q., Wang, F.-C., and Zhao, Y.-P. (2009). *Comput. Mater. Sci.* 46: 621–625.

53 Agrawal, P.M., Sudalayandi, B.S., Raff, L.M., and Komanduri, R. (2006). *Comput. Mater. Sci.* 38: 271–281.

54 WenXing, B., ChangChun, Z., and WanZhao, C. (2004). *Physica B* 352: 156–163.

55 Li, Y., Wang, S., Arash, B., and Wang, Q. (2016). *Carbon* 100: 145–150.

56 Tiwari, S.K., Sahoo, S., Wang, N., and Huczko, A. (2020). *J. Sci.: Adv. Mater. Devices* 5: 10–29.

57 Mittal, G., Dhand, V., Rhee, K.Y. et al. (2015). *J. Ind. Eng. Chem.* 21: 11–25.

58 Mittal, V. (2014). *Macromol. Mater. Eng.* 299: 906–931.

59 Yang, Y., Liu, R., Wu, J. et al. (2015). *Sci. Rep.* 5: 1–7.

60 Tiwari, S.K., Mishra, R.K., Ha, S.K., and Huczko, A. (2018). *ChemNanoMat* 4: 598–620.

61 Liao, L., Peng, H., and Liu, Z. (2014). *J. Am. Chem. Soc.* 136: 12194–12200.

62 Liu, L., Fang, Z., Gu, A., and Guo, Z. (2011). *Tribol. Lett.* 42: 59–65.

63 Tao, X., Jiazheng, Z., and Kang, X. (1996). *J. Phys. D: Appl. Phys.* 29: 2932–2937.

64 Shen, M., Luo, J., and Wen, S. (2001). *Tribol. Trans.* 44: 494–498.

65 Shinjo, K. and Hirano, M. (1993). *Surf. Sci.* 283: 473–478.

66 Zhang, R., Ning, Z., Zhang, Y. et al. (2013). *Nat. Nanotechnol.* 8: 912–916.

67 Meng, Y., Xu, J., Jin, Z. et al. (2020). *Friction* 8: 221–300.

68 Zhang, S., Ma, T., Erdemir, A., and Li, Q. (2019). *Mater. Today* 26: 67–86.

69 Vazirisereshk, M.R., Ye, H., Ye, Z. et al. (2019). *Nano Lett.* 19: 5496–5505.

70 Urbakh, M. (2013). *Nat. Nanotechnol.* 8: 893–894.

71 Socoliuc, A., Gnecco, E., Maier, S. et al. (2006). *Science* 313: 207–210.

72 Wen, Q., Qian, W., Nie, J. et al. (2010). *Adv. Mater.* 22: 1867–1871.

73 Zhang, R., Wen, Q., Qian, W. et al. (2011). *Adv. Mater.* 23: 3387–3391.

74 Satyanarayana, N., Rajan, K.S.S., Sinha, S.K., and Shen, L. (2007). *Tribol. Lett.* 27: 181–188.

75 Samad, M.A. and Sinha, S.K. (2011). *Tribol. Int.* 44: 1932–1941.

76 Yang, S.-Y., Lin, W.-N., Huang, Y.-L. et al. (2011). *Carbon* 49: 793–803.

77 Li, W., Dichiara, A., and Bai, J. (2013). *Compos. Sci. Technol.* 74: 221–227.

78 Min, C., Liu, D., Shen, C. et al. (2018). *Tribol. Int.* 117: 217–224.

79 Qi, Z., Tan, Y., Zhang, Z. et al. (2018). *RSC Adv.* 8: 38689–38700.

80 Hwang, H.J., Jung, S.L., Cho, K.H. et al. (2010). *Wear* 268: 519–525.

81 Daavari, M., Atapour, M., Mohedano, M. et al. (2021). *Surf. Interfaces* 22: 100850.

82 Cai, H., Yan, F., and Xue, Q. (2004). *Mater. Sci. Eng., A* 364: 94–100.

83 Jung, Y.C. and Bhushan, B. (2009). *ACS Nano* 3: 4155–4163.

84 Maksimkin, A.V., Chukov, D.I., Senatov, F.S., and Salimon, A.I. (2017). *Mater. Lett.* 191: 105–107.

85 Moy, D., Niu, C., Tennent, H., and Hoch, R. (2005). Fuels and lubricants containing carbon nanotubes, filed 26 October 2004 and issued 26 May 2005, patent number- US 2005/0108926A1.

86 Habeeb, J.J. and Bogovic, C.N. (2013). Reduced friction lubricating oils containing functionalized carbon nanomaterials, filed 16 July 2010 and issued 23 May 2012, patent number- EP2454352A1.

87 Chen, Q. (2010). Carbon nanotube reinforced metal composites, filed 19 May 2006 and issued 26 January 2010, patent number- US7651766B2.

88 Gulzar, M., Masjuki, H.H., Kalam, M.A. et al. (2016). *J. Nanopart. Res.* 18: 223.

89 Padgurskas, J., Rukuiza, R., Prosyčevas, I., and Kreivaitis, R. (2013). *Tribol. Int.* 60: 224–232.

90 Bogunovic, L., Zuenkeler, S., Toensing, K., and Anselmetti, D. (2015). *Tribol. Lett.* 59: 29.

91 Gupta, R.N. and Harsha, A.P. (2018). *Lubr. Sci.* 30: 175–187.

92 Kalin, M., Kogovšek, J., Kovač, J., and Remškar, M. (2014). *Tribol. Lett.* 55: 381–391.

93 Chen, L. and Zhu, D. (2017). *Ceram. Int.* 43: 4246–4251.

94 Nan, F., Xu, Y., Xu, B. et al. (2015). *Tribol. Int.* 81: 199–208.

95 Ali, M.K.A., Xianjun, H., Mai, L. et al. (2016). *Tribol. Int.* 103: 540–554.

20

Chemically Modified Carbon Nanotubes for Corrosion Protection

Iman Mohammadi[1], Seyyed Arash Haddadi[2], Mazdak Izadi[3], and Mohammad Mahdavian[2]

[1] *Tarbiat Modares University, Faculty of Engineering, Department of Materials Engineering, Jalal AleAhmad Highway, Nasr Bridge, Tehran 14115 143, Iran*
[2] *Institute for Color Science and Technology, Surface Coating and Corrosion Department, No. 55, Vafamanesh St., Lavizan Exit, Sayad Shirazi North HWY, Tehran 1668836471, Iran*
[3] *Hamedan University of Technology, Department of Materials Engineering, Shahid Fahmideh Boulevard, Hamedan, Iran*

20.1 Introduction

All metallic structures are susceptible to corrosion and should be protected using different methods, e.g. coatings, cathodic protection, and corrosion inhibitors [1]. Protective coatings can postpone corrosion deterioration through different mechanisms [2]. However, structure failure will occur even in the most efficient and well-applied coatings [3]. So, the main issue is time. To expand the coating's useful life, different additives with versatile benefits were used in industries. Among different categories of materials, nano-objects attract special attention due to their remarkable characteristics [4].

In the last decade, in both industrial and research fields, the application of carbon nanotubes (CNTs) as one of the most well-known 1D nanostructures has drawn more and more attention for the fabrication of advanced nanocomposites with enhanced electrical, mechanical, thermal, wear resistive, and anticorrosion characteristics [5]. Owing to the unique characteristics of CNTs, e.g. large specific surface area, excellent electrical conductivity, outstanding mechanical and chemical stability, hollow layered structures, and high aspect ratio, CNTs donate multifunctional properties to corrosion protective nanocomposite coatings. Single-walled carbon nanotubes (SWCNTs) or multi-walled carbon nanotubes (MWCNTs) are made of impermeable single or multilayers of carbon and hydrogen atoms, which can enhance barrier properties of corrosion protective nanocomposite coatings [5]. Despite the advantages mentioned above, all CNT-based composite coatings suffer from a lack of dispersion [6]. CNTs tend

Chemically Modified Carbon Nanotubes for Commercial Applications, First Edition.
Edited by Jeenat Aslam, Chaudhery Mustansar Hussain, and Ruby Aslam.

to entangle because of their high surface area and length-to-diameter ratio [7]. Nanocomposite coatings with poor mechanical, electrical, and corrosion mitigation performance (CMP) are achievable by directly integrating CNTs within a coating matrix. In fact, the lack of surface functional groups and polar sites on the large hydrophobic surface of CNTs accounts for aggregations formation and poor interfacial adhesion with coating matrices. Acceptable interfacial interaction between coating matrix and CNT is vital for preparing protective coatings [8]. Therefore, applying a uniformly dispersed CNTs-based nanocomposite coating with uniform dispersion is a major concern in the industrial usage of CNT.

However, based on the synthesis methods of CNTs, by manipulating synthesis conditions, some structural defects are formed in CNTs structure; and an increase in the chemical reactivity of CNTs is achieved by the carbon double bonds creation and graphene edges occurrence on the surface of CNTs. Various surface modification methods, including noncovalent functionalization, oxidation, silanization, and grafting polymer chains onto the surface of CNTs, have been introduced to enhance the compatibility of CNTs with coatings matrices.

20.2 Modification Approaches

20.2.1 Oxidized CNTs

One of the most applicable routes to modify CNTs to use them as coating filler is the oxidation process. Different approaches were used to oxidize CNTs. Oxidation methods can be categorized into gas and liquid-phase techniques. Plasma oxidation and X-ray lithography are two gas-phase methods employed to create functional oxygen-based groups on the CNTs. The oxidation level in gas-phase methods is usually considerable, and severe destruction of nanotubes is occurred [9]. Liquid-phase methods include electrochemical and chemical oxidation of CNTs using chemicals, e.g. inorganic acids, hydrogen peroxide, ozone, and potassium permanganate [10]. The surficial damages are not intensive in liquid-phase methods and are preferred in some applications compared to gas-phase methods [9]. The surface functional groups' arrangement is also different between gas and liquid-phase processes. The graphitic surface of CNTs is directly functionalized by the attachment of oxygen-containing groups. However, the most affected part of CNTs in liquid acid treating is the two ends of CNT.

In the electrochemical approach, a conductive electrolyte is prepared, and CNTs are applied to the electrode in different shapes such as paste. The anodic potential is applied, and the exposed CNTs are oxidized [11]. This method is usually utilized when CNTs-based electrodes are expected. However, CNTs oxidation is mostly carried out using acid etching methods in the field of anticorrosion composite coatings.

Functional groups, e.g. hydroxyl and carboxyl, are the main expected moieties on the CNTs after oxidation. These hydrophilic groups guarantee the consequent dispersion of CNTs in the matrix of anticorrosion coating. Based on the protection mechanism, oxidized CNTs can be used to prepare anticorrosion coatings:

1. *Barrier coatings*: formation of dense and defect-free coating on the metal surface, blocking the corrosive moieties toward the metal surface;
2. *Conductive coatings*: preparation of conductive and adhesive film on the metals and alloys with active/passive behavior, leading to the creation of uniform passive layer;
3. *Sacrificial coatings*: zinc-rich primers with active nature related to the metal substrate, resulting in cathodic protection of coated metal.

20.2.1.1 CNTs-Based Barrier Coatings

The addition of surface oxidized CNTs were done for the following reasons in the barrier coatings [12, 13]:

- The embedded CNTs with good dispersity create a uniform and defect-free nanocomposite with reduced and prolonged diffusion paths toward the metal/coating interface;
- Development of nanocomposite coatings with better mechanical properties and barrier performance due to the improved interfacial interactions between organic/inorganic matrix and carbon-based additive;
- Improvement in the coating-substrate adhesion because of functional groups presented on the uniformly dispersed CNTs;
- Reduction in water uptake of coatings due to the more hydrophobicity nature of protective films in the presence of CNTs.

MWCNTs were oxidized using an acids mixture (HNO_3–H_2SO_4) by Vu et al. to prepare anticorrosion epoxy-based coating [12]. They reported that hydroxyl and carboxyl groups were created by acid treatment on the MWCNTs. Higher crosslinking density leading to lower porosities and coating adhesion promotion were proposed as CMP of epoxy coating containing oxidized MWCNTs. As mentioned before, surface oxidation of CNTs can be used to prepare them for the next modification steps. MWCNTs were oxidized by Najmi et al. for subsequent polyaniline (PANI) grafting decorated with zinc species [13]. The coating loss of adhesion prepared using the MWCNTs-based additive was considerably improved because of the lower diffusion rate of corrosive moieties due to the denser coating formation. Akbarzadeh et al. used CNTs with different oxidation levels to enhance the protective behavior of thin films [14]. They reported that lower water uptake, higher impedance modulus ($|Z|_{0.01Hz}$), and fewer defects were detected in the coating containing CNTs with more oxidation levels. The effect of oxidized CNTs on the coating behavior is depicted in Figure 20.1.

20.2.1.2 CNTs-Based Conductive Coatings

Coatings based on conductive polymers (CPs) are the new and effective approaches in regards to corrosion protection. Substrate passivation is the main inhibitive mechanism in this group of coatings. CNTs were considered as filler in the conductive

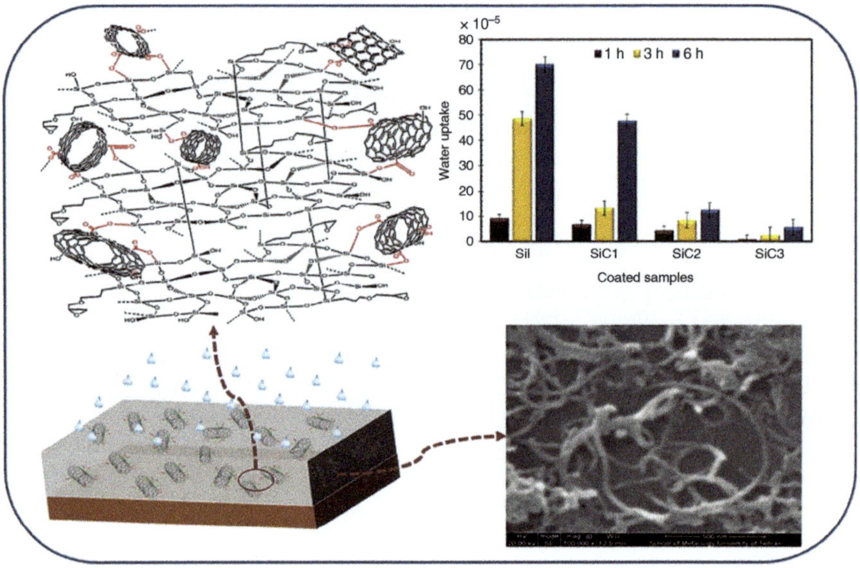

Figure 20.1 The effect of CNTs with different levels of oxidation on the structure and water uptake of silane nanocomposite coating. Source: Akbarzadeh et al. [14]/with permission of Elsevier.

coatings, e.g. PANI, polypyrrole (PPy), poly(*o*-toluidine), and polyurethane (PU) [7, 15]. The authors proposed that utilization of CNTs decreased the coating defects and promoted the passive surface film quality in this type of coating [16].

The addition of oxidized CNTs into the conductive coatings improves the protective efficiency of the system through different mechanisms [7]:

- Reducing the coating porosity, ensuring the coating cohesion, and conductance;
- Enhancing the oxygen reduction reaction, improving metal passive film formation in the coating/substrate interface;
- Improving the polymerization of CPs in the *in situ* coating preparation methods;

So, it is evident that the oxidized CNTs have considerable effects on the CMP of the conductive coatings.

20.2.1.3 CNTs-Based Sacrificial Coatings

In the sacrificial coatings, the galvanic couple formed between the metal substrate and dispersed active particles is the main reason for corrosion protection. The most common sacrificial system is zinc-rich primer. Zinc particles are in electrical contact and form a continuous network in the zinc-rich primer connected to the metal substrate. Cathodic protection of metal substrate is the main protective action in these coatings.

According to the literature, any improvement in the electrical connection of zinc particles leads to better protective behavior of primers. In regard to this point, carbon-based nano-objects such as CNTs are good candidates to enhance zinc-rich systems. The high aspect ratio of CNTs makes them efficient additives in zinc-rich primers [17]. Also, CNTs show a lower percolation threshold than the other carbon-based materials [18]. CNTs are intrinsically conductive. Furthermore, they improve the zinc particles' electrical connection increasing the protective functioning of the coating. Another positive effect of CNTs in zinc-rich coatings is mechanical properties improvement. All zinc-rich primers suffer from mechanical damage due to the high amounts of zinc particles. If CNTs with proper surface functional groups are incorporated into the coating, a strong threaded-like network will be obtained due to the good CNT-matrix interaction. CNTs can also improve the barrier performance of zinc-rich primer due to the reduction of pores and other defects. The coating-substrate adhesion improvement has also been reported for zinc-rich primers in the presence of CNTs [19].

The inhibition mechanism of zinc-rich primers is entirely different from the conductive coatings. Water uptake is the first step in the CMP of zinc-rich coatings. In the second step, zinc particles are activated by electrolyte diffusion into the organic/inorganic primer, and the substrate is cathodically protected due to the standard potential difference between zinc and coated metal. In the third step, after particles deactivation, long-term barrier protection of substrate is achieved because of zinc corrosion products deposited in the coating diffusion paths near the metal/primer interface [20]. So, the metals such as steel can be efficiently protected with zinc-rich primers. These three steps are proposed by Cubides et al. assigned to a zinc-rich primer with 60 and 70 wt% zinc particles in the presence of CNTs (Figure 20.2) [21].

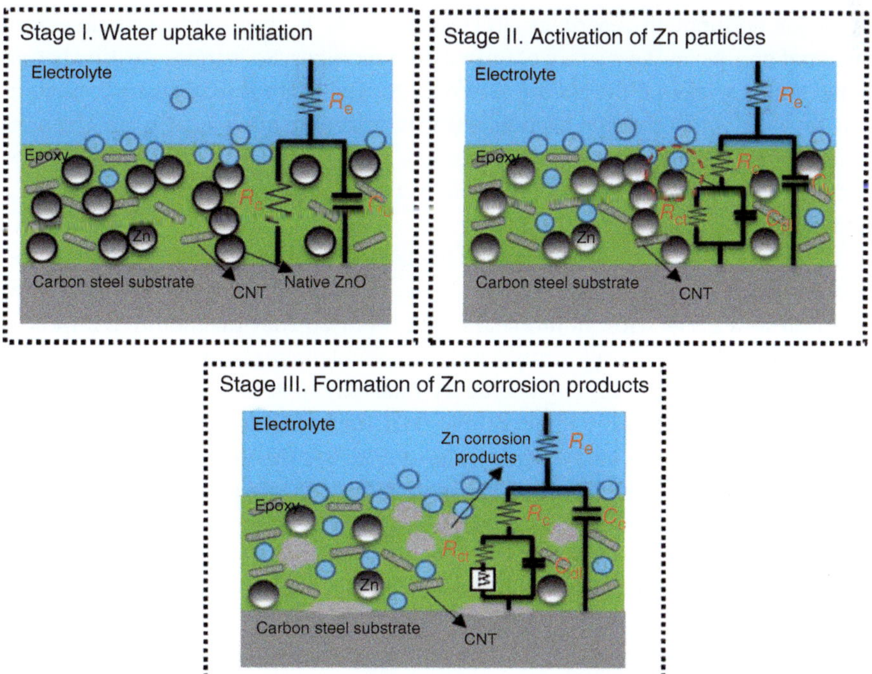

Figure 20.2 Protection mechanism in the CNT-containing epoxy zinc-rich coating: (a) Step 1: Water uptake, (b) Step 2: Zinc particle activation, and (c) Step 3: Formation of zinc corrosion products. Source: Cubides and Castaneda [21]/with permission of Elsevier.

As expected, the abovementioned positive effects of CNTs are conditional on the good dispersion of these additives into the primer matrix. CNTs treating with acids leading to surface oxidation can be considered as an effective route.

20.2.2 Silanized CNTs

Coupling agents are materials that link different additives, matrices, and substrates together. Organosilanes [$X_3Si(CH_2)_nY$] contain organofunctional (Y) reacting with organic materials and alkoxy (X) groups reacting with inorganic materials [22]. Organosilanes can be used to link surfaces containing functional moieties, e.g. hydroxyl and carboxyl (COOH), with other materials depending on the silane molecular structure.

The utilization of silane compounds can be used for functionalizing nanotubes to achieve different characteristics [23–25], including:

- Improving the CNTs-matrix interface to ensure good dispersion, leading to uniform and defect-free corrosion resistance coating;
- Inducing hydrophobic behavior on the CNTs, resulting in the reduction of electrolyte adsorption throughout the coating and consequent decrease in the corrosion rate of the coated substrate;
- Affecting the coating characteristics such as polymerization or curing degree, improving the CMP of the coating;

A major problem in regard to CNTs silanization is the shortage of necessary functional groups on the surface of untreated nanotubes. So, surface treating before silanization is vital in different applications. According to the literature, oxidation of CNTs with acid mixtures such as sulfuric acid and nitric acid is an essential pretreatment before the silanization process [26]. Hydroxyl and carboxyl groups created by oxidizing provide the appropriate condition for attaching silane coupling agents onto the CNTs.

An amino functionalized organosilane has been used by Zhou et al. to enhance the epoxy nanocomposite coating curing [25]. They modified carboxylic acid functional groups on MWCNTs with amino silane and reported that this additive improves coating crosslinking and uniformity. The hydrophobicity of polymethylmethacrylate (PMMA) coatings has been engineered by Raei Nayini et al. by embedding silane functionalized MWCNTs into the coating [23]. Different hydrophobic organosilanes were investigated in this study. Results showed that PMMA coating with a contact angle of about 140° can be obtained in the case of silane-modified MWCNTs nanocomposite coating. Ma et al. used 3-glycidoxypropyltrimethoxysilane to functionalize oxidized MWCNTs [24]. They concluded that treatment of MWCNTs with the organosilane directly affects the dispersion of the nanotubes in the coating matrix.

It is evident that surface-modified CNTs through silanization can be considered as a solution to produce engineered nanocomposite coatings with better characteristics and longer lifetimes exposed to different environments. However, there are limited reported works regarding using silanized CNTs for corrosion protection.

Table 20.1 A quick review of the investigated oxidized/silanized CNTs and their constituents with the corrosion protection approach.

Treatment method	Coating class	Probable protective mechanism in the presence of treated CNTs
Oxidation	Barrier	Coating compactness improvement, defect reduction, barrier effect, adhesion promotion, mechanical properties improvement, hydrophobicity
	Conductive	Defect reduction, active protection (promotion of passive film formation), coating compactness improvement (polymerization agent), and defect reduction
	Sacrificial	Active protection (conductance improvement), defect reduction, and mechanical properties improvement
Silanization	Barrier	Improvement in hydrophobicity and coating compactness (crosslinking agent)
	Intermediator/primer	Adhesion promotion (forming linkage between the inorganic substrate and organic coating matrix)

A quick review of the investigated oxidized/silanized CNTs and their constituents with the corrosion protection approach is provided in Table 20.1. New surface-modified CNTs are introduced every day to enhance the corrosion resistance of organic/inorganic coatings. So, finding new functionalizing materials to improve the versatile advantages of CNT will attract more attention toward corrosion protection in the future years.

20.2.3 Polymer Grafting

Grafting polymers onto CNTs not only improves the interfacial adhesion of CNTs to the polymer matrix but also donates doping primacy to modified CNTs. Functionalities of grafted polymer chains onto the surface of CNTs can be used as active sites for further chemical treatments of CNTs to fabricate sandwich-like CNT-based nanostructures [27, 28].

Covalent functionalization of CNTs influences the atomic structure and electronic network of CNTs with sp^2 carbon atoms and alters them to sp^3 carbonaceous atomic structures. Grafting polymers on CNTs is known as a promising functionalizing method for dispersibility improvement of CNTs in solvents and polymers. In this method, two approaches are proposed for grafting polymers on CNTs, including "grafting to" and "grafting from" strategies [29]. In the "grafting to" approach, commercially provided or as-synthesized polymer chains are attached to the surface of CNTs. For this purpose, the possible reactions happen between functionalities of polymer chains and the surface of CNTs such as carboxylic, carbonyl, and hydroxyl via amidation, and radical coupling. For some polymers, to increase the degree of grafting, functionalities of CNTs e.g. carboxylic and hydroxyl, created by oxidization reactions should be replaced with more reactive functional groups such as acyl chloride [27, 28].

In the "grafting from" approach, the grafting of polymers is performed by *in situ* polymerization of desired monomers on CNTs with a certain degree of reactivity. For the activation of CNTs' surface, a combination of oxidation, surface posttreatment, and treatment by initiators is required to provide enough active sites for polymerization reactions [28, 29]. In this method, desired properties such as specific polymer/CNTs ratios, the morphology of grafted polymers, and a certain degree of polymerization are approachable by tuning the synthesis routes and conditions. Based on the reactivity and chemistry of monomers, different polymerization techniques are utilized such as radical, condensation, ring opening, and ionic polymerizations [28–30]. A wide range of thermoplastic polymers, biopolymers, CPs, etc. has been polymerized and used in corrosion and other applications. Among different types of polymers, CPs, i.e. PANI, PPy, and polythiophenes family such as poly(3,4-ethylenedioxythiophene) (PEDOT), polydopamine (PDA), and their derivatives were widely used for the chemical modification of CNTs [27, 28, 31]. In Sections 20.2.3.1 and 20.2.3.2, the most recent works on polymer grafted CNTs, used in corrosion prevention will be overviewed.

20.2.3.1 CPs

CPs such as PANI, PPy, and PEDOT are excellent candidates for modification of CNTs, owing to their superior properties such as lightweight, nontoxicity, facile synthesis, doping, loading primacy, and good electrical conductivity. Unique electrochemical features and an outstanding combination of electronic/ionic conductivities of CPs have attracted much attention in the corrosion protection fields [27, 32]. Detailed corrosion mitigative mechanisms of polymer coatings filled with CPs grafted CNTs have not been interpreted completely, but most of the publications have reported the following possible mechanisms.

1. *Barrier improvement*: The presence of many entanglements and aggregations in pristine CNTs are the most important challenges for being dispersed in a polymer. Functionalization and grafting of CNTs with CPs have illustrated promising mechanical, electrical, thermal, and anticorrosion properties within polymer matrices. Grafting of polymers on CNTs leads to dispersion of individual modified CNTs and extends the modified CNTs/polymer interfaces in the final nanocomposites. Consequently, the overall barrier properties of the final nanocomposite are much higher than the unfilled coating. Moreover, the placement of polymer grafted CNTs in microscopic defects of the coatings obstacles pathways for oxygen, chloride, and water, reducing the corrosion rate [28, 33].
2. *Controlled release of dopants*: Different dopants with corrosion inhibitive properties such as NO_3^- and 4-amino-3-hydroxy-naphthalene-1-sulfonic acid can be used during the oxidative polymerization of CPs. During the reduction of CPs, the on-demand release of dopants in defects leads to the formation of inhibitive complexes and mitigates active corrosion reactions by constructing physical and chemical barriers in the coating/metal interfaces [27, 32, 34].

3. *Anodic inhibition*: As stated in the literature, the presence of CPs in polymer coatings may generate metal oxides, acting as a barrier layer and preventing corrosion reactions [32, 35].
4. *Cathodic inhibition*: The incorporation of CPs grafted CNTs in polymer coatings makes the final nanocomposite coating denser, less porous with better compactness and adhesion to the metallic substrate. Subsequently, the reduction reactions of oxygen are transferred from the metal/coating interface to CPs/coating interfaces. The restriction of cathodic reactions on the metal leads to the generation of less cathodic corrosion products, i.e. hydroxyl groups. The decrement of oxygen reduction rates postpones the cathodic delamination of coating to longer periods and prolongs the durability of the CPs grafted CNTs filled nanocomposite coatings [32, 36]. However, oxygen reduction reactions on the CPs/coating interfaces cause CPs reoxidation and reduce their active CMP. Still, the dispersion of CPs grafted CNTs in polymer coatings alters the overall conductivity of the coating and maintains the open circuit potential (OCP) of the coating close to its passive state [32].

All aforementioned mechanisms for corrosion mitigative mechanisms of CPs highlight their important rules as one of the most important categories of polymers for the chemical modification of CNTs. Consequently, the surface modification of CNTs with CPs not only enhances the mechanical and barrier performance of polymer coatings but also makes them smart to recover themselves in the case of any damage. Based on the type of CPs, polymer coatings, and metals, CNTs/CPs ratio, dopant's type, and concentration, synthesis methods (chemical or electrochemical), and electronic/ionic conductivity level, they follow at least one or more of the aforementioned corrosion mitigative mechanisms.

Among CPs, PANI in the forms of the matrix of coatings, fillers, fibers, micro/nanocarriers, and surface modifier has demonstrated good corrosion prevention due to its good stability in different environments, low price, ease in synthesis, and excellent CMP [28, 37]. PANI can be synthesized in three different states, including pernigraniline base (PB), emeraldine base (EB), and leucoemeraldine base (LB). Complete oxidation, semi-complete oxidation, and complete reduction of PANI are assigned to PB, EB, and LB, respectively. Among these states of PANI, only protonated emeraldine salt (ES) is considered as the conductive PANI. Owing to the existence of positively charged sectors on PANI chains, they can be attached to oxidized CNTs with negative charges via both electrostatic and covalent interactions [32, 38].

Akbarzadeh et al. [38] studied the CMP of PANI grafted oxidized CNTs (PCNT) in a silane coating. After oxidation of CNTs, PANI chains were grafted on the surface of oxidized CNTs using hydrogen bond, electrostatic, and π–π interactions in the presence of hydrochloric acid, as presented schematically in Figure 20.3. Different states of PANI, including ES, polystyrene (PS), and PB in different colors, including green, dark blue, and purple, were observed at pHs 4, 8, and 12, respectively. Electrochemical impedance spectroscopy (EIS) results of silane coatings (intact and scratched) displayed a bi-functional barrier-active CMP. Grgur [39] studied the

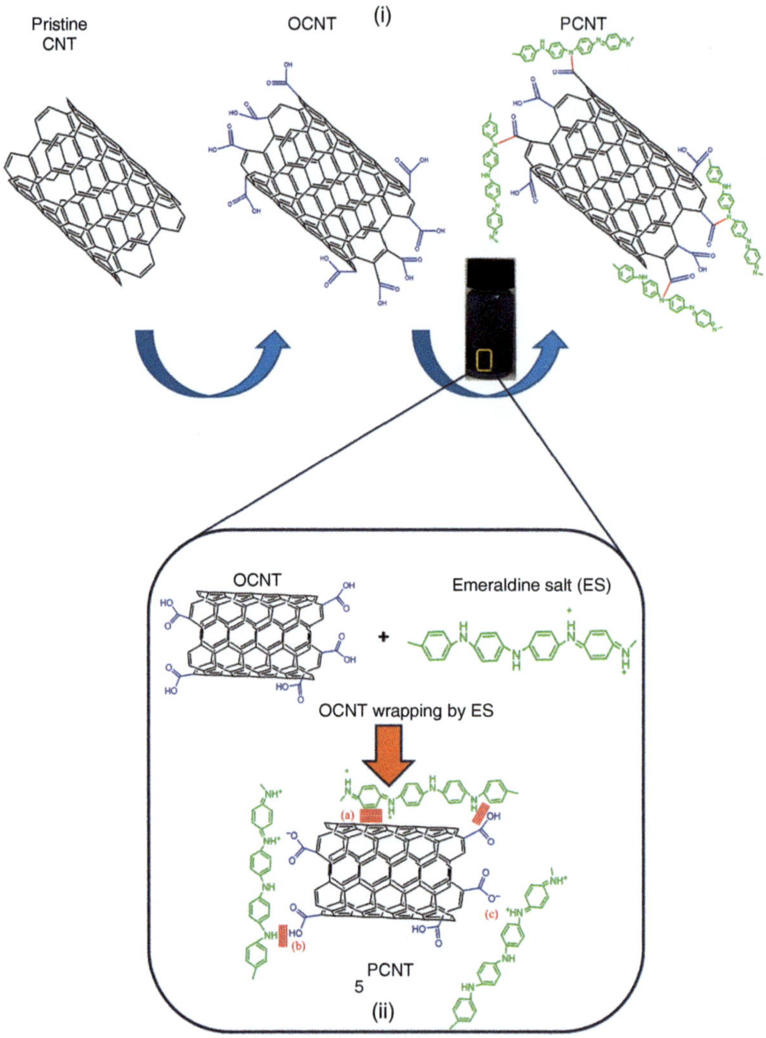

Figure 20.3 The synthesis scheme of PANI- grafted CNTs (i) and possible interactions between PANI chains and functionalities of oxidized CNTs (ii). Source: Akbarzadeh et al. [38]/with permission of American Chemical Society.

inhibition mechanism of aniline oligomers on mild steel (MS). Results showed that aniline changes the kinetics of oxygen reduction reactions on MS from a 4-electron path to a 2-electron path. As a result, the formation of hydrogen peroxide converts Fe^{2+}, generated by anodic reactions to Fe^{3+}, forming $Fe(OH)_3$ on the cathodic sites. This mechanism can explain the self-healing performance of polymer coatings containing PANI.

Qiu et al. [40] synthesized PANI-grafted CNTs nanostructures (c-PANI) via the *in situ* oxidative polymerization of aniline monomers for the fabrication of acrylic resin modified epoxy nanocomposite coatings. The unique brush-like morphology of c-PANI illustrated outstanding CMP on MS at a wide range of pH. Accordingly, owing to the formed charge transfer linkages, which originated from hydrogen bond and π–π interactions between PANI chains and functionalities of CNTs, the c-PANI/epoxy nanocomposite coating demonstrated superior barrier performance at neutral and alkaline pHs. While the protection mechanism at acidic pHs followed the passivation protection.

The long-term durability of a polyvinyl butyral (PVB) coating filled with 2 wt% PANI grafted carboxylic acid-functionalized multi-walled CNTs (PANI-CMWCNTs) was studied by Li et al. [41] on AA2024 aluminum panels. In the presence of the PANI-CMWCNTs/PVB coating, the corrosion current density of aluminum panels decreased five orders of magnitude compared to uncoated panels. In addition, the total impedance modulus ($|Z|_{0.01\,Hz}$) of the PANI-CMWCNTs/PVB coating showed three orders of magnitude increase compared to the blank PVB coating after 60 days of immersion in a 3.5 wt% NaCl electrolyte. All electrochemical observations proved the barrier enhancement and a passive layer formation over the aluminum panels as corrosion inhibitive mechanisms.

As discussed earlier, CPs grafted CNTs nanostructures can be used as nanocarriers for other active agents such as organic and inorganic corrosion inhibitors. Owing to the functionalities of both oxidized CNTs and grafted CPs chains, they can provide enough sites for the attachment of corrosion inhibitors or other types of active agents. In other words, they can be considered as potent modified nanostructures for the fabrication of sandwich-like multifunctional nanostructures using layer-by-layer assembly. This posttreatment leads to the on-demand triggered release of these active agents and promotes the self-healing/active CMP of polymer coatings filled with modified sandwich-like CPs/CNTs. In this way, Najmi et al. [42] constructed novel PANI-grafted highly-oxidized CNTs (O-MWCNT) doped with Zn^{2+}(O-MWCNT-PANI-Zn) nanostructures according to the synthesis scheme, shown in Figure 20.4. Comprehensive characterization of O-MWCNT-PANI-Zn nanostructures revealed that PANI chains were mostly grafted via covalent bonds, while Zn^{2+} were doped via electrostatic and π–cation interactions. In the solution phase, the $|Z|_{0.01\,Hz}$ value of bare MS dipped in the O-MWCNT-PANI-Zn extract improved by around 400% compared to the control solution. Impedance characterization of intact and scratched coatings indicated that the release of PANI, Zn^{2+}, and their complexes with metal accounts for the significant barrier/active CMP of the O-MWCNT-PANI-Zn/epoxy nanocomposite coating. Moreover, the incorporation of 0.15 wt% O-MWCNT-PANI-Zn nanostructures improved remarkably the coating adhesion and cathodic disbonding reduction. The combination of barrier improvement by modified CNTs, anodic and cathodic inhibitions, as well as surface passivation was mentioned as inhibition mechanisms.

As discussed earlier, PANI and its derivatives are capable of being grafted on CNTs via noncovalent interactions. π–π stacking, hydrogen bonds, and electrostatic attractions are the most probable interactions noncovalent binding mechanisms. In another study, poly(2-butylaniline) (PBA) was used by Cui et al. [43] as a dispersant agent for MWCNTs to be dispersed uniformly via π–π interactions in an epoxy matrix. A uniform suspension of MWCNTs in tetrahydrofuran (THF) was

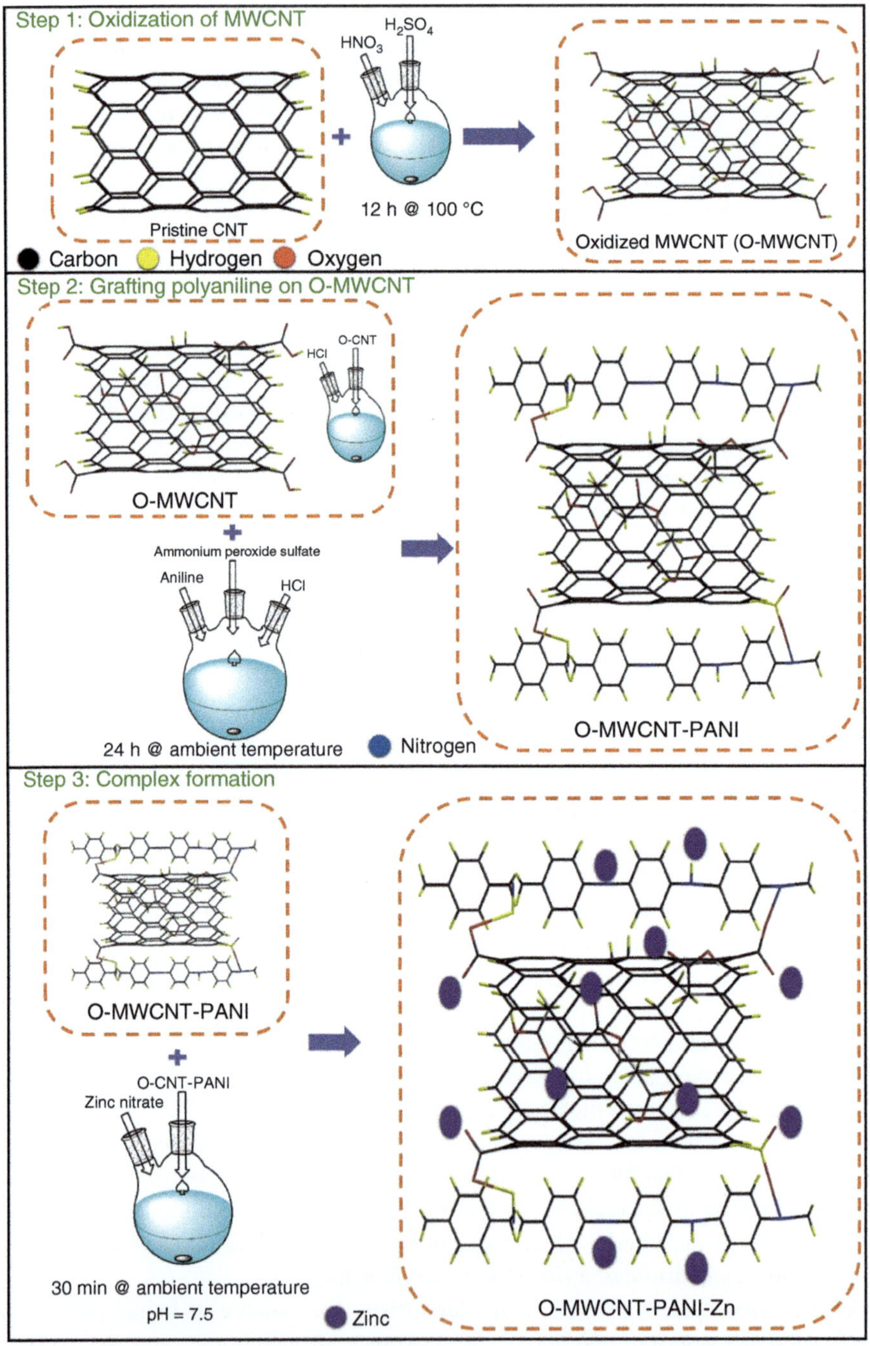

Figure 20.4 The detailed synthesis scheme of O-MWCNT-PANI-Zn nanostructures. Source: Najmi et al. [42]/with permission of Elsevier.

obtained in the presence of PBA. The dispersion of 1 wt% MWCNTs in the epoxy coating enhanced the water absorption capacity by 0.36%. Also, the long-term CMP of the MWCNTs/epoxy coating with the PBA dispersant was improved significantly, owing to the better dispersion of MWCNTs in the presence of PBA chains. As stated in the literature, PANI and its derivatives are potent polymeric candidates for modification of CNTs to enhance anticorrosion, mechanical, electrical, and thermal properties of CNTs polymer nanocomposites.

Owing to superior characteristics of PPy such as straightforward synthesis, high thermal stability, high conductivity at a wide range of pHs, good environmental stability, ion-exchange capability, and promising CMP, its application in the field of corrosion is appealing [32, 44]. In this way, one research has been conducted on the CMP of PPy/CNTs nanocomposite coating, applied *in situ* on carbon steel, by Ionita et al. [44]. Polyaminobenzene sulfonic acid-functionalized CNT (CNT-PABS) and acidized SWCNTs with carboxylic acid (CNT-CA) were used as nanofillers. Experimental characterization methods coupled with computational molecular modeling were utilized to investigate the mechanic and anticorrosion of the final nanocomposite coatings. The molecular dynamics modeling proved the random distribution of PPy chains around CNTs. The Young's modulus of final nanocomposite coatings containing about 5 wt% CNTs reached from 2.67 GPa for the neat PPy coating to 3.35–3.96 GPa for the CNT-CA/PPy coating and 4.15–4.61 GPa for the CNT-PABS/PPy coating due to the better dispersion and compatibility of CNT-PABS with PPy chains. The CMP of prepared coatings was in the order of CNT-PABS/PPy > CNT-CA/PPy > neat PPy. The better CMP of CNTs filled PPy nanocomposite coatings originated from the barrier enhancement and the electron transport hindrance of the PPy coating reinforced with CNTs.

The corrosion behavior of steel samples coated with alkyd-based paints filled with nanoalumina, polystyrene-sulfonate (PSS)-functionalized, and sulfonated MWCNTs supported PPy was evaluated in 1 M NaCl and HCl electrolytes [45]. Results revealed that the synergism between alumina@ and CNTs@PPy provide remarkable CMP in harsh chloride-containing environments, while the alkyd coating filled with alumina@PPy does not afford good CMP in harsh corrosive environments. For the alkyd coating filled with alumina@ and CNTs@PPy, an increase in the conductivity of immobile counter-ion PPy and the redox activity enhancement account for the better CMP.

As an intrinsic CP, PEDOT is the best one in this category, owing to its superior conductivity, stability, and processability. Also, PEDOT is the only CP that has been produced on a large scale for numerous applications. Shining characteristics such as mechanical flexibility, transparency, film-forming capability, and tunable conductivity make PEDOT an appropriate polymer for the fabrication of advanced nanocomposites [46]. Despite its wide applications in solar cells, electrodes, sensors, and supercapacitors, there are a few publications on the PEDOT application in the field of corrosion. The occurrence of localized corrosion on the oxide layer of aluminum is one of the main disadvantages of this metal in the field of lithium-ion batteries. Carboxylic acid-functionalized CNTs can be used as a dopant in the polymerization of CPs, not only altering the ion transport mechanisms but also improving the electrostatic repulsion capacity of a CP. Based on a recent publication [31], nano-sized PEDOT nanospheres (PEDOT-NSP) coated MWCNTs (PEDOT-NSP/MWCNT), and coaxial-structured PEDOT (PEDOT-COX) wrapped MWCNTs (PEDOT-COX/MWCNT) were used for the aluminum protection against $LiPF_6$. A Cu_2O-assisted emulsion polymerization was used for the decoration of PEDOT nanospheres (2–5 nm) on the surface of carboxylic acid-functionalized MWCNTs. A better CMP was observed for aluminum substrates coated with PEDOT-NSP/MWCNT compared to the PEDOT-COX/MWCNT and neat PEDOT coatings. The interpretation of results proved the remarkable synergistic effects between PEDOT chains with cation exchangeability and negatively charged MWCNTs with anion-repulsive properties. The treatment of MWCNTs with COO^- groups significantly reduced the diffusion of pit-causing PF_6^- anions.

PDA is one of the well-known biopolymers with relative electrical conductivity, which contains aromatic rings and different functionalities, e.g. amine, catechol, and imine groups. These functionalities provide many active sites for physical and chemical interactions of PDA chains with other functionalized nanostructures and activated surfaces [47, 48]. In addition, carbon-based nanomaterials such as CNTs are capable of interacting with PDA chains via π–π stacking interactions. PDA chains can be synthesized mainly by the polymerization of dopamine monomers (DA) via electropolymerization, enzymatic oxidation, and solution oxidation methods [33, 48]. The solution oxidation method, which is a self-polymerization route for the polymerization of DA monomers, is performed in an alkaline solution in contact with atmospheric oxygen [48]. The literature has reported that changes in the local pHs lead to the PDA release from its frameworks. Moreover, PDA grafted nanostructures are capable of being used as platforms for the fabrication of hybridized nanostructures, due to the diverse functionalities of PDA. The aforementioned functionalities are potent active sites for the synthesis and growth of secondary nanostructures and/or grafting other active agents [33, 47]. It was found that the incorporation of 2 wt% PDA grafted CNTs (PDA@CNTs) in a PU coating improved the adhesion strength and CMP of the final PDA@CNTs/PU coating. Characterization results demonstrated that the better dispersibility and greater compatibility of PDA@CNTs in the PU coatings compared to the pristine CNTs are responsible for enhanced mechanical properties and CMP [49]. The layer-by-layer assembly of PDA, chitosan (CH), and Zn^{2+} on the surface of oxidized MWCNTs (O-MWCNT-PDA-CH-Zn) was designed by Ghahremani et al. [33]. Analytical and computational studies results of O-MWCNT-PDA-CH-Zn nanocarriers revealed the successful polymerization of PDA on the surface of O-MWCNTs. CH molecules and Zn^{2+} were attached to the surface of PDA grafted O-MWCNTs via the π–π/π–cation and hydrogen bonding interactions. The efficient release of PDA, CH, and Zn^{2+}components in a saline electrolyte (3.5 wt% NaCl) was confirmed by electrochemical and microscopic analysis. The CMP of scratched O-MWCNT-PDA-CH-Zn/epoxy coatings enhanced up to 360% compared to the blank epoxy coating. After 63 days of immersion, the $|Z|_{0.01\text{Hz}}$ value of the O-MWCNT-PDA-CH-Zn filled epoxy coating reached about $\approx 1.2 \times 10^{10}$ Ω cm^2, which was much higher than that of the blank epoxy coating ($\approx 9.8 \times 10^5$ Ω cm^2).

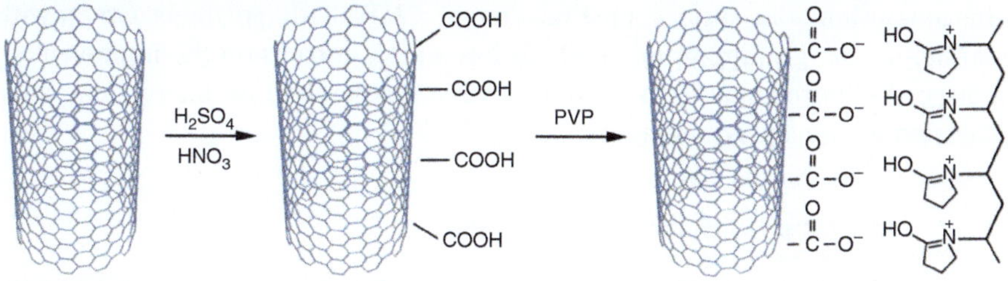

Figure 20.5 The synthesis scheme of m-MWCNTs. Source: Chen et al. [50]/with permission of Elsevier.

Characterization of coatings revealed significant enhancements of the barrier and active CMP of intact and scratched epoxy coatings with the layer-by-layer fabricated O-MWCNT-PDA-CH-Zn nanocarriers.

20.2.3.2 Other Polymers

CPs, due to their supreme CMP, have been drawn much attention compared to other polymers in corrosion mitigation applications. A wide range of other polymers such as PMMA, polystyrene (PS), poly-ε-caprolactam (PCL), polyacrylic acid (PAA), and polyvinylimidazole (PVI) were grafted chemically on CNTs via different polymerization methods [28]. However, a few publications reported the effects of non-CPs grafted CNTs or conductive/nonconductive copolymers grafted CNTs on corrosion. In a recent study [50], polyvinylpyrrolidone (PVP) grafted MWCNTs (m-MWCNTs) were used as nanofillers in a powdered fusion-bonded epoxy coating (FBE). According to the synthesis scheme illustrated in Figure 20.5, the generation of carboxylic acid functionalities on the surface of MWCNTs was performed via oxidation reactions. A "grafting-from" method was used for the modification of oxidized MWCNTs with PVP molecules. The uniform dispersion of m-MWCNTs was carried out using the combination of hot melt extrusion and ball milling methods. An electrostatic spray gun was used to spray the final nanocomposite epoxy coating on MS panels. The dispersion of 0.2 wt% m-MWCNTs improved the tensile strength by 53.7%. Up to 10 weeks, the value of coating resistance (R_c) reached from $1.98 \times 10^5\ \Omega\,cm^2$ for the neat epoxy coating to $1.18 \times 10^9\ \Omega\,cm^2$ for the m-MWCNTs filled epoxy coating.

Superior anticorrosion and barrier properties of MS were obtained by uniform dispersion of polyindole-grafted MWCNTs (f-MWCNTs/Pin) in an epoxy coating [51]. A "graft-to" synthesis route based on a radical polymerization was used for *in situ* polymerization of indole monomers. Activation and functionalization of MWCNTs were carried out via acid treatments to create –COOH groups on the MWCNTs surface. Electrochemical characterization of f-MWCNTs/Pin/epoxy nanocomposite coatings containing various contents of f-MWCNTs/Pin revealed that the coating filled with 0.25 wt% f-MWCNTs/Pin has higher CMP than those with 0.1 and 0.5 wt%. The R_c value reached from 2.683×10^4 for the blank epoxy to $1.72 \times 10^7\ \Omega\,cm^2$ for the coating filled with 0.25 wt% f-MWCNTs/Pin after 240 hours. Better dispersion of f-MWCNTs/Pin and the increase in the pathway of the electrolyte accounted for the enhanced CMP of the nanocomposite coating.

In conclusion, the surface modification of CNTs with polymers either with "grafting-to" or "grafting-from" methods has strong impacts on the improvement of dispersibility, compatibility, and molecular/intermolecular interactions after dispersion within a polymer matrix.

20.2.4 Decoration with Inorganic Materials

Due to the combined properties of two materials, the decoration of the CNTs with the inorganic nanomaterials may create novel features. As a result, these composite materials have found highly appealing prospective applications in various fields [52]. It was illustrated that the nanocrystals can be decorated on CNT's surface. The inorganic materials can interact with bulky CNTs in the form of covalent bonding [53], π–π stacking interaction [54], hydrophobic interaction, or hydrogen bonding [55], and electrostatic interaction [56]. The production and dispersion of high purity CNTs in appropriate mediums are required for decorating the CNT with proper inorganic materials. The insolubility of CNTs has been proven to be a substantial impediment to their modification and utilization. CNTs should be dispersed at the individual nanotube level to operate well in most applications, including the modification process of CNT. As a result, several approaches to increasing dispersion ability of CNT in solvents were researched. They have mainly divided into sidewall covalent functionalization and noncovalent alteration employing guest molecules as solubilizers [52]. Also, there are two main processes to decorate the inorganic materials on the surface of the CNT. The first approach is the decoration of pre-synthesized inorganic materials on the surface of the CNT. Another approach for synthesizing the composite structure is the *in situ* synthesis of the inorganic nanocrystals on the CNTs surface [52].

In the first approach, the nanostructure materials are fabricated before decorating them on the surface of the CNTs. Because of the presence of several methods that have previously been established for preparing inorganic nanocrystals with regulated morphology and size, this method makes it practical to control the morphology and size of the nanocrystals decorating on CNT. In this route, binary, ternary, or even quaternary nanocomposite structures such as Co_3O_4/CNTs, Au/CNTs, TiO_2/CNTs, Au/TiO_2/CNTs, TiO_2/Co_3O_4/CNTs, and Co/CoO/Co_3O_4/CNTs can be manufactured [57]. However, at high loading content, as-synthesized CNTs usually lack sufficient binding sites to absorb inorganic nanocrystal precursors, resulting in aggregation, low nanocrystals loading efficiency, and poor nanocrystal dispersion. By functionalizing or wrapping CNT with suitable "gluing molecules," more nanocrystals can be loaded. The CNTs surface groups play an essential role in interacting with created inorganic nanocrystals [58].

More binding sites, such as defects and functional groups (e.g. carbonyl, carboxyl groups, hydroxyl), can be produced on the surfaces of the CNTs by treating them with microwave irradiation, ultrasonication, acid, or γ-ray irradiation. CNT defects or functional groups can act as nucleation or anchoring sites for inorganic nanocrystal deposition [59]. Han et al., for example, observed that the treating of SWCNT nanoparticles with an appropriate acid might result in the homogeneous fabrication of tin oxide on them when submerged in $SnCl_2$ solution [60]. One reaction precursor may be adsorbable by the binding sites, which subsequently reacts with another reactant, such as $NaBH_4$, ethylene glycol, H_2, formaldehyde, or polyol [61, 62]. Moreover, functional groups can act as nucleation and anchoring sites and reducing agents for metal nanocrystals produced on CNT. These methods enable the production of a wide range of inorganic nanocrystal/CNT composites in a highly efficient manner. However, as previously indicated, the breakdown of the CNTs structure produced by the harsh procedure may diminish the route to decorating the nanoparticles on the surface of the CNTs.

To the *in situ* synthesis of the inorganic nanocrystals on the CNTs surface, several techniques were used to manufacture the inorganic nanocrystals, which are directly decorated on as-prepared CNTs. One of the first attempts used high-temperature sputtering, evaporation, or thermal degradation for the decoration of CNT with the inorganic nanocrystals [63]. Heating or sputtering can convert noble metals to gas. CNTs are then used as nucleation scaffolds for gaseous metal atoms, on which metallic nanocrystals are synthesized. Within the gaps between tubes in CNT bundles, for example, nanoparticle arrays of Cu, Ag, and Au, as well as ultrafine Zr, Mo, and Ti nanowires, were formed. Furthermore, nanotube curvature might be used to influence the orientation of cluster arrays and nanowires [64]. On the other hand, organometallic compounds break down quickly at high temperatures.

Other approaches for *in situ* decorating CNTs with semiconducting nanocrystals is included atomic layer deposition (ALD) and chemical vapor deposition (CVD) [65]. However, to minimize the damage created on the structure of CNTs at such high temperatures, these activities must be carried out in an oxygen-free environment. Using electroless chemical deposition, metal nanocrystals may be formed directly on CNTs. The size of the metal nanoparticles covering the CNTs may be varied by altering the reactant concentration and reaction duration. However, surfactants are often used in this procedure to scatter the CNTs so that the nanocrystals generated may be evenly placed onto the CNTs. Only metals with incredibly high reduction potential, such as noble metals, are suitable for this electroless chemical deposition technique [66, 67]. Electrochemical deposition provides a method for the decoration of several types of inorganic materials on the CNTs. CNTs should first be deposited onto the working electrodes of electrochemical cells in a standard electrochemical deposition method. Several factors such as reactant concentrations, reaction duration, and applied voltage may be changed to adjust the size and amount of loading of the inorganic nanocrystals. Other parameters, including CNT fabrication techniques, pretreatment, and CNT conductivity, all play essential roles in determining the nanocrystals size decorated on CNTs. The decorated inorganic materials have enormous size through this approach.

The decoration of the inorganic materials on the CNT can be used in various coating structures to enhance the CMP of the metallic substrates. For this purpose, the CNTs are first modified with the covalent and noncovalent bonding by using an oxidizing process and/or grafting a suitable polymer. Then, the nanocrystals are coated on CNTs and dispersed in the various coating matrix. Sivaraj and Vijayalakshmi [68] used spray pyrolysis to create zinc substituted hydroxyapatite decorated CNT thin films deposited on 316L-SS substrates. Consequently, the current approach produces a suitable biomaterial with excellent antibacterial characteristics and good CMP. Rajitha et al. [69] developed ZnO/MWCNT nanohybrids as nanohybrid fillers to enhance the protection performance of the PU coating. The results demonstrated that well-dispersed nanofillers improved the PU coating's compatibility, density, and hydrophobicity. ZnO NPs assisted CNT dispersion in the PU matrix. PU nanocomposite coating containing ZnO/CNT has shown good protective effectiveness.

He et al. [70] acidified the MWCNT using a nitration mixture to oxidize its surfaces; the SiO_2 decorated on MWCNT was then performed at room temperature through sol–gel polymerization of tetraethoxysilane (TEOS). The findings reveal that the mechanical characteristics and corrosion resistance of the SiO_2–MWCNTs/epoxy composite coating were improved due to increased barrier functionality. Guo et al. [71] used the sol–gel function to produce ZnO decorated MWCNTs (ZnO–MWCNTs). ZnO–MWCNTs were also incorporated into a chemically bonded ceramic coating to improve the CMP of the stainless steel. Because of the presence of ZnO nanoparticles on the surface of the CNTs and the excellent interaction between the ZnO and aluminum phosphate (AP) binder, there were no cracks between the ceramic matrix and the MWCNTs. Consequently, MWCNTs were coated with ceramic gel, leading to excellent bonding strength. Furthermore, electrochemical test results showed that using ZnO–MWCNTs may significantly increase the ceramic coatings CMP. In another work conducted by Cai et al. [72], sandwich-like CeO_2 decorated PDA grafted CNTs nanostructures (CeO_2-PDACNTs) were synthesized to enhance the CMP of a PU coating, applied on an aluminum alloy substrate (AA-7075). A schematic procedure to prepare CeO_2-PDACNTs is provided in Figure 20.6. Electrochemical evaluations exhibited that the corrosion resistance

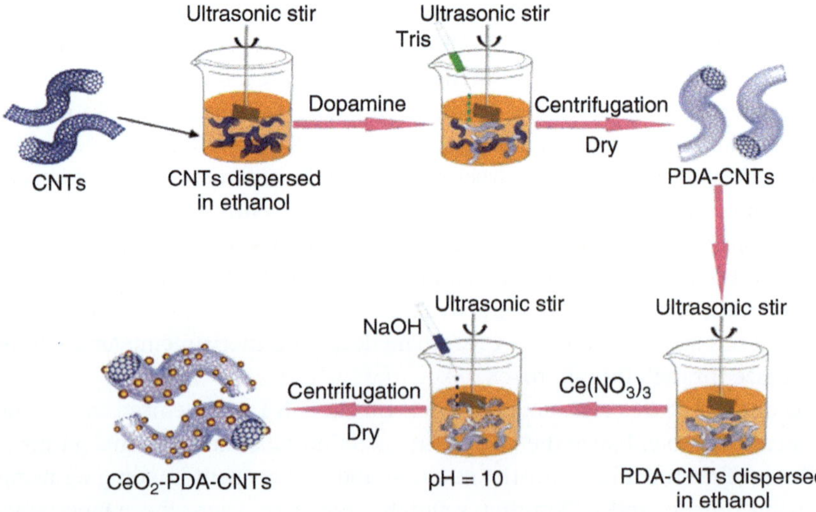

Figure 20.6 Schematic procedure to prepare CeO_2-PDACNTs. Source: Cai et al. [72]/with permission of Elsevier.

of the PU coating filled with CeO_2-PDACNTs after seven cycles of aging tests reached $10^8\ \Omega\ cm^2$, which is higher by around 3 orders of magnitude than the blank PU. The synergy of CeO_2, as a corrosion inhibitive agent, and enhanced barrier performance by modified CNTs accounted for the enhanced corrosion resistance, UV irradiation resistance, and anti-stripping ability of the PU/CeO_2-PDACNTs nanocomposite coating. Kumar et al. [73] decorated the functionalized CNT with CeO_2 nanoparticles. The produced nanocomposite was then used as a nanofiller in PU coatings. Their findings showed that steel covered with a PU coating including CNT/CeO_2 had much superior CMP than steel coated with a pure PU coating or a PU coating containing CNTs alone. Kumar et al. [74] used ultrasonic cavitation to decorate MWCNTs with –OH groups carrying TiO_2 nanoparticles. The decoration process was achieved in this study to improve the dispersion ability and loading content of CNT in the epoxy matrix. In terms of anticorrosion and mechanical performance, the resultant MWCNT/TiO_2 hybrid-epoxy nanocomposite surpasses the nanocomposite created by loading only MWCNTs or TiO_2 nanoparticles, as well as neat epoxy.

20.2.5 Loading with Corrosion Inhibitor

The incorporation of corrosion inhibitors into the structure of the coatings is an efficient method to improve the coatings' CMP. For this purpose, the corrosion inhibitor can act as the functionalized agent for CNTs. Palaniappan et al. [75] functionalized the CNT with 4,5-diphenyl-imidazole, as the corrosion inhibitors and then coated it on the surface of the nickel alloy using polyamine as a binder. Corrosion inhibition analyses, including DFT investigations, revealed that functionalized materials might provide superior nickel alloy surface CMP. Although corrosion inhibitors can enhance the active protective functionality of the coatings, due to the risk of compromising the coating's stability and self-healing activity, the corrosion inhibitors cannot be directly loaded on the surface of the CNT. In this condition, corrosion inhibitors may react with the coating matrix and leak into the environment [76]. As a result, the CMP of the coatings deteriorates.

The nanocontainers, which can encapsulate the corrosion inhibitors in their structure, have emerged as the new route for overcoming the problem in recent years [77, 78]. A nanocontainer's main function is to encapsulate the corrosion inhibitor to be controlled and released. As a consequence, the inhibitor remains trapped, preventing unwanted interactions between the passive matrix and the active component (inhibitor) as well as inhibitor leaking [79].

Due to the small size alongside the presence of the inner lumen in the structure of the CNTs, they can be used for encapsulating the corrosion inhibitors into their structure. In this manner, the corrosion inhibitors can be loaded into the inner lumen of the CNTs and released after penetration of the water into the coating structure. On the other hand, the corrosion inhibitor can be encapsulated on the surface of the CNT employing the polyelectrolyte layers. In this case, the corrosion inhibitors are released from the surface of the CNT in a quick response to the change in pH of the environment [80]. Yeole et al. [80] synthesized 2-mercaptobenzothiazole

(MBT)-loaded CNTs by a layer-by-layer deposition as an active anticorrosive pigment for the epoxy-based coatings. The three-step process for generating loaded nanotubes was employed in the report: (i) applying a PPy layer to the CNT surface; (ii) loading the corrosion inhibitor; and (iii) applying a PAA layer on the loaded inhibitor. Epoxy coatings, including CNTs, loaded with 2-MBT, as the active anticorrosive pigment, were applied to MS samples, and its anticorrosive performance was studied. Experimental results showed that the nanocontainers/epoxy-based system provided superior CMP compared to the epoxy-alone system attributed to the interaction between the corrosive environment and the synthesized nanocontainers. According to electrochemical studies and salt spray tests, combining the nanoparticles (i.e. CNTs and inhibitors) can increase the CMP for MS. The improved barrier characteristics of the coating are most likely owing to the pores being blocked by the released inhibitor from the nanocontainer.

Asaad et al. [81] present the self-healing characteristics of a new epoxy coating composed of MWCNTs as nanocontainers loaded with 5% green inhibitor-*Elaeis guineensis*/silver nanoparticles (EG/AgNPs) that served as a strong corrosion inhibitor for MS in a saltwater environment. According to the findings, the produced smart epoxy coating displayed superior self-healing capability as well as appropriate CMP when exposed to seawater. This smart coating displayed better CMP by weakening the lepidocrocite (–FeOOH) due to the release of *E. guineensis*/AgNPs. The results indicated that the synthesized pigment can be assumed as an efficient corrosion inhibitor to repair or self-heal the surface.

20.3 Commercial-Scale Application

The published materials about the modified CNTs are mainly research based in the form of papers and patents. However, there are a few companies that produced the modified CNTs in commercial-scale. Takenaka Seisakusho Co., Ltd. is a Japanese company that produces superhard, anticorrosive, CNT-composited coating named NANOTECT®. Nanochemi Group is another company established in Czech Republic, producing NANO CR COATING based on CNTs with excellent chemical resistant coating for maintenance in aggressive environments.

20.4 Summary

CNTs are well-known fillers in the field of protective coatings. Direct integration of CNTs within a coating matrix results in poor mechanical, electrical, and anticorrosion performances. In fact, the lack of surface functional groups and polar sites on the large hydrophobic surface of CNTs accounts for aggregations formation and poor interfacial adhesion with coating matrices. Various surface modification methods, including noncovalent functionalization, oxidation, silanization, and grafting polymer chains onto the surface of CNTs, have been introduced to enhance the compatibility of CNTs with coatings matrices.

Several mechanisms are proposed for enhanced CMP of nanocomposite coatings in the presence of modified CNTs. The main reason for the corrosion of coated structures is the electrolyte diffusion toward the substrate-coating interface. Also, most of the industrial electrolytes are aqueous. So, the coating lifetime can be prolonged by increasing its hydrophobicity. According to literature, CNTs reduce the coating hydrophobicity leading to a longer working lifetime. On the other hand, similar to other nanocomposite coatings, well-dispersion of nano-size additive in the matrix provides a denser coating with lower defects. It is worth mentioning that coating defects, e.g. porosities, are the main diffusion paths for most electrolytes. The third protective mechanism of CNTs is the incorporation of them into the coating as nanocarriers. The hollow lumen of CNTs can be filled with different corrosion inhibitors and fed to the coating in the presence of a diffused electrolyte. The surface of the CNTs can also be loaded with anticorrosion moieties. Finally, the conductive nature of CNTs provides another benefit in the field of protective coatings, especially sacrificial and passive layer forming systems such as zinc-rich and PANI-containing films, respectively. Dispersion of the CNTs modified with CPs in the coating matrix increases the passivating ability of the composite filler, leading to better CMP. Therefore, it is evident that CNTs have undeniable advantages in most protective coatings.

Considering the approach of recent works on developing multifunctional fillers, it is expected that surface decoration of CNTs with structures capable of carrying corrosion inhibitors, such as layered double hydroxides and metal–organic frameworks, can be of high interest in future works.

References

1 Jiang, M.-Y., Wu, L.-K., Hu, J.-M., and Zhang, J.-Q. (2015). Silane-incorporated epoxy coatings on aluminum alloy (AA2024). Part 1: Improved corrosion performance. *Corros. Sci.* 92: 118–126.

2 Ramezanzadeh, B., Ahmadi, A., and Mahdavian, M. (2016). Enhancement of the corrosion protection performance and cathodic delamination resistance of epoxy coating through treatment of steel substrate by a novel nanometric sol–gel based silane composite film filled with functionalized graphene oxide nanosheets. *Corros. Sci.* 109: 182–205.

3 Mišković-stanković, V.B., Dražić, D.M., and Teodorović, M.J. (1995). Electrolyte penetration through epoxy coatings electrodeposited on steel. *Corros. Sci.* 37: 241–252.

4 Bhanvase, B.A., Patel, M.A., and Sonawane, S.H. (2014). Kinetic properties of layer-by-layer assembled cerium zinc molybdate nanocontainers during corrosion inhibition. *Corros. Sci.* 88: 170–177.

5 Chen, J., Liu, B., Gao, X., and Xu, D. (2018). A review of the interfacial characteristics of polymer nanocomposites containing carbon nanotubes. *RSC Adv.* 8: 28048–28085.

6 Wang, X., Lu, J., and Xing, B. (2008). Sorption of organic contaminants by carbon nanotubes: influence of adsorbed organic matter. *Environ. Sci. Technol.* 42: 3207–3212.

7 Ganash, A.A. (2013). Effect of conducting composite carbon nanotube-poly (*o*-toluidine) coatings on the corrosion resistance of stainless steel material. *Polym. Compos.* 34: 1180–1185.

8 Chen, Q., Dai, L., Gao, M. et al. (2001). Plasma activation of carbon nanotubes for chemical modification. *J. Phys. Chem. B* 105: 618–622.

9 Xing, Y., Li, L., Chusuei, C.C., and Hull, R.V. (2005). Sonochemical oxidation of multiwalled carbon nanotubes. *Langmuir* 21: 4185–4190.

10 Vanyorek, L., Meszaros, R., and Barany, S. (2014). Surface and electrosurface characterization of surface-oxidized multi-walled N-doped carbon nanotubes. *Colloids Surf., A* 448: 140–146.

11 Ye, J.-S., Liu, X., Cui, H.F. et al. (2005). Electrochemical oxidation of multi-walled carbon nanotubes and its application to electrochemical double layer capacitors. *Electrochem. Commun.* 7: 249–255.

12 Vu, C.M. and Bach, Q.-V. (2021). Oxidized multiwall carbon nanotubes filled epoxy-based coating: fabrication, anticorrosive, and mechanical characteristics. *Polym. Bull.* 78: 2329–2339.

13 Najmi, P., Keshmiri, N., Ramezanzadeh, M., and Ramezanzadeh, B. (2021). Highly improving the mechanical-responses/thermal-stability of the epoxy nano-composite using novel highly-oxidized multi-walled carbon nanotubes (OMWCNT) functionalized by zinc-doped polyaniline (PANI) nanofibers. *J. Taiwan Inst. Chem. Eng.* 119: 245–258.

14 Akbarzadeh, S., Naderi, R., and Mahdavian, M. (2019). Fabrication of a highly protective silane composite coating with limited water uptake utilizing functionalized carbon nano-tubes. *Composites Part B* 175: 107109.

15 Hermas, A.-E.A., Abdel Salam, M., and Al-Juaid, S.S. (2013). In situ electrochemical preparation of multi-walled carbon nanotubes/polyaniline composite on the stainless steel. *Prog. Org. Coat.* 76: 1810–1813.

16 Hermas, A.-E.A., Salam, M.A., Al-Juaid, S.S., and Al-Thabaiti, S.A. (2009). Electrochemical coating of stainless steel with multi-walled carbon nanotubes/polyaniline composite layer. *Int. J. Nanomanuf.* 4: 166–174.

17 Li, J., Ma, P.C., Chow, W.S. et al. (2007). Correlations between percolation threshold, dispersion state, and aspect ratio of carbon nanotubes. *Adv. Funct. Mater.* 17: 3207–3215.

18 Hussain, A.K., Seetharamaiah, N., Pichumani, M., and Chakra, C.S. (2021). Research progress in organic zinc rich primer coatings for cathodic protection of metals – a comprehensive review. *Prog. Org. Coat.* 153: 106040.

19 Park, S. and Shon, M. (2015). Effects of multi-walled carbon nano tubes on corrosion protection of zinc rich epoxy resin coating. *J. Ind. Eng. Chem.* 21: 1258–1264.

20 Cubides, Y., Su, S.S., and Castaneda, H. (2016). Influence of zinc content and chloride concentration on the corrosion protection performance of zinc-rich epoxy coatings containing carbon nanotubes on carbon steel in simulated concrete pore environments. *Corrosion* 72: 1397–1423.

21 Cubides, Y. and Castaneda, H. (2016). Corrosion protection mechanisms of carbon nanotube and zinc-rich epoxy primers on carbon steel in simulated concrete pore solutions in the presence of chloride ions. *Corros. Sci.* 109: 145–161.

22 Izadi, M., Shahrabi, T., and Ramezanzadeh, B. (2018). Active corrosion protection performance of an epoxy coating applied on the mild steel modified with an eco-friendly sol–gel film impregnated with green corrosion inhibitor loaded nanocontainers. *Appl. Surf. Sci.* 440: 491–505.

23 Nayini, M.M.R., Bastani, S., and Ranjbar, Z. (2014). Synthesis and characterization of functionalized carbon nanotubes with different wetting behaviors and their influence on the wetting properties of carbon nanotubes/polymethylmethacrylate coatings. *Prog. Org. Coat.* 77: 1007–1014.

24 Ma, P.C., Kim, J.-K., and Tang, B.Z. (2007). Effects of silane functionalization on the properties of carbon nanotube/epoxy nanocomposites. *Compos. Sci. Technol.* 67: 2965–2972.

25 Zhou, T., Wang, X., Liu, X., and Lai, J. (2010). Effect of silane treatment of carboxylic-functionalized multi-walled carbon nanotubes on the thermal properties of epoxy nanocomposites. *Express Polym. Lett.* 4: 217–226.

26 Vast, L., Philippin, G., Destrée, A. et al. (2004). Chemical functionalization by a fluorinated trichlorosilane of multi-walled carbon nanotubes. *Nanotechnology* 15: 781–785.

27 Kar, K.K., Rana, S., and Pandey, J. (2015). *Handbook of Polymer Nanocomposites Processing, Performance and Application*. Springer.

28 Basheer, B.V., George, J.J., Siengchin, S., and Parameswaranpillai, J. (2020). Polymer grafted carbon nanotubes—synthesis, properties, and applications: a review. *Nano-Struct. Nano-Objects* 22: 100429.

29 Homenick, C.M., Lawson, G., and Adronov, A. (2007). Polymer grafting of carbon nanotubes using living free-radical polymerization. *Polym. Rev.* 47: 265–290.

30 Liu, P. (2005). Modifications of carbon nanotubes with polymers. *Eur. Polym. J.* 41: 2693–2703.

31 Prabakar, S.R. and Pyo, M. (2012). Corrosion protection of aluminum in $LiPF_6$ by poly(3,4-ethylenedioxythiophene) nanosphere-coated multiwalled carbon nanotube. *Corros. Sci.* 57: 42–48.

32 Deshpande, P.P., Jadhav, N.G., Gelling, V.J., and Sazou, D. (2014). Conducting polymers for corrosion protection: a review. *J. Coat. Technol. Res.* 11: 473–494.

33 Ghahremani, P., Mostafatabar, A.H., Bahlakeh, G., and Ramezanzadeh, B. (2022). Rational design of a novel multi-functional carbon-based nano-carrier based on multi-walled-CNT-oxide/polydopamine/chitosan for epoxy composite with robust pH-sensitive active anti-corrosion properties. *Carbon* 189: 113–141.

34 Bhandari, H., Choudhary, V., and Dhawan, S. (2011). Influence of self-doped poly(aniline-*co*-4-amino-3-hydroxy-naphthalene-1-sulfonic acid) on corrosion inhibition behaviour of iron in acidic medium. *Synth. Met.* 161: 753–762.

35 Wessling, B. (1994). Passivation of metals by coating with polyaniline: corrosion potential shift and morphological changes. *Adv. Mater.* 6: 226–228.

36 Zarras, P., Anderson, N., Webber, C. et al. (2003). Progress in using conductive polymers as corrosion-inhibiting coatings. *Radiat. Phys. Chem.* 68 (3–4): 387–394.

37 Gao, F., Mu, J., Bi, Z. et al. (2021). Recent advances of polyaniline composites in anticorrosive coatings: a review. *Prog. Org. Coat.* 151: 106071.

38 Akbarzadeh, S., Ramezanzadeh, M., Ramezanzadeh, B. et al. (2019). Fabrication of highly effective polyaniline grafted carbon nanotubes to induce active protective functioning in a silane coating. *Ind. Eng. Chem. Res.* 58: 20309–20322.

39 Grgur, B.N. (2014). On the role of aniline oligomers on the corrosion protection of mild steel. *Synth. Met.* 187: 57–60.

40 Qiu, G., Zhu, A., and Zhang, C. (2017). Hierarchically structured carbon nanotube–polyaniline nanobrushes for corrosion protection over a wide pH range. *RSC Adv.* 7: 35330–35339.

41 Li, A., Sun, M., Ma, Z. et al. (2020). Anticorrosion performance of polyvinyl butyral composite coatings improved by polyaniline-multiwalled carbon nanotubes/poly(methylhydrosiloxane). *Thin Solid Films* 712: 138347.

42 Najmi, P., Keshmiri, N., Ramezanzadeh, M., and Ramezanzadeh, B. (2021). Synthesis and application of Zn-doped polyaniline modified multi-walled carbon nanotubes as stimuli-responsive nanocarrier in the epoxy matrix for achieving excellent barrier-self-healing corrosion protection potency. *Chem. Eng. J.* 412: 128637.

43 Cui, M., Ren, S., Qiu, S. et al. (2018). Non-covalent functionalized multi-wall carbon nanotubes filled epoxy composites: effect on corrosion protection and tribological performance. *Surf. Coat. Technol.* 340: 74–85.

44 Ioniţă, M. and Prună, A. (2011). Polypyrrole/carbon nanotube composites: molecular modeling and experimental investigation as anti-corrosive coating. *Prog. Org. Coat.* 72: 647–652.

45 Gergely, A., Pászti, Z., Hakkel, O. et al. (2012). Corrosion protection of cold-rolled steel with alkyd paint coatings composited with submicron-structure types polypyrrole-modified nano-size alumina and carbon nanotubes. *Mater. Sci. Eng., B* 177: 1571–1582.

46 Gueye, M.N., Carella, A., Faure-Vincent, J. et al. (2020). Progress in understanding structure and transport properties of PEDOT-based materials: a critical review. *Prog. Mater Sci.* 108: 100616.

47 Chen, Y., Ren, B., Gao, S., and Cao, R. (2020). The sandwich-like structures of polydopamine and 8-hydroxyquinoline coated graphene oxide for excellent corrosion resistance of epoxy coatings. *J. Colloid Interface Sci.* 565: 436–448.

48 Batul, R., Tamanna, T., Khaliq, A., and Yu, A. (2017). Recent progress in the biomedical applications of polydopamine nanostructures. *Biomater. Sci.* 5: 1204–1229.

49 Cai, G., Hou, J., Jiang, D., and Dong, Z. (2018). Polydopamine-wrapped carbon nanotubes to improve the corrosion barrier of polyurethane coating. *RSC Adv.* 8: 23727–23741.

50 Chen, S., Wang, X., Zhu, G. et al. (2021). Developing multi-wall carbon nanotubes/fusion-bonded epoxy powder nanocomposite coatings with superior anti-corrosion and mechanical properties. *Colloids Surf., A* 628: 127309.

51 Nayak, S.R., Mohana, K.N.S., Hegde, M.B. et al. (2021). Functionalized multi-walled carbon nanotube/polyindole incorporated epoxy: an effective anti-corrosion coating material for mild steel. *J. Alloys Compd.* 856: 158057.

52 Chu, H., Wei, L., Cui, R. et al. (2010). Carbon nanotubes combined with inorganic nanomaterials: preparations and applications. *Coord. Chem. Rev.* 254: 1117–1134.

53 Coleman, K.S., Bailey, S.R., Fogden, S., and Green, M.L. (2003). Functionalization of single-walled carbon nanotubes via the Bingel reaction. *J. Am. Chem. Soc.* 125: 8722–8723.

54 Liu, L., Wang, T., Li, J. et al. (2003). Self-assembly of gold nanoparticles to carbon nanotubes using a thiol-terminated pyrene as interlinker. *Chem. Phys. Lett.* 367: 747–752.

55 Han, L., Wu, W., Kirk, F.L. et al. (2004). A direct route toward assembly of nanoparticle–carbon nanotube composite materials. *Langmuir* 20: 6019–6025.

56 Li, C., Jin, Z., Chu, H., and Li, Y. (2008). Seed-mediated growth of ZnO nanorods on multiwalled carbon nanotubes. *J. Nanosci. Nanotechnol.* 8: 4441–4446.

57 Li, J., Tang, S., Lu, L., and Zeng, H.C. (2007). Preparation of nanocomposites of metals, metal oxides, and carbon nanotubes via self-assembly. *J. Am. Chem. Soc.* 129: 9401–9409.

58 Pan, B., Cui, D., Ozkan, C.S. et al. (2008). Effects of carbon nanotubes on photoluminescence properties of quantum dots. *J. Phys. Chem. C* 112: 939–944.

59 Peng, X., Chen, J., Misewich, J.A., and Wong, S.S. (2009). Carbon nanotube–nanocrystal heterostructures. *Chem. Soc. Rev.* 38: 1076–1098.

60 Han, W.-Q. and Zettl, A. (2003). Coating single-walled carbon nanotubes with tin oxide. *Nano Lett.* 3: 681–683.

61 Xing, Y. (2004). Synthesis and electrochemical characterization of uniformly-dispersed high loading Pt nanoparticles on sonochemically-treated carbon nanotubes. *J. Phys. Chem. B* 108: 19255–19259.

62 Wu, B., Hu, D., Kuang, Y. et al. (2009). Functionalization of carbon nanotubes by an ionic-liquid polymer: dispersion of Pt and PtRu nanoparticles on carbon nanotubes and their electrocatalytic oxidation of methanol. *Angew. Chem. Int. Ed.* 48: 4751–4754.

63 Zhang, Y., Franklin, N.W., Chen, R.J., and Dai, H. (2000). Metal coating on suspended carbon nanotubes and its implication to metal–tube interaction. *Chem. Phys. Lett.* 331: 35–41.

64 Planeix, J., Coustel, N., Coq, B. et al. (1994). Application of carbon nanotubes as supports in heterogeneous catalysis. *J. Am. Chem. Soc.* 116: 7935–7936.

65 Yu, H., Quan, X., Chen, S., and Zhao, H. (2007). TiO_2-multiwalled carbon nanotube heterojunction arrays and their charge separation capability. *J. Phys. Chem. C* 111: 12987–12991.

66 Hu, X. and Dong, S. (2008). Metal nanomaterials and carbon nanotubes—synthesis, functionalization and potential applications towards electrochemistry. *J. Mater. Chem.* 18: 1279–1295.

67 Vairavapandian, D., Vichchulada, P., and Lay, M.D. (2008). Preparation and modification of carbon nanotubes: review of recent advances and applications in catalysis and sensing. *Anal. Chim. Acta* 626: 119–129.

68 Sivaraj, D. and Vijayalakshmi, K. (2018). Enhanced corrosion resistance and antibacterial activity of Zn-HA decorated MWCNTs film coated on medical grade 316L SS implant by novel spray pyrolysis technique. *J. Anal. Appl. Pyrolysis* 134: 176–182.

69 Rajitha, K., Mohana, K.N.S., Hegde, M.B. et al. (2020). Fabrication of ZnO/rGO and ZnO/MWCNT nanohybrids to reinforce the anticorrosion performance of polyurethane coating. *FlatChem* 24: 100208.

70 He, Y., Chen, C., Zhong, F., and Chen, H. (2015). Synthesis and characterization: silicon oxide-coated multiwalled carbon nanotubes and properties of composite coating research. *High Perform. Polym.* 27: 352–361.

71 Guo, Y., Bian, D., and Zhao, Y. (2019). Preparation and corrosion behavior of chemically bonded ceramic coatings reinforced with ZnO-MWCNTs composite. *Mater. Res. Express* 6: 085021.

72 Cai, G., Xiao, S., Deng, C. et al. (2021). CeO_2 grafted carbon nanotube via polydopamine wrapping to enhance corrosion barrier of polyurethane coating. *Corros. Sci.* 178: 109014.

73 Kumar, A.M., Rahman, M.M., and Gasem, Z.M. (2015). A promising nanocomposite from CNTs and nano-ceria: nanostructured fillers in polyurethane coatings for surface protection. *RSC Adv.* 5: 63537–63544.

74 Kumar, A., Kumar, K., Ghosh, P.K., and Yadav, K. (2018). MWCNT/TiO_2 hybrid nano filler toward high-performance epoxy composite. *Ultrason. Sonochem.* 41: 37–46.

75 Palaniappan, N., Cole, I., Kuznetsov, A., and Thomas, K.J. (2019). Experimental and DFT studies of carbon nanotubes covalently functionalized with an imidazole derivative for electrochemical stability and green corrosion inhibition as a barrier layer on the nickel alloy surface in a sulphuric acidic medium. *RSC Adv.* 9: 38677–38686.

76 Borisova, D., Akçakayıran, D., Schenderlein, M. et al. (2013). Nanocontainer-based anticorrosive coatings: effect of the container size on the self-healing performance. *Adv. Funct. Mater.* 23: 3799–3812.

77 Mohammadi, I., Izadi, M., Shahrabi, T. et al. (2019). Enhanced epoxy coating based on cerium loaded Na-montmorillonite as active anti-corrosive nanoreservoirs for corrosion protection of mild steel: synthesis, characterization, and electrochemical behavior. *Prog. Org. Coat.* 131: 119–130.

78 Izadi, M., Shahrabi, T., Mohammadi, I., and Ramezanzadeh, B. (2019). Synthesis of impregnated Na+-montmorillonite as an eco-friendly inhibitive carrier and its subsequent protective effect on silane coated mild steel. *Prog. Org. Coat.* 135: 135–147.

79 Shchukin, D.G. and Möhwald, H. (2007). Self-repairing coatings containing active nanoreservoirs. *Small* 3: 926–943.

80 Yeole, K., Agarwal, I., and Mhaske, S. (2016). The effect of carbon nanotubes loaded with 2-mercaptobenzothiazole in epoxy-based coatings. *J. Coat. Technol. Res.* 13: 31–40.

81 Asaad, M.A., Raja, P.B., Huseien, G.F. et al. (2021). Self-healing epoxy coating doped with *Elaeis guineensis*/silver nanoparticles: a robust corrosion inhibitor. *Constr. Build. Mater.* 312: 125396.

21

Chemically Modified Carbon Nanotubes and Sustainability

Shahidul Islam Bhat[1], Mohammad Mobin[1], Shayesta Islam[2], Mohd Irfan[3], Rabia Kouser[4], Kulsoom Koser[5], Ruby Aslam[1], and Saman Zehra[1]

[1] *Aligarh Muslim University, Corrosion Research Laboratory, Department of Applied Chemistry, Faculty of Engineering and Technology, Aligarh 202002, India*
[2] *SKUAST K, Division of Environmental Sciences, Shalimar Campus, Srinagar, Kashmir 190025, India*
[3] *Ajanta Caves and Conservation Research Laboratory, Aurangabad 431118, India*
[4] *Jamia Millia Islamia, Materials Research Laboratory, Department of Chemistry, New Delhi 110025, India*
[5] *Bio Inorganic Research Laboratory, Department of Chemistry, Jamia Millia Islamia, New Delhi 110025, India*

Abbreviations

APP	atmospheric-pressure plasma
CNTs	carbon nanotubes
HPP	high-pressure plasma
LPP	low-pressure plasma
MWCNTs	multi-walled carbon nanotubes
SWCNTs	single-walled carbon nanotubes

21.1 Introduction

Carbon nanotubes (CNTs) commonly named as being the allotropes of carbon are tubular structures composed predominantly of graphite. Their outer diameter varies from 3 to 30 nm and is formed of two layers or more [1]. Typically CNTs are categorized into two classes viz. singl-walled carbon nanotubes (SWCNTs) and mult-walled carbon nanotubes (MWCNTs). The SWCNTs are usually tapered than the MWCNTs, with the diameter ranging 1–2 nm, and acquire curved morphologies instead of straight. The SWCNTs are the nanotubes with a single wall and were reported in 1993 for the first time [1], and those possessing more than one wall are MWCNTs and were reported by Iijima in 1991 [2]. Almost all physical characteristics of CNTs are graphene derived. The carbon atoms are densely arranged in a regular sp^2 hybridized pattern which forms the basis of other sp^2 hybridized carbon bonded species such as allotropes, fullerenes, and CNTs. CNT is theoretically distinct with a cylindrical shape and the carbon atoms are bound as a rolled-up

Chemically Modified Carbon Nanotubes for Commercial Applications, First Edition.
Edited by Jeenat Aslam, Chaudhery Mustansar Hussain, and Ruby Aslam.

sheet of graphene. The property of catenation enables carbon to form structures having different properties. In MWCNTs, the distance between the layers was found to vary from 0.34 to 0.39 m from real space analysis [3]. Also in MWCNTs, the diameter on the interior side sweeps from 0.4 nm to few nanometers and on the exterior side in between 2 nm and 20 to 30 nm. Usually, both ends of MWCNTs have sealed ends, capped by half-fullerene molecules of dome shape. The diameter of SWCNT sweeps between 0.4 and 2–3 nm, and their length varies up to certain micrometers. SWCNTs usually assemble in bundles and get oriented hexagonally to attain crystal-like orientations [4].

21.2 Chemically Modified Carbon Nanotubes

The modification of a carbon-derived material via chemical means was reported by Murray et al. in 1970 [5]. The modification was achieved on attachment of the specific molecules that impart a particular specific chemical change within the substrate molecule [6]. The aforementioned changes can be achieved in different ways among which the surface and the bulk modification are the most common. The surface modification method involves polymer grafting, electrochemical induced methods, and the metal nanofiller deposition. On the other hand, the bulk modification includes covalent modification by thermally activated processes, phosphorous acid (H_3PO_3) as a reducing agent, microwave-assisted modification, and ball milling modification.

21.2.1 Surface Modification

Generally, CNTs are hydrophobic with low surface energies, making their incompatibility with polymer matrices and some polar solvents. These get aggregated into macro visible clusters due to low surface energy and strong interparticle van der Waals force of interaction that also leads to a reduction in their wettability. The extent of surface functionalization has a direct relation with wettability of carbon materials. The agglomeration tendency, weak dispersion, and weaker interfacial interactions in polymer-carbon/blend can be controlled by customizing the filler plane approximately. This functionalization/customization decreases the tendency of filler agglomeration by reducing filler–matrix interfacial adhesion via formation of chemical bonds. In view of this significant efforts were developed for functionalization of carbon fillers [7–9]. There are two broad divisions of functionalization namely chemical (covalent) and the physical (noncovalent) functionalization as a consequence of interplay among the carbon atoms of the filler and the agile molecules (flowchart given in Scheme 21.1). During chemical functionalization process distinct moieties get fastened to the filler plane by means of covalent bonds. The chemically modified filler can generate powerful bonding between the faces with semipolar or polar medium and becomes the basis for the high mechanical shear. The functionalized nanofillers were found to get dispersed in various polar solvents and were found to enhance the colors of paints, inks, and coatings [10, 11].

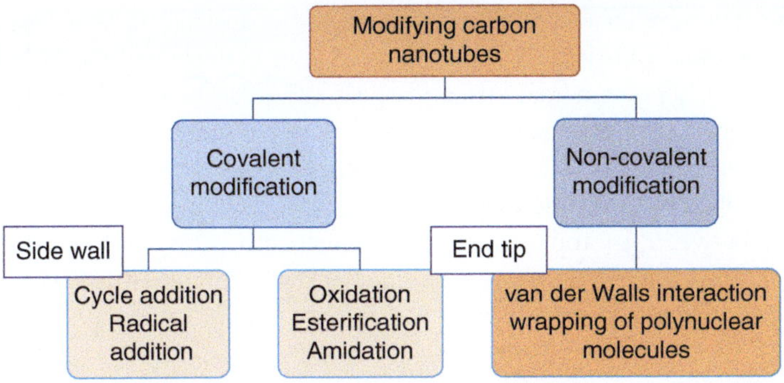

Scheme 21.1 Flow chart depicting CNT modification.

21.2.1.1 Chemical Functionalization (Covalent Functionalization)

The surface modification/functionalization of CNTs is governed by some oxidative procedure. The oxidized substance function as starting materials for another set of chemical alterations, via polymer grafting [12], esterification [13], arylation [14], silanation [15], alkylation, etc. Under moist conditions, oxidation takes place with the aid of strong acids such as H_2SO_4, HNO_3 [16], or strong oxidizing agents such as $KMnO_4$ [17], in dry medium such as ozone (O_3) [15, 18]. The surface modification (nondestructive) is performed by encapsulation/wrapping processes [19, 20]. The possible sites of CNT functionalization in terms of reaction-prone sites and process of functionalization are shown in Figure 21.1.

Surface Oxidation The surface of CNTs being inert and as a result, a large number of oxygen-containing functionalities might be hooked on the carbon surface

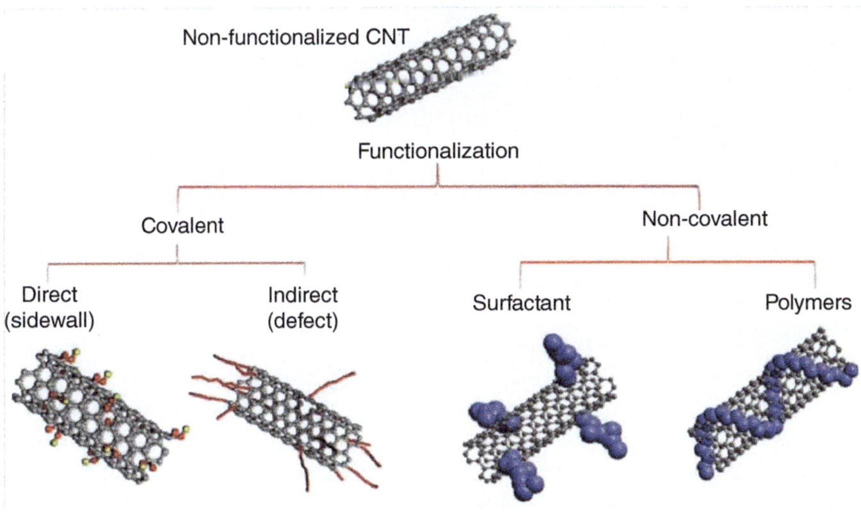

Figure 21.1 Different functionalization methods of CNTs. Source: Karimi et al. [21]/with permission of Taylor & Francis.

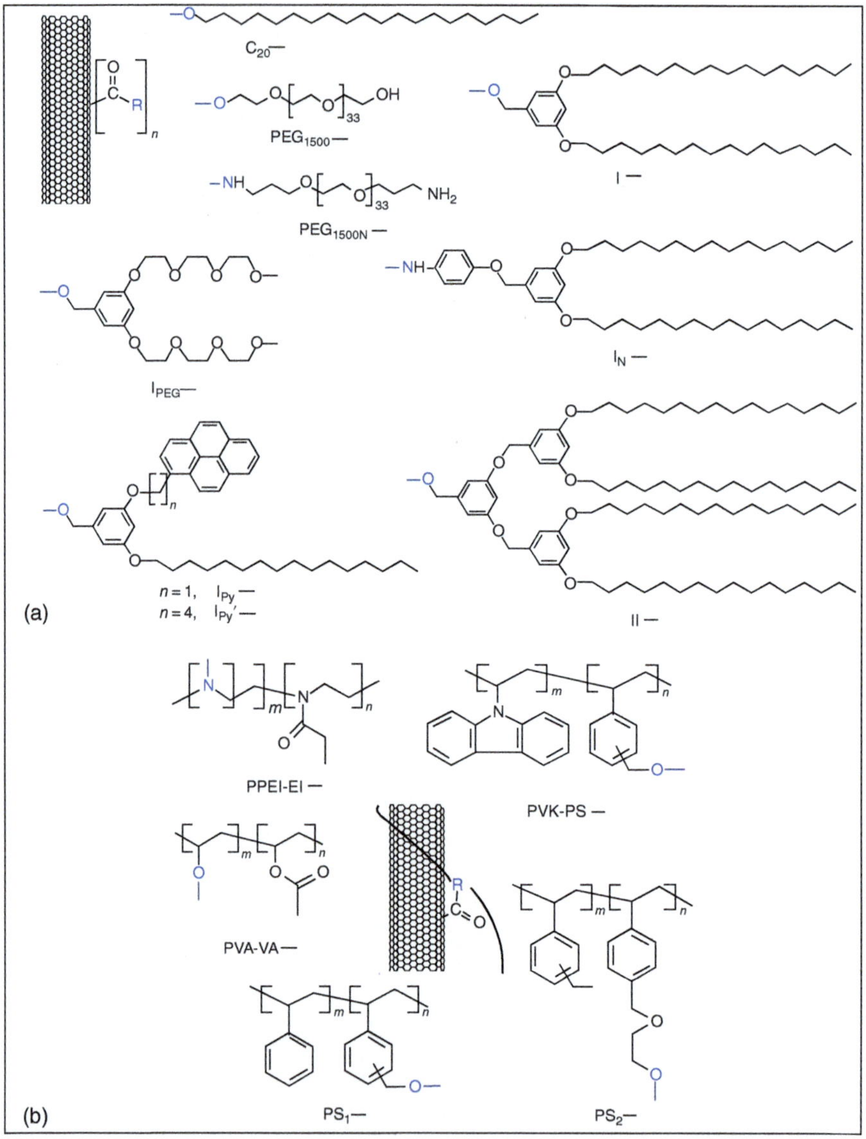

Figure 21.2 Schematic presentation of functionalization of CNTs. Source: Ya-Ping et al. [22]/with permission of American Chemical Society.

by treatment of strong oxidizing agents in aqueous medium or gaseous phase. The common oxidizing agents used for this purpose are nitric oxide/air mixtures, ozone, and air. However, one more potent practice for superficial oxidation is the plasma method, where powdered fillers are easily redesigned at precise interval of time. The attaching of the functionalities to the carbon-based substances on oxidation (Figure 21.2), in which various functional groups viz. phenolic, carboxylic quinine, hydroquinone, ketonic, and hydroxylic groups are embedded on to their surface by means of covalent bonding.

Wet Oxidation The raw CNTs are oxidized by varying processing conditions such as (different sonication/treatment time, temperature, and oxidant concentration) to attach surface functional groups. Using concentrated mineral acids for the oxidation of CNTs is considered as a preferred functionalization tool at the laboratory scale. Sahoo et al. [23] reported the oxidation of raw MWNTs with H_2SO_4/HNO_3 in (3 : 1) ratio at 90 °C for 10 minutes and their incorporation in polyurethanes. The elevation of 740% and 180% in the Young's modulus and tensile strength respectively above unblended polyurethane (PU) were attained at an optimized loading of 20 wt% of oxidized MWCNT [23]. The distribution of the CNTs into the polymer backbone via covalent connection is enhanced by the chemical oxidation. So et al. [24] assimilated both the virgin and oxidized CNTs into the polyimide matrix and found that the non-functionalized CNTs exhibited elevated values of conductivity in comparison to the functionalized ones. A similar type of study was carried out by Spitalsky et al. [25] where they introduced MWCNTs in epoxy resins, an increase in flexural modulus up to double in order of magnitude, and a two-fold conductivity decrease in DC was found. Simsek et al. [26] discussed amino-functionalized/CNTs that were reported to possess lesser electrical conductivity in contrast to those without functionalization.

Gaseous Phase Oxidation/Dry Oxidation The wet oxidation poses no attraction in industrial scale functionalization due to ill effects on health and environmental hazards. The chemicals used in dry oxidation contain highly corrosive ions and are difficult to operate. The various hazardous gases are eliminated in the current process on health and environmental grounds. The wet oxidation forms aggregates or lumps of carbon materials and upon wet treatment with losing their fluffy character. To overcome these shortcomings, different dry oxidation techniques particularly air, steam, ozone, and blend of air/few gases are used. The dry oxidation method removes the aforementioned issues and also eliminates some intermediate drying steps.

Oxidation with Air The technique of air oxidation is a tedious technique, taking several hours depending on desired extent of oxidation. The functionalization at the CNT surface occurs at an elevated temperature in comparison to the carbon black. In the first instant, the carbonaceous material gets oxidized and is followed by etching and end capping [27]. Li et al. [28] reported thermally oxidized MWCNTs in a tube-shaped furnace with sufficient supply of air. The quartz (SiO_2) tube on CNT impregnation was kept in an oven (rotated 25 rpm) and exposed eventually to the air for uniform oxidation. The samples were oxidized at 480 to 750 °C for 20 minutes and it should be ensured that the oxidation process increases the specific area and improves the intrinsic CNT morphology. The increased conductivities of air oxidized MWCNTs in comparison to ones that get oxidized was given by Park et al. [29], and near about parallel results were put forward by Yuen et al. [30].

Ozonation of CNTs On ozonation, amorphous carbon atoms are separated out from nanotube surface, and diverse functional groups such as carboxylic as alcoholic, esteric, aldehydic, and ketonic moieties are generated on the nanotube facet.

The aforementioned functionalities get attached via cyclo addition reaction (1,3 bipolar) to the CNT double bonds employing Criegee's mechanism [31]. The polar groups allow the CNTs to easily get dispersed in various solvents and polymer matrices. Vennerberg et al. [32] employed oxidation treatment in gaseous phase with MWCNTs employing a vertical bed reactor. A significant oxidation level was achieved (of the order of 8 at% of O) with a low or no delamination of the MWCNT sidewalls. It was observed that primary hydroxyl groups get linked to nanotube surface of during short O_3 exposure, however on O_3 exposure for a longer duration of time mainly leads to formation of carboxylic functionalities. The surface defects were discovered to perform a prominent part in the oxidation process. The ozonation of CNTs is usually accomplished as a gas/solid operation; however on attaining a steady CNT oxidation agglomeration of the nanofillers has always posed a problem principally when the reaction surface area is sizeable. There are reports in literature on ozonation of CNTs in the solution phase, where aqueous medium is used for sonication of CNTs and ozone is bubbled through the dispersion phase for oxidation of CNTs [33–41]. The decomposition of ozone molecule into hydroxyl radical takes place in aqueous phase as hydroxyl radical is considered as a strong oxidizing agent ($E^{\circ} = 3.06\,V$) in comparison to ozone molecule ($E^{\circ} = 2.06\,V$). Lara et al. [42], studied the MWCNT functionalization by ozone at alkaline pH in the gaseous state and established that the effectiveness of ozone oxidation in primary aqueous solution is lesser in comparison to ozone gas. At higher values of pH, OH^- ions get adsorbed on the CNT carbon atoms, restricting the hydroxyl radical ions kinetically produced by O_3 in water. Hence, the increasing pH reduces the efficiency of O_3 oxidation.

Plasma Treatment An alternate treatment for insertion of functional groups into the CNT backbone and compatibility improvement is the plasma treatment. It is considered to be advantageous in comparison to acid treatments as it is a single-step functionalization and a dry process. It has been found that the degree of functionalization can be controlled by altering certain plasma processing parameters such as process gases, time of processing, plasma power, and gas pressure [43]. The plasma treatment is of considerable interest as compared to other processes because of its green nature and shorter times of reaction. In plasma treatment, ions, free radicals, metastable species, and ions cause interconnection, ablation, and oxidation reactions relying upon gaseous nature. The types of plasmas are low-pressure plasma (LPP), hig-pressure plasma (HPP), and atmospheric-pressure plasma. In case of HPP, a high vacuum is created for the generation of gas, and this system is considered not appropriate for handling of pulverized substances. A scant literature has been found on surface handling of pulverized carbon substances by LPP techniques using various reactors such as downstream reactor, fluidized circulating bed reactor/rotary drum reactor [44–46].

21.2.1.2 Physical Modification (Non-covalent Functionalization)

The physical modification (functionalization by noncovalent methods) is an alluring tool for biomolecular immobilization of on CNTs due to its tendency to perpetuate the CNT innate characteristics while providing novel functionalities. The noncovalent interactions aid in preserving the conformational structure of the immobilized biomolecules after immobilization on CNT surface, whereas covalent methods alter the sp^2 hybridized CNT structure, hence imposing a negative impact on their electronic and mechanical properties. The CNT functionalization by noncovalent means can be attained by π–π interrelation (polymers or aromatic), CH–π types of interactions between biomolecules and the CNTs, and electrostatic interactions [47–49]. Without making any compromise on mechanical and electrical properties, noncovalent functionalization retains the CNT biocompatibility. A potent functionalization pathway not only introduces uniform surface functionalities, but it also minimizes its influence on the CNT electronic characteristics.

π–π Interactions with Aromatic Molecules The aromatic group-containing molecules are competent in achieving CNT functionalization by noncovalent methods. The functionality is achieved on formation of site specified and directional stacking of π–π interconnections linking nanotube surface and the aromatic species. Woods et al. utilized density functional theory (DFT) for adsorption of aromatic compounds such as benzene on SWCNTs and found out the adsorption was mainly achieved by the π orbital interaction of the benzene and the CNT [50]. The aforementioned besides enhancing the dissolution of CNTs in the solvents also takes a prominent part in noncovalent CNT functionalization. Some aromatic compounds, such as pyrene, anthracene, and phthalocyanine-derived products were utilized for the bimolecular immobilization onto the CNT surface.

Figure 21.3 depicts a pyrene molecule operationalized SWCNT. The attached 1-pyrenebutanoic acid, succinimidyl ester (PBSE) molecules on the sides of SWCNT were greatly impenetrable to desorption in aqueous solution as shown in Figure 21.4 and accelerate the reaction between succinimidyl ester groups and primary or secondary amines of proteins.

π–π Interactions with Polymers Polymers offer a 3D nanostructure possessing an extensive electro operative zone that can be used for noncovalent functionalization

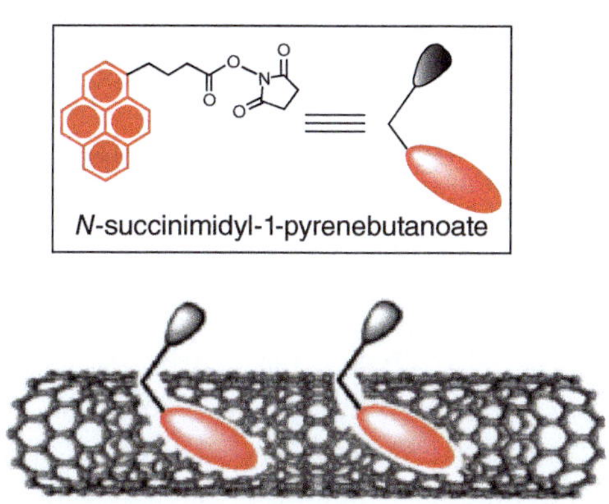

Figure 21.3 SWCNT functionalized with pyrene-based molecular tether, 1-pyrenebutanoic acid, and succinimidyl ester (PBSE). Source: Adapted from Zhao et al. [47]/with permission of.

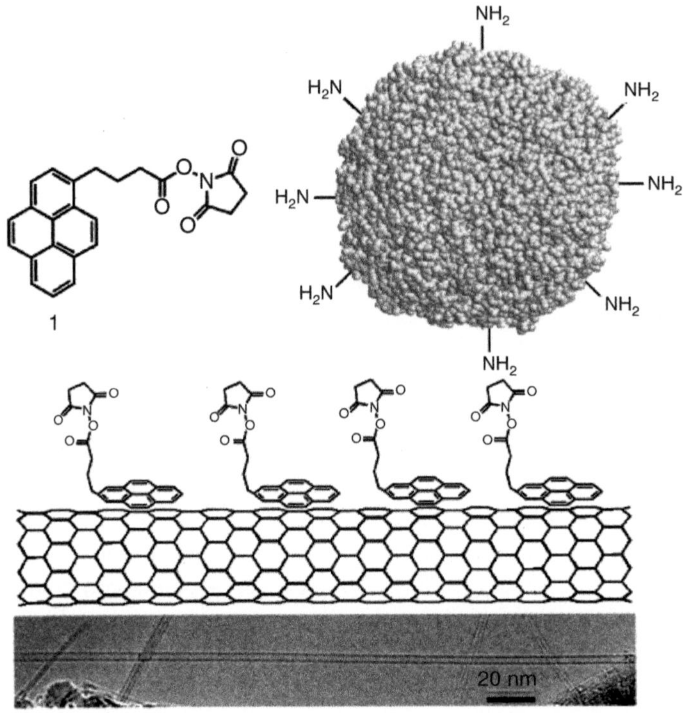

Figure 21.4 Amine groups on a protein react with the attached succinimidyl esterto form amide bonds for protein immobilization. Source: Chen et al. [51]/with permission of American Chemical Society.

for biosensing use. The biopolymers such as cellulose derivatives, glycolipids, and polypyrroles were tailored for functionalization of CNTs for various applications [52–54]. The films based on polypyrrole/CNT were developed as the complex can enhance the stability, conductivity, and the reactions of biomolecules mainly the electron transfer reactions are enhanced up to a considerable limit [55]. A 10-fold enhancement in the catalytic current in comparison to glassy bare electrode was achieved on using polypyrrole/SWCNT composite as a working electrode. The polymeric layers also aid in preserving the electro enzymatic action of working electrode for a couple of months [56].

Electrostatic and CH–π Interactions The electrostatic interactions play a significant role in biomolecular adsorption onto the CNTs. The myoglobin protein (positively charged) gets immobilized on SWCNTs functionalized by phosphonate via electrostatic interactions and hence exhibited an adequate bioelectro catalytic activity [57].The π–π stacking has also been reported between the π electrons of the SWCNTs and the matrix that originated as a result of strong electrostatic interactions (repulsions) and debundling (partial) of the SWCNT fillers.

Another most widely used pathway of noncovalent functionalization is the use of amphiphilic molecules that interacts with hydrophilic as well as hydrophobic molecules, whereby the nanotube layers interact with the hydrophobic end groups [58]. Some researchers have developed natural polysaccharides such as amylose, chitosan, and sodium alginate to wrap around side walls of CNTs by noncovalent interactions, bestowing diverse surface phenomenon to the resulting composites for interacting with biological systems [59]. Kumar et al. developed a three-electrode surface systems impregnated with CNTs/chitosan and it was found that the CNT diameter increased from 10 to 20 nm and 70 to 100 nm for noncovalent and to 50 to 80 nm for covalent functionalized systems. It was concluded that chitosan has a densely-packed structure when covalently bonded to CNTs than in the noncovalent state.

21.3 Future Scope

The problem of air/water pollution is a state of concern in the current era. The nanohybrids derived from CNTs in combination with inorganic moieties (metal oxide nanofillers) have successfully solved the aforementioned problems. The nanohybrid state leads to development of materials with improved properties such as sorption, conductivity, mechanical, catalytic, and magnetic properties. Thus, the aforementioned area will have a promising potential in the development of materials with a superior performance.

21.4 Conclusion

CNTs were considered the topic of discussion in the current era as a result of their varied structures. Their modification provides not only selectivity but specificity for the varied detection of analyte molecules. A widespread application in sensing technology, drug delivery, and catalysis. The current chapter focuses on the developments at commercial scale for fabrication of chemically modified CNTs. The chapter highlights the different commercial scale developments for fabrication of chemically modified CNTs. Various methods of chemical modification such as surface modification, plasma treatment, dry oxidation wet oxidation along with noncovalent functionalization has been discussed.

Author's Contributions

All the authors have contributed equally.

Conflict of Interest

There is no conflict of interest between the authors.

Other Sources

https://www.intechopen.com/chapters/17304
https://www.nanowerk.com/nanotechnology/introduction/introduction_to_nanotechnology_22.php
https://aip.scitation.org/doi/10.1063/5.0009644
https://www.britannica.com/science/carbon-nanotube
https://www.cheaptubes.com/carbon-nanotubes-properties-and-applications

Acknowledgments

Shahidul Islam Bhat acknowledges the UGC, New Delhi, India for financial support in the form of Dr. DS Kothari postdoc vide award number grant number (No. F.4-2/2006 (BSR)/CH/20-21/0146).

References

1 Iijima, S. and Ichihashi, T. (1993). Single-shell carbon nanotubes of 1-nm diameter. *Nature* 363: 603–605.
2 Iijima, S. (1991). Helical microtubules of graphitic carbon. *Nature* 354: 56–58.
3 Ajayan, P.M. and Ebbesen, T.W. (1997). Nanometre-size tubes of carbon. *Rep. Prog. Phys.* 60: 10–25.
4 Chico, L., Crespi, V.H., Benedict, L.X. et al. (1996). Pure carbon nanoscale devices: nanotube heterojunctions. *Phys. Rev. Lett.* 76: 971–974.
5 Murray, R.W., Ewlng, A.G., and Dust, R.A. (1987). Chemically modified electrodes. Molecular design for electroanalysis. *Anal. Chem.* 59: 379A–390A.
6 Oni, J., Diab, N., and Reiter, S. (2005). *Sens. Actuators, B* 105: 208.
7 Balasubramanian, K. and Burghard, M. (2005). Chemically functionalized carbon nanotubes. *Small* 1: 180–192.
8 Hirsch, A. (2002). Functionalization of single-walled carbon nanotubes. *Angew. Chem. Int. Ed.* 41: 1853–1859.
9 Hirsch, A. and Vostrowsky, O. (2005). Functionalization of carbon nanotubes. *Top. Curr. Chem.* 245: 193–237.
10 Rahman, U.O., Bhat, S.I., Yu, H., and Ahmad, S. (2017). Hyperbranched soya alkyd nanocomposite: a sustainable feedstock based anticorrosive nanocomposite coatings. *ACS Sustainable Chem. Eng.* 5 (11): 9725–9734.
11 Bhat, S.I. and Ahmad, S. (2018). Castor oil-TiO_2 hyperbranched poly (ester amide) nanocomposite: a sustainable, green precursor-based anticorrosive nanocomposite coatings. *Prog. Org. Coat.* 123: 326–336.
12 Liu, P. (2005). Modifications of carbon nanotubes with polymers. *Eur. Polym. J.* 41: 2693–2703.
13 Hamon, M.A., Hui, H., and Bhowmik, P. (2002). Ester-functionalized soluble single-walled carbon nanotubes. *Appl. Phys. A* 74: 333–338.

14 Stephenson, J.J., Sadana, A.K., Higginbotham, A.L., and Tour, J.M. (2006). Highly functionalized and soluble multiwalled carbon nanotubes by reductive alkylation and arylation: the billups reaction. *Chem. Mater.* 18: 4658–4661.

15 Ma, P.C., Kim, J.K., and Tang, B.Z. (2006). Functionalization of carbon nanotubes using a silane coupling agent. *Carbon* 44: 3232–3238.

16 Esumi, K., Ishigami, M., Nakajima, A. et al. (1996). Chemical treatment of carbon nanotubes. *Carbon* 34: 279–281.

17 Yu, R., Chen, L., Liu, Q. et al. (1998). Platinum deposition on carbon nanotubes via chemical modification. *Chem. Mater.* 10: 718–722.

18 Sham, M.L. and Kim, J.K. (2006). Surface functionalities of multi-wall carbon nanotubes after UV/ozone and TETA treatments. *Carbon* 44: 768–777.

19 Star, A., Stoddart, J.F., and Steuerman, D. (2001). Preparation and properties of polymer-wrapped single-walled carbon nanotubes. *Angew. Chem. Int. Ed.* 40: 1721–1725.

20 Kostarelos, K., Lacerda, L., and Pastorin, G. (2007). Cellular uptake of functionalized carbon nanotubes is independent of functional group and cell type. *Nat. Nanotechnol.* 2: 108–113.

21 Karimi, M., Solati, N., Amiri, M. et al. (2015). Carbon nanotubes part I: preparation of a novel and versatile drug-delivery vehicle. *Expert Opin. Drug Delivery* 12 (7): 1–17.

22 Sun, Y.-P., Kefu, F., Lin, Y., and Huang, W. (2002). Functionalized carbon nanotubes: properties and applications. *Acc. Chem. Res.* 35 (12): 1096–1104.

23 Sahoo, N.G., Jung, Y.C., Yoo, H.J., and Cho, J.W. (2006). Effect of functionalized carbon nanotubes on molecular interaction and properties of polyurethane composites. *Macromol. Chem. Phys.* 207: 1773–1780.

24 So, H.H., Cho, J.W., and Sahoo, N.G. (2007). Effect of carbon nanotubes on mechanical and electrical properties of polyimide/carbon nanotubes nanocomposites. *Eur. Polym. J.* 43: 3750–3756.

25 Spitalsky, Z., Krontiras, C.A., Georga, S.N., and Galiotis, C. (2007). Effect of oxidation treatment of multiwalled carbon nanotubes on the mechanical and electrical properties of their epoxy composites. *Composites, Part A* 40: 778–783.

26 Simsek, Y., Ozyuzer, L., Seyhan, A.T. et al. (2007). Temperature dependence of electrical conductivity in double-wall and multi-wall carbon nanotube/polyester nanocomposites. *J. Mater. Sci.* 42: 9689–9695.

27 Park, Y.S., Choi, Y.C., Kim, K.S., and Chung, D.C. (2001). High yield purification of multiwalled carbon nanotubes by selective oxidation during thermal annealing. *Carbon* 39: 655–661.

28 Li, C., Wang, D., Liang Wang, T.X. et al. (2004). Oxidation of multiwalled carbon nanotubes by air: benefits for electric double layer capacitors. *Powder Technol.* 142: 175–179.

29 Park, W.K., Kim, J.H., Lee, S.S. et al. (2005). Effect of carbon nanotube pre-treatment on dispersion and electrical properties of melt mixed multi-walled carbon nanotubes/poly(methyl methacrylate) composites. *Macromol. Res.* 13: 206–211.

30 Yuen, S.M., Ma, C.C.M., Lin, Y.Y., and Kuan, H.C. (2007). Preparation, morphology and properties of acid and amine modified multiwalled carbon nanotube/polyimide composite. *Compos. Sci. Technol.* 67: 2564–2573.

31 Criegee, R. (1975). Mechanism of ozonolysis. *Angew. Chem. Int. Ed.* 14: 745–752.

32 Vennerberg, D.C., Quirino, R.L., Jang, Y., and Kessler, M.R. (2014). Oxidation behavior of multiwalled carbon nanotubes fluidized with ozone. *ACS Appl. Mater. Interfaces* 6: 1835–1842.

33 Wepasnick, K., Smith, B.A., Schrote, K.E. et al. (2011). Surface and structural characterization of multi-walled carbon nanotubes following different oxidative treatments. *Carbon* 49: 24–36.

34 Mawhinney, D.B., Naumenko, V., Kuznetsova, A. et al. (2000). Infrared spectral evidence for the etching of carbon nanotubes: ozone oxidation at 298 K. *J. Am. Chem. Soc.* 122: 2383–2384.

35 Cai, L., Bahr, J.L., Yao, Y., and Tour, J.M. (2002). Ozonation of single-walled carbon nanotubes and their assemblies on rigid self-assembled monolayers. *Chem. Mater.* 14: 4235–4241.

36 Chen, Z., Ziegler, K.J., Shaver, J. et al. (2006). Cutting of single-walled carbon nanotubes by ozonolysis. *J. Phys. Chem. B* 110: 11624–11627.

37 Banerjee, S. and Wong, S.S. (2002). Rational sidewall functionalization and purification of single-walled carbon nanotubes by solution-phase ozonolysis. *J. Phys. Chem. B* 106: 12144–12151.

38 Simmons, J.M., Nichols, B.M., Baker, S.E. et al. (2006). Effect of ozone oxidation on single-walled carbon nanotubes. *J. Phys. Chem.* 110: 7113–7118.

39 Byl, O., Liu, J., and Yates, J.T. (2005). Etching of carbon nanotubes by ozone: a surface area study. *Langmuir* 21: 4200–4204.

40 Hemraj-Benny, T., Bandosz, T.J., and Wong, S.S. (2008). Effect of ozonolysis on the pore structure, surface chemistry, and bundling of single-walled carbon nanotubes. *J. Colloid Interface Sci.* 317: 375–382.

41 Peng, K., Liu, L.Q., Li, H. et al. (2011). Room temperature functionalization of carbon nanotubes using an ozone/water vapor mixture. *Carbon* 49: 70–76.

42 Lara, F.M., Perez-Mendoza, M.J., Altmajer-Vaz, D. et al. (2013). Functionalization of multiwall carbon nanotubes by ozone at basic pH. Comparison with oxygen plasma and ozone in gas phase. *J. Phys. Chem. C* 117: 11647–11655.

43 Felten, A., Bittencourt, C., Pireaux, J.J. et al. (2005). Radio-frequency plasma functionalization of carbon nanotubes surface O_2, NH_3, and CF_4 treatments. *J. Appl. Phys.* 98: 074308. https://doi.org/10.1063/1.2071455.

44 Tressaud, A., Shirasaki, T., Nanse, G., and Papirer, E. (2002). Fluorinated carbon blacks: influence of the morphology of the starting material on the fluorination mechanism. *Carbon* 40: 217–220.

45 Schukin, L.I., Kornievich, M.V., Vartapetjan, R.S., and Beznisko, S.I. (2002). Low-temperature plasma oxidation of activated carbons. *Carbon* 40: 2028–2030.

46 Donnet, J.B., Wang, W.D., and Vidal, A. (1994). Observation of plasma treated carbon black surfaces by scanning tunnelling microscopy. *Carbon* 32: 199–2006.

47 Zhao, Y.L. and Stoddart, J.F. (2009). Noncovalent functionalization of single-walled carbon nanotubes. *Acc. Chem. Res.* 42: 1161–1171.

48 Gao, C., Guo, Z., Liu, J.H., and Huang, X.J. (2012). The new age of carbon nanotubes: an updated review of functionalized carbon nanotubes in electrochemical sensors. *Nanoscale* 4: 1948–1963.

49 Chen, R.J., Bangsaruntip, S., Drouvalakis, K.A. et al. (2003). Noncovalent functionalization of carbon nanotubes for highly specific electronic biosensors. *Proc. Natl. Acad. Sci. U.S.A.* 100: 4984–4989.

50 Woods, L.M., Bădescu, Ş.C., and Reinecke, T.L. (2007). Adsorption of simple benzene derivatives on carbon nanotubes. *Phys. Rev. B* 75: 155415.

51 Chen, R.J., Zhang, Y., Wang, D., and Dai, H. (2001). Noncovalent sidewall functionalization of single-walled carbon nanotubes for protein immobilization. *J. Am. Chem. Soc.* 123: 3838–3839.

52 Zheng, W., Li, Q., Su, L. et al. (2006). Direct electrochemistry of multi-copper oxidases at carbon nanotubes noncovalently functionalized with cellulose derivatives. *Electroanalysis* 18: 587–594.

53 Chen, J., Liu, H., Weimer, W.A. et al. (2002). Noncovalent engineering of carbon nanotube surfaces by rigid, functional conjugated polymers. *J. Am. Chem. Soc.* 124: 9034–9035.

54 Cesarino, I., Moraes, F.C., Lanza, M.R., and Machado, S.A. (2012). Electrochemical detection of carbamate pesticides in fruit and vegetables with a biosensor based on acetylcholinesterase immobilised on a composite of polyaniline-carbon nanotubes. *Food Chem.* 135: 873–879.

55 Wu, T.M., Chang, H.-L., and Lin, Y.-W. (2009). Synthesis and characterization of conductive polypyrrole/multi-walled carbon nanotubes composites with improved solubility and conductivity. *Compos. Sci. Technol.* 69: 639–644.

56 Silveira, C.M., Baur, J., Holzinger, M. et al. (2010). Enhanced direct electron transfer of a multihemic nitrite reductase on single-walled carbon nanotube modified electrodes. *Electroanalysis* 22: 2973–2978.

57 Li, P., Liu, H., Ding, Y. et al. (2012). Synthesis of water-soluble phosphonate functionalized single walled carbon nanotubes and their applications in biosensing. *J. Mater. Chem.* 22: 15370–15378.

58 Britz, D.A. and Khlobystov, A.N. (2006). Noncovalent interactions of molecules with single walled carbon nanotubes. *Chem. Soc. Rev.* 35: 637–659.

59 Zhang, X., Meng, L., and Lu, Q. (2009). Cell behaviors on polysaccharide wrapped SWCNT a quantitative study of the surface properties of biomimetic nanofibrous scaffolds. *ACS Nano* 3: 3200–3206.

Index

Chemically Modified Carbon Nanotubes for Commercial Applications, First Edition.
Edited by Jeenat Aslam, Chaudhery Mustansar Hussain, and Ruby Aslam.

x

y